# Progress in Nonlinear Differential Equations and Their Applications

Volume 75

# Differential Equations, Chaos and Variational Problems

Vasile Staicu
Editor

Birkhäuser
Basel · Boston · Berlin

Editor:
Vasile Staicu
Department of Mathematics
University of Aveiro
3810-193 Aveiro
Portugal

e-mail: vasile@ua.pt

2000 Mathematics Subject Classification: 34, 35, 37, 39, 45, 49, 58, 70, 76, 80, 91, 93

Library of Congress Control Number: 2007935492

Bibliographic information published by Die Deutsche Bibliothek
Die Deutsche Bibliothek lists this publication in the Deutsche Nationalbibliografie;
detailed bibliographic data is available in the Internet at <http://dnb.ddb.de>.

ISBN 978-3-7643-8481-4 Birkhäuser Verlag AG, Basel · Boston · Berlin

© 2008 Birkhäuser Verlag AG
Basel · Boston · Berlin
P.O. Box 133, CH-4010 Basel, Switzerland
Part of Springer Science+Business Media
Printed on acid-free paper produced from chlorine-free pulp. TCF ∞
Printed in Germany

ISBN 978-3-7643-8481-4                    e-ISBN 978-3-7643-8482-1

9 8 7 6 5 4 3 2 1                         www.birkhauser.ch

# Contents

# Contents

# Editorial Introduction

This book is a collection of original papers and state-of-the-art contributions written by leading experts in the areas of differential equations, chaos and variational problems in honour of Arrigo Cellina and James A. Yorke, whose remarkable scientific carrier was a source of inspiration to many mathematicians, on the occasion of their 65th birthday.

Arrigo Cellina and James A. Yorke were born on the same day: August 3, 1941. Both received their Ph.D. degrees from the University of Maryland, where they met first in the late 1960s, at the Institute for Fluid Dynamics and Applied Mathematics. They had offices next to each other and though they were of the same age, Yorke was already Assistant Professor, while Cellina was a Graduate Student. Each one of them had a small daughter, and this contributed to their friendship.

Arrigo Cellina                    James A. Yorke

Yorke arrived at the office every day with a provision of cans of Coca Cola, his daily ration, that he put in the air conditioning fan, to keep cool. Cellina says that he was very impressed by Yorke's way of doing mathematics; Yorke could prove very interesting new results using almost elementary mathematical tools, little more than second year Calculus.

From those years, he remembers for example the article *Noncontinuable solutions of differential-delay equations* where Yorke shows, in an elementary way but with a clever use of the extension theorem, that the basic theorem of continuation of solutions to ordinary differential equations cannot be valid for functional equations (at that time very fashionable). In the article *A continuous differential equation in Hilbert space without existence*, Yorke gave the first example of the nonexistence of solutions to Cauchy problems for an ordinary differential in a Hilbert space. Furthermore, in a joint paper with one of his students, Saperstone, he proved a controllability theorem without using the hypothesis that the origin belongs to the interior of the set of controls. This is just a sample of important problems to which Yorke made nontrivial contributions.

Yorke went around always carrying in his pocket a notebook where he annotated the mathematical problems that seemed important for future investigation. In those years Yorke's collaboration with Andrezj Lasota began, which produced outstanding results in the theory of "chaos". Yorke became famous even in non-mathematical circles for his mathematical model for the spread of gonorrhoea. While traditional models were not in accord with experimental data, he proposed a simple model based on the existence of two groups of people and proved that this model fits well the experimental data. Later, in a 1975 paper entitled *Period three implies chaos* with T.Y. Lee, Yorke introduced a rigorous mathematical definition of the term "chaos" for the study of dynamical systems. From then on, he played a leading role in the further research on chaos, including its control and applications.

Yorke's goals to explore interdisciplinary mathematics were fully realized after he earned his Ph.D. and joined the faculty of the Institute for Physical Science and Technology (IPST), an institute established in 1950 to foster excellence in interdisciplinary research and education at the University of Maryland. He said: *All along the goal of myself and my fellow researchers here at Maryland has been to find the concepts that the applied scientist needs.* His chaos research group introduced many basic concepts with exotic names like *crises*, the *control of chaos, fractal basin boundary, strange non-chaotic attractors*, and the *Kaplan–Yorke dimension.* One remarkable application of Yorke's theory of chaos has been the weather prediction.

In 2003 Yorke shared with Benoit Mandelbrot of Yale University the prize for *Science and Technology of Complexity* of the Science and Technology Foundation of Japan for the *Creation of Universal Concepts in Complex Systems-Chaos and Fractals.* With this prize, Jim Yorke was recognized for his outstanding achievements in nonlinear dynamics that have greatly advanced the frontiers of science and technology.

Yorke's research has been highly influential, with some of his papers receiving hundreds of citations. He is the author of three books on chaos, of a monograph on gonorrhoea epidemiology, and of more than 300 papers in the areas of ordinary differential equations, dynamical systems, delay differential equations, applied and random dynamical systems.

He believes that a Ph.D. in mathematics is a licence to investigate the universe, and he has supervised over 40 Ph.D. dissertations in the departments of mathematics, physics and computer science.

Currently, Jim Yorke is a Distinguished University Professor of Mathematics and Physics, and Chair of the Mathematics Department of the University of Maryland.

Arrigo Cellina received a Ph.D. degree in mathematics in 1968 and went back to Italy, where he was Assistant Professor and then Full Professor at the Universities of Perugia, Florence, and Padua, at the International School for Advanced Studies (SISSA) in Trieste, and at the University of Milan. He was a member of the scientific committee and then Director (1999–2001) of the International Mathematical Summer Centre (CIME) in Florence, Italy, and also a member of the scientific council of CIM (International Centre for Mathematics) seated in Coimbra, Portugal. Presently he is Professor at the University of Milan "Bicocca" and coordinator of the Doctoral Program of this university.

In Italy, the International School for Advanced Studies (SISSA) was established in 1978, in Trieste, as a dedicated and autonomous scientific institute to develop top-level research in mathematics, physics, astrophysics, biology and neuroscience, and to provide qualified graduate training to Italian and foreign laureates, to train them for research and academic teaching.

SISSA was the first Italian school to set up post-laurea courses aimed at a Ph. D. degree (Doctor Philosophiae). Cellina was one of the professors, founders and, for several years, the Coordinator of the Sector of Functional Analysis and Applications at SISSA, from 1978 until 1996.

I was lucky to have been initiated to mathematical research on Aubin–Cellina's book *Differential inclusions* in a research seminar at the University of Bucharest. Three years later I began my Ph.D. studies on differential inclusions at SISSA, under the supervision of Arrigo Cellina. I arrived at SISSA coming from Florence where I spent a very rewarding and training period of one year as a Research Fellow of GNAFA under the supervision of Roberto Conti, and I remember that Arrigo welcomed me with a kindness equal to his erudition.

Always available to discuss and to help his students to overcome difficulties, not only of mathematical orders, Arrigo taught me a lot more than differential inclusions. I remember with great pleasure his beautiful lessons, the long hours of reflection in front of the blackboard in his office, as well as the walks along the sea or in the park of Miramare.

I remember SISSA of those days as a very exciting environment. A community of researchers worked there, while several others were visiting SISSA and gave short courses or seminars concerning their new results. The Sector of Functional Analysis and Applications was located in a beautiful place, close to the Castle of Miramare, and near the International Centre for Theoretical Physics (ICTP), with an excellent library where we could spend much of our time. Without a doubt, this has been a very fruitful and rewarding period of my life, both as a scientific and as a life experience. Cellina's contribution has been significant.

Cellina's scientific work has always been highly original, introducing entirely new techniques to attack the difficult problems he considered. He introduced the notion of *graph approximate selection* for upper semicontinuous multifunctions, thus establishing a basic connection between ordinary differential inclusions and differential inclusions. He also introduced the *fixed-point approach* to prove the existence of differential inclusions based on *continuous selections from multifunctions with decomposable values.*

The Baire category method, for the analysis of differential inclusions without convexity assumptions, has been developed starting from Cellina's seminal paper *On the differential inclusion* $x' \in [-1, 1]$, published in 1980 by the *Rendiconti dell'Academia dei Lincei*. Eventually this method recently found applications to problems of the Calculus of Variations, without convexity or quasi-convexity assumptions, as well as to implicit differential equations. This year, it found even more striking new applications to the construction of deep counterexamples in the theory of multidimensional fluid flow.

More recently, Cellina's research activity was devoted to the area of the Calculus of Variations, where he obtained important results on the validity of the Euler-Lagrange equation, on the regularity of minimizers, on necessary and sufficient conditions for the existence of minima, and on uniqueness and comparison of minima without strict convexity.

The book *Differential Inclusions*, co-authored by Cellina with J. P. Aubin and published by Springer, as well as several of his eighty papers published in first-class journals, are now classic references to their subject. Cellina also edited several volumes with lectures and seminars of CIME sessions, published by Springer in the subseries *Fondazione C.I.M.E.* of the *Lecture Notes in Mathematics* series.

Cellina mentored ten Ph.D. students: seven of them while at SISSA and three others at the university of Milan. Among his former students, many are now Professors in prestigious universities in Italy, Portugal, Chile or other countries. Several more mathematicians continue to be inspired by his ground-breaking ideas.

Cellina and Yorke during the conference in Aveiro

In June 2006, I had the privilege to organize in Aveiro (Portugal) with my colleagues from the *Functional Analysis and Applications* research group, the conference *Views on ODEs*, in celebration of the 65th birthday of Arrigo Cellina and James A. Yorke. Several friends, former students and collaborators, presently leading experts in differential equations, chaos and variational problems, gathered in Aveiro on this occasion to discuss their new results. The present volume collects thirty-two original papers and state-of-the-art contributions of participants to this conference and brings the reader to the frontier of research in these modern fields of research.

I wish to thank Professor Haim Brezis for accepting to publish this book as a volume of the series *Progress in Nonlinear Differential Equations and Their Applications*. I also thank Thomas Hempfling for the professional and pleasant collaboration during the preparation of this volume. Finally, I gratefully acknowledge partial financial support from the Portuguese Foundation for Science and Technology (FCT) under the Project POCI/MAT/55524/2004 and from the *Mathematics and Applications* research unit of the University of Aveiro.

Aveiro, October 2007                                                             Vasile Staicu

Progress in Nonlinear Differential Equations
and Their Applications, Vol. 75, 1–14
© 2007 Birkhäuser Verlag Basel/Switzerland

# Nodal and Multiple Constant Sign Solution for Equations with the $p$-Laplacian

Ravi P. Agarwal, Michael E. Filippakis,
Donal O'Regan, and Nikolaos S. Papageorgiou

*Dedicated to Arrigo Cellina and James Yorke*

**Abstract.** We consider nonlinear elliptic equations driven by the $p$-Laplacian with a nonsmooth potential (hemivariational inequalities). We obtain the existence of multiple nontrivial solutions and we determine their sign (one positive, one negative and the third nodal). Our approach uses nonsmooth critical point theory coupled with the method of upper-lower solutions.

**Mathematics Subject Classification (2000).** 35J60, 35J70.

**Keywords.** Scalar $p$-Laplacian, eigenvalues, $(S)_+$-operator, local minimizer, positive solution, nodal solution.

## 1. Introduction

Let $Z \subseteq \mathbb{R}^N$ be a bounded domain with a $C^2$-boundary $\partial Z$. We consider the following nonlinear elliptic problem with nonsmooth potential (hemivariational inequality):

$$\left\{ \begin{array}{l} -\mathrm{div}\left(\|Dx(z)\|^{p-2}Dx(z)\right) \in \partial j\left(z, x(z)\right) \text{ a.e. on } Z\,, \\ x|_{\partial Z} = 0, \quad 1 < p < \infty\,. \end{array} \right\} \tag{1.1}$$

Here $j(z, x)$ is a measurable function on $Z \times \mathbb{R}$ and $x \to j(z, x)$ is locally Lipschitz and in general nonsmooth. By $\partial j(z, \cdot)$ we denote the generalized subdifferential of $j(z, \cdot)$ in the sense of Clarke [3]. The aim of this lecture is to produce multiple nontrivial solutions for problem (1.1) and also determine their sign (positive, negative or nodal (sign-changing) solutions). Recently this problem was studied for equations driven by the $p$-Laplacian with a $C^1$-potential function (single-valued

Researcher M. E. Filippakis supported by a grant of the National Scholarship Foundation of Greece (I.K.Y.).

right hand side), by Ambrosetti-Garcia Azorero-Peral Alonso [1], Carl-Perera [2], Garcia Azorero-Peral Alonso [7], Garcia Azorero-Manfredi-Peral Alonso [8], Zhang-Chen-Li [15] and Zhang-Li [16]. In [1], [7], [8], the authors consider certain nonlinear eigenvalue problems and obtain the existence of two strictly positive solutions for all small values of the parameter $\lambda \in \mathbb{R}$ (i.e., for all $\lambda \in (0, \lambda^*)$). In [2], [15], [16] the emphasis is on the existence of nodal (sign changing) solutions. Carl-Perera [2] extend to the $p$-Laplacian the method of Dancer-Du [6], by assuming the existence of an ordered pair of upper-lower solutions. In contrast, Zhang-Chen-Li [15] and Zhang-Li [16], base their approach on the invariance properties of certain carefully constructed pseudogradient flow. Our approach here is closer to that of Dancer-Du [6] and Carl-Perera [2], but in contrast to them, we do not assume the existence of upper-lower solutions, but instead we construct them and we use a recent alternative variational characterization of the second eigenvalue $\lambda_2$ of $(-\triangle_p, W_0^{1,p}(Z))$ due to Cuesta-de Figueiredo-Gossez [5], together with a nonsmooth version of the second deformation theorem due to Corvellec [4].

## 2. Mathematical background

Let $X$ be a Banach space and $X^*$ its topological dual. By $\langle \cdot, \cdot \rangle$ we denote the duality brackets for the pair $(X, X^*)$. Let $\varphi : X \to \mathbb{R}$ be a locally Lipschitz. The generalized directional derivative $\varphi^0(x; h)$ of $\varphi$ at $x \in X$ in the direction $h \in X$, is given by

$$\varphi^0(x; h) = \limsup_{\substack{x' \to x \\ \lambda \downarrow 0}} \frac{\varphi(x' + \lambda h) - \varphi(x')}{\lambda}.$$

The function $h \to \varphi^0(x; h)$ is sublinear continuous and so it is the support function of a nonempty, convex and $w^*$-compact set $\partial\varphi(x) \subseteq X^*$ defined by

$$\partial\varphi(x) = \left\{ x^* \in X^* : \langle x^*, h \rangle \leq \varphi^0(x; h) \quad \text{for all } h \in X \right\}.$$

The multifunction $x \to \partial\varphi(x)$ is known as the generalized subdifferential or subdifferential in the sense of Clarke. If $\varphi$ is continuous convex, then $\partial\varphi(x)$ coincides with the subdifferential in the sense of convex analysis. If $\varphi \in C^1(X)$, then $\partial\varphi(x) = \{\varphi'(x)\}$. We say that $x \in X$ is a critical point of $\varphi$, if $0 \in \partial\varphi(x)$. The main reference for this subdifferential, is the book of Clarke [3].

Given a locally Lipschitz function $\varphi : X \to \mathbb{R}$, we say that $\varphi$ satisfies the nonsmooth Palais-Smale condition at level $c \in \mathbb{R}$ (the nonsmooth $PS_c$-condition for short), if every sequence $\{x_n\}_{n\geq 1} \subseteq X$ such that $\varphi(x_n) \to c$ and $m(x_n) = \inf\{\|x\| : x^* \in \partial\varphi(x_n)\} \to 0$ as $n \to \infty$, has a strongly convergent subsequence. If this is true at every level $c \in \mathbb{R}$, then we say that $\varphi$ satisfies the $PS$-condition.

**Definition 2.1.** *Let $Y$ be a Hausdorff topological space and $E_0, E, D$ nonempty, closed subsets of $Y$ with $E_0 \subseteq E$. We say that $\{E_0, E\}$ is linking with $D$ in $Y$, if the following hold:*

(a) $E_0 \cap D = \emptyset$;

(b) *for any $\gamma \in C(E, Y)$ such that $\gamma|_{E_0} = id|_{E_0}$, we have $\gamma(E) \cap D \neq \emptyset$.*

Using this geometric notion, we can have the following minimax characterization of critical values for nonsmooth, locally Lipschitz functions (see Gasinski-Papageorgiou [9], p.139).

**Theorem 2.2.** *If $X$ is a Banach space, $E_0, E, D$ are nonempty, closed subsets of $X$, $\{E_0, E\}$ are linking with $D$ in $X$, $\varphi : X \to \mathbb{R}$ is locally Lipschitz, $\sup_{E_0} < \inf_D \varphi$, $\Gamma = \{\gamma \in C(E, X) : \gamma|_{E_0} = id|_{E_0}\}$, $c = \inf_{\gamma \in \Gamma} \sup_{v \in E} \varphi(\gamma(v))$ and $\varphi$ satisfies the nonsmooth $PS_c$-condition, then $c \geq \inf_D \varphi$ and $c$ is a critical value of $\varphi$.*

**Remark 2.3.** By appropriate choices of the linking sets $\{E_0, E, D\}$, from Theorem 2.2, we obtain nonsmooth versions of the mountain pass theorem, saddle point theorem, and generalized mountain pass theorem. For details, see Gasinksi-Papageorgiou [9].

Given a locally Lipschitz function $\varphi : X \to \mathbb{R}$, we set

$$\varphi^c = \{x \in X : \varphi(x) < c\} \quad \text{(the strict sublevel set of } \varphi \text{ at } c \in \mathbb{R})$$

and $K_c = \{x \in X : 0 \in \partial\varphi(x), \varphi(x) = c\}$ (the critical points of $\varphi$ at the level $c$).

The next theorem is a nonsmooth version, of the so-called "second deformation theorem" (see Gasinski-Papageorgiou [10], p.628) and it is due to Corvellec [4].

**Theorem 2.4.** *If $X$ is a Banach space, $\varphi : X \to \mathbb{R}$ is locally Lipschitz, it satisfies the nonsmooth $PS$-condition, $a \in \mathbb{R}$, $b \in \mathbb{R} \cup \{+\infty\}$, $\varphi$ has no critical points in $\varphi^{-1}(a, b)$ and $K_a$ is discrete nonempty and contains only local minimizers of $\varphi$, then there exists a deformation $h : [0, 1] \times \varphi^{\cdot b} \to \varphi^{\cdot b}$ such that*

(a) $h(t, \cdot)|_{K_a} = Id|_{K_a}$ *for all $t \in [0, 1]$;*

(b) $h(t, \varphi^{\cdot b}) \subseteq \varphi^{\cdot a} \cup K_a$;

(c) $\varphi(h(t, x)) \leq \varphi(x)$ *for all $(t, x) \in [0, 1] \times \varphi^{\cdot b}$.*

**Remark 2.5.** In particular then $\varphi^{\cdot b} \cup K_a$ is a weak deformation retract of $\varphi^{\cdot b}$.

Let us mention a few basic things about the spectrum of $(-\triangle_p, W_0^{1,p}(Z))$, which we will need in the sequel. So let $m \in L^\infty(Z)_+$, $m \neq 0$ and consider the following weighted eigenvalue problem:

$$\left\{ \begin{array}{l} -\text{div}\left(\|Dx(z)\|^{p-2} Dx(z)\right) = \lambda m(z)|x(z)|^{p-2}x(z) \text{ a.e. on } Z\,, \\ x|_{\partial Z} = 0,\ 1 < p < \infty\,. \end{array} \right\} \tag{2.1}$$

Problem (2.1) has at least eigenvalue $\widehat{\lambda}_1(m) > 0$, which is simple, isolated and admits the following variational characterization in terms of the Rayleigh quotient:

$$\widehat{\lambda}_1(m) = \min \left[ \frac{\|Dx\|_p^p}{\int_Z m|x|^p dz} : x \in W_0^{1,p}(Z)\,, \quad x \neq 0 \right] \tag{2.2}$$

The minimum is attained on the corresponding one dimensional eigenspace $E(\lambda_1)$. By $u_1$ we denote the normalized eigenfunction, i.e., $\int_Z m|u_1|^p dz = 1$

(if $m \equiv 1$, then $\|u_1\|_p = 1$). We have $E(\lambda_1) = \mathbb{R}u_1$ and $u_1 \in C_0^1(\overline{Z})$ (nonlinear regularity theory, see Lieberman [13] and Gasinski-Papageorgiou [10], p.738). We set

$$C_+ = \left\{ x \in C_0^1(\overline{Z}) : x(z) \geq 0 \text{ for all } z \in \overline{Z} \right\}$$

$$\text{and } \text{int}C_+ = \left\{ x \in C_+ : x(z) > 0 \text{ for all } z \in Z \text{ and } \frac{\partial x}{\partial n}(z) < 0 \text{ for all } z \in \partial Z \right\}.$$

The nonlinear strong maximum principle of Vazquez [14], implies that $u_1 \in \text{int}C_+$.

Since $\widehat{\lambda}_1(m)$ is isolated, we can define the second eigenvalue of $(-\triangle_p, W_0^{1,p}(Z), m)$ by

$$\widehat{\lambda}_2^*(m) = \inf \left[ \widehat{\lambda} : \widehat{\lambda} \text{ is an eigenvalue of } (2.1), \widehat{\lambda} \neq \widehat{\lambda}_1(m) \right] > \widehat{\lambda}_1(m).$$

Also by virtue of the Liusternik-Schnirelmann theory, we can find an increasing sequence of eigenvalues $\{\widehat{\lambda}_k(m)\}_{k \geq 1}$ such that $\widehat{\lambda}_k(m) \to \infty$. These are the so-called LS-eigenvalues. We have

$$\widehat{\lambda}_2^*(m) = \widehat{\lambda}_2(m),$$

i.e., the second eigenvalue and the second LS-eigenvalue coincide. The eigenvalues $\widehat{\lambda}_1(m)$ and $\widehat{\lambda}_2(m)$ exhibit the following monotonicity properties with respect to the weight function $m \in L^\infty(Z)_+$ :

-   If $m_1(z) \leq m_2(z)$ a.e. on $Z$, and $m_1 \neq m_2$, then $\lambda_1(m_2) < \lambda_1(m_1)$ (see (2.2)).
-   If $m_1(z) < m_2(z)$ a.e. on $Z$, then $\lambda_2(m_2) < \lambda_2(m_1)$.

If $m \equiv 1$, then we write $\widehat{\lambda}_1(1) = \lambda_1$ and $\widehat{\lambda}_2(1) = \lambda_2$. Recently Cuesta-de Figueiredo-Gossez [5], produced the following alternative variational characterization of $\lambda_2$ :

$$\lambda_2 = \inf_{\gamma_0 \in \Gamma_0} \sup_{x \in \gamma_0([-1,1])} \|Dx\|_p^p \tag{2.3}$$

with $\Gamma_0 = \{\gamma_0 \in C([-1,1], S) : \gamma_0(-1) = -u_1, \gamma_0(1) = u_1\}$, $S = W_0^{1,p}(Z) \cap \partial B_1^{L^p(Z)}$ and $\partial B_1^{L^p(Z)} = \{x \in L^p(Z) : \|x\|_p = 1\}$.

Finally we recall the notions of upper and lower solution for problem (1.1).

**Definition 2.6.**

(a) *A function $\overline{x} \in W^{1,p}(Z)$ is an upper solution of (1.1), if $\overline{x}|_{\partial Z} \geq 0$ and*

$$\int_Z \|D\overline{x}\|^{p-2}(D\overline{x}, Dv)_{\mathbb{R}^N} dz \geq \int_Z uv dz$$

*for all $v \in W_0^{1,p}(Z)$, $v \geq 0$ and all $u \in L^\eta(Z)$, $u(t) \in \partial j(t, \overline{x}(z))$ a.e. on $Z$ for some $1 < \eta < p^*$.*

(b) *A function $\underline{x} \in W^{1,p}(Z)$ is a lower solution of (1.1), if $\overline{x}|_{\partial Z} \leq 0$ and*

$$\int_Z \|D\underline{x}\|^{p-2}(D\underline{x}, Dv)_{\mathbb{R}^N} dz \leq \int_Z uv dz$$

*for all $v \in W_0^{1,p}(Z)$, $v \geq 0$ and all $u \in L^\eta(Z)$, $u(z) \in \partial j(z, \underline{x}(z))$ a.e. on $Z$ for some $1 < \eta < p^*$.*

## 3. Multiple constant sign solutions

In this section, we produce multiple solutions of constant sign. Our approach is based on variational techniques, coupled with the method of upper lower solutions. We need the following hypotheses on the nonsmooth potential $j(z, x)$.

$\underline{H(j)_1}$: $j : Z \times \mathbb{R} \to \mathbb{R}$ is a function such that $j(t, 0) = 0$ and $\partial j(z, 0) = \{0\}$ a.e. on $Z$, and

(i) for all $x \in \mathbb{R}$, $z \to j(z, x)$ is measurable;

(ii) for almost all $z \in Z$, $x \to j(z, x)$ is locally Lipschitz;

(iii) for a.a. $z \in Z$, all $x \in \mathbb{R}$ and all $u \in \partial j(z, x)$, we have

$$|u| \leq a(z) + c|x|^{p-1} \quad \text{with} \quad a \in L^\infty(Z)_+, \ c > 0;$$

(iv) there exists $\theta \in L^\infty(Z)_+$, $\theta(z) \leq \lambda_1$ a.e. on $Z$, $\theta \neq \lambda_1$ such that

$$\limsup_{|x| \to \infty} \frac{u}{|x|^{p-2}x} \leq \theta(z)$$

uniformly for a.a. $z \in Z$ and all $u \in \partial j(z, x)$;

(v) there exists $\eta, \widehat{\eta} \in L^\infty(Z)_+$, $\lambda_1 \leq \eta(z) \leq \widehat{\eta}(z)$ a.e. on $Z$, $\lambda_1 \neq \eta$ such that

$$\eta(z) \leq \liminf_{x \to 0} \frac{u}{|x|^{p-2}x} \leq \limsup_{x \to 0} \frac{u}{|x|^{p-2}x} \leq \widehat{\eta}(z)$$

uniformly for a.a. $z \in Z$ and all $u \in \partial j(z, x)$;

(vi) for a.a. $z \in Z$, all $x \in \mathbb{R}$ and all $u \in \partial j(z, x)$, we have $ux \geq 0$ (sign condition).

Let $\varepsilon > 0$ and $\gamma_\varepsilon \in L^\infty(Z)_+$, $\gamma_\varepsilon \neq 0$ and consider the following auxiliary problem:

$$\left\{ \begin{array}{l} -\text{div}\big(\|Dx(z)\|^{p-2}Dx(z)\big) = \big(\theta(z) + \varepsilon\big)|x(z)|^{p-2}x(z) + \gamma_\varepsilon(z) \ \text{a.e. on } Z, \\ x|_{\partial Z} = 0. \end{array} \right\} \quad (3.1)$$

In what follows by $\langle \cdot, \cdot \rangle$ we denote the duality brackets for the pair $(W_0^{1,p}(Z), W^{-1,p'}(Z))$ $(\frac{1}{p} + \frac{1}{p'} = 1)$. Let $A : W_0^{1,p}(Z) \to W^{-1,p'}(Z)$ be the nonlinear operator defined by

$$\langle A(x), y \rangle = \int_Z \|Dx\|^{p-2}(Dx, Dy)_{\mathbb{R}^N} dz \quad \text{for all } x, y \in W_0^{1,p}(Z).$$

We can check that $A$ is monotone, continuous, hence maximal monotone. In particular then we can deduce that $A$ is pseudomonotone and of type $(S)_+$.

Also let $N_\varepsilon : L^p(Z) \to L^{p'}(Z)$ be the bounded, continuous map defined by

$$N_\varepsilon(x)(\cdot) = \big(\theta(\cdot) + \varepsilon\big)|x(\cdot)|^{p-2}x(\cdot).$$

Evidently due to the compact embedding of $W_0^{1,p}(Z)$ into $L^p(Z)$, we have that $N_\varepsilon|_{W_0^{1,p}(Z)}$ is completely continuous. Hence $x \to A(x) - N_\varepsilon(x)$ is pseudomonotone. Moreover, from the hypothesis on $\theta$ (see $H(j)_1(iv)$), we can show that there exists $\xi_0 > 0$ such that

$$\|Dx\|_p^p - \int_Z \theta|x|^p dz \geq \xi_0 \|Dx\|_p^p \quad \text{for all } x \in W_0^{1,p}(Z). \tag{3.2}$$

Therefore for $\varepsilon > 0$ small the pseudomonotone operator $x \to A(x) - N_\varepsilon(x)$ is coercive. But a pseudomonotone coercive operator is surjective (see Gasinski-Papageorgiou [10], p.336). Combining this fact with the nonlinear strong maximum principle, we are led to the following existence result concerning problem (3.1).

**Proposition 3.1.** *If $\theta \in L^\infty(Z)_+$ is as in hypothesis $H(j)_1(iv)$, then for $\varepsilon > 0$ small problem (3.1) has a solution $\overline{x} \in intC_+$.*

Because of hypothesis $H(j)_1(iv)$, we deduce easily the following fact:

**Proposition 3.2.** *If hypotheses $H(j)_1 \to (iv)$ hold and $\varepsilon > 0$ is small, then the solution $\overline{x} \in intC_+$ obtained in Proposition 3.1 is a strict upper solution for (1.1) (strict means that $\overline{x}$ is an upper solution which is not a solution).*

Clearly $\underline{x} \equiv 0$ is a lower solution for (1.1).

Let $C = [0, \overline{x}] = \{x \in W_0^{1,p}(Z) : 0 \leq x(z) \leq \overline{x}(z) \text{ a.e. on } Z\}$. We introduce the truncation function $\tau_+ : \mathbb{R} \to \mathbb{R}$ defined by

$$\tau_+(x) = \begin{cases} 0 & \text{if } x \leq 0 \\ x & \text{if } x > 0 \end{cases}.$$

We set $j_1(z, x) = j(z, \tau_+(x))$. This is still a locally Lipschitz integrand. We introduce $\varphi_+ : W_0^{1,p}(Z) \to \mathbb{R}$ defined by

$$\varphi_+(x) = \frac{1}{p}\|Dx\|_p^p - \int_Z j_+(z, x(z))dz \quad \text{for all } x \in W_0^{1,p}(Z).$$

The function $\varphi_+$ is Lipschitz continuous on bounded sets, hence locally Lipschitz. Using hypothesis $H(j)_1(iv)$ and (3.2), we can show that $\varphi_+$ is coercive. Moreover, due to the compact embedding of $W_0^{1,p}(Z)$ into $L^p(Z)$, $\varphi_+$ is weakly lower semicontinuous. Therefore by virtue of Weierstrass theorem, we can find $x_0 \in C$ such that

$$\varphi_+(x_0) = \inf_C \varphi_+. \tag{3.3}$$

Hypothesis $H(j)_1(v)$ implies that for $\mu > 0$ small we have $\varphi_+(\mu u_1) < 0 = \varphi_+(0)$. Since $\mu u_1 \in C$, it follows that $x_0 \neq 0$. Moreover, from (3.3) we have

$$0 \leq \langle A(x_0), y - x_0 \rangle - \int_Z u_0(z)(y - x_0)(z)dz \quad \text{for all } y \in C, \tag{3.4}$$

with $u_0 \in L^{p'}(Z)$, $u_0(z) \in \partial j_+(z, x_0(z)) = \partial j(z, x_0(z))$ a.e. on $Z$. For $h \in W_0^{1,p}(Z)$ and $\varepsilon > 0$, we define

$$y(z) = \begin{cases} 0 & \text{if } z \in \{x_0 + \varepsilon h \le 0\} \\ x_0(z) + \varepsilon h(z) & \text{if } z \in \{0 < x_0 + \varepsilon h \le \overline{x}\} \\ \overline{x}(z) & \text{if } z \in \{\overline{x} \le x_0 + \varepsilon h\} \end{cases}.$$

Evidently $y \in C$ and so we can use it as a test function in (3.4). Then we obtain

$$0 \le \langle A(x_0) - u_0, h \rangle. \tag{3.5}$$

Because $h \in W_0^{1,p}(Z)$ was arbitrary, from (3.5) we conclude that

$$A(x_0) = u_0 \quad \Rightarrow \quad x_0 \in W_0^{1,p}(Z) \text{ is a solution of (1.1)}. \tag{3.6}$$

Nonlinear regularity theory implies that $x_0 \in C_0^1(\overline{Z})$, while the nonlinear strong maximum principle of Vazquez [14], tell us that $x_0 \in \text{int}C_+$.

Using the comparison principles of Guedda-Veron [11], we can show that

$$\overline{x} - x_0 \in \text{int}C_+.$$

Therefore $x_0$ is a local $C_0^1(\overline{Z})$-minimizer of $\varphi$, hence $x_0$ is a local $W_0^{1,p}(Z)$-minimizer of $\varphi$ (see Gasinski-Papageorgiou [9], pp.655–656 and Kyritsi-Papageorgiou [12]). Therefore we can state the following result:

**Proposition 3.3.** *If hypotheses $H(j)_1$ hold, then there exists $x_0 \in C$ which is a local minimizer of $\varphi_+$ and of $\varphi$.*

If instead of (3.1), we consider the following auxiliary problem

$$\left\{ \begin{array}{l} -\text{div}\big(\|Dv(z)\|^{p-2}Dv(z)\big) = \big(\theta(z) + \varepsilon\big)|v(z)|^{p-2}v(z) - \gamma_\varepsilon(z) \text{ a.e. on } Z, \\ v|_{\partial Z} = 0. \end{array} \right\} \tag{3.7}$$

then we obtain as before a solution $\underline{v} \in -\text{int}C_+$ of (3.7). We can check that this $\underline{v} \in -\text{int}C_+$ is a strict lower solution for problem (1.1). Now we consider the set

$$D = \big\{ x \in v \in W_0^{1,p}(Z) : \underline{v}(z) \le v(z) \le 0 \text{ a.e. on } Z \big\}.$$

We introduce the truncation function $\tau_- : \mathbb{R} \to \mathbb{R}_-$. defined by

$$\tau_-(x) = \begin{cases} x & \text{if } x < 0 \\ 0 & \text{if } x \ge 0 \end{cases}.$$

Then $j_-(z, x) = j(z, \tau(x))$ and $\varphi_-(x) = \frac{1}{p}\|Dx\|_p^p - \int_Z j_-(z, x(z))dz$ for all $x \in W_0^{1,p}(Z)$. We consider the minimization problem $\inf\limits_D \varphi_-$. Reasoning as with $\varphi_+$ on $C$, we obtain:

**Proposition 3.4.** *If hypotheses $H(j)_1$ hold, then there exists $v_0 \in D$ which is a local minimizer of $\varphi_-$ and of $\varphi$.*

Propositions 3.3 and 3.4, lead to the following multiplicity theorem for solutions of constant sign for problem (1.1).

**Theorem 3.5.** *If hypotheses $H(j)_1$ hold, then problem (1.1) has at least two constant sign smooth solutions $x_0 \in intC_+$ and $v_0 \in -intC_+$.*

**Remark 3.6.** Since $x_0, v_0$ are both local minimizers of $\varphi$, from the mountain pass theorem, we obtain a third critical point $y_0$ of $\varphi$, distinct from $x_0, v_0$. However, at this point we can not guarantee that $y_0 \neq 0$, let alone that it is nodal. This will be done in the next section under additional hypotheses.

## 4. Nodal solutions

In this section we produce a third nontrivial solution for problem (1.1) which is nodal (i.e., sign-changing). Our approach was inspired by the work of Dancer-Du [6]. Roughly speaking the strategy is the following: Continuing the argument employed in Section 3, we produce a smallest positive solution $y_+$ and a biggest negative solution $y_-$. In particular $\{y_\pm\}$ is an ordered pair of upper-lower solutions. So, if we form the order interval $[y_-, y_+]$ and we argue as in Section 3, we can show that problem (1.1) has a solution $y_0 \in [y_-, y_+]$ distinct from $y_-, y_+$. If we can show that $y_0 \neq 0$, then clearly $y_0$ is a nodal solution of (1.1). To show the nontriviality of $y_0$, we use Theorem 2.4 and (2.3).

We start implementing the strategy, by proving that the set of upper (resp. lower) solutions for problem (1.1), is downward (resp. upward) directed. The proof relies on the use of the truncation function

$$\xi_\varepsilon(s) = \begin{cases} -\varepsilon & \text{if } s < \varepsilon \\ s & \text{if } s \in [-\varepsilon, \varepsilon] \\ \varepsilon & \text{if } s > \varepsilon \end{cases} .$$

Note that

$$\frac{1}{\varepsilon}\xi_\varepsilon\big((y_1 - y_1)^-(z)\big) \to \chi_{\{y_1 < y_2\}}(z) \quad \text{a.e. on } Z \text{ as } \varepsilon \downarrow 0.$$

So we have the following lemmata

**Lemma 4.1.** *If $y_1, y_2 \in W^{1,p}(Z)$ are two upper solutions for problem (1.1) and $y = \min\{y_1, y_2\} \in W^{1,p}(Z)$, then $y$ is also an upper solution for problem (1.1).*

**Lemma 4.2.** *If $v_1, v_2 \in W^{1,p}(Z)$ are two lower solutions for problem (1.1) and $v = \max\{v_1, v_2\} \in W^{1,p}(Z)$, then $v$ is also a lower solution for problem (1.1).*

In Section 3 we used zero as a lower solution for the "positive" problem and as an upper solution for the "negative" problem. However, this is not good enough for the purpose of generating a smallest positive and a biggest negative solution, as described earlier. For this reason, we strengthen the hypotheses on $j(z, x)$ as follows:

$\underline{H(j)_2}$: $j : Z \times \mathbb{R} \to \mathbb{R}$ is a function such that $j(t, 0) = 0$ a.e. on $Z$, $\partial j(z, 0) = \{0\}$ a.e. on $Z$, hypotheses $H(j)_2(i) \to (iv)$ and $(vi)$ are the same as hypotheses $H(j)_1(i) \to (iv)$ and $(vi)$ and

(iv) there exists $\widehat{\eta} \in L^\infty(Z)_+$, such that

$$\lambda_1 < \liminf_{x \to 0} \frac{u}{|x|^{p-2}x} \le \limsup_{x \to 0} \frac{u}{|x|^{p-2}x} \le \widehat{\eta}(z)$$

uniformly for a.a. $z \in Z$ and all $u \in \partial j(z, x)$.

Using this stronger hypothesis near origin, we can find $\mu_0 \in (0,1)$ small such that $\underline{x} = \mu_0 u_1 \in \mathrm{int}C_+$ is a strict lower solution and $\overline{v} = \mu_0(-u_1) \in -\mathrm{int}C_+$ is a strict upper solution for problem (1.1). So we can state the following lemma:

**Lemma 4.3.** *If hypotheses $H(j)_2$ hold, then problem (1.1) has a strict lower solution $\underline{x} \in \mathrm{int}C_+$ and a strict upper solution $\overline{v} \in -\mathrm{int}C_+$.*

We consider the order intervals

$$[\underline{x}, \overline{x}] = \left\{ x \in W_0^{1,p}(Z) : \underline{x}(z) \le x(z) \le \overline{x}(z) \ \text{a.e. on} \ Z \right\}$$

$$\text{and} \ \ [\underline{v}, \overline{v}] = \left\{ v \in W_0^{1,p}(Z) : \underline{v}(z) \le v(z) \le \overline{v}(z) \ \text{a.e. on} \ Z \right\}.$$

Using Lemmata 4.1 and 4.2 and Zorn's lemma, we prove the following result:

**Proposition 4.4.** *If hypotheses $H(j)_2$ hold, then problem (1.1) admits a smallest solution in the order interval $[\underline{x}, \overline{x}]$ and a biggest solution in the order interval $[\underline{v}, \overline{v}]$.*

Now let $\underline{x}_n = \varepsilon_n u_1$ with $\varepsilon_n \downarrow 0$ and let $E_+^n = [\underline{x}_n, \overline{x}]$. Proposition 4.4 implies that problem (1.1) has a smallest solution $x_*^n$ in $E_+^n$. Clearly $\{x_*^n\}_{n \ge 1} \subseteq W_0^{1,p}(Z)$ is bounded and so by passing to a suitable subsequence if necessary, we may assume that

$$x_*^n \xrightarrow{w} y_+ \ \text{in} \ W_0^{1,p}(Z) \quad \text{and} \quad x_*^n \to y_+ \ \text{in} \ L^p(Z) \ \text{as} \ n \to \infty.$$

Arguing by contradiction and using hypothesis $H(j)_2(v)$, we can show that $y_+ \neq 0$ and of course $y_+ \ge 0$. Here we use the strict monotonicity of the principal eigenvalue on the weight function (see Section 2). Moreover, by Vazquez [14], we have $y_+ \in \mathrm{int}C_+$ and using this fact it is not difficult to check that $y_+$ is in fact the smallest positive solution of problem (1.1).

Similarly, working on $E_-^n = [\underline{v}, \overline{v}_n]$ with $\overline{v}_n = \varepsilon_n(-u_1)$, $\varepsilon_n \downarrow 0$, we obtain $y_- \in -\mathrm{int}C_+$ the biggest negative solution of (1.1). So we can state the following proposition:

**Proposition 4.5.** *If hypotheses $H(j)_2$ hold, then problem (1.1) has a smallest positive solution $y_+ \in \mathrm{int}C_+$ and a biggest negative solution $y_- \in -\mathrm{int}C_+$.*

According to the scheme outlined in the beginning of the section, using this proposition, we can establish the existence of a nodal solution. As we already mentioned, a basic tool to this end, is equation (2.3). But in order to be able to use (2.3), we need to strengthen further our hypothesis near the origin. Also we need to restrict the kind of locally Lispchitz functions $j(z, x)$, we have. Namely, let $f : Z \times \mathbb{R} \to \mathbb{R}$ be a measurable function such that for every $r > 0$ there exists $a_r \in L^\infty(Z)_+$ such that

$$|f(z, x)| \le a_r(z) \ \text{for a.a.} \ z \in Z \quad \text{and all} \ |x| \le r.$$

We introduce the following two limit functions:

$$f_1(z, x) = \liminf_{x' \to x} f(z, x') \quad \text{and} \quad f_2(z, x) = \limsup_{x' \to x} f(z, x').$$

Both functions are $\mathbb{R}$-valued for a.a. $z \in Z$. In addition we assume that they are sup-measurable, meaning that for every $x : Z \to \mathbb{R}$ measurable function, the functions $z \to f_1(z, x(z))$ and $z \to f_2(z, x(z))$ are both measurable. We set

$$j(z, x) = \int_0^x f(z, s)ds. \tag{4.1}$$

Evidently $(z, x) \to j(z, x)$ is jointly measurable and for a.a. $z \in Z$, $x \to j(z, x)$ is locally Lipschitz. We have

$$\partial j(z, x) = \big[f_1(z, x), f_2(z, x)\big] \quad \text{for a.a. } z \in Z, \quad \text{for all } x \in \mathbb{R}.$$

Clearly $j(z, 0) = 0$ a.e. on $Z$ and if for a.a. $z \in Z$, $f(z, \cdot)$ is continuous at $0$, then $\partial j(z, 0) = \{0\}$ for a.a. $z \in Z$. The hypotheses on this particular nonsmooth potential function $j(z, x)$ are the following:

<u>$H(j)_3$</u>: $j : Z \times \mathbb{R} \to \mathbb{R}$ is defined by (4.1) and

    (i) $(z, x) \to f(z, x)$ is measurable with $f_1, f_2$ sup-measurable;

    (ii) for a.a. $z \in Z$, $x \to f(z, x)$ is continuous at $x = 0$;

    (iii) $|f(z, x)| \leq a(z) + c|x|^{p-1}$ a.e. on $Z$, for all $x \in \mathbb{R}$, with $a \in L^\infty(Z)_+$, $c > 0$;

    (iv) there exists $\theta \in L^\infty(Z)_+$ satisfying $\theta(z) \leq \lambda_1$ a.e. on $Z$, $\theta \neq \lambda_1$ and

$$\limsup_{|x| \to \infty} \frac{f_2(z, x)}{|x|^{p-2}x} \leq \theta(z)$$

    uniformly for a.a. $z \in Z$;

    (v) there exists $\widehat{\eta} \in L^\infty(Z)_+$ such that

$$\lambda_2 < \liminf_{x \to 0} \frac{f_1(z, x)}{|x|^{p-2}x} \limsup_{x \to 0} \frac{f_2(z, x)}{|x|^{p-2}x} \leq \widehat{\eta}(z)$$

    uniformly for a.a. $z \in Z$;

    (vi) for a.a. $z \in Z$ and all $x \in \mathbb{R}$, we have $f_1(z, x)x \geq 0$ (sign condition).

From Proposition 4.5, we have a smallest positive solution $y_+ \in \text{int}C_+$ and a biggest negative solution $y_- \in -\text{int}C_+$ for problem (1.1). We have

$$A(y_\pm) = u_\pm \quad \text{with} \quad u_\pm \in L^{p'}(Z), \; u_\pm(z) \in \partial j(z, x_\pm(z)) \quad \text{a.e. on } Z.$$

We introduce the following truncations of the functions $f(z,x)$ :

$$\widehat{f}_+(z,x) = \begin{cases} 0 & \text{if } x < 0 \\ f(z,x) & \text{if } 0 \le x \le y_+(z) \ , \\ u_+(z) & \text{if } y_+(z) < x \end{cases}$$

$$\widehat{f}_-(z,x) = \begin{cases} u_-(z) & \text{if } x < y_-(z) \\ f(z,x) & \text{if } y_-(z) \le x \le 0 \ , \\ 0 & \text{if } 0 < x \end{cases}$$

$$\widehat{f}(z,x) = \begin{cases} u_-(z) & \text{if } x < y_-(z) \\ f(z,x) & \text{if } y_-(z) \le x \le y_+(z) \ , \\ u_+(z) & \text{if } y_+(z) < x \end{cases}$$

Using them, we define the corresponding locally Lipschitz potential functions, namely $\widehat{j}_+(z,x) = \int_0^x \widehat{f}_+(z,s)ds$, $\widehat{j}_-(z,x) = \int_0^x \widehat{f}_-(z,s)ds$ and $\widehat{j}(z,x) = \int_0^x \widehat{f}(z,s)ds$ for all $(z,x) \in Z \times \mathbb{R}$.

Also, we introduce the corresponding locally Lipschitz Euler functionals defined on $W_0^{1,p}(Z)$. So we have

$$\widehat{\varphi}_+(x) = \frac{1}{p}\|Dx\|_p^p - \int_Z \widehat{j}_+\big(z,x(z)\big)dz, \ \widehat{\varphi}_-(x) = \frac{1}{p}\|Dx\|_p^p - \int_Z \widehat{j}_-\big(z,x(z)\big)dz$$

and $\widehat{\varphi}(x) = \frac{1}{p}\|Dx\|_p^p - \int_Z \widehat{j}\big(z,x(z)\big)dz$ for all $x \in W_0^{1,p}(Z)$.

Finally, we set

$$T_+ = [0,y_+], \ T_- = [y_-,0] \ \text{and} \ T = [y_-,y_+]\,.$$

We can show that the critical points of $\varphi_+$ (resp. of $\varphi_-,\varphi$) are in $T_+$ (resp. in $T_-,T$). So the critical points of $\widehat{\varphi}_+$ (resp. $\widehat{\varphi}_-$) are $\{0,y_+\}$ (resp. $\{0,y_-\}$). Moreover,

$$\widehat{\varphi}_+(y_+) = \inf \widehat{\varphi}_+ < 0 = \widehat{\varphi}_+(0) \quad \text{and} \quad \widehat{\varphi}_-(y_-) = \inf \widehat{\varphi}_- < 0 = \widehat{\varphi}_-(0)\,.$$

Clearly $y_+,y_-$ are local $C_0^1(\overline{Z})$-minimizers of $\widehat{\varphi}$ and so they are also local $W_0^{1,p}(Z)$-minimizers. Without any loss of generality, we may assume that they are isolated critical points of $\widehat{\varphi}$. So we can find $\delta > 0$ small such that

$$\widehat{\varphi}(y_-) < \inf\big[\widehat{\varphi}(x) : x \in \partial B_\delta(y_-)\big] \le 0,$$

$$\widehat{\varphi}(y_+) < \inf\big[\widehat{\varphi}(x) : x \in \partial B_\delta(y_+)\big] \le 0,$$

where $\partial B_\delta(y_\pm) = \{x \in W_0^{1,p}(Z) : \|x - y_\pm\| = \delta\}$. Assume without loss of generality that $\widehat{\varphi}(y_-) \le \widehat{\varphi}(y_+)$.

If we set $S = \partial B_\delta(y_+)$, $T_0 = \{y_-,y_+\}$ and $T = [y_-,y_+]$, then we can check that the pair $\{T_0,T\}$ is linking with $S$ in $W_0^{1,p}(Z)$. So by virtue of Theorem 2.2, we can find $y_0 \in W_0^{1,p}(Z)$ a critical point of $\widehat{\varphi}$ such that

$$\widehat{\varphi}(y_\pm) < \widehat{\varphi}(y_0) = \inf_{\overline{\gamma} \in \Gamma} \max_{t \in [-1,1]} \widehat{\varphi}\big(\overline{\gamma}(t)\big) \tag{4.2}$$

where $\overline{\Gamma} = \{\overline{\gamma} \in C([-1,1], W_0^{1,p}(Z)) : \overline{\gamma}(-1) = y_-, \overline{\gamma}(1) = y_+\}$. Note that from (4.2) we infer that $y_0 \ne y_\pm$.

We will show that $\widehat{\varphi}(y_0) < \widehat{\varphi}(0) = 0$ and so $y_0 \neq 0$. Hence $y_0$ is the desired nodal solution. To establish the nontriviality of $y_0$, it suffices to construct a path $\overline{\gamma}_0 \in \overline{\Gamma}$ such that

$$\widehat{\varphi}(\gamma_0(t)) < 0 \quad \text{for all } t \in [0,1] \ (\text{see } (4.2)) \, .$$

Using (2.3), we can produce a continuous path $\gamma_0$ joining $-\varepsilon u_1$ and $\varepsilon u_1$ for $\varepsilon > 0$ small. Note that if $S_c = C_0^1(\overline{Z}) \cap \partial B_1^{L^p(Z)}$ and $S = W_0^{1,p}(Z) \cap \partial B_1^{L^p(Z)}$ are equipped with the relative $C_0^1(\overline{Z})$ and $W_0^{1,p}(Z)$ topologies respectively, then

$$C\bigl([-1,1], S_c\bigr) \quad \text{is dense in } C\bigl([-1,1], S\bigr) \, .$$

Also we have

$$\widehat{\varphi}|_{\gamma_0} < 0 \, . \tag{4.3}$$

Using Theorem 2.4, we can generate the continuous path

$$\gamma_+(t) = h(t, \varepsilon u_1) \, , \quad t \in [0,1] \, ,$$

with $h(t,x)$ the deformation of Theorem 2.4. This path joins $\varepsilon u_1$ and $y_+$. Moreover, we have

$$\widehat{\varphi}|_{\gamma_+} < 0 \, . \tag{4.4}$$

In a similar fashion we produce a continuous path $\gamma_-$ joining $y_-$ with $-\varepsilon u_1$ such that

$$\widehat{\varphi}|_{\gamma_-} < 0 \, . \tag{4.5}$$

Concatinating $\gamma_-, \gamma_0$ and $\gamma_+$, we produce a path $\overline{\gamma}_0 \in \overline{\Gamma}$ such that

$$\widehat{\varphi}|_{\overline{\gamma}_0} < 0 \ (\text{see } (4.3),(4.4) \text{ and } (4.5)) \, .$$

This proves that $y_0 \neq 0$ and so $y_0$ is a nodal solution. Nonlinear regularity theory implies that $y_0 \in C_0^1(\overline{Z})$.

Therefore we can state the following theorem on the existence of nodal solutions

**Theorem 4.6.** *If hypotheses $H(j)_3$ hold, then problem (1.1) has a nodal solution $y_0 \in C_0^1(\overline{Z})$.*

Combining Theorems 3.5 and 4.6, we can state the following multiplicity result for problem (1.1).

**Theorem 4.7.** *If hypotheses $H(j)_3$ hold, then problem (1.1) has at least three nontrivial solutions, one positive $x_0 \in \text{int}C_+$, one negative $v_0 \in -\text{int}C_+$ and the third $y_0 \in C_0^1(\overline{Z})$ nodal.*

## Acknowledgment

Many thanks to our TₑX-pert for developing this class file.

# References

[1] A. Ambrosetti, J. G. Azorero, J. P. Alonso, *Multiplicity results for some nonlinear elliptic equations*, J. Funct. Anal. **137** (1996), 219–242.

[2] S. Carl, K. Perera, *Sign-Changing and multiple solutions for the p-Laplacian*, Abstr. Appl. Anal. **7** (2003) 613–626.

[3] F. H. Clarke, *Optimization and Nonsmooth Analysis*, Willey, New York, 1983.

[4] J.-N. Corvellec, *On the second deformation lemma*, Topol. Math. Nonl. Anal., **17** (2001), 55–66.

[5] M. Cuesta, D. de Figueiredo, J.-P. Gossez, *The beginning of the Fucik spectrum of the p-Laplacian*, J. Diff. Eqns., **159** (1999), 212–238.

[6] N. Dancer, Y. Du, *On sign-changing solutions of certain semilinear elliptic problems*, Appl. Anal., **56** (1995), 193–206.

[7] J. G. Azorero, I. P. Alonso, *Some results about the existence of a second positive solution in a quasilinear critical problem*, Indiana Univ. Math. Jour., **43** (1994), 941–957.

[8] J. G. Azorero, J. Manfredi, I. P. Alonso, *Sobolev versus Hölder local minimizers and global multiplicity for some quasilinear elliptic equations*, Comm. Contemp. Math., **2** (2000), 385–404.

[9] L. Gasinski, N. S. Papageorgiou, *Nonsmooth Critical Point Theory and Nonlinear Boundary Value Problems*, Chapman and Hall/CRC Press, Boca Raton, 2005.

[10] L. Gasinski, N. S. Papageorgiou, *Nonlinear Analysis*, Chapman and Hall/CRC Press, Boca Raton, 2006.

[11] M. Guedda, L. Veron, *Quasilinear elliptic equations involving critical Sobolev exponents*, Nonlin. Anal., **13** (1989), 879–902.

[12] S. Kyritsi, N. S. Papageorgiou, *Multiple solutions of constant sign for nonlinear nonsmooth eigenvalue problems and resonance*, Calc. Var., **20** (2004), 1–24.

[13] G. Lieberman, *Boundary regularity for solutions of degenerate equations*, Nonlin. Anal., **12** (1988), 1203–1219.

[14] J. Vazquez, *A strong maximum principle for some quasilinear elliptic equations*, Appl. Math. Optim., **12** (1984), 191–202.

[15] Z. Zhang, J. Chen, S. Li, *Construction of pseudogradient vector field and sign-changing multiple solutions involving p-Laplacian*, J. Diff. Eqns, **201** (2004), 287–303.

[16] Z. Zhang, S. Li, *On sign-changing and multiple solutions of the p-Laplacian*, J. Funct. Anal., **197** (2003), 447–468.

Ravi P. Agarwal
Department of Mathematical Sciences,
Florida Institute of Technology,
Melbourne 32901-6975, FL, USA
e-mail: agarwal@fit.edu

Michael E. Filippakis
Department of Mathematics,
National Technical University,
Zografou Campus, Athens 15780, Greece
e-mail: mfilip@math.ntua.gr

Donal O'Regan
Department of Mathematics,
National University of Ireland,
Galway, Ireland
e-mail: oregan@nuigalway.ie

Nikolaos S. Papageorgiou
Department of Mathematics,
National Technical University,
Zografou Campus, Athens 15780, Greece
e-mail: npapg@math.ntua.gr

Progress in Nonlinear Differential Equations
and Their Applications, Vol. 75, 15–28
© 2007 Birkhäuser Verlag Basel/Switzerland

# A Young Measures Approach to Averaging

Zvi Artstein

*Dedicated to Arrigo Cellina and James Yorke*

**Abstract.** Employing a fast time scale in the Averaging Method results in a limit dynamics driven by a Young measure. The rate of convergence to the limit induces quantitative estimates for the averaging. Advantages that can be drawn from the Young measures approach, in particular, allowing time-varying averages, are displayed along with a connection to singularly perturbed systems.

## 1. Introduction

The Averaging Method suggests that a time-varying yet small perturbation on a long time interval, can be approximated by a time-invariant perturbation obtained by "averaging" the original one. The method has been introduced in the 19th Century as a practical device helping computations of stellar motions. Its rigorous grounds have been affirmed in the middle of the 20th Century. Many applications, including to fields beyond computations, make the field very attractive today. For a historical account and many applications consult Lochak and Meunier [10], Sanders and Verhulst [14], Verhulst [17], and references therein.

In this paper we make a connection between the averaging method and another useful tool, namely, probability measure-valued maps, called Young measures. These were introduced by L.C. Young as generalized curves in the calculus of variations; other usages are as relaxed controls, worked out by J. Warga, limits of solutions of partial differential equations, and many more. For an account of some of the possible applications consult the monographs and surveys Young [19], Warga [18], Valadier, [16], Pedregal [12, 13], Balder [6], and references therein. For a connection to singular perturbations extending, in particular, the Levinson-Tikhonov scope, see Artstein [3].

Incumbent of the Hettie H. Heineman Professorial Chair in Mathematics. Research supported by the Israel Science Foundation.

The qualitative consequences of the averaging method played a role in all the aforementioned applications of Young measures. The purpose of this note is to show that the Young measures approach can contribute to the considerations of averaging, including to the quantitative estimates the theory offers.

In the next section we explain how Young measures arise in the averaging considerations. A general estimate based on the distance in the sense of Young measures is displayed in Section 3. Applications to the classical averaging, along with some examples, are given in Section 4. Averaging considerations relative to subsequences, resulting, in particular, in time-varying averages, is a feature Young measures help to clarify; it is displayed in Section 5 along with a comment on the connection to singularly perturbed systems.

## 2. The connection

In this section we provide the basic definitions of Young measures and explain how they arise in considerations of averaging. We start actually with the latter, namely, provide the motivation first.

Averaging of ordinary differential equations is concerned with an equation which depends on a small positive parameter $\varepsilon$ and given by

$$\frac{dx}{dt} = \varepsilon f(t, x, \varepsilon), \quad x(0) = x_0. \tag{2.1}$$

We assume, throughout, continuity of $f(t, x, \varepsilon)$ in $x$ and measurability in $t$ (continuity in $\varepsilon$ is not needed in general; it is explicitly assumed below when used). In many applications one has to carry out a change of variables in order to arrive to the form (2.1); in fact, the form (2.1) already depicts the small perturbation; see Verhulst [17] for an elaborate discussion. Of interest is the limit behaviour of solutions of (2.1) as $\varepsilon \to 0$. A typical result assures, under appropriate conditions, that the solution, say $x(\cdot)$, of (2.1) (it depends on $\varepsilon$) is close to the solution, say $x_0(\cdot)$, of the averaged equation, namely, the equation

$$\frac{dx}{dt} = \varepsilon f^0(x), \quad x(0) = x_0; \tag{2.2}$$

here the time-invariant right hand side of (2.2) is the limit average of the original equation, namely,

$$f^0(x) = \lim_{T \to \infty, \ \varepsilon \to 0} \frac{1}{T} \int_0^T f(t, x, \varepsilon) dt, \tag{2.3}$$

assuming, of course, that the limit exists. (The order of convergence between $T$ and $\varepsilon$ in (2.3) may play a role; we do not address this issue in this general discussion.) Furthermore, the theory assures that the two solutions, $x(\cdot)$ and $x_0(\cdot)$, are uniformly close on an interval of length of order $\varepsilon^{-1}$, say uniformly on $[0, \varepsilon^{-1}]$. Estimating the order of approximation is a prime goal of the theory. Discussions and examples can be found in Arnold [1], Bogoliubov and Mitropolsky [8], Guckenheimer and Holmes [9], Lochak and Meunier [10], Sanders and Verhulst [14], Verhulst [17]. We provide some concrete examples later on.

A standard approach to verifying the approximation and establishing the order of approximation is via differential or integral inequalities, e.g., Gronwall inequalities, carefully executed so to produce the appropriate estimates. We offer another approach which starts with a change of time scales, namely, $s = \varepsilon t$. In the "fast" time variable $s$ equations (2.1) and (2.2) take the form

$$\frac{dx}{ds} = f(s/\varepsilon, x, \varepsilon), \ x(0) = x_0 \tag{2.4}$$

and, respectively,

$$\frac{dx}{ds} = f^0(x), \ x(0) = x_0. \tag{2.5}$$

Verifying an approximation estimate for solutions of (2.1) and (2.2) uniformly on $[0, \varepsilon^{-1}]$ amounts to verifying the same estimate for solutions of (2.4) and (2.5) uniformly on $[0, 1]$.

When attempting to apply limit considerations to the right hand side of (2.4) a difficulty arises, namely, to determine the limit, as $\varepsilon \to 0$, of the function $f(\frac{s}{\varepsilon}, x, \varepsilon)$, as a function of $s$ for a fixed $x$. Indeed, the point-wise limit may not exist, while weak limits, although resulting in the desired average, are not easy to manipulate when quantitative estimates are sought. What we suggest is to employ the *Young measures limit*, as follows.

The best way to explain the idea is via a concrete example. Suppose that the right hand side in (2.4) is the function $\sin(\frac{s}{\varepsilon})$. As $\varepsilon \to 0$ the function oscillates more and more rapidly. What the Young measure limit captures is the distribution of the values of the function. Indeed, on any fixed small interval, say $[s_1, s_2]$ in $[0, 1]$, when $\varepsilon$ is small the values of $\sin(\frac{s}{\varepsilon})$ are distributed very closely to the distribution of the values of the sin function over one period; namely, the distribution is $\mu_0(d\xi) = \pi^{-1}(1 - \xi^2)^{-\frac{1}{2}}d\xi$ which is a probability measure over the space of values of the mapping $\sin(\cdot)$. A way to depict the limit is to identify it with the probability measure-valued map, say $\mu(\cdot)(d\xi)$ which assigns to each $s \in [0, 1]$ the probability distribution $\mu_0(d\xi)$ just defined. In the example, the same probability distribution is assigned to all $s$ in the interval. The general definition of a Young measure allows probability measure-valued maps which may not be constant over the interval. Later we take advantage of this possibility when allowing time-varying averages.

A probability measure on $R^n$ is a $\sigma$-additive mapping, say $\mu$, from the Borel subsets of $R^n$ into $[0, 1]$ such that $\mu(R^n) = 1$. The space of probability measures is endowed with the weak convergence of measures, namely, $\mu_i$ converge to $\mu_0$ if $\int h(\xi)\mu_i(d\xi)$ converge to $\int h(\xi)\mu_0(d\xi)$ for every bounded and continuous mapping $h(\cdot) : R^n \to R$. Here $\xi$ is an element of $R^n$. The space of probability measures on $R^n$ is denoted $\mathcal{P}(R^n)$. In the next section we recall the Prohorov metric; it makes the space $\mathcal{P}(R^n)$ with the weak convergence of measures a complete metric space. On this space see Billingsley [7].

A measurable mapping $\mu(\cdot) : [0, 1] \to \mathcal{P}(R^n)$ is called a Young measure, the measurability being with respect to the weak convergence. A Young measure, say $\mu(\cdot)$, is associated with a measure, marked in this paper in bold face font, say $\boldsymbol{\mu}$, on $[0, 1] \times R^n$ defined on rectangles $E \times B$ by $\boldsymbol{\mu}(E \times B) = \int_E \mu(s)(B)ds$. The resulting measure is also a probability measure (since the base space has Lebesgue measure one, otherwise we get a probability measure multiplied by the Lebesgue measure of that base). The convergence in the space of Young measures is now derived from the convergence on $\mathcal{P}([0, 1] \times R^n)$, and likewise the Prohorov metric. A useful consequence is that the space of Young measures with values supported on a compact subset of $R^n$ is a compact set in the space of Young measures.

An $R^n$-valued function, say $f(s)$, is identified with the Young measure whose values are the Dirac measures supported on the singletons $\{f(s)\}$. The convergence of functions in the sense of Young measures, say of $\{f_i(\cdot)\}$, is taken to be the convergence of the associated Young measures. More on the basic theory of Young measures and their convergence see Balder [6], Valadier [16].

The application of the Young measure convergence to the averaging problem is via the convergence, as $\varepsilon \to 0$, in the sense of Young measures of the functions $f(\frac{s}{\varepsilon}, x, \varepsilon)$. We shall also consider convergence of $f(\frac{s}{\varepsilon_j}, x, \varepsilon_j)$ for a subsequence $\varepsilon_j \to 0$. The limit in the general case is a Young measure, say $\mu_0(s, x)(d\xi)$ (here $x$ is the parameter carried over from the function $f(\frac{s}{\varepsilon}, x, \varepsilon)$, and $d\xi$ is an infinitesimal element in $R^n$). The resulting limit differential equation is defined by

$$\frac{dx}{ds} = E\big(\mu_0(s, x)(d\xi)\big), \quad x(0) = x_0, \tag{2.6}$$

where $E(\mu_0(s, x)(d\xi))$ is the expectation with respect to $\xi$ of the measure, namely, it is equal to $\int_{R^n} \xi \mu_0(s, x)(d\xi)$. Thus, the differential equation (2.6) is an ordinary differential equation whose right hand side is determined via an average of values. When the measure $\mu_0(s, x)(d\xi)$ is a Dirac measure, namely a function, the equation reduces to the form in (2.4). We abuse rigorous terminology and refer to $\mu_0(s, x)$ as the right hand side of the differential equation (2.6). It is easy to see that when the convergence holds when $\varepsilon \to 0$ (rather than for a subsequence $\varepsilon_j \to 0$), the limit Young measure is constant-valued, see Remark 5.3.

It should be pointed out that, throughout the derivations, it is the expectation of the Young measure which plays a role, and not the Young measure itself. Considering the entire Young measure does not, however, restrict the scope of the applications and, in turn, helps in the analysis.

It has been known for a long time that, under appropriate conditions, if the right hand side, say $f_i(s, x)$, of a differential equation converges in the sense of Young measures to, say, $\mu_0(s, x)$, then the corresponding solutions converge uniformly on bounded intervals. This may be considered a qualitative aspect of averaging. It was exploited in many frameworks. One such application is to relaxed controls, see Warga [18], Young [19]. Applications more related to the averaging principle were to systems with oscillating parameter and to singularly perturbed systems, see Artstein [2, 3], Artstein and Vigodner [5]. In the present paper the direction of the contribution is reversed, namely, we exploit the Young measures convergence in order to derive quantitative estimates within the averaging method.

## 3. A quantitative estimate

In this section we present an estimate of the distance between solutions of two ordinary differential equations driven by Young measures. The estimate is based on local averages of the equations. Standard ordinary differential equations with $R^n$-valued right hand side form a particular case. The application of the estimate to the averaging method is carried out in the next section. There are other applications of the general estimates, e.g., to relaxed controls; they are not examined in the present paper.

The Prohorov distance $\rho(\cdot, \cdot)$ among probability measures is the basis of the quantitative estimates in this paper. For measures, say $\mu$ and $\nu$ on a metric space $M$, the Prohorov distance $\rho(\mu, \nu)$ is defined by

$$\rho(\mu, \nu) = \inf \left\{ \eta : \text{ for all } B, \ \mu(B) \leq \nu(B^\eta) + \eta \text{ and } \nu(B) \leq \mu(B^\eta) + \eta \right\} \quad (3.1)$$

(here $B^\eta$ is the $\eta$-neighborhood of the Borel set $B$ in $M$); see Billingsley [7]. Notice that when the two measures are Dirac, namely, each supported on a point, then the Prohorov distance coincides with the Euclidean distance. We apply the definition to measures in the space $\mathcal{P}(R^n)$ of probability measures in $R^n$ and to Young measures (as defined in the previous section) in $\mathcal{P}([0, 1] \times R^n)$.

**Lemma 3.1.** *Let $\mu$ and $\nu$ be two probability measures on $R^n$ whose supports are included in the ball of radius $r$. Then $|E(\mu(d\xi)) - E(\nu(d\xi))| \leq \alpha\rho(\mu, \nu)$ with $\alpha$ depending only on $r$.*

*Proof.* Recall that in general, when $h(\xi) : R^n \to [0, \infty)$ is a nonnegative measurable function which is bounded, say by $r$, and $\mu$ is a probability measure on $R^n$ then

$$\int_{R^n} h(\xi)\mu(d\xi) = \int_0^r \mu\Big(\{\xi : h(\xi) \geq \lambda\}\Big)d\lambda, \quad (3.2)$$

where $d\lambda$ being the Lebesgue measure, see, e.g., Billingsley [7, page 223]. The $i$-th coordinate of the expectation $E(\mu(d\xi))$ is the integral $\int_{R^n} \xi_i\mu(d\xi)$ (with $\xi_i$ being the $i$-th coordinate of $\xi$). Clearly, the integration can be carried out when $\xi_i$ is replaced by 0 outside of the ball of radius $r$, and when replacing $\xi_i$ by $\xi_i + r$ the integration is of a positive bounded function. Employing (3.2) we deduce that the difference between the $i$-th coordinates of $E(\mu(d\xi))$ and $E(\nu(d\xi))$ is

$$\int_0^r \Big(\mu(\{\xi : \xi_i + r \geq \lambda\}) - \nu(\{\xi : \xi_i + r \geq \lambda\})\Big)d\lambda. \quad (3.3)$$

The inequalities in (3.1) establishing the Prohorov distance imply, in particular, that $\mu(\{\xi : \xi_i + r \geq \lambda\}) \leq \nu(\{\xi : \xi_i + r + \rho(\mu, \nu) \geq \lambda\}) + \rho(\mu, \nu)$ and likewise with the roles of $\mu$ and $\nu$ exchanged. Plugging these inequalities in (3.3) implies the existence of the desired coefficient $\alpha$; this completes the proof. $\square$

The estimate in this section refers to solutions of two ordinary differential equations given by

$$\frac{dx}{ds} = E\big(\mu_i(s,x)(d\xi)\big), \quad x(0) = x_0, \tag{3.4}$$

where for every $x$ the mappings $\mu_i(\cdot,x)$ for $i = 1, 2$ are Young measures defined on $s \in [0,1]$, with values being probability distributions on $R^n$ (Compare with (2.6)). Lipschitz continuity of $\mu_i(s,x)$ in the variable $x$ is understood with respect to the Prohorov metric on $\mathcal{P}(R^n)$.

Recall that we use bold face, say $\boldsymbol{\mu}$, to indicate the Young measure $\mu(\cdot)$ defined on $s \in [0,1]$. A Young measure $\boldsymbol{\mu}$ is bounded if there is a bound, say $r$, such that all the supports of $\mu(s)$ for all $s$ are contained in the ball of radius $r$ in $R^n$. When $I$ is a subinterval of $[0,1]$ we write $\boldsymbol{\mu}(I)$ for the normalized value of $\boldsymbol{\mu}(I \times R^n)$ namely, we divide the latter by the length of $I$; then $\boldsymbol{\mu}(I)$ is a probability measure (analogous to $\mu(s)$ for a single $s$). Notice that $E(\boldsymbol{\mu}(I)(d\xi))$ is the average of the expectations of $\mu(s)$ for $s \in I$; it coincides with the average of the function on the interval when the measures $\mu(s)$ are Dirac measures.

The first estimate utilizes the Prohorov distance on $\mathcal{P}(R^n)$ and information about the averages of the Young measures. We later connect it to the Prohorov distance on the space of Young measures.

**Theorem 3.2.** *For each $x$ let $\mu_1(s,x)$ and $\mu_2(s,x)$ be Young measures on $s \in [0,1]$ uniformly bounded, say by $r$. Suppose that both mappings are Lipschitz in the variable $x$ with Lipschitz constant $K$. Let $(\Delta, \eta)$ be a pair of nonnegative numbers such that whenever an interval $I$ in $[0,1]$ is of length $\Delta$ then $|E(\mu_1(x)(I)(d\xi)) - E(\mu_2(x)(I)(d\xi))| \leq \eta$. Then the distance between the solutions $x_1(s)$ and $x_2(s)$ of (3.4), with $\boldsymbol{\mu}_1$ and, respectively, $\boldsymbol{\mu}_2$ being the right hand side, is uniformly bounded on $[0,1]$ by $c\max(\Delta, \eta)$ where $c$ is a constant which depends only on $r$ and $K$.*

*Proof.* In view of the Lipschitz condition of each $\mu_i(s,x)$ and Lemma 3.1 we deduce that both $E(\mu_i(s,x)(d\xi))$ are uniformly Lipschitz in the $x$ variable; denote the Lipschitz constant by $K_1$.

Rather than estimating directly the distance $|x_1(s) - x_2(s)|$ we start by estimating $|x_1(s) - x_2(s)|e^{-sK_1}$; let $s_0$ be a point in $[0,1]$ at which the maximum of the latter term is attained, and denote this maximum by $m_0$.

Since we are dealing with solutions of the differential equation (3.4) for $i = 1, 2$, the equalities $x_i(s) = x_0 + \int_0^s E(\mu_i(\sigma, x_i(\sigma))(d\xi))ds$ hold, and we can use the triangle inequality and deduce that $m_0$ is bounded by

$$e^{-s_0 K_1}\left| \int_0^{s_0} \Big( E\big(\mu_1(\sigma, x_1(\sigma))(d\xi)\big) - E\big(\mu_2(\sigma, x_1(\sigma))(d\xi)\big) \Big) d\sigma \right|$$
$$+ e^{-s_0 K_1}\left| \int_0^{s_0} \Big( E\big(\mu_2(\sigma, x_1(\sigma))(d\xi)\big) - E\big(\mu_2(\sigma, x_2(\sigma))(d\xi)\big) \Big) d\sigma \right|. \tag{3.5}$$

We estimate the two terms in (3.5) separately.

We claim that the first term in (3.5) is bounded by $c_1(\Delta + \eta)$ with $c_1$ depending only on $r$ and $K$. To show this we divide the interval $[0, s_0]$ to sub intervals of length $\Delta$, with the exception, possibly, of the last one which may be shorter. Consider the piecewise constant map $z(\cdot)$ which on each interval of the partition takes the value $x_1(s')$ where $s'$ is the middle point of the interval. Then for $i = 1, 2$

$$e^{-s_0 K_1} \left| \int_0^{s_0} \left( E\Big(m_i\big(\sigma, x_1(\sigma)\big)(d\xi)\Big) - E\Big(m_i\big(\sigma, z(\sigma)\big)(d\xi)\Big) \right) d\sigma \right| \leq c_2 \Delta \quad (3.6)$$

with $c_2$ depending only on $K_1$ and $r$; indeed, the estimate follows from the claim stated in the beginning of the proof and the observation that $|x(s) - z(s)| \leq r\Delta$. The condition on $(\Delta, \eta)$ implies that

$$e^{-s_0 K_1} \left| \int_0^{s_0} \left( E\Big(\mu_1\big(\sigma, z(\sigma)\big)(d\xi)\Big) - E\Big(\mu_2\big(\sigma, z(\sigma)\big)(d\xi)\Big) \right) d\sigma \right| \leq c_3(\Delta + \eta), \quad (3.7)$$

with $c_3$ depending only on $r$. Combining the two inequalities in (3.6) and the one in (3.7) verifies the estimate concerning the first term in (3.5).

We resort again to the observation in the beginning of the proof and conclude that the second term in (3.5) is bounded by

$$e^{-s_0 K_1} \int_0^{s_0} |x_1(\sigma) - x_2(\sigma)| \, d\sigma. \quad (3.8)$$

Multiplying the integrand in (3.8) by $e^{-\sigma K_1} e^{\sigma K_1}$ and using the particular choice of $s_0$ as the point where $m_0$ is attained we can bound (3.8) by

$$e^{-s_0 K_1} m_0 \int_0^{s_0} K_1 e^{\sigma K_1} d\sigma \leq m_0 \big(1 - e^{-K_1}\big), \quad (3.9)$$

namely, a bound on the second term in (3.5). Combined with the previous estimate we get

$$m_0 \leq c_1(\Delta + \eta) + m_0 \big(1 - e^{-K_1}\big). \quad (3.10)$$

The last inequality clearly establishes the desired estimate, however, for the quantity $m_0$. Since $|x_1(s) - x_2(s)| \leq e^{K_1} m_0$ the proof is concluded. $\qquad \square$

The preceding result utilizes the distance between averages of the two Young measures. Being close in the Prohorov distance is not a necessary condition for the averages to be close; it induces, however, an estimate on how close the averages are, as follows.

**Lemma 3.3.** *Let $\mu_1(\cdot)$ and $\mu_2(\cdot)$ be two bounded Young measures on $[0, 1]$, say bounded by $r$, and let $\delta = \rho(\boldsymbol{\mu}_1, \boldsymbol{\mu}_2)$. Then whenever an interval $I$ in $[0, 1]$ is of length $\delta^{\frac{1}{2}}$ then $|E(\mu_1(I)(d\xi)) - E(\mu_2(I)(d\xi))| \leq c\delta^{\frac{1}{2}}$ with $c$ depending only on $r$.*

*Proof.* From the definition of the Prohorov distance we deduce (in fact, for every interval) that $|\boldsymbol{\mu}_1(I \times R^n) - \boldsymbol{\mu}_2(I \times R^n)| \leq 2\delta$. Dividing the two terms by $\delta^{\frac{1}{2}}$ and invoking Lemma 3.1 yield the result. $\qquad \square$

**Corollary 3.4.** *For every $x$ let $\mu_1(s, x)$ and $\mu_2(s, x)$ be Young measures, uniformly bounded, say by $r$. Suppose that both mappings are Lipschitz in the variable $x$ with Lipschitz constant $K$. Suppose that $\rho(\mu_1(x), \mu_2(x)) \leq \delta$ for every $x$. Then the distance between the solutions $x_1(s)$ and $x_2(s)$ of (3.4) with $\mu_1$ and, respectively, $\mu_2$ being the right hand side, is uniformly bounded on $[0, 1]$ by $c\delta^{\frac{1}{2}}$ with $c$ being a constant which depends only on $r$ and $K$.*

*Proof.* Follows from Theorem 3.2 and Lemma 3.3. □

As mentioned already, the results in this section apply also when the right hand side of the equation is an $R^n$-valued function. An elaboration on this particular case with further applications will be documented separately.

# 4. The application to averaging

In this section, following the arguments in Section 2, we translate the general estimates obtained in the previous section to estimates within the averaging method in ordinary differential equations.

The equation we examine is (2.1), copied here with the specification of the time span.

$$\frac{dx}{dt} = \varepsilon f(t, x, \varepsilon), \quad x(0) = x_0, \quad t \in [0, \varepsilon^{-1}]. \tag{4.1}$$

The change of time scales $s = \varepsilon t$ transforms the equation to

$$\frac{dx}{ds} = f(s/\varepsilon, x, \varepsilon), \quad x(0) = x_0, \quad s \in [0, 1]. \tag{4.2}$$

The following assumption is a bit stronger than the existence of the averaged equation as reflected in (2.2) and (2.3); it captures, however, all the applications of the classical averaging of ordinary differential equations. In the next section we introduce a generalization that goes beyond the classical averaging.

**Assumption 4.1.** As $\varepsilon \to 0$ the functions $f(\frac{s}{\varepsilon}, x, \varepsilon)$ converge in the sense of Young measures to a Young measure on $[0, 1]$, say $\mu_0(s, x)$, which is constant-valued, namely, does not depend on $s$. We denote the common value by $\mu_0(x)$.

Under Assumption 4.1 we identify the averaged equation as

$$\frac{dx}{dt} = \varepsilon E\big(\mu_0(x)(d\xi)\big), \quad x(0) = x_0, \quad t \in [0, \varepsilon^{-1}]. \tag{4.3}$$

The averaging method establishes the distance (hopefully small) between solutions of the time varying equation (4.1) and the time-invariant one (4.3), uniformly on the interval $[0, \varepsilon^{-1}]$. Notice that in the $s$-scale the time-invariant equation is $\frac{dx}{ds} = E(\mu_0(x)(d\xi))$.

**Lemma 4.2.** *Suppose that $f(x, t, \varepsilon)$ is uniformly bounded, say by $r$, and Lipschitz in $x$ with a Lipschitz constant independent of $t$ and $\varepsilon$. Under Assumption 4.1 the probability measure $\mu_0(x)$ is bounded by $r$ and Lipschitz in $x$.*

*Proof.* It is easy to see that the probability measure $\mu_0(x)$ is, in fact, the limit as $\varepsilon \to 0$ of the distributions of the values $f(\frac{s}{\varepsilon}, x, \varepsilon)$ over $[0, 1]$ with the Lebesgue measure. The latter distributions are uniformly Lipschitz in $x$, and the limit, clearly, inherits the Lipschitz property. □

**Theorem 4.3.** *Suppose that $f(x, t, \varepsilon)$ is uniformly bounded, say by $r$, and Lipschitz in $x$ with a Lipschitz constant independent of $t$ and $\varepsilon$. Suppose that Assumption 4.1 holds. Let $(\Delta(\varepsilon), \eta(\varepsilon))$ be a pair of nonnegative numbers such that for every $s \in [0, 1 - \Delta(\varepsilon)]$*

$$\left| \frac{1}{\Delta(\varepsilon)} \int_s^{s+\Delta(\varepsilon)} \left( f(\sigma/\varepsilon, x, \varepsilon)d\sigma - E\big(\mu_0(x)(d\xi)\big) \right) d\sigma \right| \le \eta(\varepsilon). \tag{4.4}$$

*Then the distance between the solutions of (4.1) and (4.3) is uniformly bounded on $[0, \varepsilon^{-1}]$ by $c \max(\Delta(\varepsilon), \eta(\varepsilon))$ where $c$ is a constant which depends only on $r$ and $K$. In particular, if $\delta(\varepsilon)$ is a bound on the Prohorov distance between $f(\frac{s}{\varepsilon}, x, \varepsilon)$ and $\mu_0(x)$ for all $x$, then the mentioned distance is bounded by $c\delta^{\frac{1}{2}}$ where $c$ is a constant which depends only on $r$ and $K$.*

*Proof.* The two statements amount to applying Theorem 3.2 and Corollary 3.4 in the framework of the present section, on, however, the time scale $s$. □

We display now some illustrations of the previous result. The first two are paraphrases of the well known periodic averaging, see Sanders and Verhulst [14, Theorem 3.2.10], Guckenheimer and Holmes [9, Theorem 4.1.1], (but notice that in our version the period may depend on $\varepsilon$).

**Example 4.4.** Consider the system

$$\frac{dx}{dt} = \varepsilon f\big(p(\varepsilon)t, x\big), \quad x(0) = x_0, \tag{4.5}$$

with $p(\varepsilon)$ positive and bounded away from 0 and $f(t, x)$ periodic with period $T$. Suppose also that $f(t, x)$ is uniformly bounded and uniformly Lipschitz in $x$. Then the distance between the solution of (4.5) and the solution of the autonomous averaged equation

$$\frac{dx}{dt} = \varepsilon \frac{1}{T} \int_0^T f(t, x)dt, \quad x(0) = x_0, \tag{4.6}$$

is of order $\varepsilon$, uniformly on the intervals $[0, \varepsilon^{-1}]$. Indeed, it is easy to see that the conditions of Theorem 4.3 hold with the Young measure being the distribution of $f(t, x)$ over a period (leading indeed to the averaged equation (4.6)) and with $(\Delta(\varepsilon), \eta(\varepsilon)) = (\varepsilon p(\varepsilon)T, 0)$. The boundedness away of $p(\varepsilon)$ from zero guarantees the desired estimate.

**Example 4.5.** Consider equation (4.5) with, however, allowing $p(\varepsilon)$ approaching zero as $\varepsilon \to 0$. For instance, let $p(\varepsilon) = |\log \varepsilon|^{-1}$. The natural averaged system is again (4.6). The pair measuring the averages is again $(\Delta(\varepsilon), \eta(\varepsilon)) = (\varepsilon p(\varepsilon)T, 0)$. Thus, the order of approximation in this case is $\varepsilon|\log \varepsilon|$, which converge to zero as $\varepsilon \to 0$. When $\varepsilon p(\varepsilon)$ does not approach zero as $\varepsilon \to 0$ the approximation is of order 1. It is easy to see that any order between $\varepsilon$ and 1 can be materialized.

**Example 4.6.** Consider the system

$$\frac{dx}{dt} = \varepsilon\Big(f_1\big(p_1(\varepsilon)t, x\big) + \ldots + f_k\big(p_k(\varepsilon)t, x\big)\Big), \quad x(0) = x_0, \tag{4.7}$$

with each $p_j(\varepsilon)$ positive and bounded away from 0 and $f_j(t,x)$ periodic with period $T$. Suppose also that $f_j(t,x)$ are uniformly bounded and uniformly Lipschitz in $x$. It is clear that the conditions of Theorem 4.3 hold with the Young measure being the cumulated distribution of $f_j(t,x)$, $j = 1, \ldots, k$ over a period, leading to the averaged equation

$$\frac{dx}{dt} = \varepsilon \frac{1}{T} \int_0^T \big(f_1(t,x) + \ldots + f_k(t,x)\big) dt, \quad x(0) = x_0. \tag{4.8}$$

It is also easy to see that the Prohorov distance between the right hand side of (4.7) (in, however, the $s$ time scale) and the limit Young measure is of order $\varepsilon$. By Theorem 4.3 we deduce that the rate of approximation in $\varepsilon^{\frac{1}{2}}$. If, for instance, we allow $p(\varepsilon)$ to be $|\log \varepsilon|^{-1}$ away from 0, the Prohorov distance becomes of order $\varepsilon|\log \varepsilon|$ and the order of approximation is therefore $\varepsilon^{\frac{1}{2}}|\log \varepsilon|$.

## 5. Time-varying averages

Underlying the classical averaging is Assumption 4.1, guaranteeing that one equation captures the limit behaviour as $\varepsilon \to 0$ of the time-varying system. The Young measures approach allows a relatively simple and rigorous examination of a situation where more than one limit equation exists. We examine this possibility now. We consider the system (4.1) and its fast scale version (4.2).

**Definition 5.1.** *A Young measure, say $\mu(s,x)$, is a limit average of the right hand side of the time-varying equation if it is a limit, in the sense of Young measures, of a subsequence $f(\frac{s}{\varepsilon_j}, x, \varepsilon_j)$, as $\varepsilon_j \to 0$.*

The next result guarantees that limit averages exist; recall that in the classical averaging method the existence of the averaged equation is an assumption.

**Lemma 5.2.** *Suppose that $f(x,t,\varepsilon)$ is uniformly bounded, say by $r$, continuous in $\varepsilon$ and Lipschitz in $x$ with a Lipschitz constant independent of $t$ and $\varepsilon$. Then every sequence $\varepsilon_i \to 0$ has a subsequence, say $\varepsilon_j$, such that $f(\frac{s}{\varepsilon_j}, x, \varepsilon_j)$ converges in the sense of Young measures. Any such limit average, say $\mu(s,x)$, is bounded by $r$ and Lipschitz in the $x$ variable. The family of all such limit averages is compact and connected in the space of Young measures.*

*Proof.* The boundedness by $r$ is clear. Existence of subsequences which converge and the compactness follow since the family of Young measures on $[0,1]$ with values in a compact set is compact. The continuity in $\varepsilon$ implies that as a Young measure $f(\frac{s}{\varepsilon}, x, \varepsilon)$ depends continuously on $\varepsilon$; this easily implies that the family of limit averages is connected. What is a bit subtle is the Lipschitz dependence on $x$ (the Lipschitz dependence of the entire Young measure $\boldsymbol{\mu}(x)$ on $x$ is easy, but we need the Lipschitz property of $\mu(s,x)$ separately for almost every each $s$). To this end let $f(\frac{s}{\varepsilon_j}, x, \varepsilon_j)$ converge to $\mu(s,x)$. For almost every $s_0 \in [0,1]$ the probability measure $\mu(s_0, x)$ is the limit of $\mu(I, x)$ for $I$ of length tending to zero with $s_0 \in I$ (for the definition of $\mu(I)$ see the paragraph preceding Theorem 3.2). The value $\mu(I, x)$ is, in turn, the limit of the distribution of $f(\frac{s}{\varepsilon_j}, x, \varepsilon_j)$ for $s \in I$, thus it is Lipschitz in $x$. A simple limit argument shows that the limit $\mu(s_0, x)$ is also Lipschitz in $x$. $\qquad \square$

**Remark 5.3.** Notice that unlike the case in Section 4, the limit Young measures may not be constant-valued. If, however, $f(\frac{s}{\varepsilon}, x, \varepsilon)$ converge as $\varepsilon \to 0$, say to $\mu(s,x)$, the latter is constant-valued. Indeed, it is easy to see that if $f(\frac{s}{\varepsilon}, x, \varepsilon_j)$ converge to $\mu(s,x)$ then on, say $s \in [0, \frac{1}{2}]$, the sequence generated by $\frac{1}{2}\varepsilon_j$ converges to $\mu(2s, x)$. Such relations clearly imply that $\mu(s,x)$ does not depend on $s$.

The preceding observations combined with the estimates of Section 3 enable to determine approximation rates for subsequences of solutions. We state now a global result, followed by an illustration.

We consider equation (4.1) with $f(t, x, \varepsilon)$ bounded and Lipschitz in $x$, and denote the family of its limit averages by $\mathcal{E}$.

**Theorem 5.4.** *Suppose that $f(t, x, \varepsilon)$ is uniformly bounded, continuous in $\varepsilon$ and Lipschitz in $x$ with a Lipschitz constant independent of $t$ and $\varepsilon$. Let $\delta(\varepsilon)$ be the Prohorov distance between $f(\frac{s}{\varepsilon}, x, \varepsilon)$ and the family $\mathcal{E}$ of limit averages of the equation. Then the distance between the solution of*

$$\frac{dx}{dt} = \varepsilon f(t, x, \varepsilon), \quad x(0) = x_0, \tag{5.1}$$

*and the solution of*

$$\frac{dx}{dt} = \varepsilon E\big(\mu(\varepsilon t, x)(d\xi)\big), \quad x(0) = x_0, \tag{5.2}$$

*for one of the Young measures in $\mathcal{E}$, is of order $\delta(\varepsilon)^{\frac{1}{2}}$, uniformly on $[0, \varepsilon^{-1}]$.*

*Proof.* The statement amounts to applying Corollary 3.4 to equation (5.1) in, however, the fast time scale and to equation (5.2) when its right hand side is the Young measure closest to $f(\frac{s}{\varepsilon}, x, \varepsilon)$ in $\mathcal{E}$. $\qquad \square$

**Example 5.5.** Consider a system

$$\frac{dx}{dt} = \varepsilon f(t, x), \quad x(0) = x_0, \tag{5.3}$$

with right hand side which on intervals $[N_{j-1}, N_j]$, $j = 1, 2, \dots$ with $N_0 = 0$, is equal to one of, say, three prescribed periodic functions, $f_i(t, x)$ with periods $T_i$, for $i = 1, 2, 3$. Furthermore, the functions appear cyclically, namely, on $[N_{j-1}, N_j]$ the right hand side is $f_{i(j)}(t, x)$ where $i(j) = j \mod 3$. Consider now the case where, say, $N_1 = 2$ and $N_{j+1} = N_j^2$ for $j \geq 1$. The limit averages are easy to determine. Indeed, when $\varepsilon_j^{-1} = N_j$ the distance of $f(\frac{s}{\varepsilon_j}, x)$ from the constant-valued Young measure whose distribution is equal to the distribution of $f_{i(j)}(t, x)$ over a period, is less than or equal to $\varepsilon_j T_{i(j)}$. Denote this distribution by $\mu_{i(j)}(x)$. When $\varepsilon_{j+1} \leq \varepsilon \leq \varepsilon_j$ the associated Young measure takes the value $\mu_{i(j)}(x)$ on part of the interval, say $s \in [0, s']$ and takes the value $\mu_{i(j+1)}(x)$ on the rest of the interval; the dividing point is $s' = (\varepsilon^{-1} - \varepsilon_j^{-1})(\varepsilon_{j+1}^{-1} - \varepsilon_j^{-1})^{-1}$. Respectively, the solution of (5.3) is of order $\varepsilon$ away from the solution of the autonomous equation on $[0, \varepsilon^{-1} s']$ and continues to be of order $\varepsilon$ away from the solution of the corresponding autonomous equation on $[\varepsilon^{-1} s', \varepsilon^{-1}]$. The order $\varepsilon$ for the approximation is justified by Theorem 3.2 applied separately on the two subintervals.

**Remark 5.6.** We conclude by relating the previous considerations to singularly perturbed systems. Consider the system

$$\begin{aligned} \frac{dx}{dt} &= \varepsilon f(x, y) \\ \frac{dy}{dt} &= g(x, y) \,, \end{aligned} \tag{5.4}$$

with $x \in R^n$ and $y \in R^m$ and when the desired time interval is $t \in [0, \varepsilon^{-1}]$. Thus, the $x$-equation in (5.4) is of the form (2.1) with the time dependency provided by the trajectory $y(t)$, which in turn is coupled with $x$. This is, indeed, the general form used in the averaging problem, see Arnold [1], Lochak and Meunier [10], Verhulst [17]. When the fast time scale $s = \varepsilon t$ is introduced, the system becomes

$$\begin{aligned} \frac{dx}{ds} &= f(x, y) \\ \varepsilon \frac{dy}{ds} &= g(x, y) \,, \end{aligned} \tag{5.5}$$

namely, a singularly perturbed equation; the desired time interval here is $s \in [0, 1]$. The classical approach to the analysis of (5.5) is via the Levinson-Tikhonov method, assuming that for a fixed $x$ the fast solution converges to an asymptotically stale point $y(x)$. Further regularity conditions identify the limit behaviour as $(x(s), y(x(s)))$ with $x(s)$ the solution of $\frac{dx}{ds} = f(x, y(x))$. See O'Malley [11], Tikhonov et al. [15]. Such an analysis covers only a degenerate form of averaging. Removing the assumption of the convergence of the fast dynamics to a stable equilibrium is offered in Artstein and Vigodner [5], Artstein [3], where the equilibria are replaced by invariant measures; the resulting differential equation is driven by a Young measure, exactly as in the present paper. In the special case where $g(x, y)$ in (5.4) does not depend on the $x$ variable we get exactly the form (2.1), with time dependency induced by the solution $y(t)$ of $\frac{dx}{dt} = g(y)$. Indeed, Example 5.5 is a simplified version of an example analyzed in detail in Artstein [4] within the framework of singular perturbations.

# References

[1] V. I. Arnold, *Geometrical Methods in the Theory of Ordinary Differential Equations, second edition.* Springer, New York, 1988.

[2] Z. Artstein, *Chattering variational limits of control systems.* Forum Mathematicum 5 (1993), 369–403.

[3] Z. Artstein, *On singularly perturbed ordinary differential equations with measure-valued limits.* Mathematica Bohemica 127 (2002), 139–152.

[4] Z. Artstein, *Distributional convergence in planar dynamics and singular perturbations.* J. Differential Equations 201 (2004), 250–286.

[5] Z. Artstein and A. Vigodner, *Singularly perturbed ordinary differential equations with dynamic limits.* Proceedings of the Royal Society of Edinburgh 126A (1996), 541–569

[6] E. J. Balder, *Lectures on Young measure theory and its applications to economics.* Rend. Istit. Mat. Univ. Trieste 31 (2000), supplemento 1, 1–69.

[7] P. Billingsley, *Convergence of Probability Measures.* Wiley, New York, 1968.

[8] N. N. Bogoliubov and Y. A. Mitropolsky, *Asymptotic Methods in the Theory of Non-linear Oscillations.* English Translation, Gordon and Breach, New York, 1961.

[9] J. Guckenheimer and P. Holmes, *Nonlinear Oscillations, Dynamical Systems, and Bifurcations of Vector Fields.* Applied Mathematical Sciences 42. Springer-Verlag, New York, 1983.

[10] P. Lochak and C. Meunier, *Multiphase Averaging for Classical Systems. With Applications to Adiabatic Theorems.* Applied Mathematical Sciences, 72. Springer-Verlag, New York, 1988.

[11] R. E. O'Malley, Jr., *Singular Perturbation Methods for Ordinary Differential Equations.* Springer-Verlag, New York, 1991.

[12] P. Pedregal, *Parameterized Measures and Variational Principles.* Birkhäuser Verlag, Basel, 1997.

[13] P. Pedregal, *Optimization, relaxation and Young measures.* Bull. Amer. Math. Soc. 36 (1999), 27–58.

[14] J. A. Sanders and F. Verhulst, *Averaging Methods in Nonlinear Dynamical Systems.* Springer-Verlag, New York, 1985.

[15] A. N. Tikhonov, A. B. Vasiléva and A. G. Sveshnikov, *Differential Equations.* Springer-Verlag, Berlin, 1985.

[16] M. Valadier, *A course on Young measures.* Rend. Istit. Mat. Univ. Trieste 26 (1994) supp., 349–394.

[17] F. Verhulst, *Methods and Applications of Singular Perturbations.* Texts in Applied Mathematics 50, Springer, New York, 2005.

[18] J. Warga, *Optimal Control of Differential and Functional Equations.* Academic Press, New York, 1972.

[19] L. C. Young, *Lectures on the Calculus of Variations and Optimal Control Theory.* Saunders, New York, 1969.

Zvi Artstein
Department of Mathematics
The Weizmann Institute of Science
Rehovot 76100, Israel
e-mail: zvi.artstein@weizmann.ac.il

Progress in Nonlinear Differential Equations
and Their Applications, Vol. 75, 29–47
© 2007 Birkhäuser Verlag Basel/Switzerland

# Viability Kernels and Capture Basins for Analyzing the Dynamic Behavior: Lorenz Attractors, Julia Sets, and Hutchinson's Maps

Jean-Pierre Aubin and Patrick Saint-Pierre

*To Arrigo and Jim, for a happy mathematical continuation*

**Abstract.** Some issues on chaotic solutions, to Lorenz systems, for instance, are related to the concepts of viability kernels of subsets under continuous time systems, or in the case of Julia or Cantor sets, for instance, under discrete time systems.

It happens that viability kernels of subsets, capture basins of targets and the combination of the twos provide tools for the analysis of the local behavior around equilibria (local stable and unstable manifolds), the asymptotic behavior and the fluctuation of evolutions between two areas of a domain, etc.

Since algorithms and softwares do exist for computing the viability kernels and the capture basins, as well as evolutions viable in the viability kernel until they converge to a target in finite time, we are able to localize the attractor, to compute local stable and unstable manifolds, heteroclinic points, etc.

**Mathematics Subject Classification (2000).** Primary 37F50; Secondary 28A80.

**Keywords.** Lorenz attractor, fractals, fluctuation, viability kernel, chaos à la Saari, Julia sets, Mandelbrot function, Hutchison maps.

## 1. Introduction

Existence and uniqueness of solutions to differential equations, was and still is identified by many scientists with the mathematical description of *determinism*, after the 1796 book *L'Exposition du système du monde* and the 1814 book *Essai philosophique sur les probabilités* by Pierre Simon de Laplace [1749–1827]: "*We must regard the present state of the universe as the effect of its anterior state and not as the cause of the state which follows. An intelligence which, at a given instant, would know all the forces of which the nature is animated and the respective situation of the beings of which it is made of, if by the way it was wide enough to subject these data to analysis, would embrace in a unique formula the movements of the largest bodies of the universe and those of the lightest atom: Nothing would be uncertain for it, and the future, as for the past, would present at its eyes."*

Does it imply what is meant by *"predictability"*? In 1770, even before Laplace, Paul Henri Thiry, baron d'Holbach [1694–1778] wrote in his book *Système de la nature*: *"Finally, if everything in nature is linked to everything, if all motions are born from each other although they communicate secretely to each other unseen from us, we must hold for certain that there is no cause small enough or remote enough which sometimes does not bring about the largest and the closest effects on us. The first elements of a thunderstorm may gather in the arid plains of Lybia, then will come to us with the winds, make our weather heavier, alter the moods and the passions of a man of influence, deciding the fate of several nations."*

Later, in a mathematical perspective, Henri Poincaré [1854–1912] wrote in *"La science et l'hypothèse"* (1908): *"If we knew exactly the laws of Nature and the situation of the universe at the initial moment, we could predict exactly the situation of that same universe at a succeeding moment. But even if it was the case that the natural laws had no longer any secret for us, we could still know the situation approximately. If that enabled us to predict the succeeding situation with the same approximation, that is all we require, and we should say that the phenomenon has been predicted, that it is governed by the laws. But it is not always so; it may happen that small differences in the initial conditions produce very great ones in the final phenomena. A small error in the former will produce an enormous error in the latter. Prediction becomes impossible and we obtain a fortuitous phenomenon."* Jumping ahead in time, with the advent of computers, while studying a simplified meteorological model made of a system of three differential equations, Edward Lorenz discovered in the beginning of the 1960's that for certain parameters where the system has three repelling equilibria, the "limit set" was quite strange, "chaotic" in the sense that evolutions approach one equilibrium while circling around it, then suddenly leave away toward another equilibrium around which it turns again, and so on. In other words, this behavior is strange. "chaotic", in the sense that the limit set of an evolution is not a trajectory of a periodic solution. Lorenz presented in 1979 a lecture to the American Association for the Advancement of Sciences entitled: *Predictability: Does the flap of a butterfly's wing in Brazil set off a tornado in Texas?*

Chaos was thus "born again", resurrected with another name, soon to become famous. Some nonlinear *differential equations* produce indeed *chaotic* behavior, quite *sensitive* to initial conditions. Was chance rooted in deterministic system? Is uncertainty observed in living systems living systems consistent with differential equations?

However, for many problems arising in biological, cognitive, social and economic sciences, I believe we face a completely *orthogonal situation*, governed by *differential inclusions*, but producing evolutions as *regular or stable* (in a very loose sense) as possible for the sake of adaptation and viability required for life.

*Myopic, conservative, lazy and opportunistic* agents, from molecules to (wo)men, have some *contingent* freedom to choose among some *regulons* to govern the evolution, in order to protect themselves against *tychastic* uncertainty, obeying no statistical regularity, produced by *tyches*, from the Greek goddess Tyche, meaning and embodying a concept of "chance".

Differential inclusions and dynamical games are useful tools for providing mathematical metaphors for this kind of evolution. Their chaotic properties have to be studied ... for avoiding them.

Some issues on chaotic evolution are related to *viability kernels* and *capture basins* of subsets under continuous time systems (attractor of the Lorenz, for instance) or discrete time systems (Julia and Mandelbrot sets, Cantor and fractal nature of inverses of Hutchinson maps, etc.).

Viability kernels of subsets, capture basins of targets and the combination of those two provide other tools for the analysis of the local behavior around equilibria (*local stable and unstable manifolds*), the asymptotic behavior (*localizing the attractor in the intersection of the forward and backward viability kernels*), the *fluctuation basin, heteroclines*, etc.

The viability kernel of a constrained subset under an evolutionary system is the (possibly empty) set of initial states from which at least one evolution remains (viable) forever in the constrained set and the capture basin of a target is the set of initial states from which at least one evolution reaches the target **in finite time**. We refer to the viability literature for the mathematical properties of viability kernels and capture basins. We stress here the numerical results are provided by the Viability Kernel and Capture Basin Algorithms introduced in [23, Saint-Pierre] (See also [10, Cardaliaguet, Quincampoix & Saint-Pierre], [22, Quincampoix & Saint-Pierre]). They allow us to

1. compute the viability kernel of a constrained set or the capture basin of a target under a differential equation, or even, a control system,
2. compute the evolutions viable in the constrained set forever or until they reach the target in finite time. Indeed, starting from an initial state in the viability kernel, standard algorithms for computing solutions (the so-called shooting methods) do not take into consideration the necessary conditions for imposing the viability of the solution. Since the initial state lies only in an approximation of the viability kernel, the absence of these corrections does not allow us to "tame" evolutions which quickly leave the constrained set, above all for systems which are sensitive to initial states, as the Lorenz system.

These algorithms *handle subsets instead of functions*, and are part of the emerging field of *"set-valued numerical analysis"*, to which belongs by the way *interval arithmetic* (see [19, 20, Moore]). Since fluctuation basins can be expressed in terms of viability kernels and capture basins, these algorithms provide new numerical results that can be of interest.

For systems with high sensitivity to initial states, this is quite important, because, even starting from an initial state in the attractor, approximations provided with very precise schemes of solutions which should be viable in the attractor may actually leave it (and converge to it, but from the outside).

Although the results we present in this paper hold true for control systems, we test them only on the Lorenz system for the sake of comparison in the continuous time case, on the Julia sets in the discrete time case.

## 2. Viability concepts

**Definition 2.1 (Control Systems).** *Let $X$ and $\mathcal{U}$ be finite dimensional spaces,*

1. *$f : (x, u) \in X \times \mathcal{U} \mapsto f(x, u) \in X$,*
2. *$U : x \in X \rightsquigarrow U(x) \subset \mathcal{U}$ with which we associate the set-valued map $F : x \in X \rightsquigarrow F(x) := f(x, U(x)) \subset X$.*

*The evolutionary system $\mathcal{S} : X \rightsquigarrow \mathcal{C}(0, \infty; X)$ maps any $x \in X$ to the set $\mathcal{S}(x)$ of evolutions $x(\cdot)$ starting from $x$ and governed by the control system (or by a differential inclusion)*

$$x'(t) = f\big(x(t), u(t)\big) \text{ where } u(t) \in U\big(x(t)\big) \text{ or } x'(t) \in F(x(t)) \qquad (2.1)$$

*The backward (or negative) evolutionary system $\overleftarrow{\mathcal{S}}$ is associated with*

$$x'(t) = -f\big(x(t), u(t)\big) \text{ where } u(t) \in U(x(t)) \text{ or } x'(t) \in -F(x(t))$$

There are several ways for describing continuity of the evolutionary system $x \rightsquigarrow \mathcal{S}(x)$ with respect to the initial state, regarded as stability property.

**Definition 2.2.** *The evolutionary system $\mathcal{S}$ is said to be upper semicompact from $X$ to $\mathcal{C}(0, \infty; X)$ if for any $x_n \in X$ converging to $x$ in $X$ and for any evolution $x_n(\cdot) \in \mathcal{S}(x_n)$ starting at $x_n$, there exists a subsequence of $x_n(\cdot)$ converging to a evolution $x(\cdot) \in \mathcal{S}(x)$ uniformly on compact intervals.*

The main (and difficult) theorem states that the set of solutions depends continuously upon the initial states in the upper semicompact sense. This is the case for Marchaud differential inclusions:

**Definition 2.3.** *We say that $F$ is a* Marchaud *map if*

$$\begin{cases} (i) & \text{the graph and the domain of } F \text{ are nonempty and closed} \\ (ii) & \text{the values } F(x) \text{ of } F \text{ are convex} \\ (iii) & \exists\, c > 0 \text{ such that } \forall\, x \in X, \\ & \|F(x)\| := \sup_{v \in F(x)} \|v\| \leq c(\|x\| + 1) \end{cases} \qquad (2.2)$$

We recall the statement of the Stability Theorem 3.5.2 of [2, Aubin]:

**Theorem 2.4.** *If $F : X \rightsquigarrow X$ is Marchaud, the solution map $\mathcal{S}$ is an upper semicompact evolutionary system.*

The basic viability problem is reformulated in terms of the definition of viability kernels:

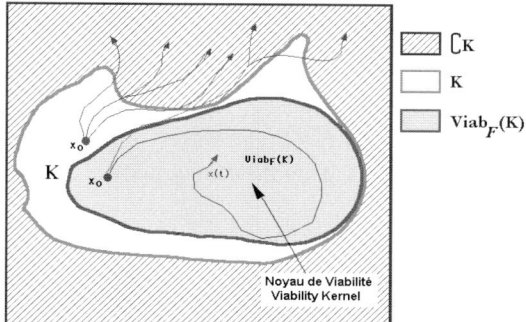

FIGURE 1. **Viability Kernels.** Let $K \subset X$ being regarded as a constrained environnement. The *viability kernel* Viab$(K)$ of $K$ is the set of initial states $x_0 \in K$ such that one evolution $x(\cdot) \in \mathcal{S}(x_0)$ starting at $x_0$ is *viable* in $K$

$$\mathrm{Viab}_{\mathcal{S}}(K) := \{x_0 \in K \mid \exists x(\cdot) \in \mathcal{S}(x_0) \text{ such that } \forall t \geq 0, \ x(t) \in K\}$$

When targets to be reached in finite time are involved, we introduce the concept of capture basins of targets:

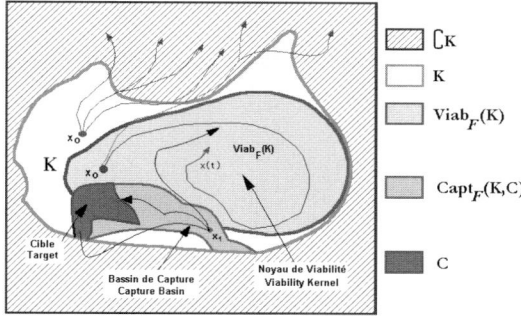

FIGURE 2. **Capture Basins of Targets.** Let $C \subset K \subset X$ be two subsets, $C$ being regarded as a target, $K$ as a constrained environnement. The *viable-capture basin* Capt$(K, C)$ of $C$ in $K$ is the set of initial states $x_0 \in K$ such that $C$ is reached in finite time before possibly leaving $K$ by at least one solution $x(\cdot) \in \mathcal{S}(x_0)$ of the control system starting at $x_0$

$$\mathrm{Capt}_{\mathcal{S}}(K, C) :=$$
$$\{x \in K \mid \exists x(\cdot) \in \mathcal{S}(x), \exists t^* > 0 \text{ such that } x(t^*) \in C \text{ and } \forall t \in [0, t^*], \ x(t) \in K\}$$

For more details on viability theory, we refer to [2, Aubin] and to the forth-coming book [5, Aubin, Bayen, Bonneuil & Saint-Pierre]. We provide here basic only definitions and some results which may be relevant to fluctuations of evolutions.

### 2.1. Lorenz systems

Lorenz introduced the following variables

1. $x$, proportional to the intensity of convective motion,
2. $y$, proportional to the temperature difference between ascending and descending currents,
3. $z$, proportional to the distortion (from linearity) of the vertical temperature profile.

Their evolution is governed by the following system of differential equations:

$$\begin{cases} (i) & x'(t) = \sigma y(t) - \sigma x(t) \\ (ii) & y'(t) = rx(t) - y(t) - x(t)z(t) \\ (iii) & z'(t) = x(t)y(t) - bz(t) \end{cases} \tag{2.3}$$

where $\sigma > b + 1$.

The vertical axis $(0, 0, z)_{z \in \mathbb{R}}$ is a symmetry axis, which is also the viability kernel of the hyperplane $(0, y, z)$ under the Lorenz system, from which the solutions are $(0, 0, ze^{-bt})$.

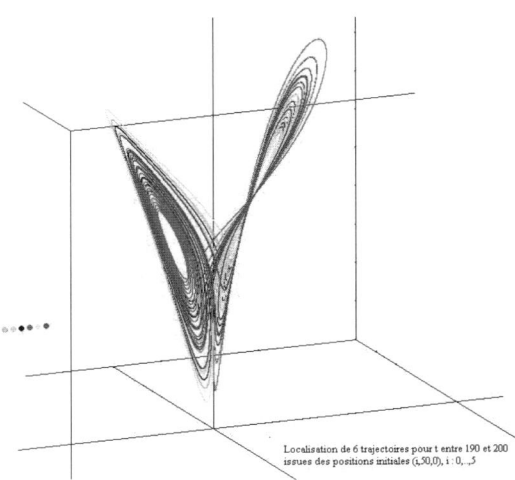

Localisation de 6 trajectoires pour t entre 190 et 200
issues des positions initiales (i,50,0), i : 0, ..5

FIGURE 3. Trajectories of six evolutions starting from initial conditions $(i, 50, 0)$, $i = 0, \ldots, 5$. Only the part of the trajectories from step times ranging between 190 and 200 are shown for clarity. If the normalized Rayleigh number $r \in {]}0, 1{[}$, then 0 is an asymptotically stable equilibrium. If $r = 1$, the equilibrium 0 is "neutrally stable"

When $r > 1$, the equilibrium $0$ becomes unstable and two more equilibria appear:

$$e_1 := \left(\sqrt{b(r-1)}, \sqrt{b(r-1)}, r-1\right) \quad \text{and}$$

$$e_2 := \left(-\sqrt{b(r-1)}, -\sqrt{b(r-1)}, r-1\right).$$

They are stable when $1 < r^\star := \dfrac{\sigma(\sigma + b + 3)}{\sigma - b - 1}$ and unstable when $r > r^\star$.

The Viability Kernel Algorithm provides the forward and backward viability kernels under the Lorenz system:

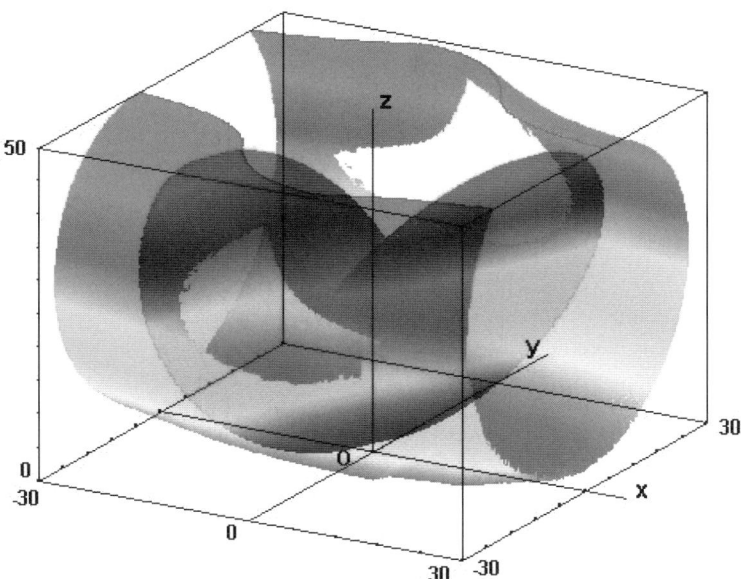

FIGURE 4. **Examples of viability kernel of a cube $K$ in $\mathbb{R}^3$ of the forward and backward Lorenz system.** We take $\sigma = 10$, $b = \frac{8}{3}$ and $r = 28$. Whenever the backward viability kernel is contained in the interior of $K$, the backward viability kernel is contained in the forward viability kernel. The color scale provides the third coordinates

The following figure provides views of the Backward Viability Kernel

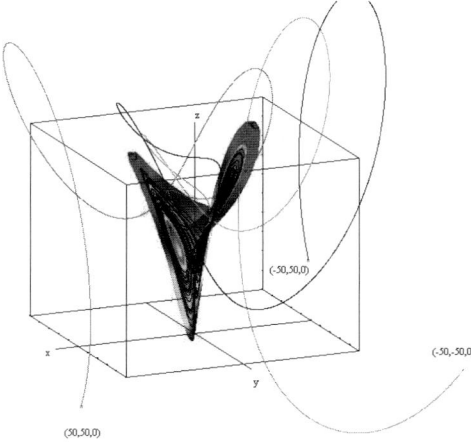

FIGURE 5. The trajectories of three evolutions converging to the backward viability kernel are shown

The Viability Kernel Algorithm "tames" evolutions in the viability kernel in the following sense:

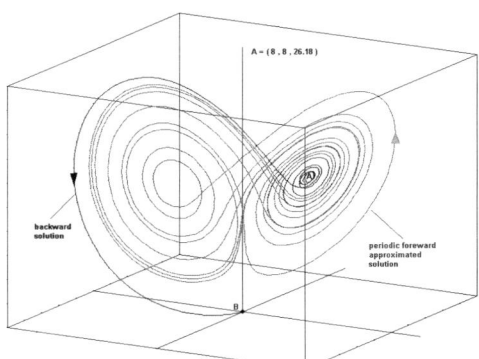

FIGURE 6. **Taming Evolutions in the Viability Kernels.** The viability kernel algorithm involves a correction procedure allowing them to compute evolutions viable in the viability kernel. They "trap and tame" viable evolutions even though they are sensitive to initial conditions, whereas usual solvers without lacking such correction procedure cannot maintain them viable so that solutions escape quickly from the kernel or basin

## 3. Attractors and their localization

We recall the définirons of limit sets of evolutions and of attractors:

Let $x(\cdot) \in \mathcal{C}(0, \infty; X)$ be an evolution. We say that the (closed) subsets

$$\begin{cases} \omega(x(\cdot)) := \bigcap_{T > 0} cl(x(\mathbb{R}_+)) \\ \alpha(x(\cdot)) := \overleftarrow{\omega}(x(\cdot)) := \bigcap_{T > 0} cl(x(\mathbb{R}_-)) \end{cases}$$

of the cluster points when $t \to \infty$ and $t \to -\infty$ are respectively the $\omega$-*limit set* of $x(\cdot)$ and the $\alpha$-*limit set* of the evolution $x(\cdot)$.

**Definition 3.1.** *Let* $\mathcal{S}^K(x)$ *denote the set of evolutions starting at* $x \in K$ *viable in* $K$. *We denote by*

$$\text{Attr}_{\mathcal{S}}(K) := \bigcup_{x(\cdot) \in \mathcal{S}^K(x)\ x \in K} \{\omega(x(\cdot))\} \quad and$$

$$\text{Attr}_{\overleftarrow{\mathcal{S}}}(K) := \bigcup_{x(\cdot) \in \mathcal{S}^K(x)\ x \in K} \{\alpha(x(\cdot))\}$$

*the* $\omega$-*attractor or simply* attractor *and the* $\alpha$-*attractor or* backward attractor *of the subset* $K$ *under* $\mathcal{S}$ *respectively.*

If $\mathcal{S}$ is upper semicompact, the $\omega$-limit set $\omega(x(\cdot))$ of an evolution $x(\cdot) \in \mathcal{S}(x)$ is always forward and backward viable under $\mathcal{S}$:

$$\omega(x(\cdot)) = \text{Viab}_{\mathcal{S}}(\omega(x(\cdot))) = \text{Viab}_{\overleftarrow{\mathcal{S}}}(\omega(x(\cdot)))$$

The forward and backward attractors of $K$ under $\mathcal{S}$, as well as their closures are respectively subsets viable and backward viable under the evolutionary system:

$$\text{Attr}_{\mathcal{S}}(K) = \text{Viab}_{\mathcal{S}}(\text{Attr}_{\mathcal{S}}(K)) = \text{Viab}_{\overleftarrow{\mathcal{S}}}(\text{Attr}_{\mathcal{S}}(K))$$

and

$$\text{Attr}_{\overleftarrow{\mathcal{S}}}(K) = \text{Viab}_{\mathcal{S}}(\text{Attr}_{\overleftarrow{\mathcal{S}}}(K)) = \text{Viab}_{\overleftarrow{\mathcal{S}}}(\text{Attr}_{\overleftarrow{\mathcal{S}}}(K))$$

They are consequently contained in the intersection of the viability kernel of $K$ and the backward viability kernel of $K$:

$$\text{Attr}_{\mathcal{S}}(K) \cup \text{Attr}_{\overleftarrow{\mathcal{S}}}(K) \subset \text{Viab}_{\mathcal{S}}(K) \cap \text{Viab}_{\overleftarrow{\mathcal{S}}}(K)$$

**Theorem 3.2 (Localization of the attractor).** *Assume that the evolutionary system is upper semicompact. Then*

$$\text{Attr}_{\mathcal{S}}(K \backslash \text{Viab}_{\overleftarrow{\mathcal{S}}}(K)) \subset \text{Viab}_{\mathcal{S}}(K) \cap \partial \text{Viab}_{\overleftarrow{\mathcal{S}}}(K).$$

*Consequently, if* $\text{Viab}_{\overleftarrow{\mathcal{S}}}(K) \subset \text{Int}(K)$, *then*

$$\text{Attr}_{\mathcal{S}}(K) \subset \text{Viab}_{\overleftarrow{\mathcal{S}}}(K) \subset \text{Inv}_{\mathcal{S}}(K).$$

Consider two subsets $K$ and $L \subset K$. we can also prove that

$$\text{Attr}_{\mathcal{S}}(K) \subset \text{Attr}_{\mathcal{S}}(L) \cup \overline{\complement\text{Capt}_{\mathcal{S}}(K, \text{Viab}_{\mathcal{S}}(L))}. \tag{3.1}$$

As a consequence, we can localize the Lorenz attractor in its backward viability kernel.

**3.1. Localization of the Lorenz attractor**

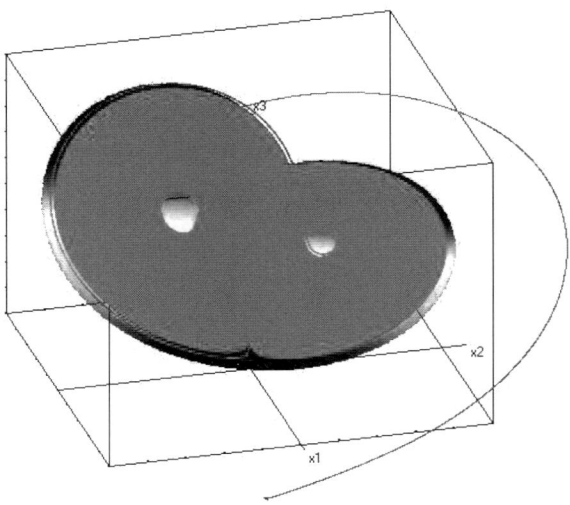

FIGURE 7. Usually, the Lorenz limit set is illustrated (but not computed) by computing solutions to the systems and drawing their trajectories. By the way, these trajectories do not belong to the limit set (except when they start from the limit set), just approach it. In our example, we draw the trajectory of only one solution and we superpose it to the viability kernel under the backward system. The "eyes" around around the two nonzero equilibria are disjoint from the attractor

## 4. Fluctuation basins

Let $K_1 \subset K$ and $K_2 \subset K$ be two closed subsets covering $K$: $K = K_1 \cup K_2$. The *fluctuation basin* $\text{Fluct}(K_1, K_2)$ between $K_1$ and $K_2$ is the subset of initial states $x \in K$ from which all evolutions $x(\cdot) \in \mathcal{S}(x)$ viable in $K$ fluctuate back and forth between $K_1$ to $K_2$ in the sense that the evolution leaves successively $K_1$ and $K_2$ in finite time. We refer to [7, Aubin & Saint-Pierre] the following characterization of the fluctuation basin:

**Theorem 4.1.** *The fluctuation basin is equal to the complement*

$$\text{Fluct}(K_1, K_2) := \complement \left( \text{Capt}_\mathcal{S} \left( K_1 \cup K_2, \text{Viab}_\mathcal{S}(K_1) \cup \text{Viab}_\mathcal{S}(K_2) \right) \right)$$

*of the capture basin of the union* $\text{Viab}_\mathcal{S}(K_1) \cup \text{Viab}_\mathcal{S}(K_2)$. *If we assume furthermore that* $K_i \subset \text{Int}(K_i)$, $i = 1, 2$, *then*

$$\text{Fluct}(K_1, K_2) := \complement \left( \text{Viab}_\mathcal{S}(K_1) \cup \text{Viab}_\mathcal{S}(K_2) \right).$$

As an example, one can compute the fluctuation basin under the Lorenz system by the Viability Kernel Algorithm:

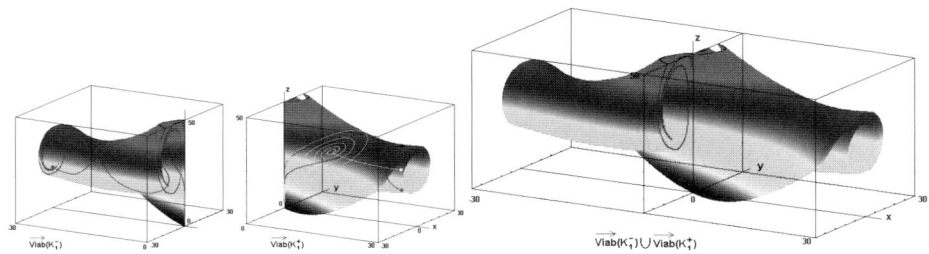

FIGURE 8. **Fluctuation under the Lorenz system.** Forward Viability Kernel of $K_1^-$ and $K_1^+$. Viability kernels of $K_1^-$ and $K_1^+$ are separated on the left and united on the right. Evolutions starting from each of these viability kernels remain in them, and cannot fluctuate. Starting outside the viability kernel of $K_1^-$, the evolutions reach $K_1^+$ in finite time. Either they reach it in the viability kernel of $K_1^+$, and they do not fluctuate anymore, or they reach the capture basin of $K_1^-$ viable in $K$, and they fluctuate once more. They fluctuate back and forth from $K_1^-$ to $K_1^+$ if they do not belong to the capture basin of the union of the two viability kernels

## 5. Local behavior around equilibria

**Definition 5.1 (Stable and unstable local manifolds).** *Let $E$ be the intersection of the forward and backward viability kernels of some compact subset $C$. A subset $L$ between $E$ and $C$ is a* local stable manifold *of $E$ in $C$ if from any initial states $x \in L$ starts at least one evolution $x(\cdot) \in \mathcal{S}(x)$ viable in $L$ such that its $\omega$-limit set $\omega(x(\cdot)) \subset E$ is contained in $E$ (when $E := \{e\}$ is reduced to an equilibrium, such an evolution converges to this equilibrium). A* local unstable manifold *of $E$ is a local stable manifold for the backward system.*

These definitions can be couched in termes of viability kernels:

**Theorem 5.2.** *The* local stable manifold *of $E$ in $C$ is equal to the (forward) viability kernel $\mathrm{Viab}_{\mathcal{S}}(C)$ and the* local unstable manifold *of $E$ in $C$ is equal to the backward viability kernel $\mathrm{Viab}_{\overleftarrow{\mathcal{S}}}(C)$.*
*For any state $x$ belonging to $C \backslash (\mathrm{Viab}_{\mathcal{S}}(C) \cup \mathrm{Viab}_{\overleftarrow{\mathcal{S}}}(C))$, all evolutions passing through $x$ arrived in $C$ a finite time earlier and will leave $C$ in finite time without meeting the union of the local forward and backward manifolds.*

The Viability Kernel Algorithm allows us to compute the stable and unstable manifolds around the trivial equilibrium:

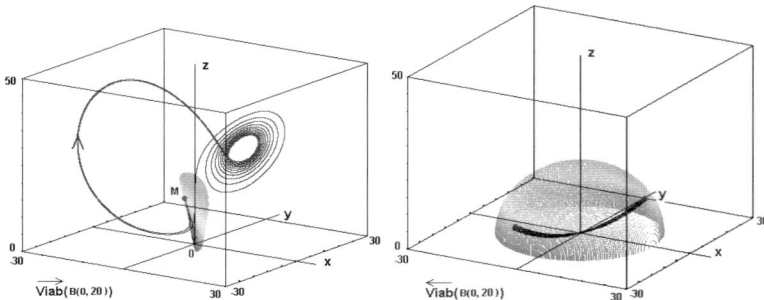

FIGURE 9. **Local Manifolds around the Trivial Equilibrium.** Forward and backward viability kernels of a neighborhood of the origin (trivial equilibrium) are the two-dimensional local stable and one-dimensional unstable manifolds.

The trajectory of an evolution (red) computed with the viability kernel algorithm remains viable in the stable manifold and converges to the origin, whereas another approximation of the evolution (grey) starting from the same initial state computed with a high precision Runge-Kutta method leaves the stable manifold due to the sensitivity to initial conditions

and the nontrivial equilibrium:

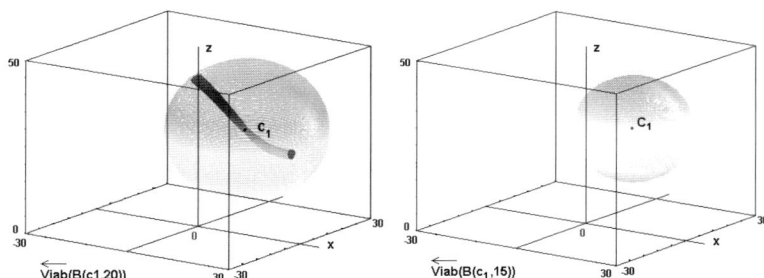

FIGURE 10. **Local Manifolds around a Nontrivial Equilibrium.** Forward and backward viability kernels of a neighborhood of a nontrivial equilibrium $e_1$ are the one-dimensional local stable manifold and the singleton reduced to the equilibrium.

The equilibrium is thus given by the viability kernel algorithm, and not computed analytically.

For every non-equilibrium point $x$ in this neighborhood, the evolution passing through $x$ arrived in it a finite time earlier and will leave it in finite time. Evolutions may cross this neighborhood, but for any only a finite duration. This explains why these balls (the "eyes") are disjoint from the attractor

The Viability Kernel Algorithm allows us to compute also and heterocline joining the trivial equilibrium to any of the nontrivial equilibrium:

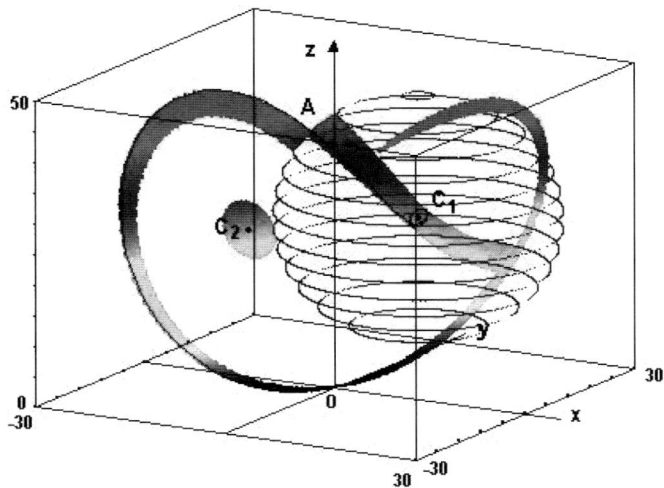

FIGURE 11. **Heteroclinic Evolutions.** Let $C_1 \subset K$ and $C_2 \subset K$ be two subsets. Then through any state in the intersection

$$\mathrm{Capt}_{\overleftarrow{\mathcal{S}}}(K, \mathrm{Viab}_{\overleftarrow{\mathcal{S}}}(C_1)) \cap \mathrm{Capt}_{\mathcal{S}}(K, \mathrm{Viab}_{\mathcal{S}}(C_2))$$

passes an *heteroclinic evolution* $x(\cdot)$ linking the $\alpha$-limit set $\alpha(x(\cdot)) \subset \mathrm{Attr}_{\overleftarrow{\mathcal{S}}}(C_1)$ to the $\omega$-limit set $\mathrm{Attr}_{\mathcal{S}}(C_2)$: There exist subsequence $t_n^- \to -\infty$ and $t_n^+ \to +\infty$ such that $x(t_n^+)$ converges to $C_2$ and $x(t_n^-)$ converges to $C_1$

Through any state in the intersection $\mathrm{Viab}_{\overleftarrow{\mathcal{S}}}(\overline{K \backslash B(e_1, 15)}) \cap \mathrm{Viab}_{\mathcal{S}}(B(e_1, 15))$ passes an evolution connecting the trivial equilibrium to the nontrivial equilibrium. Knowing such an element, one can compute the heterocline:

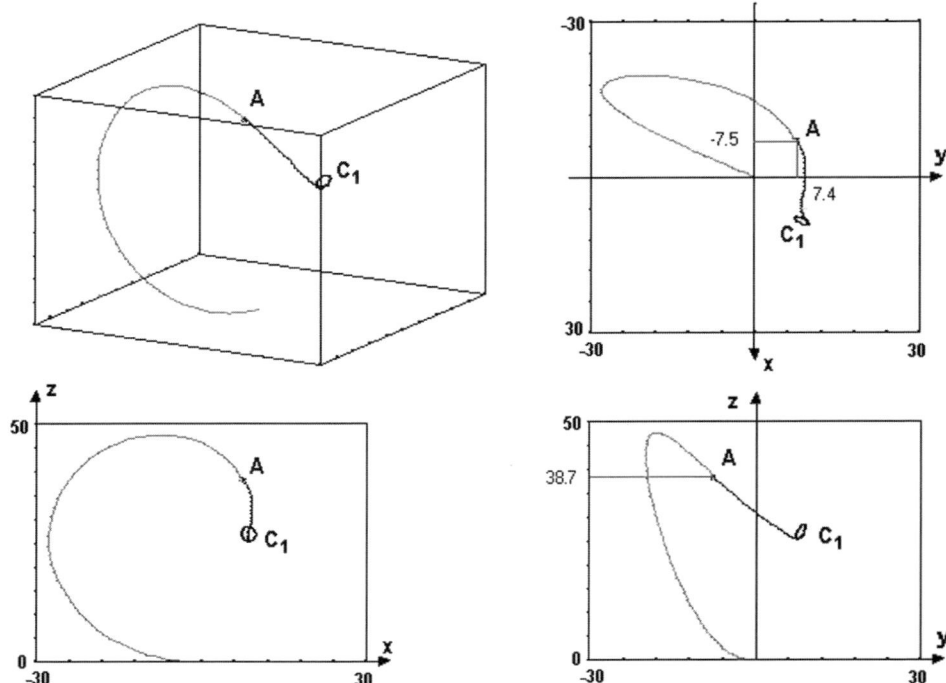

FIGURE 12. **The heteroclinic evolution from** $0$ **to** $e_1$. The heteroclinic evolution linking the trivial equilibrium $0$ to the non trivial equilibrium $e_1$ is obtained from a point in the intersection
$\operatorname{Viab}_{\overleftarrow{S}}(\overline{K \backslash B(e_1, 15)}) \cap \operatorname{Viab}_{S}(B(e_1, 15))$ by piecing together the evolution of the forward system viable in viability kernel of a neighborhood
of $e_1$ and the evolution of the backward system viable in its complement

## 6. Fractal properties of viability kernels under discrete systems

Viability kernels under some discrete systems enjoy fractal properties as well as
high sensitivity. Some of them are Cantor subsets. We begin with Julia sets:

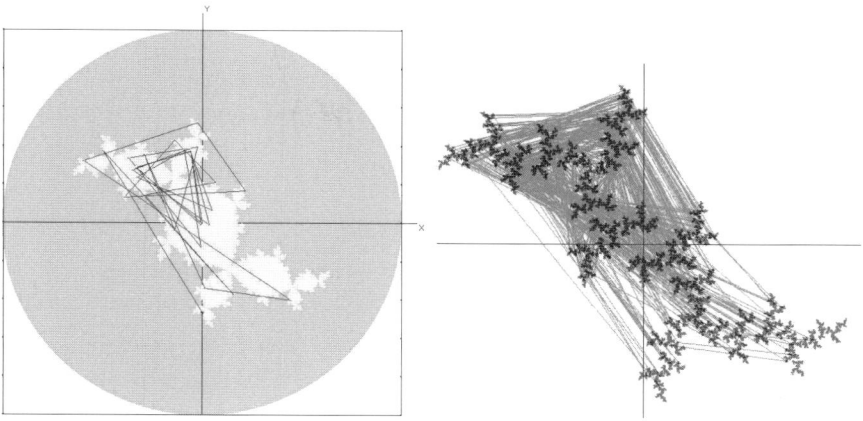

FIGURE 13. **Julia Sets and Fractals.** Pierre Fatou [1878–1929] and Gaston Julia [1893–1978] studied in depth the iterates of complex function $z \mapsto z^2 + u$, or, equivalently, of the map

$$(x, y) \mapsto \varphi(x, y) := (x^2 - y^2 + a, 2xy + b).$$

The subset $K_u := \mathrm{Viab}_\varphi(B(0, 1))$ is the *filled-in Julia set* and its boundary $J_u := \partial K_u$ the *Julia set*.

Examples of a filled-in Julia sets and of a Julia set with empty interior, called Fatou dust

Contrary to "shooting methods", the viability kernel algorithm provides the exact filled-in Julia sets and the viable iterates.

Actually, Theorem 6.1 provides a characterization of the Julia set which can be used by the viability kernel algorithm to compute it:

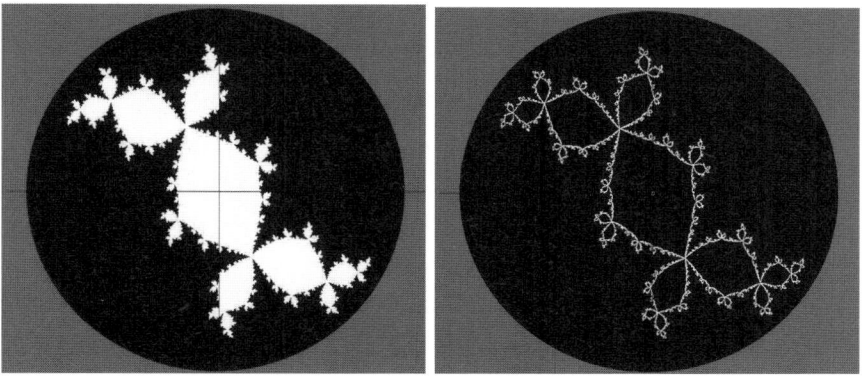

FIGURE 14

**Theorem 6.1 (Julia sets and Filled-in Julia sets).** *The boundary*

$$\partial \mathrm{Viab}_\varphi(K) = \mathrm{Viab}_\varphi(K \setminus C)$$

*of a viability kernel is the viability kernel of the complement of a subset $C \subset$* $\mathrm{Viab}_\varphi(K)$

*if and only*

*if the boundary $\partial \mathrm{Viab}_{\mathcal{S}}(K)$ is viable and the interior $\mathrm{Int}(\mathrm{Viab}_{\mathcal{S}}(K))$ captures $C$.*

Hence the viability kernel algorithm computes also the Julia set.

We turn our attention to viability kernels under disconnecting discrete systems.

**Definition 6.2 (Hutchinson Map).** *A set-valued map $\Phi$ is said to be* disconnecting *on a subset $K$ if there exists a finite number $p$ of functions $\alpha_i : K \mapsto X$ such that*

$$\forall\, x \in K, \quad \Phi^{-1}(x) := \bigcup_{i=1}^{p} \alpha_i(x)$$

*and such that there exist constants $\lambda_i \in ]0,1[$ satisfying: for each subset $C \subset K$,*

$$\left\{ \begin{array}{lll} (i) & \forall\, i = 1, \ldots, p, & \alpha_i(C) \subset C \ (\alpha_i \text{ is antiextensive}) \\ (ii) & \forall\, i \neq j, & \alpha_i(C) \cap \alpha_j(C) = \emptyset \\ (iii) & \forall\, i = 1, \ldots, p, & \mathrm{diam}(\alpha_i(C)) \leq \lambda_i \mathrm{diam}(C) \end{array} \right.$$

*If the functions $\alpha_i : K \mapsto K$ are contractions with Lipschitz constants $\lambda_i \in ]0,1[$, then $\Phi^{-1}$ is called an* Hutchinson map *(introduced in 1981 by John Hutchinson and also called an* iterated function system *by Michael Barnsley).*

**Definition 6.3 (Cantor Sets).** *A subset $K$ is said to be*

1. perfect *if it is closed and if each of its elements is a limit of other elements of $K$,*
2. totally disconnected *if it contains no nonempty open subset,*
3. a Cantor set *if it is non-empty compact, totally disconnected and perfect.*

Classical results can be reformulated in the following way:

**Theorem 6.4.** *The viability kernel of a compact set under a disconnecting map is an uncountable Cantor set.*

The first point to check is that the Cantor set is a cantor set and to compute it with the Viability Kernel Algorithm:

**Theorem 6.5.** *The* Cantor Ternary Set *is the viability kernel of the interval $[0,1]$ under the* Cantor Ternary Map $\Phi$ *defined on $K := [0,1] \subset \mathbb{R}$ by*

$$\Phi(x) := (3x, 3(1-x))$$

*The Cantor Ternary Set is a self similar, symmetric, uncountable Cantor set with fractal dimension $\frac{\log 2}{\log 3}$ and satisfies $\mathbf{C} = \alpha_1(\mathbf{C}) \cup \alpha_2(\mathbf{C})$ and $\alpha_1(\mathbf{C}) \cap \alpha_2(\mathbf{C}) = \emptyset$.*

The interval $[0, 1]$ is viable under the the the Verhulst logistic differential equation $x'(t) = rx(t)(1 - x(t))$ whereas its viability kernel is a Cantor set under its discrete analogue $x_{n+1} = rx_{n+1}(1 - x_{n+1})$ when $r > 4$, also called the logistic system or the quadratic map. This is due to the fact that under for $r > 4$, the quadrating map is disconnecting:

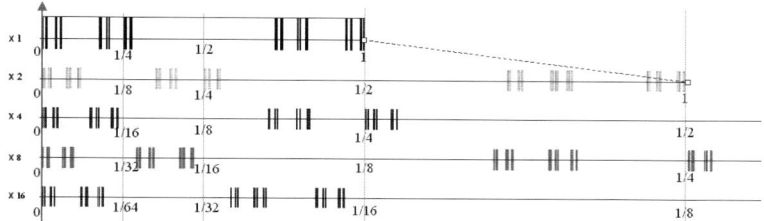

FIGURE 15. **The Quadratic Map.** The quadratic map $\Phi(x) := \{rx(1 - x)\}$ is disconnecting for $r > 4$. The viability kernel of the interval $[0, 1]$ under $\Phi(x) := \{5x(1 - x)\}$ is a uncountable, symmetric Cantor set

The two-dimensional version of the Cantor set is the Sierpinski Gasket:

FIGURE 16. **The Sierpinski Gasket.** The Sierpinski Gasket is the viability kernel under the discrete map

$$\Phi(x, y) := \left\{ (2x, 2y), (2x - 1, 2y), \left( 2x - \frac{1}{2}, 2y - 1 \right) \right\}$$

Since this map is disconnecting, the Sierpinski Gasket is a self similar, uncountable Cantor set with fractal dimension $\frac{\log 3}{\log 2}$

# References

[1] D. Aubin, *A Cultural History of Catastrophes and Chaos: Around the Institut des Hautes Études Scientifiques, France, 1958–1980*, Doctoral Thesis, Princeton University, UMI # 9817022, 1998

[2] J.-P. Aubin, *Viability Theory,* Birkhäuser, Boston, Basel, Berlin, 1991

[3] J.-P. Aubin, *Mutational and morphological analysis: tools for shape regulation and morphogenesis,* Birkhäuser, 1999

[4] J.-P. Aubin, *Viability Kernels and Capture Basins of Sets under Differential Inclusions,* SIAM J. Control, **40** (2001), 853–881

[5] J.-P. Aubin, A. Bayen, N. Bonneuil, and P. Saint-Pierre, *Viability, Control and Game Theories: Regulation of Complex Evolutionary Systems Under Uncertainty,* Springer-Verlag (in preparation)

[6] J.-P. Aubin and H. Frankowska, *Set-Valued Analysis,* Birkhäuser, 1990

[7] J.-P. Aubin and P. Saint-Pierre, *Fluctuations between Subsets of Evolutions Governed by Chaotic Systems,* Mediterranean Journal of Mathematics, **1** (2004), 123–149

[8] C. Bonnati, A. Pumarino, and M. Viana, *Lorenz attractors with arbitrary expanding directions,* C. R. Acad. Sci., **235** (1997), 883–888

[9] A. B. Cambel, *Applied Chaos Theory – A Paradigm for Complexity,* Academic Press, 1993

[10] P. Cardaliaguet, M. Quincampoix and P. Saint-Pierre, *Set-valued numerical methods for optimal control and differential games,* In: *Stochastic and differential games. Theory and numerical methods,* Annals of the International Society of Dynamical Games, 177–247 Birkhäuser (1999)

[11] R. L. Devaney and L. Keen, *Chaos and Fractals: The Mathematics Behind the Computer Graphics,* American Mathematical Society, 1988

[12] C. Doss-Bachelet, J.-P. Francoise, and C. Piquet, *Bursting oscillations in two coupled Fitzhugh-Nagumo systems,* Complex Us, 2003

[13] J. Gleick, *Chaos: Making a New Science,* Penguin Books, 1987

[14] C. Grebogi, E. Ott, and J. A. Yorke, *Chaos, Strange Attractors, and Fractal Basin Boundaries in Nonlinear Dynamical Systems,* "Dynamical Systems", Science, **238** (1987), 585

[15] J. Guckenheimer and R. F. Williams, *Structural stability of Lorenz attractors,* Inst. Hautes Etudes Sci. Publ. Math. **50** (1979), 59–72

[16] S. P. Hasting and W. C. Troy, *A shooting approach to the Lorenz equations,* Bull. Amer. Math. Soc. **27** (1992), 298–303.

[17] E. N. Lorenz, *Deterministic nonperiodic flow,* J. Atmos. Sci. **20** (1987) , 130–141

[18] E. N. Lorenz, *The Essence of Chaos,* University of Washington Press, 1993

[19] R. E. Moore, *Methods and applications of interval analysis,* Studies in Applied Mathematics, SIAM, Philadelphia, 1979

[20] R. E. Moore, *Interval Analysis,* Prentice-Hall, 1966

[21] E. Ott, *Chaos in Dynamical Systems,* Cambridge University Press, 1993

[22] M. Quincampoix and P. Saint-Pierre, *An algorithm for viability kernels in Hölderian case: Approximation by discrete viability kernels*, J. Math. Syst. Est. and Control, **8** (1998), 17–29

[23] P. Saint-Pierre, *Approximation of the viability kernel*, Applied Mathematics & Optimisation, **29** (1994), 187–209

[24] Shi Shuzhong, *Viability theorems for a class of differential-operator inclusions*, J. of Diff. Eq., **79** (1989), 232–257

[25] C. Sparrow, *The Lorenz equations: bifurcations, chaos and strange attractors*, Springer, 1982

**Acknowledgment**

Many thanks to Agathe Tyche for typing this artcile.

Jean-Pierre Aubin
LASTRE (Laboratoire d'Applications des Systèmes Tychastiques Régulés)
14, rue Domat
FR-75005 Paris
France
e-mail: aubin.jp@gmail.com
URL: http://lastre.asso.fr

Patrick Saint-Pierre
12, rue de la Roue
FR-92140 Clamart
France
e-mail: patrick.saint.pierre@gmail.com

Progress in Nonlinear Differential Equations
and Their Applications, Vol. 75, 49–60
© 2007 Birkhäuser Verlag Basel/Switzerland

# Generalized Steiner Selections Applied to Standard Problems of Set-Valued Numerical Analysis

Robert Baier

*Dedicated to Arrigo Cellina and James Yorke*

**Abstract.** Generalized Steiner points and the corresponding selections for set-valued maps share interesting commutation properties with set operations which make them suitable for the set-valued numerical problems presented here. This short overview will present first applications of these selections to standard problems in this area, namely representation of convex, compact sets in $\mathbb{R}^n$ and set operations, set-valued integration and interpolation as well as the calculation of attainable sets of linear differential inclusions. Hereby, the convergence results are given uniformly for a dense countable representation of generalized Steiner points/selections. To achieve this aim, stronger conditions on the set-valued map $F$ have to be taken into account, e.g., the Lipschitz condition on $F$ has to be satisfied for the Demyanov distance instead of the Hausdorff distance. To establish an overview on several applications, not the strongest available results are formulated in this article.

**Mathematics Subject Classification (2000).** 54C65; 93B03, 93C05, 28B20.

**Keywords.** generalized Steiner selections, set-valued quadrature methods and interpolation, linear differential inclusions, attainable sets, Lipschitz and absolutely continuous selections, set operations.

## 1. Preliminaries

In this section, some basic notations for convex sets are introduced. $B_r(m)$ denotes the closed Euclidean ball with radius $r$ and center $m$ in $\mathbb{R}^n$, $B_1$, $S_{n-1}$ the unit ball resp. sphere, $\|\cdot\|$ the Euclidean norm in $\mathbb{R}^n$ and $\mathcal{K}_c(\mathbb{R}^n)$ the set of all convex, compact, nonempty subsets of $\mathbb{R}^n$. $\delta^*(l, C)$ and $Y(l, C)$ are the support function resp. the supporting face of $C \in \mathcal{K}_c(\mathbb{R}^n)$ in direction $l \in \mathbb{R}^n$, where $Y(l, C)$

coincides with the subdifferential of the support function. Unique supporting points are denoted by $y(l, C)$.

In this paragraph, some well-known set operations are briefly recalled. The Minkowski sum of two sets $C, D \in \mathcal{K}_c(\mathbb{R}^n)$, the scalar multiplication with $\lambda \in \mathbb{R}$ and the image under a linear matrix $A \in \mathbb{R}^{p \times n}$ are defined as usual:

$$C + D = \bigcup_{\substack{c \in C \\ d \in D}} \{c + d\}, \quad \lambda C = \bigcup_{c \in C} \{\lambda c\} \quad \text{and} \quad AC = \bigcup_{c \in C} \{Ac\}. \qquad (1.1)$$

The Demyanov difference from [8, 16] is defined as

$$C \dot{-} D = \overline{\mathrm{co}} \bigcup_{l \in T_C \cap T_D} \{y(l, C) - y(l, D)\},$$

where $T_C \subset S_{n-1}$ defines the set of directions $l$ with $Y(l, C) = \{y(l, C)\}$. The Demyanov distance $d_D(C, D)$ could be calculated as the maximal norm element $\|C \dot{-} D\|$ of the Demyanov difference and is stronger than the Hausdorff distance. It plays a major role in this article, since it could also be expressed by the norm of the differences of generalized Steiner points.

Within the set of all Borel probability measures on the Borel $\sigma$-algebra $\mathcal{B}$ onto $B_1$, a smooth measure $\beta$ is defined by a density function $\theta \in \mathcal{C}^1(B_1)$ and

$$\beta(A) = \int_A \beta(dp) = \int_A \theta(p) \, dp,$$

where $A \in \mathcal{B}$ is a Borel-measurable subset of $B_1$. This class of measures is shortly denoted by $\mathcal{SM}$, the so-called smooth measures.

Atomic measures from $\mathcal{AM}$ are concentrated in a single point $l \in S_{n-1}$, i.e.,

$$\alpha_{[l]}(A) = \begin{cases} 0, & \text{if } l \notin A, \ A \in \mathcal{B}, \\ 1, & \text{if } l \in A, \ A \in \mathcal{B}. \end{cases}$$

Measures with finite support in $S_{n-1}$ (class $\mathcal{FM}$) are convex combination of measures in $\mathcal{AM}$ (cf. [4]). $\mathcal{CM}$ denotes either the family of measures $\mathcal{AM}$ or $\mathcal{FM}$.

Generalized Steiner points and selections are introduced for smooth measures in $\mathcal{SM}$ by Dentcheva in [9–11]. They are generalization of the well-known Steiner center $\mathrm{St}\,(U)$ (take the smooth measure with uniform density in the next definition) as mentioned in [9], cf. also the references given therein.

**Definition 1.1.** *The generalized Steiner (GS-) point of a set $C \in \mathcal{K}_c(\mathbb{R}^n)$ for a measure $\gamma \in \{\mathcal{FM}, \mathcal{SM}\}$ is defined as*

$$\mathrm{St}_\gamma(C) := \int_{B_1} \mathrm{St}\,(Y(p, C)) \, \gamma(dp).$$

Definition 1.1 equals the definition given in [9] (cf. [4, Lemma 3.3]), where the norm-minimal element of $Y(p, C)$ is used instead of the Steiner center. However, the definition above from [4] generalizes the GS-point from smooth measures to measures with finite support. For atomic measures $\alpha_{[l]}, \alpha_{[\xi]} \in \mathcal{AM}$ with $l, \xi \in S_{n-1}$ and $\lambda \in [0,1]$, the following formulas (cf. [4, Lemma 3.3]) apply for the GS-point:

$$\mathrm{St}_{\alpha_{[l]}}(C) = \mathrm{St}\left(Y\left(l, C\right)\right), \quad \mathrm{St}_{\lambda\alpha_{[l]}+(1-\lambda)\alpha_{[\xi]}}(C) = \lambda\,\mathrm{St}_{\alpha_{[l]}}(C) + (1-\lambda)\,\mathrm{St}_{\alpha_{[\xi]}}(C)$$

For simpler notation, we set

$$\mathrm{St}_{\alpha_{[l]}}(C) = \begin{cases} \mathrm{St}(C), & \text{if } l = 0_{\mathbb{R}^n}, \\ \mathrm{St}_{\alpha_{[\eta]}}(C), & \text{if } l \neq 0_{\mathbb{R}^n} \text{ and } \eta = \frac{1}{\|l\|} \cdot l. \end{cases}$$

## 2. Representation and arithmetics of sets

GS-points form a dense, non-minimal representation of a convex compact set, i.e.,

$$C = \overline{\bigcup_{\alpha \in \mathcal{FM}} \left\{\mathrm{St}_{\alpha}(C)\right\}} = \overline{\bigcup_{\beta \in \mathcal{SM}} \left\{\mathrm{St}_{\beta}(C)\right\}} \tag{2.1}$$

(see [9, Lemma 5.4] for measures in $\mathcal{SM}$ resp. [4, Corollary 3.5] for the class $\mathcal{FM}$).

**Proposition 2.1.** *For $C \in \mathcal{K}_c(\mathbb{R}^n)$, there exists a sequence $(\beta_m)_{m \in \mathbb{N}} \subset \mathcal{SM}$ with*

$$C = \overline{\bigcup_{m \in \mathbb{N}} \left\{\mathrm{St}_{\beta_m}(C)\right\}}.$$

*The sequence of measures could also be chosen from $\mathcal{FM}$.*

*Proof.* This follows for smooth measures from [10, Theorem 3.4]. (2.1) allows to choose a measure $\alpha_{m,N} \in \mathcal{FM}$ for $N \in \mathbb{N}$ and each $m \in \mathbb{N}$ with

$$\|\mathrm{St}_{\alpha_{m,N}}(C) - \mathrm{St}_{\beta_m}(C)\| \leq \frac{1}{N}.$$

The following union gives the representation stated in the proposition:

$$\overline{\bigcup_{N \in \mathbb{N}} \bigcup_{m \in \mathbb{N}} \left\{\mathrm{St}_{\alpha_{m,N}}(C)\right\}} \qquad \qquad \square$$

GS-points commute with the arithmetical operations for sets in $\mathcal{K}_c(\mathbb{R}^n)$, cf. [9, Remarks after Theorem 3.6] for measures in $\mathcal{SM}$ resp. [4, Lemma 4.1] for $\mathcal{FM}$.

**Proposition 2.2.** *Let $C, D \in \mathcal{K}_c(\mathbb{R}^n)$ and $\gamma \in \mathcal{CM}$. Then,*

$$\mathrm{St}_{\gamma}\left(\lambda C + \mu D\right) = \lambda\,\mathrm{St}_{\gamma}(C) + \mu\,\mathrm{St}_{\gamma}(D) \quad (\lambda, \mu \geq 0),$$

$$\mathrm{St}_{\gamma}\left(RC\right) = R\,\mathrm{St}_{\widetilde{\gamma}}(C) \qquad\qquad (R \text{ orthogonal matrix}). \tag{2.2}$$

*Hereby, $\widetilde{\gamma}(B) = \gamma(R \cdot B)$ for all sets $B \in \mathcal{B}$.*

**Example 2.3.** Let

$$U = \mathrm{co}\left\{\begin{pmatrix} 2 \\ 2 \end{pmatrix}, \begin{pmatrix} 0 \\ 2 \end{pmatrix}, \begin{pmatrix} -2 \\ -2 \end{pmatrix}, \begin{pmatrix} 0 \\ -2 \end{pmatrix}\right\}, \quad V = \mathrm{co}\left\{\begin{pmatrix} 1 \\ 1 \end{pmatrix}, \begin{pmatrix} -1 \\ 1 \end{pmatrix}, \begin{pmatrix} -1 \\ -1 \end{pmatrix}, \begin{pmatrix} 1 \\ -1 \end{pmatrix}\right\}$$

and $W = U + V$. Figure 1 shows that $U$ and $V$ are represented in the left picture by 8 GS-points $\mathrm{St}_{\alpha_{[l^i]}}(C)$ (8 small crosses), $i = 1, \ldots, 8$. By Proposition 2.2 the

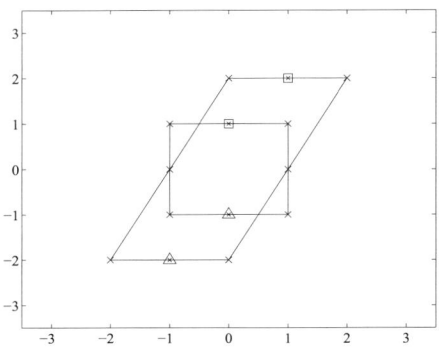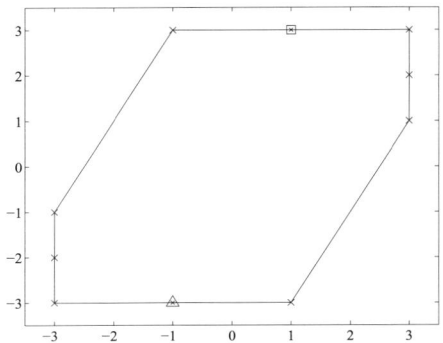

FIGURE 1. Minkowski sum $W$ (right) of the summands $U, V$ (left)

GS-points of $U$ and $V$ in common directions $l^i$ are added (see how the GS-points marked by squares resp. triangles add to form the GS-point of $W$ in the same direction in the right picture).

One could not expect that (2.2) holds for a general matrix $R \in \mathbb{R}^{n \times n}$. Nevertheless, this property could be fulfilled for special classes of sets.

**Definition 2.4.** *The tuple $(M, U)$ with $M \in \mathbb{R}^{n \times m}$ and $U \subset \mathbb{R}^m$ fulfills the GSCL-property (commutation of GS-points under linear maps), if*

$$\mathrm{St}_{\alpha_{[l]}}\left(MU\right) = M\,\mathrm{St}_{\alpha_{[M^\top l]}}\left(U\right) \quad \text{for each } l \in S_{n-1}.$$

**Lemma 2.5.** *Let $M \in \mathbb{R}^{n \times m}$ and $p^0 \in \mathbb{R}^m$. Then, $(M, \{p^0\})$, $(M, B_1(0_{\mathbb{R}^m}))$ and $(M, [-1, 1]^m)$ fulfill the GSCL-property.*

*Proof.* Clearly, all GS-points of singletons coincide with the only element of the set, so that the case $U = \{p^0\}$ is simple to prove.

For a set $U$ symmetric to the origin (i.e., $U = (-1) \cdot U$), one has for $\eta \in S_{m-1}$:

$$Y\left(-\eta, U\right) = -Y\left(\eta, U\right), \quad \mathrm{St}\left(U\right) = 0_{\mathbb{R}^m} \quad \text{and} \quad \mathrm{St}_{\alpha_{[-\eta]}}\left(U\right) = -\,\mathrm{St}_{\alpha_{[\eta]}}\left(U\right).$$

If $l \in S_{n-1}$ and $\eta := M^\top l = 0_{\mathbb{R}^m}$, then $Y\left(M^\top l, U\right) = U$, $MU = (-1) \cdot MU$ so that

$$M\,\mathrm{St}_{\alpha_{[M^\top l]}}\left(U\right) = M\,\mathrm{St}\left(U\right) = 0_{\mathbb{R}^n} = \mathrm{St}\left(MU\right) = \mathrm{St}\left(Y\left(l, MU\right)\right) = \mathrm{St}_{\alpha_{[l]}}\left(MU\right).$$

If $\eta \neq 0_{\mathbb{R}^m}$, then $Y\left(\eta, B_1(0_{\mathbb{R}^m})\right) = \{y(\eta, B_1(0_{\mathbb{R}^m}))\}$ and $M\,\mathrm{St}_{\alpha_{[\eta]}}\left(B_1(0_{\mathbb{R}^m})\right)$ equals

$$M\,\mathrm{St}\left(Y\left(\eta, B_1(0_{\mathbb{R}^m})\right)\right) = My(\eta, B_1(0_{\mathbb{R}^m})) = \mathrm{St}_{\alpha_{[l]}}\left(MB_1(0_{\mathbb{R}^m})\right).$$

Let $v \in \mathbb{R}^m$. Then, $M \operatorname{St}_{\alpha_{[\eta]}} \left( \operatorname{co}\{-v, v\} \right)$ coincides with

$$M \operatorname{St} \left( Y\left(\eta, \operatorname{co}\{-v,v\}\right) \right) = M \cdot \begin{cases} v & \text{if } \eta^\top v > 0, \\ 0_{\mathbb{R}^m} & \text{if } \eta^\top v = 0, \\ -v & \text{if } \eta^\top v < 0 \end{cases} = \begin{cases} Mv & \text{if } l^\top Mv > 0, \\ 0_{\mathbb{R}^n} & \text{if } l^\top Mv = 0, \\ -Mv & \text{if } l^\top Mv < 0 \end{cases}$$

$$= \operatorname{St} \left( Y\left(l, \operatorname{co}\{-Mv, Mv\}\right) \right) = \operatorname{St}_{\alpha_{[l]}} \left( M \operatorname{co}\{-v, v\} \right).$$

The assertion follows from $[-1,1]^m = \sum\limits_{i=1}^{m} \operatorname{co}\{-e^i, e^i\}$ with unit vectors $e^i \in \mathbb{R}^m$. $\qquad\square$

An immediate consequence of Proposition 2.2 is the representation of the set operations in (1.1) with $A$ being orthogonal (cf. [4, Corollary 4.4]) as well as for the Demyanov difference/distance, cf. [4, Theorems 4.5, 4.6 and Corollary 4.8].

**Theorem 2.6.** *Let $C, D \in \mathcal{K}_c(\mathbb{R}^n)$. Then, there exists $(\gamma_m)_{m \in \mathbb{N}} \subset \mathcal{CM}$ with*

$$C \doteq D = \overline{\bigcup_{m \in \mathbb{N}} \left\{ \operatorname{St}_{\gamma_m}(C) - \operatorname{St}_{\gamma_m}(D) \right\}}, \quad \operatorname{d}_D(C, D) = \sup_{m \in \mathbb{N}} \| \operatorname{St}_{\gamma_m}(C) - \operatorname{St}_{\gamma_m}(D) \|.$$

## 3. Regularity of set-valued maps

In this paper, a set-valued map $F : I \rightrightarrows \mathbb{R}^n$ is given with images in $\mathcal{K}_c(\mathbb{R}^n)$ and to each measure $\gamma \in \mathcal{CM}$ the generalized Steiner (GS-) selection corresponds:

$$t \mapsto \operatorname{St}_\gamma \left( F(t) \right).$$

It is interesting that the regularity of the set-valued map $F$ carries over to the uniform regularity of its GS-selection and vice versa, if the regularity is in some sense uniform. The first result states the Castaing representation by GS-selections characterizing the measurability of $F$ (i.e., each preimage of an open set lies in $\mathcal{B}$).

**Theorem 3.1.** *Let $F : I \rightrightarrows \mathbb{R}^n$ be measurable with images in $\mathcal{K}_c(\mathbb{R}^n)$. Then, $\operatorname{St}_\gamma \left( F(\cdot) \right)$ is measurable for each $\gamma \in \mathcal{CM}$ and there exists $(\gamma_m)_{m \in \mathbb{N}} \subset \mathcal{CM}$ with*

$$F(t) = \overline{\bigcup_{m \in \mathbb{N}} \left\{ \operatorname{St}_{\gamma_m}\left( F(t) \right) \right\}} \quad (t \in I).$$

*Proof.* For smooth measure, this result could be found in [10, Theorem 3.4].
For atomic measures, proceed as in the proof of [10, Theorem 3.4] and choose the same measures $(\beta_m)_{m \in \mathbb{N}} \subset \mathcal{SM}$ with densities $(\theta_m)_{m \in \mathbb{N}} \subset \mathcal{C}^1(B_1)$. Given an accuracy $\varepsilon > 0$, each point $(t, y) \in \operatorname{graph} F$, i.e., $y \in F(t)$, could be approached as

$$\| y - \operatorname{St}_{\beta_m}\left( F(t) \right) \| \leq \tilde{\varepsilon} := \frac{\varepsilon}{4\sqrt{n}}.$$

Apply [7, Proposition 3.4.5] to construct countable, dense sequences $(g_{m,k})_{k \in \mathbb{N}}$ of simple, measurable functions in $L_1(B_1, \mathcal{B}, \beta_m)$. Let $s_{m,i}$ be the function with values of the $i$-th coordinate of $\operatorname{St}_{\beta_m}\left( F(\cdot) \right)$. By [7, Proposition 3.4.2], there exists a simple, measurable map $h_{m,i}$ with $\|s_{m,i} - h_{m,i}\|_{L_1} < \tilde{\varepsilon}$. Following the proof of [7, Proposition 3.4.5], one could choose $g_{m,k,i}$ with $\|h_{m,i} - g_{m,k,i}\|_{L_1} < \tilde{\varepsilon}$. Due to the construction, $g_{m,k,i}$ is a finite sum of terms $a_{m,k,i,j} \cdot \chi_{A_{m,k,i,j}}(\cdot)$ with $a_{m,k,i,j}$ being an $\tilde{\varepsilon}$-approximation of the values $s_{m,i}$ on $A_{m,k,i,j}$. Hence, one could replace $a_{m,k,i,j}$ by $s_{m,i}(\xi_{m,k,i,j})$ with $\xi_{m,k,i,j} \in A_{m,k,i,j}$ so that the resulting sum coincides with the measure with finite support in $\bigcup_{j=1}^{N(m,k,i)} \{\xi_{m,k,i,j}\}$. Since one could approach $(t, y) \in \operatorname{graph} F$ within accuracy $\varepsilon$, the Castaing representation is proved.

For each $l \in S_{n-1}$, the Borel measurability of the GS-selection $\operatorname{St}_{\alpha_{[l]}}(F(\cdot))$ follows from the one of marginal map $t \mapsto Y(l, F(t))$ by [6, Theorem 3.4]. Indeed, the proof of [4, Lemma 3.2] could be modified by focussing on the time $t$ instead of the direction $l$. □

**Proposition 3.2 ( [4, Proposition 5.1]).** *Let* $F : I \rightrightarrows \mathbb{R}^n$ *be a set-valued map with images in* $\mathcal{K}_c(\mathbb{R}^n)$. *Then,* $F$ *is D-Lipschitz, i.e.,* $\mathrm{d}_D(F(t), F(\tau)) \leq L \cdot |t - \tau|$, *if and only if for each measure* $\gamma \in \mathcal{SM}$, *the GS-selection* $\operatorname{St}_\gamma(F(\cdot))$ *is uniformly Lipschitz continuous with constant* $L$. $\mathcal{SM}$ *could be replaced by* $\mathcal{AM}$ *or* $\mathcal{FM}$.

E.g., the maps $F(t) = r(t)U$ with $U \in \mathcal{K}_c(\mathbb{R}^n)$, $r(t) \geq 0$ or $A(t)B_1$ with $A(\cdot) \in \mathcal{C}(I)$, $A(t)$ invertible, are D-Lipschitz. If the Lipschitz continuity of $F$ is demanded only w.r.t. the Hausdorff distance, the GS-selections for $\mathcal{SM}$ are still Lipschitz (cf. [10, Theorem 4.1]), but with constants depending on the measure.

**Example 3.3 ( [4, Example 5.2]).** Let $I = [-\frac{\pi}{2}, \frac{3\pi}{2}]$ and consider the set-valued map $F(t) = \operatorname{co}\{\binom{0}{0}, \binom{\cos(t)}{\sin(t)}\}$ on $I$. Then, $F$ is Lipschitz continuous w.r.t. Hausdorff distance with constant 1.
Consider $\theta \in \mathcal{C}^1(\mathbb{R}^n)$, the normal Dirac sequence $(\theta_m)_{m \in \mathbb{N}}$ from [14, Chapter 7.1C] and their measures $\beta_m \in \mathcal{SM}$. If $L > 0$, the Lipschitz constants $L_{\beta_m}$ with

$$L_{\beta_m} = L \cdot \left( n \cdot \max_{l \in S_{n-1}} \theta_m(l) + \max_{p \in B_1} \|\nabla \theta_m(p)\| \right), \quad \nabla \theta_m(p) = m^n \cdot \nabla \theta(m \cdot p),$$

tend to $+\infty$, since the second maximum is positive and bounded uniformly in $m$.

In [4] the bounded variation of $F$ is discussed and results on the uniform bounded variation of the GS-selections are obtained. Before discussing the case of absolutely continuity, we recall the definition of the Aumann integral in [2]:

$$\int_I F(t)\, dt = \left\{ \int_I f(t)\, dt \mid f \in \mathcal{L}_1(I) \text{ and } f \text{ be a selection of } F \right\}$$

**Proposition 3.4.** *Let* $F$ *be an indefinite integral as in [1], i.e., there exists a measurable, integrably bounded* $G : I \rightrightarrows \mathbb{R}^n$ *with images in* $\mathcal{K}_c(\mathbb{R}^n)$, $F_0 \in \mathcal{K}_c(\mathbb{R}^n)$ *and*

$$F(t) := F_0 + \int_{t_0}^t G(s)\, ds \quad (t \in I).$$

*Then, each GS-selection of* $F$ *is absolutely continuous for all* $\gamma \in \mathcal{CM}$ *with*

$$\operatorname{St}_\gamma(F(t)) = \operatorname{St}_\gamma(F_0) + \int_I \operatorname{St}_\gamma(G(t))\, dt. \tag{3.1}$$

*Proof.* The measurability of $\mathrm{St}_\gamma\left(G(\cdot)\right)$ follows from Theorem 3.1, the integrability by the integrably boundedness of $G(\cdot)$. Equation (3.1) can be proved by Proposition 2.2 and [4, Propositions 6.2 and 6.4]. $\qquad\square$

Especially, the proposition yields a dense, countable representation of the Aumann-integral by Lebesgue integrals of GS-selections with measures in $\mathcal{CM}$.

## 4. Set-valued interpolation and quadrature methods

**Proposition 4.1 (piecewise linear interpolation in [20]).** *Let $I = [t_0, T]$ and $F : I \Rightarrow \mathbb{R}^n$ be a set-valued map with images in $\mathcal{K}_c(\mathbb{R}^n)$ which is D-Lipschitz with constant $L$. Then, the piecewise linear interpolation*

$$P_1(F; t) := P_1(t) := \frac{t_{i+1} - t}{h} F(t_i) + \frac{t - t_i}{h} F(t_{i+1}) \quad (t \in [t_i, t_{i+1}])$$

*with step-size $h = \frac{T - t_0}{N}$, $N \in \mathbb{N}$, and grid points $t_i = t_0 + ih$, $i = 0, \ldots, N$, yields*

$$\mathrm{d}_D\left(F(t), P_1(t)\right) = \sup_{m \in \mathbb{N}} \|\,\mathrm{St}_{\gamma_m}\left(F(t)\right) - \mathrm{St}_{\gamma_m}\left(P_1(t)\right)\| \le \frac{L}{4} h$$

*with a suitable sequence $(\gamma_m)_{m \in \mathbb{N}} \subset \mathcal{CM}$.*

*Proof.* Propositions 2.2, 3.2 and Theorem 2.6 yield:

$$\mathrm{d}_D(F(t), P_1(t)) = \mathrm{d}_D\left(\frac{t_{i+1} - t}{h} F(t) + \frac{t - t_i}{h} F(t), \frac{t_{i+1} - t}{h} F(t_i) + \frac{t - t_i}{h} F(t_{i+1})\right)$$

$$\le \frac{t_{i+1} - t}{h} \mathrm{d}_D(F(t), F(t_i)) + \frac{t - t_i}{h} \mathrm{d}_D(F(t), F(t_{i+1}))$$

$$\le \frac{t_{i+1} - t}{h} \cdot L \cdot |t - t_i| + \frac{t - t_i}{h} \cdot L \cdot |t - t_{i+1}|$$

$$= \frac{2L}{h}(t_{i+1} - t)(t - t_i) \le \frac{L}{4} \cdot h \qquad\qquad \square$$

Hence, $F(t)$ could be densely approximated by a countable number of piecewise linear interpolants of GS-selections, since $\mathrm{St}_{\gamma_m}\left(P_1(F; \cdot)\right) = P_1(\mathrm{St}_{\gamma_m}\left(F(\cdot)\right); \cdot)$.

Given some weights $b_\mu \ge 0$ and nodes $c_\mu \in [0, 1]$, $\mu = 1, \ldots, s$, a set-valued quadrature formula (cf. [5, 12] and references therein) is given by

$$Q(F) := (T - t_0) \sum_{\mu=1}^{s} b_\mu F\left(t_0 + c_\mu(T - t_0)\right)$$

for a set-valued function $F : I \Rightarrow \mathbb{R}^n$ with images in $\mathcal{K}_c(\mathbb{R}^n)$. The iterated version of this quadrature formula is given as

$$Q_N(F) := h \sum_{i=0}^{N-1} \sum_{\mu=1}^{s} b_\mu F(t_i + c_\mu h) \tag{4.1}$$

for the equi-distant step-size $h = \frac{T - t_0}{N}$, $N \in \mathbb{N}$, and nodes $t_i = t_0 + ih$, $i = 0, \ldots, N$.

In the next proposition, the assumption of bounded variation in [4, Proposition 6.6] is replaced by the stronger condition of Lipschitz continuity to shorten the exposition. In the proof, Propositions 2.2, 3.2, 3.4 and Theorem 2.6 are used.

**Proposition 4.2.** *Let $F : I \rightrightarrows \mathbb{R}^n$ have images in $\mathcal{K}_c(\mathbb{R}^n)$ and be D-Lipschitzian with constant $L$. Consider an iterated set-valued quadrature method (4.1) with $\sum_{\mu=1}^{s} b_\mu = 1$ and $N \in \mathbb{N}$. Then, there exists $(\gamma_m)_{m\in\mathbb{N}} \subset \mathcal{CM}$ with*

$$d_D \left( \int_I F(t)\, dt, Q_N(F) \right) = \sup_{m\in\mathbb{N}} \| \int_I \mathrm{St}_{\gamma_m}\left( F(t) \right) dt - Q_N\left( \mathrm{St}_{\gamma_m}\left( F(\cdot) \right) \right) \| \le Lh$$

Consequently, Proposition 4.2 shows that the integral of each GS-selection $\mathrm{St}_{\gamma_m}\left( F(\cdot) \right)$ is uniformly approximated by the corresponding point-wise iterated quadrature formula of order $\mathcal{O}(h)$, since $\mathrm{St}_{\gamma_m}\left( Q_N(F(\cdot)) \right) = Q_N\left( \mathrm{St}_{\gamma_m}\left( F(\cdot) \right) \right)$.

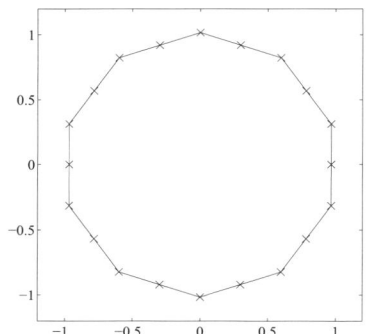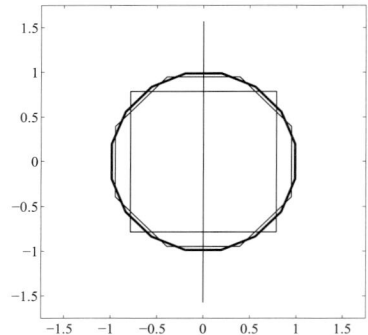

FIGURE 2. Approximations of the Aumann integral

**Example 4.3 ( [19, Example before Theorem 2]).** Consider $F : [0, 2\pi] \rightrightarrows \mathbb{R}^n$ with $F(t) = \frac{1}{4} \cdot \binom{\sin(t)}{\cos(t)}[-1, 1]$ and $\delta^*(l, F(t)) = \frac{1}{4} \cdot |\sin(t)l_1 + \cos(t)l_2|$ for $l \in S_1$.
Then, $\int_0^{2\pi} F(t)\, dt = B_1$ and

$$\mathrm{St}_{\alpha_{[l]}}\left( F(t) \right) = \begin{cases} \binom{\sin(t)}{\cos(t)}, & \text{if } \sin(t)l_1 + \cos(t)l_2 > 0, \\ \binom{0}{0}, & \text{if } \sin(t)l_1 + \cos(t)l_2 = 0, \\ -\binom{\sin(t)}{\cos(t)}, & \text{if } \sin(t)l_1 + \cos(t)l_2 < 0. \end{cases}$$

Clearly, $\mathrm{St}_{\alpha_{[l]}}\left( F(\cdot) \right)$ has bounded variation uniformly in $l \in S_1$, since it is piece-wise Lipschitz with maximal two jumps depending on $l$ in $I$ (the jump height is independent from $l$). Hence, a set-valued iterated quadrature method converges at least with order $\mathcal{O}(h)$ with the weakened form of Proposition 4.2 in [4]. Figure 2 shows the iterated Riemann sum $Q_N(F) = h \sum_{i=0}^{N-1} F(t_i)$ for $N = 10$ (left picture, the GS-points are marked by crosses) and the approximating sequence of the convex hulls $Q_{N_i}(F)$ with $N_i = 2^i$, $i = 0, 1, \ldots, 4$ (on the right, $i = 4$ emphasized).

TABLE 1. Approximate convergence order for iter. Riemann sum

| $i$ | $N$ | $M$ | $\Delta_N$ | $p_N$ | $i$ | $N$ | $M$ | $\Delta_N$ | $p_N$ |
|---|---|---|---|---|---|---|---|---|---|
| 0 | 1 | 3 | 1.83540766 | — | 4 | 16 | 24 | 0.18109238 | 1.01091 |
| 1 | 2 | 3 | 1.83540766 | 0.00000 | 5 | 32 | 40 | 0.09037697 | 1.00270 |
| 2 | 4 | 8 | 0.78379807 | 1.22755 | 6 | 64 | 72 | 0.04516741 | 1.00067 |
| 3 | 8 | 16 | 0.36493295 | 1.10285 | 7 | 128 | 136 | 0.02258107 | 1.00017 |

Table 1 shows the approximate convergence order for the iterated Riemann sum. $N$ is the number of subintervals, $M$ is the resulting number of different GS-points, $\Delta_N \approx d_D(Q_N(F), \int_I F(t)\,dt)$ is an approximation of the Demyanov distance to the reference set (iterated trapezoidal rule with $N_{\mathrm{ref}} = 100000$) and $p_N$ is the estimated order of convergence which tends to the expected order 1.

## 5. Linear differential inclusions

Consider the linear differential inclusion (LDI) with absolutely continuous solutions $x(\cdot)$ and given integrable matrix functions $A : I \to \mathbb{R}^{n \times n}$, $B : I \to \mathbb{R}^{n \times m}$, a starting set $X_0 \in \mathcal{K}_c(\mathbb{R}^n)$ and a control region $U \in \mathcal{K}_c(\mathbb{R}^m)$.

$$x'(t) \in A(t)x(t) + B(t)U \quad \left(\text{a.e. } t \in I = [t_0, T]\right), \tag{5.1}$$

$$x(t_0) \in X_0, \tag{5.2}$$

The following representation of the attainable set $\mathcal{A}(T, t_0, X_0)$ (the set of all end points $x(T)$ of absolutely continuous solutions) is well-known and is recalled in the next lemma, cf., e.g., [17].

**Lemma 5.1.** *Given the problem (LDI) in (5.1)–(5.2) and $l \in S_{n-1}$, the reachable set can be represented with the fundamental matrix solution $\Phi(\cdot, \cdot)$ as*

$$\mathcal{A}(T, t_0, X_0) = \Phi(T, t_0)X_0 + \int_I \Phi(T, \tau)B(\tau)U \, d\tau,$$

$$Y\left(l, \mathcal{A}(T, t_0, X_0)\right) = \Phi(T, t_0)Y\left(\Phi(T, t_0)^\top l, X_0\right)$$

$$+ \int_I \Phi(T, \tau)B(\tau)Y\left(B(\tau)^\top \Phi(T, \tau)^\top l, U\right) d\tau. \tag{5.3}$$

*Proof.* The second equality follows from [15, §2, Theorem 1] applied to the subdifferential $\partial \delta^*(l, F(t)) = Y\left(l, F(t)\right)$ with $F(t) = \Phi(T, t)B(t)U$. $\square$

**Corollary 5.2.** *Given the problem (LDI) in (5.1)–(5.2), $\gamma \in \mathcal{CM}$ and $t \in I = [t_0, T]$, the GS-point of the reachable set evaluates as*

$$\mathrm{St}_\gamma \left( \mathcal{A}(t, t_0, X_0) \right) = \mathrm{St}_\gamma \left( \Phi(t, t_0) X_0 \right) + \int_I \mathrm{St}_\gamma \left( \Phi(t, \tau) B(\tau) U \right) d\tau \quad (t \in I) .$$

*If furthermore, $X_0$ and $U$ are singletons, Euclidean balls or unit cubes in $\mathbb{R}^n$ resp. $\mathbb{R}^m$ and $\eta(t; l) = \Phi(t, t_0)^\top l$, $\zeta(t, \tau; l) = B(\tau)^\top \Phi(t, \tau)^\top l$, then*

$$\mathrm{St}_{\alpha_{[l]}} \left( \mathcal{A}(t, t_0, X_0) \right) = \Phi(t, t_0) \, \mathrm{St}_{\alpha_{[\eta(t;l)]}} \left( X_0 \right)$$

$$+ \int_I \Phi(t, \tau) B(\tau) \, \mathrm{St}_{\zeta(t, \tau; l)} \left( U \right) d\tau . \tag{5.4}$$

*Proof.* Clearly, Lemma 5.1 can be applied together with Lemma 2.5, since $(\Phi(t, t_0), X_0)$ and $(\Phi(t, \tau) B(\tau), U)$ fulfill the GSCL-property. $\qquad\square$

Equation (5.4) means that the GS-selection $u(\cdot) = \mathrm{St}_{\zeta(T, \cdot; l)} \left( U \right)$ is the optimal control for the optimal control problem (OCP)

$$\begin{aligned} \max \quad & l^\top x(T) \\ \text{s.t.} \quad & x'(t) = A(t) x(t) + B(t) u(t) \quad (\text{a.e. } t \in I) , \\ & x(t_0) = \mathrm{St}_{\alpha_{[\eta(T;l)]}} \left( X_0 \right) \end{aligned} \tag{5.5}$$

The corresponding solution $x(\cdot)$ is nothing else than an extremal solution of (LDI), where the proof was considerably simple and does not use the maximum principle. Although it should be noted that mild assumptions are available in [13] on which the strong convexity of the attainable set follows, one should observe that in any case there could not appear a multivalued situation in (5.4) as present in (5.3).

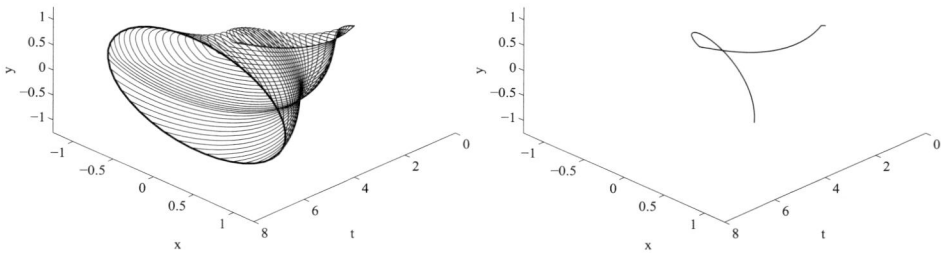

FIGURE 3. Euler's solutions with $N = 100$ for Example 5.3

**Example 5.3.** Consider (LDI) on $I = [0, 2\pi]$ with

$$A(t) = \begin{pmatrix} 0 & 1 \\ -1 & 0 \end{pmatrix} , \quad B(t) = \frac{1}{4} \cdot \begin{pmatrix} 0 \\ 1 \end{pmatrix} , \quad U = [-1, 1], \quad X_0 = \{0_{\mathbb{R}^2}\}$$

The attainable set $\mathcal{A}(2\pi, 0, \{0_{\mathbb{R}^2}\})$ equals the Aumann integral in Example 4.3. Figure 3 shows 40 Euler grid functions with $N = 100$ subintervals (left picture) approximating optimal solutions of the ODE in (OCP), each one corresponds to a different direction $l \in S_{n-1}$. The opening of the solution funnel is bounded by the attainable set $\mathcal{A}(2\pi, 0, \{0_{\mathbb{R}^2}\})$. In the right picture, one trajectory originating from a GS-selection is depicted which has two kinks due to jumps in the control.

## 6. Conclusions

Using the Demyanov difference in the regularity concepts of set-valued maps, error estimates which compare GS-points in common outer normals are possible (in contrast to [5, 12, 19, 20]). This overview did not present the weakest versions of available results. As examples it should be mentioned that the class of sets which fulfill the GSCL-property is broader than mentioned in Lemma 2.5. One could introduce concepts of bounded variation as in [4] to establish order of convergence $\mathcal{O}(h)$ under weaker assumptions than in Propositions 4.1 and 4.2. For the generalizations for higher order in interpolation resp. quadrature methods and existence proofs of smooth dense solutions, special classes of "smooth" set-valued maps need to be studied in a forthcoming paper. Furthermore, the application of GS-points to set-valued Runge-Kutta methods (cf. [3, 18]) needs further investigation.

**Acknowledgment**

The author would like to thank Elza Farkhi for many suggestions and her support.

## References

[1] Z. Artstein, *On the calculus of closed set-valued functions*. Indiana Univ. Math. J. **24**, no. 5 (1974), 433–441.

[2] R. Aumann, *Integrals of set-valued functions*. J. Math. Anal. Appl. **12** (1965), 1–12.

[3] R. Baier, *Selection Strategies for Set-Valued Runge-Kutta Methods*. Lecture Notes in Comp. Sci. **3401** (2005), 149–157.

[4] R. Baier and E. Farkhi, *Regularity and Integration of Set-Valued Maps Represented by Generalized Steiner Points*. Set-Valued Anal. **15**, no. 2 (2006), 185–207.

[5] R. Baier and F. Lempio, *Computing Aumann's integral*. In A. Kurzhanski and V. Veliov, eds., *Modeling Techniques for Uncertain Systems, Proceedings of a Conferences held in Sopron, Hungary, July 6-10, 1992*. Volume **18** of Progress in Systems and Control Theory, Birkhäuser, Basel, 1994, 71–92.

[6] T. Benavides, G. Acedo and H.-K. Xu, *Random fixed points of set-valued operators*. Proc. Amer. Math. Soc. **124**, no. 3 (1996), 831–838.

[7] D. Cohn, *Measure Theory*. Birkhäuser, Boston, 1980.

[8] V. Demyanov and A. Rubinov, *Constructive nonsmooth analysis*, volume **7** of Approximation and Optimization. Peter Lang, Frankfurt am Main, 1995.

[9] D. Dentcheva, *Differentiable Selections and Castaing Representations of Multifunctions*. J. Math. Anal. Appl. **223** (1998), 371–396.

[10] D. Dentcheva, *Regular Castaing Representations of Multifunctions with Applications to Stochastic Programming*. SIAM J. Optim. **10** (2000), 732–749.

[11] D. Dentcheva, *Continuity of Multifunctions Characterized by Steiner Selections.* Nonlinear Anal. **47** (2001), 1985–1996.

[12] T. Donchev and E. Farkhi, *Moduli of smoothness of vector valued functions of a real variable and applications.* Numer. Funct. Anal. Optim. 11, no. **5** & **6** (1990), 497–509.

[13] H. Frankowska and C. Olech, *R-convexity of the integral of the set-valued functions.* In *Contributions to analysis and geometry. Conference held at the Johns Hopkins University, Baltimore, Maryland, April 24-25, 1980*, D. Clark, G. Pecelli and R. Sacksteder, eds., John Hopkins Univ. Press, Baltimore, MD, 1981, 117–129.

[14] F. Hirsch and G. Lacombe, *Elements of Functional Analysis.* Volume **192** of Graduate Texts in Mathematics. Springer, New York, 1999.

[15] A. Ioffe and V. Levin, *Subdifferentials of convex functions.* Trans. Moscow Math. Soc. **26** (1972), 1–72.

[16] A. Rubinov and I. Akhundov, *Difference of compact sets in the sense of Demyanov and its application to non-smooth analysis.* Optimization **23**, no. 3 (1992), 179–188.

[17] L. Sonneborn and F. van Vleck, *The bang-bang principle for linear control problems.* SIAM J. Control, Ser. A, **2** (1965), 151–159.

[18] V. Veliov, *Second Order Discrete Approximation to Linear Differential Inclusions.* SIAM J. Numer. Anal. **29**, no. 2 (1992), 439–451.

[19] V. Veliov, *Discrete approximations of integrals of multivalued mappings.* C. R. Acad. Bulgare Sci. **42**, no. 12 (1989), 51–54.

[20] A. Vitale, *Approximation of convex set-valued functions.* J. Approx. Theory **26** (1979), 301–316.

Robert Baier
University of Bayreuth
Department of Mathematics
Chair of Applied Mathematics
D-95440 Bayreuth
Germany
e-mail: robert.baier@uni-bayreuth.de

Progress in Nonlinear Differential Equations
and Their Applications, Vol. 75, 61–77
© 2007 Birkhäuser Verlag Basel/Switzerland

# On the Euler-Lagrange Equation for a Variational Problem

Stefano Bianchini

*Dedicated to Arrigo Cellina and James Yorke*

**Abstract.** In this paper we prove the existence of a solution in $L^\infty_{\text{loc}}(\Omega)$ to the Euler-Lagrange equation for the variational problem

$$\inf_{\bar u + W^{1,\infty}_0(\Omega)} \int_\Omega \big( \mathbf{I}_D(\nabla u) + g(u) \big)\, dx\,, \tag{0.1}$$

with $D$ convex closed subset of $\mathbb{R}^n$ with non empty interior. By means of a disintegration theorem, we next show that the Euler-Lagrange equation can be reduced to an ODE along characteristics, and we deduce that the solution to Euler-Lagrange is different from 0 a.e. and satisfies a uniqueness property. Using these results, we prove a conjecture on the existence of variations on vector fields [3].

**Mathematics Subject Classification (2000).** 49K20.

**Keywords.** Extended valued functions, Euler-Lagrange equation, Hamilton-Jacobi equation, disintegration of a measure.

## 1. Introduction

We consider the existence of a solution to the Euler-Lagrange equation for the minimization problem

$$\inf \left\{ g(u), u \in \bar u + W^{1,\infty}_0(\Omega), \nabla u \in D \right\}, \tag{1.1}$$

where $g : \mathbb{R} \mapsto \mathbb{R}$ strictly monotone increasing and differentiable, $\Omega$ open set with compact closure in $\mathbb{R}^n$, and $D$ convex closed subset of $\mathbb{R}^n$. Under the assumption that $\nabla \bar u \in D$ a.e. in $\Omega$, there is a unique solution $u$ to (1.1) and we can actually give an explicit representation of $u$ is terms of a Lax-type formula. The solution is clearly Lipschitz continuous because $\nabla u \in \partial D$ a.e. in $\Omega$.

The Euler-Lagrange equation for (1.1) can be written as

$$\operatorname{div}\big(\pi(x)\big) = g'\big(u(x)\big), \quad \pi(x) \cdot \nabla u(x) = \max\big\{\pi(x) \cdot d, d \in D\big\}, \tag{1.2}$$

where $\pi$ is a measurable function. The first equation is considered in the distribution sense, and the second relation follows by using the subdifferential to the convex function

$$\mathbf{I}_D(x) = \begin{cases} 0 & x \in D \\ +\infty & x \notin D \end{cases}$$

in the standard formulation of the Euler-Lagrange equations. It means that the vector $\pi(x)$ lies in the convex support cone of $\partial D$ at the point $\nabla u(x)$.

In [5], the authors prove that under the assumption $D = B(0,1)$ (in which case $u$ is basically the solution to the Eiconal equation), there is a solution to the Euler-Lagrange equation (1.2), which can be rewritten as

$$\operatorname{div}\big(p(x)\nabla u(x)\big) = g'\big(u(x)\big), \quad p \geq 0. \tag{1.3}$$

The main point in the proof is that in the region $\Omega \setminus J$, where $J$ is the singularity set of $u$, the solution $u$ is $C^{1,1}$, and thus the above equation can be reduced to an ODE for $p$ along the characteristics. We recall that in this case $u$ is locally semi convex, so that $\nabla u$ has many properties of monotone functions (see for example [1] for a survey on monotone functions).

Simple examples show that such differentiability properties do not hold for general sets $D$. However, using some weaker continuity property of $\nabla u$, in [4] the author proves that the Euler-Lagrange equation holds and in the case the dual $D^*$ of $D$ is strictly convex, an explicit representation formula of the solution can be given.

In this paper we extend the results of [4] to the general case, i.e., under the only assumption that $D$, $D^*$ are convex. The proof is based on the following steps.

First, by relaying on the explicit form of the solution (Section 2), one defines the set valued function $\mathcal{B}(x) \subset \partial\Omega$ as the set of boundary data such that

$$u(y) - u(x) = |y - x|_{D^*}, \quad y \in \partial\Omega,$$

where $|\cdot|_{D*}$ is the pseudo norm generated by $D^*$, the dual of $D$ (Section 3). If $D^*$ is strictly convex, it is known that $\mathcal{B}(x)$ is single valued $\mathcal{H}^n$ a.e., and we denote this function by $b(x)$ (Section 3.1).

Then we select a $L^\infty$ function $b(x) \in \mathcal{B}(x)$ with good properties: the principal one is that it can be approximated $\mathcal{H}^n$ a.e. by functions $b_i(x)$ generated by strictly convex sets $D_i^*$. This allows to pass many properties of $b_i$ to the limit, in particular that if we define the vector field

$$d(x) = \frac{b(x) - x}{|b(x) - x|_{D*}}$$

then $\operatorname{div} d$ is a bounded measure and it is the limit of the vectors $d_i(x)$ constructed by considering the strictly convex sets $D_i^*$ (Section 3.2).

Using the estimates inherited by the sequence $d_i$, one can prove the following theorem (Section 4):

**Theorem 1.1.** *The Lebesgue measure can be disintegrated as*

$$\mathcal{H}^n|_\Omega = \int \nu_{(y,z)} d\mu(y,z) , \tag{1.4}$$

*where $\nu_y$ is a probability measure supported only on the set of lines $[x, b(x)]$, and $(y, z)$ parameterizes all segments $[x, b(x)]$.*

*Moveover we have the explicit expression of $\nu_{(y,z)}$: $\nu_{(y,z)}(s) = \alpha(s, y, z)\mathcal{H}^1(s)$ with $\alpha$ satisfying the ODE*

$$\frac{d}{ds}\alpha(s, y, z) = \alpha(s, y, z)(\mathrm{div}d)_{a.c.} . \tag{1.5}$$

*with $(\mathrm{div}d)_{a.c.}$ the absolutely continuous part of $\mathrm{div}d$.*

The proof is divided in several steps: the problem is not in the existence of the disintegration, but in the form of the measure $\nu_{(y,z)}$. The idea is to use an estimate on the divergence of $d$ to obtain that the measures $\nu_{(y,z)}$ are absolutely continuous w.r.t. $\mathcal{H}^1$ (Section 4.1).

As a corollary, using an estimate of $(\mathrm{div}d)_{a.c.}$ along the segment $x + td(x)$, one obtains that $\alpha > 0$ $\mathcal{H}^n$ a.e. (Section 4.2).

The final step is to use this disintegration of $\mathcal{H}^n|_\Omega$ to prove that the weak solution to the Euler Lagrange equation

$$\mathrm{div}\big(p(x)d(x)\big) = g'\big(u(x)\big), \quad d(x) = \partial D\big(\nabla u(x)\big) \tag{1.6}$$

satisfies an ODE on each segment $x + td(x)$. This allows to find the explicit form of the solution as

$$p(s, y, z) = \frac{1}{\alpha(s, y, z)} \int_{-\infty}^{s} \alpha(t, y, z)g\big(u(t)\big)dt . \tag{1.7}$$

This shows that $p > 0$ $\mathcal{H}^n$ a.e. in $\Omega$, so that we can prove the following theorem:

**Theorem 1.2.** *There exists a solution for the Euler Lagrange equation (1.6) strictly positive $\mathcal{H}^n$ a.e. in $\Omega$.*

As noticed in [4], a consequence of this positivity is that a conjecture stated in [3] is true (Section 6).

## 2. Preliminaries

We consider the following variation problem

$$\inf_{\bar{u}+W_0^{1,\infty}} \int_\Omega \big(\mathbf{I}_D(\nabla u) + g(u)\big)dx , \tag{2.1}$$

with $g : \mathbb{R} \mapsto \mathbb{R}$ strictly monotone increasing and differentiable, $\Omega$ open set with compact closure in $\mathbb{R}^n$. The function $\mathbf{1}_A$ is the indicative function of a set $A \subset \mathbb{R}^n$,

$$\mathbf{1}_A(x) = \begin{cases} 0 & x \in A \\ +\infty & x \notin A \end{cases} \tag{2.2}$$

Moreover, to have a finite infimum in (2.1), we assume that the function $\bar{u}$ satisfies

$$\nabla \bar{u} \in D. \tag{2.3}$$

As a consequence, the infimum is finite and it is attained.

To avoid degeneracies, in the following we assume that $D$ is a bounded convex closed subset of $\mathbb{R}^n$, with non empty interior, and without loss of generality we suppose that

$$B(0, r) = \left\{ x \in \mathbb{R}^n, |x| \leq r \right\} \subset D. \tag{2.4}$$

We then denote the dual convex set $D^*$ by

$$D^* = \left\{ d \in \mathbb{R}^n : d \cdot \ell \leq 1 \ \forall \ell \in D \right\}, \tag{2.5}$$

where the scalar product of two vectors $x, y \in \mathbb{R}^n$ is $x \cdot y$. The set $D^*$ is closed, convex and $D^{**} = D$. We will write the support set at $\bar{\ell} \in \partial D$ as

$$\delta D(\bar{\ell}) = \left\{ d \in D^* : d \cdot \bar{\ell} = \sup_{\ell \in D} d \cdot \ell \right\}. \tag{2.6}$$

Let $| \cdot |_D$ be the pseudo-norm given by the Minkowski functional

$$|x|_D = \inf \left\{ k \in \mathbb{R} : x \in kD \right\} = \sup \left\{ d \cdot x, d \in D^* \right\}, \tag{2.7}$$

and define the dual pseudo-norm by

$$|x|_{D^*} = \inf \left\{ k \in \mathbb{R} : x \in kD^* \right\} = \sup \left\{ \ell \cdot x, \ell \in D \right\}. \tag{2.8}$$

Note that due to convexity the triangle inequality holds,

$$|x + y|_{D^*} \leq |x|_{D^*} + |y|_{D^*}, \quad x, y \in R^n, \tag{2.9}$$

and that $| \cdot |_D, | \cdot |_{D^*}$ are the Legendre transforms of $\mathbf{1}_{D^*}, \mathbf{1}_D$ respectively.

In the following, we denote with $\mathcal{H}^{n-1}$ the $n-1$ dimensional Hausdorff measure [2], Definition 2.46 of page 72: for any $\Omega' \subset \Omega$,

$$|\Omega'|_{\mathcal{H}^{n-1}} = \mathcal{H}^{n-1}(\Omega')$$

$$= \kappa \sup_{\delta > 0} \left( \inf \left\{ \sum_{i \in I} |\mathrm{diam}(B_i)|^{n-1}, \mathrm{diam}(B_i) \leq \delta, \Omega \subset \bigcup_{i \in I} B_i \right\} \right), \tag{2.10}$$

where $\kappa$ is the constant such that $\mathcal{H}^{n-1}$ is equivalent to the Lebesgue measure on $n-1$ dimensional planes:

$$\kappa = \frac{\pi^{\frac{n-1}{2}}}{\Gamma\left(1 + \frac{n-1}{2}\right)}, \quad \Gamma(\alpha) = \int_0^\infty t^{\alpha-1} e^{-t} dt.$$

We recall that $\mathcal{H}^n$ is the $n$ dimensional Lebesgue measure $\mathcal{L}^n$, [2], Theorem 2.53 of page 76.

If $f : X \mapsto Y$ is a measurable map between the measure space $(X, \mathcal{S}, \mu)$ into the measurable space $(Y, \mathcal{T})$, we define the push forward measure $f\sharp\mu$ as ( [2], Definition 1.70 of page 32)

$$f\sharp\mu(T) = \mu\big(f^{-1}(T)\big), \quad T \in \mathcal{T}. \tag{2.11}$$

The first proposition is the explicit representation of the solution by a Hopf-Lax type formula.

**Proposition 2.1.** *The solution of* (2.1) *is given explicitly by*

$$u(x) = \max\Big\{u(\bar{x}) - |\bar{x} - x|_{D^*}, \bar{x} \in \partial\Omega, \alpha x + (1 - \alpha)\bar{x} \in \Omega \ \forall \alpha \in (0, 1)\Big\}. \tag{2.12}$$

*Moreover, $u$ is Lipschitz continuous and $\nabla u \in D$ a.e..*

The proof of this proposition is standard, and can be found for example in [4], Proposition 2.1. The basic ideas are that the function defined by (2.12) has derivative still in $D$ and is the lowest possible function such that $\nabla u \in D$.

## 3. Regularity estimates

Before studying the Euler-Lagrange equation for the variational problem (2.1), we introduce some important functions and prove some basic regularity estimates.

Define the set valued functions

$$\mathcal{B}(x) = \big\{\bar{x} \in \partial\Omega : u(x) = u(\bar{x}) - |\bar{x} - x|_{D^*}\big\} \subset \partial\Omega, \tag{3.1}$$

$$x \mapsto \mathcal{D}(x) = \left\{\frac{\bar{x} - x}{|\bar{x} - x|_{D^*}}, \quad \bar{x} \in \mathcal{B}(x)\right\} \subset \partial D^*. \tag{3.2}$$

Thus $\mathcal{D}(x)$ is the set of directions where $u$ has the maximal growth in the norm $|\cdot|_{D^*}$. It is easy to prove that both sets $\mathcal{B}(x)$, $\mathcal{D}(x)$ are closed not empty subset of $\partial\Omega$, $\partial D^*$, respectively (see the proof of [4], Proposition 2.1). The normalization in (3.2) implies that

$$u(x + td) = u(x) + t \tag{3.3}$$

for all $x \in \Omega$, $d \in \mathcal{D}(x)$. We can say that $\mathcal{B}(x)$ is the set where the half lines $x + td(x)$, with $d \in \mathcal{D}(x)$ and $t \geq 0$, intersect $\partial\Omega$.

The first result is the upper continuity of the set valued map $\mathcal{D}(x)$.

**Proposition 3.1.** *The function $\mathcal{D}(x)$ is closed graph and upper semicontinuous: more precisely for all $y \in \Omega$, for all $\epsilon > 0$ there exists $\delta > 0$ such that*

$$\mathcal{D}(x) \subset \mathcal{D}(y) + B(0, \epsilon) \tag{3.4}$$

*for $x \in y + B(0, \delta)$.*

The same result can be said for the the function $\mathcal{B}(x)$.

*Proof.* Fixed the point $y$, by rescaling we can restrict to the set of points distant 1 from $y$

$$D^*(y,1) = \{x : |x - y|_{D^*}\},$$

and we can assume that $u(y) = 0$. By the explicit formula of solutions, the set $\mathcal{D}(y)$ is given by

$$\mathcal{D}(y) = \{z - y : |z - y|_{D^*} = 1, \ u(z) = 1\},$$

so that it follows from Lipschitz continuity that for all $\epsilon$ there is a $\delta$ such that

$$u(z) < 1 - \epsilon \qquad \forall z : |z - y|_{D^*} = 1, \text{dist}(z, \mathcal{D}(y)) > \delta.$$

We thus have that for all $x$ such that $|x - y|_{D^*} \leq \epsilon/2$

$$u(x) \geq -\epsilon/2 > u(z) - 1 + \epsilon/2.$$

Thus the set $\mathcal{D}(z)$ for such a $z$ has a distance from $\mathcal{D}(y)$ less than $\mathcal{O}(\delta + \epsilon)$. The closed graph property follows from the fact that each $\mathcal{D}(x)$ is closed. □

We next prove that the set valued map $\mathcal{B}$ is measurable: we repeat the computations of [4], Proposition 3.3. We recall that if $F$ is a set valued function, then

$$F^{-1}(A) = \{x : F(x) \cap A \neq \emptyset\}. \tag{3.5}$$

**Lemma 3.2.** *The function $\mathcal{B}(x)$ is measurable, i.e., the inverse image of open sets are Borel measurable.*

*Proof.* We have to prove that for all open set $O$ in $\partial\Omega$, the inverse image

$$\left\{x : \mathcal{B}(x) \in O\right\}$$

is Borel. Take a sequence of closed set $\bar{O}_i \subset O$, $i \in \mathbb{N}$, on the boundary $\partial\Omega$ such that $\cup_i \bar{O}_i = O$. The measurability of $\mathcal{B}^{-1}(\bar{O}_i)$ is trivial for the function

$$u_0 = \max\{u(\bar{x}) - |\bar{x} - x|_{D^*}, \bar{x} \in \bar{O}_i, \alpha x + (1 - \alpha)\bar{x} \in \Omega \ \forall \alpha \in [0,1]\},$$

since it coincides with the whole $\Omega$. Then, one only observes that

$$\mathcal{B}^{-1}(\bar{O}_i) = \{x : u_0(x) = u(x)\},$$

where $u(x)$ is the solution to the variational problem. Since $u_0$, $u$ are Lipschitz function, it follows that $\mathcal{B}^{-1}(\bar{O}_i)$ is a closed set in $\Omega$, hence $\mathcal{B}^{-1}(O) = \cup_i \mathcal{B}^{-1}(\bar{O}_i)$ is Borel. □

We finally prove the following relation among the derivative of the Lipschitz function $u$ and the function $\mathcal{D}$.

**Lemma 3.3.** *If $x$ is a point of differentiability of $u$, then*

$$\nabla u \in \partial D^*(d) = \{\ell \in D : \ell \cdot d = 1\}, \tag{3.6}$$

*where $d \in \mathcal{D}(x)$. In particular, $d \in \partial D(\nabla u(x))$, so that*

$$\mathcal{D}(x) \subset \partial D(\nabla u(x)). \tag{3.7}$$

*Proof.* This follows from the equation $u(x + td) = u(x) + 1$ for all $d \in \mathcal{D}(x)$, which gives

$$\nabla u \cdot d = 1 \qquad \Longrightarrow \qquad \nabla u \in \partial \mathcal{D}^*(d)$$

by definition of $\partial \mathcal{D}^*$ and the fact that $\nabla u \in D$ (Proposition 2.1).          □

### 3.1. The strictly convex case

In this generality we cannot say much on the functions $\mathcal{B}$, $\mathcal{D}$. Following [4], in this section we make the following assumption.

**Assumption.** *The conjugate set $D^*$ is strictly convex.*

This implies that $D$ is differentiable. Using (3.7), it follows that

**Corollary 3.4.** *The functions $\mathcal{B}$ and $\mathcal{D}$ are single valued in each differentiability point of $u$.*

We can thus consider the set of lines

$$\Sigma(x) = \bigcup_{d \in \mathcal{D}(x)} \{ x + td : t \in \mathbb{R}, u(x + td) = u(x) + t \} . \tag{3.8}$$

The set $\mathcal{B}(x)$ is the set of end points for $t \geq 0$, while by considering the end points for $t \leq 0$, we define the function

$$a(x) = \{ x - td : t \geq 0, d \in \mathcal{D}, u(x - td) \\ = u(x) - t, u(x - (t + \epsilon)d) > u(x) - (t + \epsilon) \; \forall \epsilon > 0 \} . \tag{3.9}$$

From the strict convexity of $D^*$, it follows in fact that $a(x)$ is single valued: if $\mathcal{D}(x)$ contains two different directions, then $a(x) = x$. This is clearly not the case when $D^*$ is not strictly convex.

As a corollary of the explicit form of the solution and the above definitions, we have

**Corollary 3.5.** *If $D^*$ is strictly convex, the function $a(x)$ is single valued and, on the differentiability set of $u$, the functions $\mathcal{B}(x)$, $\mathcal{D}(x)$ are single valued.*

*Moreover, the solution $u$ can be written as*

$$u(x) = \min \left\{ u(\bar{x}) + |x - \bar{x}|_{D^*}, \bar{x} \in \bigcup_{y \in \Omega} a(y), \alpha x + (1 - \alpha)\bar{x} \in \Omega \; \forall \alpha \in (0, 1) \right\},$$

$$u(x) = \max \left\{ u(\bar{x}) - |\bar{x} - x|_{D^*}, \bar{x} \in \bigcup_{y \in \Omega} \mathcal{B}(y), \alpha x + (1 - \alpha)\bar{x} \in \Omega \; \forall \alpha \in (0, 1) \right\}. \tag{3.10}$$

In the set $S$ where $\mathcal{B}(x)$, $\mathcal{D}(x)$ are single valued, we will use the notation

$$\mathcal{B}(x)\Big|_S = b(x), \quad \mathcal{D}(x)\Big|_S = d(x). \tag{3.11}$$

An important consequence of the continuity property of $\mathcal{B}$ (Proposition 3.1) and Corollary 3.4 is that we have some stability of the vector $d$ w.r.t. perturbation of the boundary data, of the set $D$ and approximation by smooth vector fields.

**Proposition 3.6.** *The function $x \mapsto d(x)$ is continuous w.r.t. the inherited topology on the differentiability set of $u$. Moreover, it holds*

1. *If $u_i(\partial\Omega) \to u(\partial\Omega)$ in $L^\infty(\partial\Omega)$, then $d_i(x) \to d(x)$ in $L^p(\Omega)$, for all $p \in [1,\infty)$, where $d_i = \delta D(\nabla u)$.*
2. *If $\rho_\epsilon$ is a convolution kernel, then $\rho_\epsilon * d$ converges to $d$ in $L^p(\Omega)$, for all $p \in [1,\infty)$.*
3. *If $D_i$ is a sequence of convex sets converging to $D$ w.r.t. the Hausdorff distance, with $D_i^*$ strictly convex and $D \subset D_i$, then the vector field $d_i(x)$ corresponding to the solution $u_i$ to*

$$\inf_{\bar{u}+W_0^{1,\infty}(\Omega)} \int_\Omega \left(\mathbf{I}_{D_i}(\nabla u) + g(u)\right) dx\,, \tag{3.12}$$

*converges to the vector field $d$ corresponding to $u$ in $L^p(\Omega)$, for all $p \in [1,\infty)$.*

The key argument in the proof is the convergence of $\mathcal{D}(x)$, which follows from the upper continuity of $\mathcal{D}(x)$.

### 3.2. The general case

We now consider a sequence of strictly convex sets $D_i$ converging to $D$: natural candidates are the sets $D_i$ obtained by the inf-convolution,

$$D_i = D\square B(0, 1/i) = \left\{x : \exists x_1 \in D, x_2 \in B(0, 1/i), x = x_1 + x_2\right\}. \tag{3.13}$$

By construction, $D_i$ is smooth and its Legendre transform is

$$|x|_{D_i^*} = |x|_{D^*} + \frac{1}{i}|x|_B\,. \tag{3.14}$$

We thus have that $D_i^*$ is strictly convex, and $D_i^* \subset D^*$. By computations similar to the proof of Proposition 3.1, it follows that for all $x \in \Omega$, $\epsilon > 0$,

$$\mathcal{D}_i(x) \subset \mathcal{D}(x) + B(0, \epsilon)\,, \tag{3.15}$$

if $i = i(\epsilon, x)$ sufficiently large.

We now show that the functions $d_i(x)$ converges a.e. to a precise measurable selection of the set valued function $\mathcal{D}(x)$.

**Lemma 3.7.** *Define the selection $\mathfrak{B}(x) \subset \mathcal{B}(x)$ as*

$$\mathfrak{B}(x) = \left\{b \in \mathcal{B}(x), |b - x| \text{ is minimal}\right\}. \tag{3.16}$$

*Then $\mathfrak{B}(x)$ is single valued $\mathcal{H}^n$ a.e. in $\Omega$, if $\Omega$ is a polyhedron.*

*Proof.* This follows simple because if $\mathfrak{B}(x)$ contains two vectors with directions on different faces of $D^*$, then the set is $n - 1$ rectifiable, with the same proof given in [4], Lemma 6.2 and Proposition 6.4.

Conversely, the set where $\mathcal{B}(x)$ contains two vectors on the same face, is contained in the set

$$\{x \in \Omega : \exists x_1, x_2 \in \partial\Omega, x_1 \neq x_2, |x_1 - x| = |x_2 - x|\}.$$

which for a polyhedron is contained in the planes bisecting the angles of the faces of $\partial\Omega$: it is thus of measure 0.                                    $\square$

In the following we assume thus that $\Omega$ is a polyhedron. The general case can be treated by covering the open set $\Omega$ with a countable number of polyhedra, and solving the Euler-Lagrange equation in each polyhedron with suitable boundary conditions.

**Proposition 3.8.** *Assume that $u_i|_{\partial\Omega} \to u|_{\partial\Omega}$ in $C^0(\partial\Omega, \mathbb{R})$. Then, for all $x \in \Omega$, for all $\epsilon$ there exists $\bar{i}$ such that*

$$\mathcal{B}_i(x) \subset \mathcal{B}(x) + B(0, \epsilon).\tag{3.17}$$

*Proof.* The proof follows the same analysis of Proposition 3.1: in fact, for $i \gg 1$ the set $\mathcal{B}_i(x)$ is a subset of $\mathcal{B}(x) + B(0, \epsilon)$.                    $\square$

In particular we have a corollary:

**Corollary 3.9.** *Let $d(x)$ be the vector field*

$$d(x) = \frac{b(x) - x}{|b(x) - x|_{D^*}} \in L^\infty(\Omega, \partial D^*).\tag{3.18}$$

*Then the function $d_i \in L^\infty(\Omega, \partial D_i^*)$ converges to $d$ $\mathcal{H}^n$ a.e..*

Also in this case, we can consider the set of lines

$$\Sigma(x) = \bigcup_{d \in \mathcal{D}(x)} \{x + td : t \in \mathbb{R}, u(x + td) = u(x) + t\}.\tag{3.19}$$

This set reduces to a segment $\mathcal{H}^n$ a.e.. We can introduce the function

$$x \mapsto a(x) = \left\{x + td(x), t = \inf\left\{s : d(x + sd(x)) = d(x)\right\}\right\}.\tag{3.20}$$

Since $d$ is Borel measurable, then also $a(x)$ is: moreover, $a(a(x)) = x$. We will denote by $J$ the set of the initial points of $\Sigma$:

$$J = \bigcup_{x \in \Omega} a(x).\tag{3.21}$$

Following the same proof of [4], Proposition 5.1, we can prove that

**Proposition 3.10.** *The divergence of the vector field $d(x)$ is a positive locally finite Radon measure, satisfying*

$$\operatorname{div} d(x) + \frac{\varrho}{\operatorname{dist}(\Omega', \partial\Omega)} \geq 0\tag{3.22}$$

*for all $x \in \Omega' \subset\subset \Omega$, with $\rho$ depending on on $D$. Moreover, we have the estimate*

$$|\text{div } d|\big(B(x,r)\big) \leq \text{dist}(\partial D^*, 0)^{-1}|\partial B(0,r)| + \frac{2\varrho|B(x,r)|}{\text{dist}\big(B(x,r), \partial\Omega\big)}, \quad B(x,r) \subset\subset \Omega. \tag{3.23}$$

*Finally, the singular part is strictly positive in $\Omega$.*

*Proof.* The idea of the proof is that the above estimates holds for the approximating sequence $d_i$, so that we can pass to the limit $\text{div} d_i \rightharpoonup \text{div} d$. $\qquad\square$

To end this section, we deduce also a uniform estimate on the push forward of the $n-1$ dimension Lebesgue measure by the map $x \mapsto x + td(x)$. These estimates are the extension to this case of the estimates of [4], Lemma 5.4.

Let $\bar{x}$ be a Lebesgue point of $d$: without any loss of generality we assume

$$\bar{x} = 0, \qquad d(0) = e_1 = (1,0),$$

where we denote with $e_i$ the unit vector along the $i$-th coordinate axis.

Consider the Borel measurable sets

$$Z = Z(0) = B(0,r) \cap \{e_1 \cdot y = 0\} \cap \{-e_1 \cdot a(y), e_1 \cdot b(y) \geq 2h\}, \tag{3.24}$$

$$\begin{aligned} Z(s) &= \{x = y + sd(y), y \in S(0)\}, \\ \mathcal{Z}(-h,h) &= \{x = y + sd(y), y \in S(0), \quad t \in [-h,h]\}. \end{aligned} \tag{3.25}$$

We now show that $(\mathbb{I} + td)\sharp\mathcal{H}^{n-1}|_{\mathcal{Z}}$ remains equivalent to $\mathcal{H}^{n-1}$.

**Proposition 3.11.** *We have the following estimates: there exists $C = C(h,r,D)$ such that*

$$\frac{1}{C}\mathcal{H}^{n-1}\big(Z(s)\big) \leq \mathcal{H}^{n-1}\big(Z(t)\big) \leq C\mathcal{H}^{n-1}\big(Z(s)\big), \tag{3.26}$$

*for all $-h \leq s \leq t \leq h$. Moreover*

$$\frac{1}{C}\mathcal{H}^n\big(\mathcal{Z}(s,t)\big) \leq \mathcal{H}^{n-1}\big(Z(0)\big)(t-s) \leq C\mathcal{H}^n\big(\mathcal{Z}(s,t)\big). \tag{3.27}$$

*The constant $C(h,r,D)$ can be estimated by*

$$C(h,r,D) = \frac{1}{\mathfrak{e}}\exp\Big\{\frac{\varrho}{\mathfrak{e}h}s\Big\}, \qquad \mathfrak{e} = \inf_{y \in Z}\{e_i \cot d(y)\}, \tag{3.28}$$

*with $\rho$ the constant entering in (3.22).*

*Proof.* By the construction of $Z$, it follows that the vector field $d_i(x)$ converges to $d$ for all points $x \in C(-h,h)$. Similarly, if we define $A = \{y - 3hd(y)/2, y \in Z\}$,

$$\tilde{u}_i = \inf\{u(a) + |x - a|_{D^i}, a \in A\},$$

then the vector field $\tilde{d}_i$ corresponding to the function $\tilde{u}_i$ converges to $d$ for all $x \in \mathcal{Z}(-h,h)$. By Lusin and Egoroff theorems, we can find a compact set $Z_k$ where $d$ is continuous and $d_i$ converges to $d$ uniformly. We thus have that $(\mathbb{I} + td_i)Z_k$ is compact, and if $O(t)$ is an open set containing $Z(t)$ such that $\mathcal{H}^{n-1}(O \setminus Z(t)) \leq \epsilon$, then $(\mathbb{I} + td_i)Z_k \subset O(t)$, $i, k \ll 1$, and

$$\mathcal{H}^{n-1}(O(t)) \geq \mathcal{H}^{n-1}((\mathbb{I} + td_i)Z_k) \geq \frac{1}{C}\mathcal{H}^{n-1}(Z_k).$$

We have used the uniform estimate on the divergence of $d_i$, which can be obtained by Proposition 3.10. By passing to the limit we obtain the first inequality of (3.26).

To obtain the second one, one can repeat the computation starting from the set $\cup_{x \in Z_k} a(x)$, where now $A(x)$ is slightly perturbed in order to have the convergence of the $d_i$: more precisely, the set $A(x)$ is the set $z - (t - s + h)d(z)$, with $z \in Z(t)$. $\qquad \square$

Using the above lemma we show that the set $J = \cup_{x \in \Omega} a(x)$ is negligible, by using an argument similar to the one in [4, 5].

**Proposition 3.12.** *The set $J$ has Lebesgue measure 0.*

## 4. Disintegration of the Lebesgue measure

Consider now the Borel maps $x \mapsto b(x)$, $x \mapsto d(x)$, and the Borel measure $m$ on $\mathbb{R}^n \times \mathbb{R}^n \times \mathbb{R}^n$ defined by

$$\int_{\mathbb{R}^n \times \mathbb{R}^n \times \mathbb{R}^n} \phi(x, y, z)dm = \int_\Omega \phi(x, b(x), d(x))dx, \qquad (4.1)$$

for all continuous function $\phi : \mathbb{R}^n \times \mathbb{R}^n \times \mathbb{R}^n \mapsto \mathbb{R}$. We can write equivalently

$$m = (\mathbb{I}, b, d)\sharp\mathcal{H}^n|_\Omega. \qquad (4.2)$$

Denote with $\pi_2$ the projection on the second set of coordinates, i.e.,

$$\pi_2 \quad : \quad \begin{matrix} \mathbb{R}^n \times \mathbb{R}^n \times \mathbb{R}^n & \mapsto & \mathbb{R}^n \\ (x, y, z) & \mapsto & \pi_2(x, y, z) = (y, z) \end{matrix}$$

We recall the following disintegration theorem [2], Theorem 2.28 of page 57:

**Theorem 4.1.** *Let $m$ be a positive Radon measure on $\mathbb{R}^{n_1} \times \mathbb{R}^{n_2}$ such that, if $\pi$ is the projection on $\mathbb{R}^{n_2}$, the measure $\mu = \pi\sharp m$ is Radon. Then there exist finite positive Radon measures $\nu_{x_2}$ on $\mathbb{R}^{n_1}$ such that $x_2 \mapsto \nu_{x_2}$ is measurable, $\nu_y(\mathbb{R}^{n_1}) = 1$ for $\mu$ a.e., and for all measurable functions $f \in L^1(\mathbb{R}^{n_1} \times \mathbb{R}^{n_2}, m)$, the function $x_1 \mapsto f(x_1, x_2)$ is measurable for $\mu$ a.e.*

$$x_2 \mapsto \int f(x_1, x_2)d\nu_y(x_1) \in L^1(\mathbb{R}^{n_2}, \mu) \qquad (4.3)$$

*and the following disintegration formula holds:*

$$\int_{\mathbb{R}^{n_1} \times \mathbb{R}^{n_2}} f(x_1, x_2)dm(x_1, x_2) = \int_{\mathbb{R}^{n_2}} \left( \int_{\mathbb{R}^{n_1}} f(x_1, x_2)d\nu_{x_2}(x_1) \right)d\mu(x_2). \qquad (4.4)$$

*This decomposition is unique, in the sense that if there is another $\mu$ measurable map $\nu'_{x_2}$ such that (4.3) and (4.4) hold for every Borel function Borel function*

$f$ with compact support, and such that $\nu'_{x_2}(\mathbb{R}^{n_1}) \in L^1_{\text{loc}}(\mathbb{R}^{n_2}, \mu)$, then $\nu'_{x_2} = \nu_{x_2}$ for $\mu$ a.e..

In our case, since $|m|(\mathbb{R}^{3n}) = |\Omega|$, the measure $m$ is clearly Radon, and also the projection $\mu = \pi_2 \sharp m$,

$$\int_{\mathbb{R}^{2n}} \phi(y, z) d\mu = \int_{\Omega} \phi(b(x), d(x)) dx,$$

is clearly a Radon measure. Moreover, $\mu$ is concentrated in the set

$$\Upsilon = \{(y, z) : \exists x \in b^{-1}(y), d(x) = z\}. \tag{4.5}$$

because

$$d^{-1}(\mathbb{R}^{2n} \setminus \Upsilon) = \emptyset,$$

it follows that $\mu$ is concentrated on $\Upsilon$.

*Remark* 4.1. We use the disintegration w.r.t. $(b(x), d(x))$ because we need not only to control the direction $d$, but also the end point $b(x)$: in fact the coordinates $(b(x), d(x))$ parameterizes the set $\cup_{x \in \Omega} \Sigma(x)$, and to all $(b, d)$ there corresponds at most one segment $x + td(x)$.

Applying the above theorem to our case, we obtain the following proposition:

**Proposition 4.2.** *There exists probability measures $\nu_{y,z}$ such that for all functions $f \in L^1(\mathbb{R}^{3n}, m)$ it holds*

$$\int_{\mathbb{R}^{3n}} f(x, y, z) dm(x, y, z) = \int_{\mathbb{R}^{2n}} \left( \int_{\mathbb{R}^n} f(x, y, z) d\nu_{(y,z)}(x) \right) d\mu(y, z), \tag{4.6}$$

*where $\mu = \pi_2 \sharp m$. Moreover $\mu$ is concentrated on $\Upsilon$, and $\nu_{(y,z)}$ is concentrated on the lines $(b, d)^{-1}(y, z)$.*

*Proof.* The support of the measure $\mu$ and $\nu_{(x,y)}$ is a consequence of the construction of $\mu$ and the fact that $(b, d)^{-1}(y, z)$ is a straight line for $(y, z) \in \Upsilon$. □

The problem here is that we do not know much about the structure of $\nu_{(y,x)}$, apart from the the validity of the disintegration formula: in fact, there are examples where the disintegration of the Lebesgue measure w.r.t. sets of straight lines does not generate measures $\nu$ absolutely continuous w.r.t. $\mathcal{H}^1$.

### 4.1. Absolute continuity of the measures $\nu_{y,z}$

We first recall that in [4], Theorems 5.7 and 5.8 show that, in the case of strictly convex sets $D_i^*$, the disintegration of the Lebesgue measure satisfies

$$d\nu_{(y,z),i}(s) = \alpha(s, y, z) ds,$$

with $\alpha(s, y, z)$ uniformly bounded in all sets $\Omega' \subset\subset \Omega$. If $\phi = \phi(x_1)$, then

$$\int d\nu_{(y,z),i} \phi(x_1 + sd(x_1)) \leq C(h) \int \phi(x_1) dx_1$$

for a.e. $(y, z)$ such that $e_1 \cdot y \geq \sup\{x_1 \in \text{supp}\phi\} + h$, $e_1 \cdot z \geq 1/2$. In fact, the last condition means that $e_i \cdot d(x) \geq 1/2$, i.e. the integral on the line $x + sd(x)$ is equivalent to the integral in $dx_1$, and the other assures that we are away from the boundary.

**Lemma 4.3.** *If $(b_i, d_i) \to (b, d)$ a.e., then $m_i = (\mathbb{I}, b_i, d_i) \sharp \mathcal{H}^n|_\Omega \rightharpoonup m$ and $\mu_i = (b_i, d_i) \sharp \mathcal{H}^n|_\Omega \rightharpoonup \mu$.*

The proof is a straightforward application of the definition of push forward of a measure.

We are now able to show that $\nu_{(y,z)}$ is a.c. w.r.t. Lebesgue.

**Proposition 4.4.** *The measures $\nu_{(y,z)}$ is a.c. w.r.t. $\mathcal{H}^1$.*

*Proof.* It is sufficient to prove the statement on the sets $e_1 \cdot y \geq \sup\{x \in \operatorname{supp}\phi\} + h$, $e_1 \dot{z} \geq 1/2$, since by changing direction or constant $h$ we can cover all $\Upsilon$ with a countable number of sets satisfying these assumptions.

For this set, taking as a test function $\psi(x, y, z) = \psi(y, z)\phi(x_1)$, we thus obtain

$$\int_\Omega \psi\big(b_i(x), d_i(x)\big)\phi(x_1)dx$$
$$= \int_{\mathbb{R}^{2n}} d\mu_i \psi(y, z) \int_\mathbb{R} d\nu_{(y,z),i}(s)\phi(s) \leq C \int_{\mathbb{R}^{2n}} d\mu_i \psi(y) \int_\mathbb{R} \phi(x_1)dx_1 \,.$$

By the weak convergence, we obtain (the integral $\int \phi(t)dt$ is constant w.r.t. $(y, z)$)

$$\int_{\mathbb{R}^{2n}} d\mu(y, z)\psi(y, z) \int_\mathbb{R} d\nu_{y,z}(s)\phi(s)$$
$$= \int_\Omega \psi\big(b(x), d(x)\big)\phi(x_1)dx \leq C \int_{\mathbb{R}^{2n}} d\mu(y, z)\psi(y, z) \int_\mathbb{R} \phi(x_1)dx_1 \,.$$

It thus follows that, by the freedom in choosing $\psi$, that for all $\phi$ for $\mu$ a.e.

$$\int_\mathbb{R} \phi(s)d\nu_{(y,z)}(s) \leq C \int_\mathbb{R} \phi(s)ds \,,$$

so that $\nu_y$ is absolutely continuous w.r.t. $\mathcal{H}^1$:

$$\nu_y \leq C\mathcal{H}^1 \qquad \mu \text{ a.e.} \,. \tag{4.7}$$

$\square$

The next step is to study the relation between $d\nu_{(y,z)}(s) = \alpha(s, y, z)ds$ and the divergence of $d(x)$.

### 4.2. Divergence disintegration

By definition of divergence, we have that

$$\int_\Omega \phi(x)d\xi = -\int_\Omega d \cdot \nabla\phi dx \qquad \phi \in C_c^1(\Omega, \mathbb{R}) \,. \tag{4.8}$$

We use again the approximation given by the $d_i$.

By [4], Theorem 5.8, for the $d_i$ we can write for all $\psi(x,y,z) = \psi(y,z)\phi(x_1)$, where $\psi$ is selected in such a way that $\mathrm{supp}\{\psi(y,z)\phi(x_1)\} \subset\subset \Omega$,

$$\int_{\mathbb{R}^n} \phi(x_1)d\Big(\mathrm{div}\big(\psi(b_i(x),d_i(x))d_i(x)\big)\Big) = -\int_{\mathbb{R}^n} \psi\big(b_i(x),d_i(x)\big)\big(e_1\cdot d_i(x)\big)\phi'(x_1)dx\,.$$
(4.9)

Note that $\psi$ does not need to be smooth, since we are working with $d_i$.

Passing to the limit and using the weak convergence of $\mathrm{div}d_i$, we can write for the vector $d$ the same estimate:

$$\int_{\mathbb{R}^n} \phi(x_1)d\Big(\mathrm{div}\big(\psi(b(x),d(x))d(x)\big)\Big) = -\int_{\mathbb{R}^n} \psi\big(b(x),d(x)\big)\big(e_1\cdot d(x)\big)\phi'(x_1)dx\,.$$
(4.10)

The first part of the integral defines a linear function on $\psi(y)$ (if the divergence is a measure) for all $\phi$ fixed: in fact we have the uniform estimate on the divergence measure given by (3.23).

If we thus consider the functional

$$\psi \mapsto \int_{\mathbb{R}^n} \phi(x_1)\mathrm{div}\Big(\psi\big(b(x),d(x)\big)d(x)\Big)dx\,,$$

we have that it can be written as

$$\int_{\mathbb{R}^n} \phi(x_1)\mathrm{div}\Big(\psi\big(b(x),d(x)\big)d(x)\Big)dx = \int d\tilde{\mu}_\phi(y,z)\psi(y,z)$$

$$= \int d\tilde{\mu}(y,z)c_\phi(y,z)\psi(y,z)\,,$$

where here and in the following we will use the notation

$$\tilde{m} = (\mathbb{I},b,d)\sharp|\mathrm{div}d|\,,\qquad \tilde{m} = \pi_2\sharp\tilde{\mu} = (b,d)\sharp|\mathrm{div}d|\,.$$

We have used the fact that $|\tilde{\mu}_\phi| \le \tilde{\mu}$.

Note that since $(b,d)$ is Borel, we obtain Borel measures. By restricting the divergence to $\Omega' \subset\subset \Omega$, we have a finite measure by Proposition 3.10, so that the Radon property holds and we can apply again the disintegration theorem.

By taking a countable dense set in $C_c$, we have that the dependence w.r.t. $\phi$ is linear for all $\phi$ for $\tilde{\mu}$ a.e., and moreover, from the bounded estimate, we have that $c_\phi \le \|\phi\|$. We thus can extend it to the whole $C_c$ and use the representation theorem to write

$$\int_{\mathbb{R}^n} \phi(x_1)\mathrm{div}\Big(\psi\big(b(x),d(x)\big)d(x)\Big)dx = \int d\tilde{\mu}\psi(y,z)\int d\tilde{\nu}_{(y,z)}(s)\phi(s)\,.\qquad(4.11)$$

Using the disintegration of the r.h.s. of (4.8), we have the equality

$$\int d\tilde{\mu}\psi(y,z)\int d\tilde{\nu}_{(y,z)}(s)\phi(s) = -\int d\mu\psi(y,z)\int d\nu_{(y,z)}(s)\phi'(s)\,.\qquad(4.12)$$

By decomposing

$$d\tilde{\mu} = k(y,z)d\mu + d\tilde{\mu}_s\,,$$

we have that for all $\phi$ (at least in a countable dense subset of $C_c$)

$$\int ds\alpha(s,y,z)\phi'(s) = -k(y,z)\int d\tilde{\nu}_y(s)\phi(s)\,. \tag{4.13}$$

This gives the following representation of $\alpha(s,y,z)$.

**Proposition 4.5.** *The density $\alpha(s,y,z)$ can be written as*

$$\alpha(s,y,z) = k(y,z)\tilde{\nu}_{(y,z)}(-\infty, s)\,, \tag{4.14}$$

*where $\tilde{\nu}_{(y,z)}$ is the measure corresponding to the disintegration*

$$\mathrm{div}d = \int \tilde{\nu}_{(y,z)}d\tilde{\mu}(y,z)\,, \tag{4.15}$$

*and $k(y,z) = d\tilde{\mu}/d\mu$, $\mu = (b,d)\sharp\mathcal{H}^n|_{\Omega}$.*

This representation can be improved to show that only the absolutely continuous part of the divergence enters in the computation of $\alpha(s,y,z)$.

### 4.3. Explicit representation of the density $\alpha(s,y,z)$

We recall that by Proposition 3.11, the following estimate holds:

$$\int \phi(x_1)\mathrm{div}(\chi_A d)dx \leq \int \phi(x_1)dx_1\,, \tag{4.16}$$

where the constant $C$ depends on the compact support of $\phi$, the set $A$ and the distance from the initial points $A$ and the boundary $\partial\Omega$. This shows that we can cover the set where the singular part of $\mathrm{div}d$ is concentrated by cylinders with total Lebesgue measure as small as we want.

It thus follows that

$$\tilde{m} = (\mathrm{div}d)_{\mathrm{a.c.}}m + \tilde{m}_s\,.$$

so that we can obtain

**Theorem 4.6.** *The Lebesgue density satisfies the ODE*

$$\frac{d}{ds}\alpha(s,y,z) = \alpha(s,y,z)(\mathrm{div}d)_{\mathrm{a.c.}}\,. \tag{4.17}$$

A more precise estimate and using the same argument of [4], Theorem 5.7, it follows that

$$\alpha(s,y,z) \leq \varrho\max\left\{\frac{|b(x) - x|}{\epsilon'}, \frac{|x - a(x)|}{\epsilon}\right\}\,, \tag{4.18}$$

where $x$ is the point corresponding to $(s,y,z)$.

By studying the ODE (4.14), on can easily prove the following corollary:

**Corollary 4.7.** *The function $\alpha(s,y,z)$ is different from $0$ $\mathcal{H}^n$ a.e. in $\Omega$, and uniformly bounded in $\Omega' \subset\subset \Omega$.*

This concludes the analysis of the disintegration of the Lebesgue measure.

## 5. Existence of a solution to the Euler-Lagrange equation

After this machinery, it is quite easy to solve the Euler Lagrange equation, and thus to prove that the solution is positive inside the domain. In fact, the Euler Lagrange equation can be written thus as

$$\operatorname{div}\big(p(x)d(x)\big) = g'\big(u(x)\big),$$

with $p \geq 0$. By weak compactness, there exists a solution to the Euler Lagrange equation: this result is proved in [4], Theorem 4.3.

We disintegrate the Euler Lagrange equation: for all $\phi \in C_c^1(\Omega)$,

$$\int_\Omega \big(p(x)d(x)\nabla\phi(x) + g(x)\phi(x)\big)dx$$
$$= \int d\mu(y,z) \left( \int_{\mathbb{R}} \big(p(y,s)\phi_s(y,s) + g(y,s)\phi(y,s)\big)\alpha(s,y,z)ds \right).$$

We thus obtain that on each line the function $p$ satisfies

$$p(s,y,z) = p_0(y,z) + \frac{1}{\alpha(s,y,z)} \int_{-\infty}^s \alpha(t,y,z)g\big(u(t)\big)dt, \qquad (5.1)$$

where the initial data $p_0(y,z)$ can be chosen arbitrary(but in a measurable way) if the segment $x + td(x)$ has both ends on the boundary, otherwise $p_0(y,z) = 0$. Clearly this solution is different from 0 a.e. on $\Omega$, because $c(s,y,z) \neq 0$ a.e..

It thus follows

**Theorem 5.1.** *The Euler Lagrange equation corresponding to the variational problem (2.1) has a solution which is different from 0 a.e. in $\Omega$.*

The possibility of choosing $p_0(y,z)$ allows in the general case to join the polyhedra in such a way that non divergence on the boundary occurs, so that one can construct the solution to Euler-Lagrange equation for general $\Omega$.

## 6. A conjecture of Bertone-Cellina

In this section we consider the following conjecture. Let $\Omega$ be an open bounded set in $\mathbb{R}^n$, and $D$ a convex closed bounded set in $\mathbb{R}^n$. Let $u \in W^{1,\infty}(\Omega)$ such that $\nabla u \in D$ a.e.. The conjecture stated in [3] is the following:

1. either there exists a function $\eta \in W_0^{1,\infty}(\Omega)$, $\eta \neq 0$, such that $\nabla\eta + \nabla u \in D$;
2. or there exists a divergence free vector $\pi \in (L_{\text{loc}}^1(\Omega))^n$ such that $\pi \neq 0$ and

$$\pi(x) \cdot \nabla u(x) = \max_{k \in D}\big\{\pi \cdot k\big\} \qquad (6.1)$$

a.e. in $\Omega$.

The following proposition has been proved in [3, 4]:

**Proposition 6.1.** *If there exists $\pi$ satisfying point (2), then $\eta = 0$.*

To prove the other implication, consider in fact the two minimization problem

$$\inf_{\bar{u}+W_0^{1,\infty}(\Omega)} \int_\Omega \big(\mathbf{I}_D(\nabla u) - u\big)dx\,, \qquad \inf_{\bar{u}+W_0^{1,\infty}(\Omega)} \int_\Omega \big(\mathbf{I}_D(\nabla u) + u\big)dx\,. \qquad (6.2)$$

Since we assume that there are no variations, then the two solutions coincide with $u$, so in particular there are two positive functions $p^-(x)$, $p^+(x)$ belonging to $L^\infty_{\mathrm{loc}}(\Omega)$ and satisfying the Euler-Lagrange equation

$$\mathrm{div}\big(p^-(x)d(x)\big) = -1\,, \quad \mathrm{div}\big(p^+(x)d(x)\big) = 1\,. \qquad (6.3)$$

For the second minimization problem, we have to reverse the directions on $x+td(x)$, by setting

$$a(x) = \big\{x + td(x), t \in \mathbb{R}\big\} \cap \partial\Omega\,, \quad b(x) = \big\{x - td(x), t \in \mathbb{R}\big\} \cap \partial\Omega\,.$$

Clearly, it is equivalent to consider the minimization problem

$$\inf_{-\bar{u}+W_0^{1,\infty}(\Omega)} \int_\Omega \big(\mathbf{I}_{-D}(\nabla v) - v\big)dx\,,$$

and setting $u = -v$.

By adding the two equations in (6.3), it follows that $\pi(x) = p^+(x)d(x) + p^-(x)d(x)$ satisfies point (2), if we can prove that it is different from 0: this follows from Theorem 5.1.

# References

[1] G. Alberti and L. Ambrosio, *A geometric approach to monotone functions in $\mathbb{R}^n$*, Math. Z., **230** (1999), 259–316.

[2] L. Ambrosio, N. Fusco and D. Pallara, *Functions of bounded variation and free discontinuous prblems*, Oxford Mathematical Monographs, 2000.

[3] S. Bertone and A. Cellina, *On the existence of variations, control, optimization and calculus of variations* (to appear).

[4] S. Bianchini, *On the Euler-Lagrange equation for a variational problem*, DCDS (to appear).

[5] A. Cellina and S. Perrotta, *On the validity of the maximum prnciple and of the Euler-Lagrange equation*, SIAM J. Control Optim., **36** (1998), 1987–1998.

[6] L. Schwartz, *Theorie des distributions*, 3rd Edition, Hermann Paris, 1966.

Stefano Bianchini
SISSA-ISAS
Via Beirut 2-4
I-34014 Trieste
ITALY
URL: http://www.sissa.it/~bianchin/

Progress in Nonlinear Differential Equations
and Their Applications, Vol. 75, 79–103
© 2007 Birkhäuser Verlag Basel/Switzerland

# Singular Limits for Impulsive Lagrangian Systems with Dissipative Sources

Alberto Bressan

*Dedicated to Arrigo Cellina and James Yorke*

**Abstract.** Consider a mechanical system, described by finitely many Lagrangian coordinates. Assume that an external controller can influence the evolution of the system by directly assigning the values of some of the coordinates. If these assignments are implemented by means of frictionless constraints, one obtains a set of ordinary differential equations where the right hand side depends also on the time derivatives of the control functions. Some basic aspects of the mathematical theory for these equations are reviewed here. We then consider a system with an additional dissipative term, which vanishes on a stable submanifold $\mathcal{N}$. As the coefficient of the source term approaches infinity, we show that the limiting impulsive dynamics on the reduced state space $\mathcal{N}$ can be modelled by two different systems, depending on the order in which two singular limits are taken. These results are motivated by the analysis of impulsive systems with non-holonomic constraints.

## 1. Introduction

Consider a mechanical system described by $N$ Lagrangian variables $q^1, \ldots, q^N$. Its kinetic energy $T = T(q, \dot{q})$ will be given by a positive definite quadratic form of the time derivatives $\dot{q}^i$, say

$$T(q, \dot{q}) = \frac{1}{2} \sum_{i,j=1}^{N} A_{ij}(q)\, \dot{q}^i \dot{q}^j \,. \tag{1.1}$$

This work was completed with the support of our TEX-pert.

If the system is affected by external forces having components $Q_i = Q_i(t, q, \dot{q})$, its motion is determined by the equations

$$\frac{d}{dt}\frac{\partial T}{\partial \dot{q}^i} = \frac{\partial T}{\partial q^i} + Q_i(t, q, \dot{q}) \qquad i = 1, \ldots, N.$$

In a common situation, an external controller can apply additional forces, whose components $\phi_i(q, u)$ depend continuously on the state $q$ of the system and on the value $u = u(t)$ of the control function. In this case, the evolution of the system is described by

$$\frac{d}{dt}\frac{\partial T}{\partial \dot{q}^i} = \frac{\partial T}{\partial q^i} + Q_i(t, q, \dot{q}) + \phi_i(q, u) \qquad i = 1, \ldots, N.$$

This leads to a control system in standard form, where the right hand side depends continuously on the control $u$.

In a quite different but still realistic situation, a controller can prescribe the values of the last $m$ coordinates $q^{n+1}, \ldots, q^{n+m}$ as functions of time, say

$$q^{n+i}(t) = u_i(t) \qquad i = 1, \ldots, m. \tag{1.2}$$

We now make the crucial assumption that the identities (1.2) are achieved by implementing $m$ frictionless constraints. Here **frictionless** means that the forces produced by the constraints make zero work in connection with any virtual displacement of the remaining free coordinates $q^1, \ldots, q^n$. Calling $\Phi_i(t)$ the components of the additional forces, used to implement the constraints (1.2), the motion is now determined by the equations

$$\frac{d}{dt}\frac{\partial T}{\partial \dot{q}^i} = \frac{\partial T}{\partial q^i} + Q_i(t, q, \dot{q}) + \Phi_i(t) \qquad i = 1, \ldots, n + m. \tag{1.3}$$

The assumption that the constraints are frictionless is expressed by the identities

$$\Phi_1(t) = \cdots = \Phi_n(t) = 0. \tag{1.4}$$

In this case, the evolution of the remaining free coordinates $q^1, \ldots, q^n$, together with their conjugate momenta

$$p_i = \frac{\partial T}{\partial \dot{q}^i}, \qquad i = 1, \ldots, n, \tag{1.5}$$

is described by a first order system of $2n$ differential equations:

$$\begin{cases} \dot{q}^i = \phi_i(t, q(t), p(t), u(t), \dot{u}(t)), \\ \dot{p}_i = \psi_i(t, q(t), p(t), u(t), \dot{u}(t)), \end{cases} \qquad i = 1, \ldots, n, \tag{1.6}$$

where upper dots denote derivatives w.r.t. time. The presence of the time derivatives $\dot{u}_i$ of the control functions accounts for the impulsive character of (1.6).

The theory of control of Lagrangian systems by means of of moving constraints was initiated independently by Aldo Bressan and by Charles-Michel Marle, around 1980. The memoir [9] was motivated by problems of optimal control for the ski or the swing, later studied in [10]. In [18] one can find a more general geometric approach, also including some mechanical applications. A subsequent paper by Cardin and Favretti [11] clarified the relations between the two approaches.

The impulsive control of mechanical systems leads to a rich mathematical theory, different in many aspects from the standard theory of control.

## 1.1. Basic form of the equations.

Assume that the external forces $Q_i$ in (1.3) depend at most linearly on the first derivatives $\dot{q}^1$. Then, as shown in [9], the right hand sides of (1.6) are polynomials of degree $\leq 2$ w.r.t. the derivatives $\dot{u}_i$. Renaming the variables $x = (x_1, \ldots, x_{2n}) = (q^1, \ldots q^n, p_1, \ldots, p_n)$, we thus obtain a system of the form

$$\dot{x} = \tilde{f}(t, x, u) + \sum_{i=1}^{m} \tilde{g}_i(t, x, u)\, \dot{u}_i + \sum_{i,j=1}^{m} \tilde{h}_{ij}(t, x, u)\, \dot{u}_i \dot{u}_j \,. \tag{1.7}$$

To achieve a further simplification, one can introduce the additional state variables $x_0 = t$ and $x_{n+1} = u_1, \ldots, x_{n+m} = u_m$, with equations

$$\dot{x}_0 = 1\,, \qquad \dot{x}_{n+j} = \dot{u}_j \qquad j = 1, \ldots, m\,.$$

This removes the explicit dependence of the vector fields $\tilde{f}, \tilde{g}_i, \tilde{h}_{ij}$ on the variables $t$ and $u$:

$$\dot{x} = f(x) + \sum_{i=1}^{m} g_i(x)\, \dot{u}_i + \sum_{i,j=1}^{m} h_{ij}(x)\, \dot{u}_i \dot{u}_j \,. \tag{1.8}$$

In many important cases, all the coefficients $h_{ij}$ of the quadratic terms vanish identically, and the right hand side of (1.8) is an affine function of the components $\dot{u}_i$, namely

$$\dot{x} = f(x) + \sum_{i=1}^{m} g_i(x)\, \dot{u}_i \,. \tag{1.9}$$

The form of the basic equations plays a key role in determining the class of control functions for which the corresponding trajectory of the system can be uniquely defined. Systems of the form (1.9) were called "fit for jumps" in [9]. In this case, since the derivative of the control enters linearly in the equations, solutions can here be defined also in connection with a control function having jumps at certain points. On the other hand, if we insert a control having a jump in (1.8), a product like $\dot{u}_i \dot{u}_i$ will formally contain the square of a Dirac delta distribution. Therefore, if the vector field $h_{ii}$ does not vanish, the state of system will instantly reach infinity. In this case, the model is clearly not well posed.

An analytic characterization of systems "fit for jumps" was first derived in [9]. This property also admits an elegant geometric characterization, in terms of orthogonal geodesic curves. In the case of a scalar control, this characterization was obtained in Theorem 5.1 of [9]. For general vector-valued controls, the analysis in [24] showed how "fitness for jumps" is related to another fundamental property of foliations on a Riemann manifold, studied in [25, 26].

Of particular interest is the case where all vector fields $g_i$ in (1.9) commute, i.e., their Lie brackets $[g_i, g_j] \doteq (Dg_j)\, g_i - (Dg_i)\, g_j$ vanish identically. By a suitable change of coordinates one can then remove the presence of the derivatives $\dot{u}_i$ from the equations [29]. The evolution is thus described by a standard (non-impulsive) control system.

### 1.2. Construction of trajectories for non-smooth controls.

When the control functions $u = (u_1, \ldots, u_m)$ are absolutely continuous, one could simply define $v_i \doteq \dot{u}_i$, $i = 1, \ldots, n$, and use $v = (v_1, \ldots, v_m)$ as our basic control function. Using these new variables, (1.8) becomes a control system in standard form, namely

$$\dot{x} = f(x) + \sum_{i=1}^{m} g_i(x) \, v_i + \sum_{i,j=1}^{m} h_{ij}(x) \, v_i v_j \, .$$

This approach, however, is not of much interest. In most applications, the dynamics of the system and the constraints on the control functions are naturally formulated using the coordinates themselves as controls, rather then their time derivatives. Moreover, in several optimization problems, the optimal control $u^{opt}(\cdot)$ is a discontinuous function of time. Restricting the search to absolutely continuous controls would be fruitless. When studying the impulsive system in its original form (1.8), a major issue is how to define solutions for controls which are not Lipschitz continuous. A natural approach is the following. Let the functions $f$, $g_i$, and $h_{ij}$ in (1.8) be smooth, and consider the initial data

$$x(0) = \bar{x} \, . \tag{1.10}$$

Then the Cauchy problem (1.8), (1.10) has a unique solution $t \mapsto x(t, u)$ for any $\mathcal{C}^1$ control $u(\cdot)$. In order to construct a solution corresponding to a more general (possibly discontinuous) control function $u(\cdot)$, one can approximate $u$ by a sequence of $\mathcal{C}^1$ control functions $u^k$ and take the limit of the corresponding trajectories. The key problem here is to identify suitable topologies on the space of controls and on the space of trajectories which render continuous the control-to-trajectory map: $u(\cdot) \mapsto x(\cdot, u)$. This problem has been the object of many investigations, in the context of stochastic differential equations [29], and for control systems [2, 4–7, 17, 19]. Observe that, if the convergence $u^k(\cdot) \to u(\cdot)$ implies the convergence of the sequence of trajectories $x(\cdot, u^k)$, one can then uniquely define the trajectory $x(\cdot, u)$ as the limit

$$x(\cdot, u) \doteq \lim_{k \to \infty} x(\cdot, u^k) \, ,$$

in a suitable topology.

A related problem is to characterize the $\mathbf{L}^1$ closure of the set of all trajectories which correspond to smooth controls. As shown in [8], this can be done in terms of a differential inclusion with closed, convex right hand side:

$$\frac{dx}{ds} \in F\big(x(s)\big) \, , \tag{1.11}$$

where $s \mapsto t(s)$ provides a reparametrization of the time variable. Trajectories of (1.11) can be interpreted as a kind of "generalized solutions" of the impulsive control system (1.8). This characterization is useful in many applications. For example, some problems of optimal control, or asymptotic stabilization, are best studied by looking directly at the differential inclusion (1.11).

## 1.3. Stabilization to a constant state.

We recall that the impulsive system (1.8) is said to be **locally stabilizable** at a state $\bar{x}$ if, given any $\varepsilon > 0$ one can find $\delta > 0$ such that the following holds. Given any initial state $x^\dagger$ with $|x^\dagger - \bar{x}| \leq \delta$, one can find a $\mathcal{C}^1$ control $u(\cdot)$ such that the corresponding trajectory satisfies

$$\left| x(t, u) - \bar{x} \right| \leq \varepsilon \qquad \text{for all } t \geq 0 \,. \tag{1.12}$$

If a control $u(\cdot)$ can be found such that, in addition to (1.12),

$$\lim_{t \to \infty} x(t, u) = \bar{x} \,, \tag{1.13}$$

then we say that the system is **locally asymptotically stabilizable** at $\bar{x}$. Notice that here we restrict the attention to $\mathcal{C}^1$ controls. This is natural, because the more general trajectories of (1.8) are always defined as limits of solutions corresponding to smooth controls.

For results on the (asymptotic) stabilization of a Lagrangian system, by means of moving constraints, we refer to the forthcoming paper [BR5]. The key idea here is to reduce the problem to a stabilization problem for the corresponding differential inclusion (1.11). In turn, the weak stability of a differential inclusion can be analyzed by well established techniques [12, 15, 27, 28].

## 1.4. Optimization problems.

The problems of maximizing the amplitude of oscillation of a swing, or minimizing the time taken by a skier to reach the end of a trail, were among the initial motivations for research on this subject [10]. A variety of optimal control problems can be naturally formulated in the context of mechanical systems controlled by moving constraints. As shown in [5], certain cases can be reduced to an optimization problem for a standard (non-impulsive) control system. Other cases can be studied in terms of a related differential inclusion. Further results on this type of optimization problems can be found in [20–22]. For a general survey of the mathematical theory of Lagrangian systems controlled by moving constraints, we refer to [3]. A key tool in the analysis of impulsive systems is the reparametrization of the graph of the control function $t \mapsto u(t)$. Given a function $u(\cdot)$ with bounded variation (BV), one can consider a Lipschitz continuous curve $\gamma$, parametrized as $s \mapsto \big(t(s), u(s)\big)$, which contains the graph of $u$. Under suitable conditions, the impulsive equations (1.9) reduce to a standard system of O.D.E.'s, in terms of this new variable $s$. This approach relies on the basic concept of "graph completion", which was introduced in [4].

**Definition 1.1.** *A* **graph completion** *of a BV function* $u : [0, T] \mapsto I\!\!R^m$ *is a Lipschitz continuous path* $\gamma = (\gamma_0, \gamma_1, \ldots, \gamma_m) : [0, S] \mapsto [0, T] \times I\!\!R^m$ *such that*

(i) $\gamma(0) = (0, u(0))$, $\gamma(S) = (T, \ u(T))$,
(ii) $\gamma_0(s_1) \leq \gamma_0(s_2)$ *for all* $0 \leq s_1 < s_2 \leq S$,

(iii) *for each* $t \in [0, T]$ *there exists some* $s$ *such that*
$\gamma(s) = (t, u(t))$.

Notice that the path $\gamma$ provides a continuous parametrization of the graph of $u$ in the $(t, u)$ space. At a time $\tau$ where $u$ has a jump, the curve $\gamma$ must include an arc joining the left and right points $(\tau, u(\tau-))$, $(\tau, u(\tau+))$.

**Definition 1.2.** *Let $\gamma$ be a graph completion of the control function $u$. Let $y(\cdot, \gamma)$ be the unique Carathéodory solution of the Cauchy problem*

$$\frac{d}{ds} y(s) = f(y(s)) \, \dot{\gamma}_0(s) + \sum_{i=1}^m g_i(y(s)) \, \dot{\gamma}_i(s), \qquad y(0) = \bar{x}. \qquad (1.14)$$

*Then the (possibly multivalued) function*

$$t \mapsto x(t, \gamma) = \{y(s, \gamma) \, ; \ \gamma_0(s) = t\}$$

*is called the **generalized trajectory** of (1.9)–(1.10) determined by the graph completion $\gamma$.*

Observe that, by definition, the path $\gamma$ is absolutely continuous, hence the Carathéodory solution of (1.14) is well defined.

It can be shown that the trajectory $x(\cdot, \gamma)$ depends on the path $\gamma$ itself, but not on the way it is parametrized. In particular, let $\tilde{\gamma} : [0, \widetilde{S}] \mapsto [0, T] \times I\!R^m$ be another graph completion of $u$ such that

$$\tilde{\gamma}(s) = \gamma(\phi(s)) \qquad s \in [0, \widetilde{S}]$$

for some absolutely continuous, strictly increasing $\phi : [0, \widetilde{S}] \mapsto [0, S]$. Then the generalized trajectories $x(\cdot, \tilde{\gamma})$ and $x(\cdot, \gamma)$ coincide.

It is important to understand how the trajectory of the system (1.9) depends on the choice of graph completion. The main results in this direction, proved in [4], are as follows.

Let $\gamma : [0, S] \mapsto I\!R^{1+m}$ and $\tilde{\gamma} : [0, \widetilde{S}] \mapsto I\!R^{1+m}$ be any two graph completions of the same control function $u \in BV$. Define their distance as

$$\Delta(\gamma, \tilde{\gamma}) \doteq \inf_{\phi} \ \max_{s \in [0, S]} \left| \gamma(s) - \tilde{\gamma}(\phi(s)) \right|, \qquad (1.15)$$

where the infimum is taken over all continuous, strictly increasing, surjective maps $\phi : [0, S] \mapsto [0, \widetilde{S}]$. In addition, we recall that the Hausdorff distance between two compact sets $A, B \subset I\!R^N$ is

$$d_H(A, B) \doteq \max \left\{ \max_{a \in A} d(a, B), \ \max_{b \in B} d(b, A) \right\}.$$

The distances of a point from a set are here defined as

$$d(a, B) = \inf_{x \in B} |x - a|, \qquad d(b, A) = \inf_{x \in A} |x - b|.$$

We then have

**Theorem 1.3.** *Let the vector fields $f, g_i$ in (1.9) be Lipschitz continuous. Let $u_n$ : $[0,T] \mapsto I\!\!R^m$ be a sequence of control functions. For each $n \geq 0$, let $\gamma_n$ be a graph-completion of $u_n$. Assume that the total variation of the maps $\gamma_n$ remains uniformly bounded, and that*

$$\Delta(\gamma_n, \gamma_0) \to 0 \qquad as \quad n \to \infty. \qquad (1.16)$$

*Then the graphs of the corresponding (possibly multivalued) trajectories $x(\cdot, \gamma_n)$ converge to the graph of $x(\cdot, \gamma_0)$ in the Hausdorff metric.*

For a proof, see [4]. The following result, also proved in [4], relates the generalized solution obtained from a graph completion to the limit of more regular solutions, corresponding to smooth control functions.

**Theorem 1.4.** *Let $\gamma : [0, S] \mapsto I\!\!R^{1+m}$ be a graph completion of a control $u$ : $[0,T] \mapsto I\!\!R^m$. Let $(u_n)_{n\geq 1}$ be a sequence of Lipschitz continuous controls with uniformly bounded total variation, which approximates $\gamma$ in the following sense: Setting $\gamma_n(s) \doteq (s, u_n(s))$, one has*

$$\Delta(\gamma, \gamma_n) \to 0 \qquad as \quad n \to \infty. \qquad (1.17)$$

*Then the generalized solution $x(\cdot, \gamma)$ of (1.9)–(1.10) corresponding to the graph completion $\gamma$ satisfies*

$$x(t, \gamma) = \lim_{n\to\infty} x(t, u_n) \qquad (1.18)$$

*at every time $t$ where $x(\gamma, t)$ is single valued, hence almost everywhere.*

Besides the theory of impulsive control, the idea of graph completion was also used in [13, 14, 16] to achieve a definition of non-conservative products, and non-conservative solutions to hyperbolic systems in one space dimension.

The remainder of this paper is organized as follows. In Section 2 we discuss some elementary mechanical examples, which motivate the impulsive control model. Following [9], in Section 3 we sketch the derivation of the basic equation of motion, assuming that the controller always implements frictionless constraints.

Section 4 contains the main new results of this paper. We consider an impulsive control system of the form

$$\dot{x} = f(x) + \sum_{i=1}^{m} g_i(x)\dot{u}_i + \frac{1}{\varepsilon}\Phi(x). \qquad (1.14)$$

where $f, g_i, \Phi$ are smooth vector fields on $I\!\!R^n$, while $\varepsilon > 0$ is a small parameter. We assume that the source term has dissipative character. More precisely, $\Phi$ vanishes on a submanifold $\mathcal{N} \subset I\!\!R^n$, and every trajectory of the O.D.E.

$$\dot{x} = \Phi(x)$$

converges to some point in $\mathcal{N}$, as $t \to \infty$. As the "relaxation time" $\varepsilon$ approaches zero, we study the dynamics of the system (1.14), corresponding to a possibly discontinuous control $u(\cdot)$. Since the trajectory of (1.14) is defined as limit of trajectories corresponding to an approximating sequence of smooth controls $u_n(\cdot) \to u(\cdot)$, in the end one has to take two limits: $\varepsilon \to 0$ and $n \to \infty$. Depending on the order in which these limits are taken, we show that two different dynamics are obtained, for a system defined on the reduced state space $\mathcal{N}$.

## 2. Mechanical examples

A mechanical system can be affected by an external controller in two basically different ways. On one hand, the controller can influence the evolution of the system by applying additional external forces. This leads to a standard control problem, where the time derivatives of the state variables depend continuously on the control function [23].

In other situations, also physically realistic, the controller acts on the system by directly assigning the values of some of the coordinates. The remaining coordinates are then be determined by solving a set of differential equations with constraints, which also depend explicitly on the time derivative of the control function. We illustrate these two cases with simple examples.

**Example 2.1.** Consider a child riding on a swing, pushed by his mother. In this case, the equations of motion are the same as for a forced pendulum, say of length $r_0$ and mass $m$. In addition to the gravity acceleration $g$, the child is subject to a force $F = u(t)$ exerted by the parent. This force represents a control, and its value can be prescribed (within certain bounds) as function of time. Let $\theta$ be the angle formed by the swing with a vertical line. The motion is then described by the controlled differential equation

$$mr_0\ddot{\theta} = -mgr_0 \sin\theta + u(t)\,. \tag{2.1}$$

**Example 2.2.** Now consider an older boy riding on the swing. We assume that, by standing up or kneeling down, he can change at will the radius of oscillation (within certain bounds). This new system can be described in terms of two variables: the angle $\theta$ and the radius of oscillation $r$. The kinetic energy is given by

$$T(r, \theta, \dot{r}, \dot{\theta}) = \frac{m}{2}\left(\dot{r}^2 + r^2\dot{\theta}^2\right)\,, \tag{2.2}$$

while the potential energy is

$$V(r, \theta) = -mgr\cos\theta\,. \tag{2.3}$$

The control implemented by the boy amounts to assigning the radius of oscillation as a function of time, i.e., $r = u(t)$, for some control function $u$. Calling $L = T - V$ the associated Lagrangian function, the evolution of the remaining coordinate $\theta = \theta(t)$ is now determined by the equation

$$\frac{d}{dt}\frac{\partial L}{\partial \dot{\theta}} = \frac{\partial L}{\partial \theta}\,, \tag{2.4}$$

which in this case yields

$$2mr\dot{\theta}\dot{r} + mr^2\ddot{\theta} = -mgr\sin\theta\,. \tag{2.5}$$

Denoting by $\omega = \dot{\theta}$ the angular velocity, and recalling that $r = u$, we obtain the impulsive system

$$\dot{\theta} = \omega\,, \qquad \dot{\omega} = -\frac{g\sin\theta}{u} - \frac{2\omega}{u}\,\dot{u}\,. \qquad (2.6)$$

Observe that the right hand side of the second equation depends linearly on the time derivative of the control function. In this special case, we can remove the dependence on $\dot{u}$ by a change of variable. Namely, calling $p = mr^2\dot{\theta} = mu^2\dot{\theta}$ the angular momentum, from (2.6) one obtains

$$\dot{\theta} = \frac{p}{mu^2}\,, \qquad \dot{p} = -mgu\sin\theta\,, \qquad (2.7)$$

where the right hand sides do not depend on $\dot{u}$.

**Example 2.3.** A bead of mass $m$ slides without friction along a bar, while the bar can be rotated around the origin, on a vertical plane. This system can be described by two lagrangian parameters: the distance $r$ of the bead from the origin, and the angle $\theta$ formed by the bar and a vertical line. The kinetic and potential energy are still given by (2.2)-(2.3). In the present case, however, we assign the angle $\theta = u(t)$ as a function of time, while the radius $r$ is the remaining free coordinate. The evolution of the radius is now determined by the Lagrange equation

$$\frac{d}{dt}\frac{\partial L}{\partial \dot{r}} = \frac{\partial L}{\partial r}\,, \qquad (2.8)$$

where $L = T - V$. Using (2.2)–(2.3), and setting $\theta = u(t)$, from (2.8) we obtain the second order equation

$$\ddot{r} = r\dot{u}^2(t) + g\cos u(t) \qquad (2.9)$$

Observe that in this case the right hand side of the equation contains the square of the derivative of the control.

## 3. The equations of motion

Consider the Lagrangian system at (1.3), subject to frictionless constraints as in (1.2), (1.4). In this section, following [9], we derive a set of equations describing the evolution of the free coordinates $q^1, \ldots, q^n$. Remarkably, this can be done without explicitly computing the remaining components of the forces $\Phi_{n+1}, \ldots, \Phi_{n+m}$. Indeed, the variables $q^{n+1}, \ldots, q^{n+m}$ are already assigned by (1.2). Of course, their time derivatives

$$\dot{q}^{n+1} = \dot{u}_1(t)\,, \quad \ldots \quad, \dot{q}^{n+m} = \dot{u}_m(t)$$

are also determined. We now show that the evolution of components $q^1, \ldots, q^n$ can be derived from the first $n$ equations in (1.3), taking (1.4) into account.

In connection with the quadratic form (1.1) for the kinetic energy, introduce the conjugate moments

$$p_i = p_i(q, \dot{q}) \doteq \frac{\partial T}{\partial \dot{q}^i} = \sum_{i=1}^{n+m} A_{ij}(q)\,\dot{q}^j\,. \qquad (3.1)$$

Moreover, define the Hamiltonian function

$$H(q,p) \;=\; \frac{1}{2} \sum_{i,j=1}^{n+m} B^{ij}(q)\, p_i p_j \,, \tag{3.2}$$

where $B^{ij}$ are the components of the $(n+m) \times (n+m)$ inverse matrix $B = A^{-1}$. In other words,

$$\sum_{j=1}^{n+m} B^{ij} A_{jk} \;=\; \begin{array}{ll} 1 & \text{if} \quad i = k\,, \\ 0 & \text{if} \quad i \neq k\,. \end{array} \tag{3.3}$$

Next, we solve the system of Hamiltonian equations for the first $n$ variables

$$\begin{cases} \dot{q}^i &=\; \frac{\partial H}{\partial p_i}(q,p) \\ \dot{p}_i &=\; -\frac{\partial H}{\partial q^i}(q,p) + Q_i(t,q,\dot{q}) \end{cases} \qquad i = 1,\ldots,n\,. \tag{3.4}$$

Notice that (3.4) is a system of $2n$ equations for $q^1,\ldots,q^n, p_1,\ldots,p_n$, where the right hand side also depends on the remaining components $q^i, p_i$, $i = n+1,\ldots,n+m$. We can remove this explicit dependence by inserting the values

$$\begin{cases} q^{n+i} &= u_i(t)\,, & \dot{q}^{n+i} = \dot{u}_i(t) & i = 1,\ldots,m\,, \\ p_j &= p_j(p_1,\ldots,p_n, \dot{q}^{n+1},\ldots,\dot{q}^{n+m}) & j = n+1,\ldots,n+m\,. \end{cases} \tag{3.5}$$

In the second line of (3.5), in order to express $p_j$ as a linear combination of $p_1,\ldots,p_n, \dot{q}^{n+1},\ldots,\dot{q}^{n+m}$, we proceed as follows. Let $C = (C_{ij})$ be the inverse of the $m \times m$ submatrix $(B^{ij})_{i,j=n+1,\ldots,n+m}$, so that

$$\sum_{i=n+1}^{n+m} C_{ji} B^{ik} \;=\; \begin{array}{ll} 1 & \text{if} \quad j = k\,, \\ 0 & \text{if} \quad j \neq k\,, \end{array} \qquad j,k \in \{n+1,\ldots,n+m\}\,. \tag{3.6}$$

Recalling that $p = A\dot{q}$, $\dot{q} = Bp$, we multiply by $C_{ji}$ both sides of the identity

$$\dot{q}^i = \sum_{k=1}^{n} B^{ik} p_k + \sum_{k=n+1}^{n+m} B^{ik} p_k \,,$$

and sum over $i = n+1,\ldots,n+m$. By (3.6), this yields

$$p_j = \sum_{i=n+1}^{n+m} C_{ji} \dot{q}^i - \sum_{i=n+1}^{n+m} \sum_{k=1}^{n} C_{ji} B^{ik} p_k \qquad j = n+1,\ldots,n+m\,. \tag{3.7}$$

Inserting in (3.4) the values $p_{n+1},\ldots,p_{n+m}$ given by (3.7), we obtain a closed system of $2n$ equations for the $2n$ variables $q^1,\ldots,q^n, p_1,\ldots,p_n$. More precisely, fix an index $i \in \{1,\ldots,n\}$. Recalling the definition of the Hamiltonian function at (3.2), we obtain

$$\begin{aligned} \dot{q}^i &= \frac{\partial}{\partial p_i} \left\{ \frac{1}{2} \sum_{j,k=1}^{n+m} B^{jk}(q)\, p_j p_k \right\} \\ &= \sum_{j=1}^{n} B^{ij} p_j + \sum_{j=n+1}^{n+m} B^{ij} p_j \\ &= \sum_{j=1}^{n} B^{ij} p_j + \sum_{j=n+1}^{n+m} B^{ij} \left( \sum_{\ell=n+1}^{n+m} C_{j\ell}\, \dot{q}^\ell - \sum_{\ell=n+1}^{n+m} \sum_{k=1}^{n} C_{j\ell} B^{\ell k} p_k \right). \end{aligned} \tag{3.8}$$

Next, using again (3.4) and (3.7) we compute

$$
\begin{aligned}
\dot{p}_i &= -\tfrac{\partial}{\partial q^i}\left\{\tfrac{1}{2}\sum_{j,k=1}^{n+m} B^{jk}(q)\, p_j p_k\right\} + Q_i(t,q,\dot{q}) \\
&= -\left(\tfrac{1}{2}\sum_{j,k=1}^{n} + \sum_{j=1}^{n}\sum_{k=n+1}^{n+m} + \tfrac{1}{2}\sum_{j,k=n+1}^{n+m}\right)\frac{\partial B^{jk}}{\partial q^i}\, p_j p_k + Q_i(t,q,\dot{q}) \\
&= -\tfrac{1}{2}\sum_{j,k=1}^{n}\frac{\partial B^{jk}}{\partial q^i}\, p_j p_k \\
&\quad -\sum_{j=1}^{n}\sum_{k=n+1}^{n+m}\frac{\partial B^{jk}}{\partial q^i}\, p_j \left(\sum_{h=n+1}^{n+m} C_{kh}\dot{q}^h - \sum_{h=n+1}^{n+m}\sum_{\ell=1}^{n} C_{kh} B^{h\ell} p_\ell\right) \\
&\quad -\tfrac{1}{2}\sum_{j,k=n+1}^{n+m}\frac{\partial B^{jk}}{\partial q^i}\left(\sum_{h=n+1}^{n+m} C_{jh}\dot{q}^h - \sum_{h=n+1}^{n+m}\sum_{\ell=1}^{n} C_{jh} B^{h\ell} p_\ell\right) \\
&\quad \times \left(\sum_{r=n+1}^{n+m} C_{kr}\dot{q}^r - \sum_{r=n+1}^{n+m}\sum_{\ell=1}^{n} C_{kr} B^{r\ell} p_\ell\right) \\
&\quad + Q_i(t,q,\dot{q})\,.
\end{aligned}
$$
(3.9)

Recalling that $\dot{q}^{n+i} = \dot{u}_i$, and that the matrices $C(q) = (C_{ij}(q))$ are invertible, a direct inspection of the above equations reveals that:

(i) The right hand side of (3.8) is always an affine function of the derivatives $\dot{u}_1,\ldots,\dot{u}_m$.

(ii) The right hand side of (3.9) is an affine function of the derivatives $\dot{u}_1,\ldots,\dot{u}_m$ if and only if each $Q_i$ is an affine function of $\dot{q}^{n+1},\ldots,\dot{q}^{n+m}$, and moreover

$$
\frac{\partial B^{jk}(q)}{\partial q^i} \equiv 0 \qquad \text{for all } i \in \{1,\ldots,n\}\,, \quad j,k \in \{n+1,\ldots,n+m\}\,.
$$
(3.10)

Following [9], systems whose equations of motion are affine w.r.t. the time derivatives of the control will be called **fit for jumps**. In the special case where the derivatives $\dot{u}_i$ do not appear at all in the equations, we say that the system is **strongly fit for jumps.** From the above analysis we thus obtain

**Theorem 3.1.** *The Lagrangian system controlled by moving constraints (1.2)–(1.4) is "fit for jumps" if and only if the components of external forces $Q_i$ are affine functions of the derivatives $\dot{q}^j$, $j = n+1,\ldots,n+m$, and the identities (3.10) hold.*

**Theorem 3.2.** *The system (1.2)–(1.4) is "strongly fit for jumps" provided that the external forces $Q_i$ depend only on the variables $t,q^i$, and moreover the identities (3.10) hold, together with*

$$
B^{ij}(q) \equiv 0 \qquad i \in \{1,\ldots,n\}\,,\; j \in \{n+1,\ldots,n+m\}\,.
$$
(3.11)

## 4. Relaxation limits

Consider an impulsive system with dynamics depending linearly on the derivative of the control:

$$
\dot{x} = f(x) + \sum_{i=1}^{m} g_i(x)\dot{u}_i + \frac{1}{\varepsilon}\Phi(x)\,.
$$
(4.1)

Here $f, g_i, \Phi$ are smooth vector fields on $\mathbb{R}^n$, while $x \in \mathbb{R}^n$ is the state of the system, $u \in \mathbb{R}^m$ is the control function and $\varepsilon > 0$ is a small parameter, i.e., the "relaxation time". In the following we denote by $t \mapsto x(t, y)$ the solution to the Cauchy problem

$$\dot{x} = \Phi(x), \qquad\qquad x(0) = y. \qquad\qquad (4.2)$$

Our basic assumptions are

(A1) The set of equilibrium points

$$\mathcal{N} \doteq \{x \in I\!\!R^n \,;\; \Phi(x) = 0\} \qquad\qquad (4.3)$$

is a smooth manifold of dimension $r < n$. For each $x \in \mathcal{N}$, one can split $I\!\!R^n$ as a sum of a null and a stable subspace:

$$I\!\!R^n = N_x \oplus S_x, \qquad\qquad (4.4)$$

of dimension $r$ and $n - r$ respectively. Both subspaces in (4.4) are invariant under the linear mapping $D\Phi(x)$ (the differential if $\Phi$ at the point $x$). Moreover, $N_x$ coincides with the tangent space to $\mathcal{N}$ at the point $x$, so that

$$D\Phi(x)\,v = 0 \qquad\qquad v \in N_x,$$

while the restriction of $D\Phi(x)$ to the subspace $S_x$ has eigenvalues with strictly negative real part.

(A2) The equilibrium manifold $\mathcal{N}$ is globally attractive, i.e., every solution of (4.2) eventually approaches the manifold $\mathcal{N}$. For each $y \in I\!\!R^n$, the map

$$y \mapsto \pi(y) \doteq \lim_{t \to \infty} x(t, y) \qquad\qquad (4.5)$$

is a continuous projection of $I\!\!R^n$ onto $\mathcal{N}$.

Our main goal is to describe the limits of trajectories of (4.1), as the relaxation parameter $\varepsilon \to 0$.

As a preliminary, we define the $m + 1$ vector fields $\tilde{f}, \tilde{g}_i : \mathcal{N} \mapsto I\!\!R^n$ by means of a suitable projection. Recalling the decomposition (4.4), at each point $x \in \mathcal{N}$ any vector $\mathbf{v} \in I\!\!R^n$ can be decomposed as a sum of a vector in $N_x$ and a vector in $S_x$, say

$$\mathbf{v} = \pi_N \mathbf{v} + \pi_S \mathbf{v}.$$

For $x \in \mathcal{N}$ we can thus define

$$\tilde{f}(x) = \pi_N f(x), \qquad\qquad \tilde{g}_i(x) \doteq \pi_N g_i(x). \qquad\qquad (4.6)$$

This yields a "reduced" impulsive control system on $\mathcal{N}$

$$\dot{x} = \tilde{f}(x) + \sum_{i=1}^{m} \tilde{g}_i(x)\dot{u}_i \qquad\qquad x \in \mathcal{N}. \qquad\qquad (4.7)$$

By construction, the right hand side of (4.7) is tangent to $\mathcal{N}$. Hence, for every initial data $\bar{x} \in \mathcal{N}$, it defines an evolution inside $\mathcal{N}$.

Given an initial data

$$u(0) = \bar{u}\,, \qquad\qquad x(0) = \bar{x}\,, \qquad\qquad (4.8)$$

one might expect that (4.7) describes the limit of trajectories of (4.1) as $\varepsilon \to 0$. Actually, some care must be taken on the order in which limits are taken.

Let $u : [0, T] \mapsto I\!\!R^m$ be a BV function. In the case where $u$ is discontinuous and the vector fields $g_i$ do not commute, to uniquely determine the solution of (4.1), (4.8) one needs to assign a graph completion $\gamma$ of $u$. As recalled in Definitions 1.1 and 1.2, the map $\gamma : s \mapsto \big(\gamma_0(s),\, \gamma_1(s), \ldots,\, \gamma_m(s)\big)$ provides a Lipschitz continuous parametrization of a curve in $I\!\!R^{1+m}$ which contains the graph of $u$.

If $u(\cdot)$ has a jump at a time $\tau$, to compute the jump in the corresponding generalized trajectory $x(\cdot, \gamma)$ we look at the non-trivial interval

$$[s^-, s^+] = \big\{s\,;\; \gamma_0(s) = \tau\big\}\,.$$

The value $x(\tau+)$ of the trajectory corresponding to the graph completion $\gamma$ is then computed by solving the Cauchy problem

$$y(s^-) = x(\tau-)\,, \qquad \frac{d}{ds}y(s) = \sum_{i=1}^{m} g_i\big(y(s)\big)\dot{\gamma}_i(s) \qquad s \in [s^-, s^+]\,, \quad (4.9)$$

and setting

$$x(\tau+) = y(s^+)\,. \qquad\qquad (4.10)$$

Observe that the O.D.E. in (4.9) is of Carathéodory type, having a right hand side which is Lipschitz continuous w.r.t. $y$ and measurable w.r.t. $s$. Indeed, the components $\gamma_i$ are uniformly Lipschitz continuous functions of the parameter $s$.

Next, consider a sequence of smooth controls $u^\nu$ converging to $u$ in $\mathbf{L}^1$ and for a.e. $t$, and whose graphs converge to the graph completion $\gamma$ in the distance defined in [?]. Call $t \mapsto x_\varepsilon^\nu(t)$ the solution to

$$\dot{x} = f(x) + \sum_{i=1}^{m} g_i(x)\dot{u}_i^\nu + \frac{1}{\varepsilon}\Phi(x)\,, \qquad\qquad (4.11)$$

with initial data (4.8). Then set

$$x^\nu = \lim_{\varepsilon \to 0} x_\varepsilon^\nu\,, \qquad\qquad x_\varepsilon = \lim_{\nu \to \infty} x_\varepsilon^\nu\,.$$

We now have

**Theorem 4.1.** *Let the hypothesis (A1) hold. Assume that $u$ is continuous, $\bar{x} \in \mathcal{N}$ and the smooth approximations $u^\nu$ converge to $u$ uniformly on $[0, T]$. Then the following two limits are well defined for every $t \in [0, T]$ and coincide:*

$$x(t) = \lim_{\nu \to \infty}\Big(\lim_{\varepsilon \to 0} x_\varepsilon^\nu(t)\Big) = \lim_{\varepsilon \to 0}\Big(\lim_{\nu \to \infty} x_\varepsilon^\nu(t)\Big)\,. \qquad (4.12)$$

*The limit trajectory $x : [0, T] \mapsto \mathcal{N}$ is a solution of the reduced impulsive control system (4.7).*

**Theorem 4.2.** *Let the hypotheses* $(A1)$, $(A2)$ *hold. Let* $u \in BV$, *and consider a sequence of smooth controls* $u^\nu$ *converging to* $u$ *in* $\mathbf{L}^1$ *and for a.e.* $t \in [0, T]$, *and whose graphs converge to the graph completion* $\gamma$ *of* $u$, *in the distance* (1.15). *Then the two limits*

$$x^\flat(t) = \lim_{\nu \to \infty} \left( \lim_{\varepsilon \to 0} x_\varepsilon^\nu(t) \right), \qquad x^\sharp(t) = \lim_{\varepsilon \to 0} \left( \lim_{\nu \to \infty} x_\varepsilon^\nu(t) \right) \qquad (4.13)$$

*are well defined in* $\mathbf{L}^1$ *and for a.e.* $t \in [0, T]$. *However, they do not coincide, in general.*

*The first limit coincides with the solution of* (4.7) *corresponding to the graph completion* $\gamma$ *with initial data*

$$x(0) = \pi(\bar{x}) . \qquad (4.14)$$

*The second limit is obtained by solving* (4.7) *outside the jump points, while, at each time* $\tau$ *where* $u$ *has a jump, if* $t(s) = \tau$ *for* $s \in [s^-, s^+]$, *then*

$$x(\tau+) = \pi\big(y(s^+)\big) , \qquad (4.15)$$

*Here* $y(s^+)$ *is defined as in* (4.9)–(4.10), *while* $\pi : I\!\!R^n \mapsto \mathcal{N}$ *is the projection defined at* (4.5).

We shall first give a proof of Theorem 2, in several steps. Afterwards, we will show how to recover Theorem 1 by a minor modification of the previous arguments.

**1.** By the assumption (A1), there exists a family of coordinates $z = (z_1, \ldots, z_n)$ with the following properties. The set of equilibrium points has the representation

$$\mathcal{N} \doteq \big\{ x(z); \ \in I\!\!R^n; \ z_{r+1} = \cdots = z_n = 0 \big\} . \qquad (4.16)$$

The projection $\pi : I\!\!R^n \mapsto \mathcal{N}$ takes the form

$$\pi\big(x(z_1, \ldots, z_n)\big) = x(z_1, \ldots, z_r, 0, \ldots, 0) . \qquad (4.17)$$

Moreover, in the new coordinates, the O.D.E. (4.2) takes the form

$$\begin{cases} \dot{z}_j &= 0 \qquad j = 1, \ldots, r , \\ \dot{z}_j &= \phi_j(z) \quad i = r+1, \ldots, n , \end{cases} \qquad (4.18)$$

**2.** It will be convenient to construct a Lyapunov function on a neighborhood of the manifold $\mathcal{N}$. This is a smooth function $V$ such that

$$\begin{aligned} V(x) &= 0 \qquad \text{if} \quad x \in \mathcal{N} , \\ V(x) &> 0 \qquad \text{if} \quad x \notin \mathcal{N} , \end{aligned} \qquad (4.19)$$

$$\nabla V(x) \cdot \Phi(x) < 0 \qquad x \notin \mathcal{N} . \qquad (4.20)$$

The existence of such a function follows again from the assumption (A1).

**3.** Given the continuous control $u(\cdot)$, let $t \mapsto x(t) \in \mathcal{N}$ be the solution of the reduced Cauchy problem (4.7)–(4.8). Using the adapted coordinates $z_1, \ldots, z_j$, for each $\varepsilon > 0$ and $\nu \geq 1$ we call $z^{\nu,\varepsilon} = (z_1^{\nu,\varepsilon}, \ldots, z_n^{\nu,\varepsilon})$ the solution to

$$\begin{cases} \dot{z}_j^{\nu,\varepsilon} &= f^j(z) + \sum_{i=1}^m g_i^j(z) \dot{u}_i^{\nu} & j = 1, \ldots, r, \\ \dot{z}_j^{\nu,\varepsilon} &= f^j(z) + \sum_{i=1}^m g_i^j(z) \dot{u}_i^{\nu} + \frac{1}{\varepsilon} \phi_j(z) & j = r+1, \ldots, n, \end{cases} \tag{4.21}$$

with initial data

$$z(0) = \bar{z} \tag{4.22}$$

corresponding to (4.8).

**4.** Our next goal is to show that each $x^{\nu} = \lim_{\varepsilon \to 0} x_{\varepsilon}^{\nu}$ provides a solution to the Cauchy problem

$$\dot{x} = \tilde{f}(x) + \sum_{i=1}^m \tilde{g}_i(x) \dot{u}_i^{\nu}, \qquad x(0) = \pi(\bar{x}). \tag{4.23}$$

Define $\tau_{\varepsilon} = \varepsilon^{1/2}$. We claim that

$$\lim_{\varepsilon \to 0} x^{\nu}(\tau_{\varepsilon}) = \pi(\bar{x}) \tag{4.24}$$

Indeed, let $t \mapsto w(t)$ be the solution to

$$\dot{w} = \Phi(w), \qquad w(0) = \bar{x},$$

and call $y_{\varepsilon}(t) = x_{\varepsilon}^{\nu}(\varepsilon t)$, so that

$$\frac{d}{dt} y_{\varepsilon} = \varepsilon \left[ f(y_{\varepsilon}) + \sum_i g_i(y_{\varepsilon}) \dot{u}_i^{\nu} \right] + \Phi(y_{\varepsilon}).$$

Using the assumption (A2) we obtain

$$\lim_{\varepsilon \to 0} x_{\varepsilon}^{\nu}(\tau_{\varepsilon}) = \lim_{\varepsilon \to 0} y_{\varepsilon}(\varepsilon^{-1/2}) = \lim_{\varepsilon \to 0} w(\varepsilon^{-1/2}) = \pi(\bar{x}),$$

proving (4.24).

**5.** For a fixed $\nu \geq 1$, call $y^{\nu}(\cdot)$ the solution to the Cauchy problem (4.23). Since $u^{\nu}(\cdot)$ is smooth, for some constant $C_{\nu}$ we can assume

$$\left| \dot{u}^{\nu}(t) \right| \leq C_{\nu} \qquad t \in [0, T].$$

Let $\mathcal{N}_{\delta}$ be a compact neighborhood of the image set $\{y^{\nu}(t); \ t \in [0, T]\}$. Using an a priori bound on the vector fields $f, g_i$ restricted to $\mathcal{N}_{\delta}$, for some constant $M$ we can thus assume

$$\left| f(x) + \sum_{i=1}^m g_i(x) \dot{u}_i(t) \right| \leq M \qquad t \in [0, T], \ x \in \mathcal{N}_{\delta}. \tag{4.25}$$

**6.** Let any $\epsilon_0 > 0$ be given. Then there is $\delta_0 > 0$ such that

$$V(z) < \delta_0 \qquad \Longrightarrow \qquad \sum_{j=r+1}^n |z_j| \leq \epsilon_0. \tag{4.26}$$

Moreover, by (4.20) there is $c_0 > 0$ such that

$$V(x) \geq \delta_0 \qquad \Longrightarrow \qquad \nabla V(x) \cdot \Phi(x) < -c_0 \,. \tag{4.27}$$

Together, (4.25) and (4.27) imply that, for all $\varepsilon > 0$ sufficiently small,

$$\nabla V(x) \cdot \dot{x}_\varepsilon^\nu \leq \frac{1}{\varepsilon} \nabla V(x) \cdot \Phi(x) + |\nabla V(x)| \cdot M < 0 \,. \tag{4.28}$$

This implies that the trajectory $x_\varepsilon^\nu$ cannot exit from the set $\{x \,;\, V(x) \leq \delta_0\}$.

**7.** The behavior of $x_\varepsilon^\nu$ on the initial interval $[0, \tau_\varepsilon]$, with $\tau_\varepsilon = \varepsilon^{1/2}$, was described in step 4. To study the trajectory $t \mapsto x^\nu(t)$ on the remaining interval $[\tau_\varepsilon, T]$, we use the formulation (4.21) in terms of the $z$-coordinates. By the previous steps, given any $\delta_0 > 0$, for all $\varepsilon > 0$ sufficiently small we have

$$V\left(x_\varepsilon^\nu(t)\right) \leq \delta_0 \qquad t \in [\tau_\varepsilon, T] \,. \tag{4.29}$$

In particular, by (4.26), given any $\varepsilon_0 > 0$ we have

$$\sum_{j=r+1}^{n} \left|z_j^\nu(t)\right| \leq \epsilon_0 \qquad t \in [\tau_\varepsilon, T] \,. \tag{4.30}$$

for all $\varepsilon > 0$ small enough. Since $\epsilon_0 > 0$ can be taken arbitrarily small, and the functions $f$ and $g_i \dot{u}_i$ in (4.21) are uniformly Lipschitz continuous, as $\varepsilon \to 0$ we have the uniform convergence $z^{\nu,\varepsilon} \to z^\nu$, where the limit trajectory provides a solution to

$$\begin{cases} \dot{z}_j^\nu = f^j(z) + \sum_{i=1}^{m} g_i^j(z) \dot{u}_i^\nu & j = 1, \ldots, r \,, \\ \dot{z}_j^\nu = 0 & j = r+1, \ldots, n \,, \end{cases} \tag{4.31}$$

with initial data

$$z(0) = \pi(\bar{z}) = (\bar{z}_1, \ldots, \bar{z}_r, 0, \ldots, 0) \,. \tag{4.32}$$

To conclude, we now only need to observe that (4.31)-(4.32) correspond precisely to (4.7), (4.14), in the $z$ coordinates.

**8.** Finally, assume that the sequence $u^\nu$ converges to the graph-completion $\gamma$ of $u$, in the sense of graphs. An application of the results in [4], here stated in Theorem 1.4, implies that the corresponding trajectories $x^\nu(\cdot)$ converge to $x^\flat(\cdot)$ pointwise for a.e. $t \in [0, T]$, and also in $\mathbf{L}^1\left([0, T]\right)$. This completes the proof of the first part of Theorem 4.2.

**9.** In the remainder of the proof, we consider the second limit in (4.13). By the convergence assumption on the approximating controls $u^\nu(\cdot)$, for every fixed $\varepsilon > 0$ the limit

$$x_\varepsilon = \lim_{\nu \to \infty} x_\varepsilon^\nu \tag{4.33}$$

is described in terms of the graph completion $\gamma : [0, S] \mapsto [0, T] \times I\!R^m$ of the control function $u(\cdot)$. Call $s \mapsto \hat{x}_\varepsilon(s)$ the solution of

$$\frac{d}{ds}\hat{x}(s) = \left( f(\hat{x}(s)) + \frac{1}{\varepsilon}\Phi(\hat{x}(s)) \right) \cdot \frac{d\gamma_0(s)}{ds} + \sum_{i=1}^{m} g_i(\hat{x}(s)) \frac{d\gamma_i(s)}{ds} \qquad s \in [0, S],$$

(4.34)

with initial data (4.8). Since the map $s \mapsto t(s) = \gamma_0(s)$ from $[0, S]$ onto $[0, T]$ is nondecreasing, for almost every $t$ there exists a unique $s = \sigma(t)$ such that $t = t(\sigma(t))$. We then have the representation

$$x_\varepsilon(t) = \hat{x}_\varepsilon(\sigma(t)).$$

(4.35)

**10.** It now remains to describe the limit of the maps $\hat{x}_\varepsilon$, as $\varepsilon \to 0$. Consider the set

$$J \doteq \bigcup_{t \in [0,T]} ]s^-(t), s^+(t)[,$$

(4.36)

where

$$s^-(t) = \min \left\{ s ; \gamma_0(s) = t \right\}, \qquad s^+(t) = \max \left\{ s ; \gamma_0(s) = t \right\}.$$

Since $s^-(t) = s^+(t)$ for all but countably many times $t$, the set $J$ is actually the union of countably many open intervals, say

$$J = \bigcup_{k \geq 1} ]s_k^-, s_k^+[.$$

(4.37)

Our goal is to prove that as $\varepsilon \to 0$, the maps $s \mapsto \hat{x}_\varepsilon(s)$ converge to the unique map $s \mapsto \hat{x}^\sharp(s)$ providing a fixed point to the integral transformation $w \mapsto \mathcal{T}w$ defined as

$$\mathcal{T}w(s) \doteq \pi(\bar{x}) + \int_{[0,s]\setminus J} \left( \tilde{f}(w(r))\dot{\gamma}_0(r) + \sum_{i=1}^{m} \tilde{g}_i(w(r))\dot{\gamma}_i(r) \right) dr$$
$$\begin{cases} \left\{ \pi(y_k(s_k^+)) - w(s_k^-) \right\} \\ \int_{s_k^-}^{s} \sum_{i=1}^{m} g_i(w(r))\dot{\gamma}_i(r)\, dr & \text{if } s \in ]s_k^-, s_k^+[, \\ 0 & \text{if } s \notin J. \end{cases}$$

(4.38)

for all $\sigma \in [0, S]$. Some words of explanation are in order. In (4.38), the vector fields $\tilde{f}, \tilde{g}_i$ are the projections of $f, g_i$, as defined at (4.6). For a given $k$, the function $y_k : [s_k^-, s_k^+] \mapsto I\!R^n$ is defined as the solution to the Cauchy problem

$$\frac{d}{ds}y(s) = \sum_{i=1}^{m} g_i(y(s))\frac{d\gamma_i(s)}{ds}, \qquad y(s_k^-) = x(s_k^-).$$

(4.39)

Notice that, according to (4.38), the map $s \mapsto \hat{x}^\sharp(s)$ is right continuous. At each of the points $s = s_k^+$, it jumps from $y(s_k^+)$ to $\pi(y(s_k^+))$.

**12.** In this step we prove the following lemma.

**Lemma 1.** *Given $\epsilon_0 > 0$, a time $\tau < T$ and a compact set $\Gamma \subset \mathbb{R}^n$, there exists $\eta \in \,]0, \epsilon_0]$ such that, for all $\varepsilon > 0$ small enough, the following holds. Fix any $x^* \in \Gamma$ and call $s \mapsto x_\varepsilon(s)$ the solution to (4.34) with initial data*

$$x_\varepsilon\big(s^+(\tau)\big) = x^*. \tag{4.40}$$

*Then*

$$\left| x_\varepsilon\big(s^+(\tau) + \eta\big) - \pi(x^*) \right| \;\leq\; \epsilon_0\,. \tag{4.41}$$

*Proof of Lemma 1.* For notational convenience, set $s^+ \doteq s^+(\tau)$. Of course, this implies $\tau = t(s^+)$. Consider any sequence $\eta_\nu \to 0$ and define

$$\varepsilon_\nu \;\doteq\; \big[t(s^+ + \eta_\nu) - t(s^+)\big]^2 \;>\; 0\,. \tag{4.42}$$

For $\varepsilon \leq \varepsilon_\nu$, the value $x_\varepsilon\big(s^+(\tau) + \eta_\nu\big)$ is more clearly estimated by a time-rescaling technique. Set $y_\varepsilon(\xi) \doteq x_\varepsilon(s^+ + \varepsilon\xi)$. We now look at the solution to the rescaled problem

$$\begin{aligned}
\frac{dy_\varepsilon}{d\xi}(\xi) &= \varepsilon \frac{dx_\varepsilon}{ds}(s^+ + \varepsilon\xi) \\
&= \big[\varepsilon f(y_\varepsilon) + \Phi(y_\varepsilon)\big]\frac{d\gamma_0}{ds}(s^+ + \varepsilon\xi) + \sum_{i=1}^m \varepsilon\, g_i(y_\varepsilon)\frac{d\gamma_i}{ds}(s^+ + \varepsilon\xi),
\end{aligned} \tag{4.43}$$

for $\xi \in [0, \eta_\nu/\varepsilon]$, with initial data

$$y_\varepsilon(0) = x^*\,. \tag{4.44}$$

We now observe that

$$\begin{aligned}
\varepsilon \cdot \int_0^{\eta_\nu/\varepsilon} &\left( |f(y_\varepsilon(\xi))| \cdot \left|\frac{d\gamma_0}{ds}(s^+ + \varepsilon\xi)\right| + \sum_{i=1}^m |g_i(y_\varepsilon(\xi))| \cdot \left|\frac{d\gamma_i}{ds}(s^+ + \varepsilon\xi)\right| \right) d\xi \\
&\leq C \cdot \int_{s^+}^{s^+ + \eta_\nu} \left( \left|\frac{d\gamma_0(s)}{ds}\right| + \sum_{i=1}^m \left|\frac{d\gamma_i(s)}{ds}\right| \right) ds \;\to\; 0
\end{aligned} \tag{4.45}$$

as $\eta_\nu \to 0$. On the other hand, if $\varepsilon \leq \varepsilon_\nu$, as $\eta_\nu \to 0$ we have

$$\frac{1}{\varepsilon}\int_0^{\eta_\nu/\varepsilon} \left|\frac{d\gamma_0(s^+ + \varepsilon\xi)}{d\xi}\right| d\xi \;=\; \frac{t(s^+ + \eta_\nu) - t(s^+)}{\varepsilon} \;=\; \frac{\varepsilon_\nu^{1/2}}{\varepsilon} \;\to\; \infty\,. \tag{4.46}$$

Let $w(\cdot)$ be the solution to

$$\frac{d}{dt}w(t) = \Phi\big(w(t)\big)\,, \qquad\qquad w(0) = x^*\,. \tag{4.47}$$

The limits in (4.45) and (4.46) now show that, as $\nu \to \infty$, as long as $\varepsilon \in\, ]0, \varepsilon_\nu]$ we have

$$\lim_{\nu\to\infty} x_\varepsilon(s^+ + \eta_\nu) = \lim_{\nu\to\infty} y_\varepsilon(\eta_\nu/\varepsilon) = \lim_{\nu\to\infty} w(\varepsilon_\nu^{1/2}/\varepsilon) = \lim_{t\to\infty} w(t) = \pi(x^*)\,.$$

Choosing $\eta = \eta_\nu$ sufficiently small, by a standard compactness argument the conclusion of the lemma holds uniformly for all initial data $x^* \in \Gamma$. $\qquad\square$

**13.** In connection with the given graph completion $s \mapsto \gamma(s)$, we now prove another key estimate. Roughly speaking, the next lemma says that, if an interval $[a, b]$ does not contain large subintervals $[s^-, s^+]$ where $t(s) = \gamma_0(s)$ is constant, then the value of the Lyapunov function $V(x_\varepsilon(s))$ in (4.19)-(4.20) remains small throughout $[a, b]$. This happens because, when $\gamma_0$ is strictly increasing, by (4.27) the large term $\varepsilon^{-1}\Phi(x_\varepsilon)\,\dot\gamma_0$ forces $V$ to decrease.

**Lemma 2.** *For any compact set $K \subset \mathbb{R}^n$ there exists a constant $C_K$ such that the following holds. For any $\varepsilon_0 > 0$, let $[a, b] \subseteq [0, S]$ be such that, for every subinterval $[s', s''] \subseteq [a, b]$,*

$$s'' - s' > \varepsilon_0 \qquad \Longrightarrow \qquad \gamma_0(s'') > \gamma_0(s') \,. \tag{4.48}$$

*Given any $\delta > 0$, if $s \mapsto x_\varepsilon(s)$ is a solution of (4.34) with $\varepsilon > 0$ sufficiently small (depending on $K$ and $\delta$) and if*

$$V(x_\varepsilon(a)) \le \delta \,, \qquad\qquad x_\varepsilon(a) \in K \,, \tag{4.49}$$

*then*

$$V(x_\varepsilon(s)) \le \delta + C_K \cdot \varepsilon_0 \qquad\qquad \text{for all } s \in [a, b] \,. \tag{4.50}$$

*Proof of Lemma 2.* By assumption, there exist values $a = \sigma_0 < \sigma_1 < \cdots < \sigma_N = b$, and a constant $c_1 > 0$, such that

$$\sigma_i - \sigma_{i-1} \le \varepsilon_0 \,, \qquad t(\sigma_i) - t(\sigma_{i-1}) \ge c_1 \qquad i = 1, \ldots, N \,. \tag{4.51}$$

For a given $\sigma \in [a, b]$, define

$$\sigma^* \doteq \max \left\{ s \le \sigma \,;\; V(x_\varepsilon(s)) \le \delta \right\} \,.$$

Then, for $s \in [\sigma^*, \sigma]$, by (4.27) we have

$$\frac{d}{ds} V(x_\varepsilon(s)) \;=\; \nabla V(x_\varepsilon(s)) \cdot \frac{dx_\varepsilon(s)}{ds} \;\le\; -\frac{c_0}{\varepsilon}\frac{d\gamma_0(s)}{ds} + C \cdot \left( \frac{d\gamma_0(s)}{ds} + \sum_{i=1}^{m} \left| \frac{d\gamma_i(s)}{ds} \right| \right) \,.$$

Here the constant $C$ depends only on an upper bound on the vector fields $f$ and $g_i$ and on the gradient $\nabla V$, valid on a suitable compact neighborhood of $K$. For $\varepsilon > 0$ small enough, we can assume $c_0/2\varepsilon > C$, hence

$$\frac{d}{ds} V(x_\varepsilon(s)) \le -\frac{c_0}{2\varepsilon}\frac{d\gamma_0(s)}{ds} + C \cdot \sum_{i=1}^{m} \left| \frac{d\gamma_i(s)}{ds} \right| \,. \tag{4.52}$$

To fix the ideas, assume

$$s_{h-1} \le \sigma^* < \sigma_h < \cdots < \sigma_k < \sigma \le s_{k+1} \,.$$

Let $L_\gamma$ be a Lipschitz constant for the function $\gamma$, so that

$$\sum_{i=1}^{m} \left| \frac{d\gamma_i(s)}{ds} \right| \le L_\gamma \,. \tag{4.53}$$

For every $\varepsilon > 0$ small enough so that

$$\frac{c_0}{2\varepsilon} c_1 \geq C L_\gamma \varepsilon_0 ,$$

we now have

$$V\big(x_\varepsilon(\sigma_h)\big) \leq V\big(x_\varepsilon(\sigma^*)\big) + C L_\gamma (\sigma_h - \sigma^*) \leq \delta + C L_\gamma \varepsilon_0 .$$

Moreover, by induction on $j = h, h+1, \ldots, k-1$ we find

$$\begin{aligned}
V\big(x_\varepsilon(\sigma_j)\big) &\leq V\big(x_\varepsilon(\sigma_{j-1})\big) - \tfrac{c_0}{2\varepsilon}(\sigma_j - \sigma_{j-1}) + C L_\gamma, (\sigma_j - \sigma_{j-1}) \\
&\leq \big(\delta + C L_\gamma \varepsilon_0\big) - \tfrac{c_0}{2\varepsilon} c_1 + C L_\gamma \varepsilon_0 \\
&\leq \delta + C L_\gamma \varepsilon_0 ,
\end{aligned}$$

and finally

$$V\big(x_\varepsilon(\sigma)\big) \leq V\big(x_\varepsilon(\sigma_k)\big) + C L_\gamma (\sigma - \sigma_k) \leq \big(\delta + C L_\gamma \varepsilon_0\big) + C L_\gamma \varepsilon_0 .$$

This establishes the Lemma, with $C_K \doteq 2C L_\gamma$.    $\square$

**14.** In this step we construct a sequence of functions $w_N : [0, S] \mapsto I\!\!R^n$, converging to a solution of (4.38). For a given integer $N \geq 1$, consider the finite union of open intervals

$$J_N \doteq \bigcup_{k \leq N} \, ]s_k^-, s_k^+[ .$$

Then $s \mapsto w_N(s)$ is defined as the unique piecewise Lipschitz continuous function with jumps at the points $s_1^+, s_2^+, \ldots, s_N^+$, such that $w(0) = \pi(\bar{x})$ and moreover

$$\frac{dw(s)}{ds} = \tilde{f}\big(w(s)\big) \frac{d\gamma_0(s)}{ds} + \sum_{i=1}^m \tilde{g}_i\big(w(s)\big) \frac{d\gamma_i(s)}{ds} \qquad x \notin J_N , \qquad (4.54)$$

$$\frac{dw(s)}{ds} = \sum_{i=1}^m g_i\big(w(s)\big) \frac{d\gamma_i(s)}{ds} \qquad x \in J_N , \qquad (4.55)$$

$$w\big(s_k^+\big) = \pi\left( \lim_{s \uparrow s_k^+} w(s) \right) \qquad k = 1, \ldots, N . \qquad (4.56)$$

where $\pi : I\!\!R^n \mapsto \mathcal{N}$ is the projection defined at (4.5). The existence of the functions $w_N$ is straightforward. Moreover, by the local Lipschitz continuity of the projection map $\pi : I\!\!R^n \mapsto \mathcal{N}$ and of the vector fields $f, g_i$, it is clear that the maps $w_N(\cdot)$ converge to the function $\hat{x}^\sharp(\cdot)$ defined as the fixed point of the integral transformation $w \mapsto \mathcal{T}w$ in (4.38).

Notice that the convergence holds in $\mathbf{L}^1\big([0, S]\big)$, and also pointwise for a.e. $s \in [0, S]$. In the special case where $J$ consists of finitely many open intervals, one simply has $J = J_N$ and $w^\sharp = w_N$, where $N$ is the number of intervals.

**15.** Consider again the solutions $\hat{x}_\varepsilon(\cdot)$ of (4.34), with initial data $\hat{x}_\varepsilon(0) = \bar{x}$. To complete the proof of the theorem, we need to show that, as $\varepsilon \to 0$, the functions $\hat{x}_\varepsilon(\cdot)$ converge to $\hat{x}^\sharp(\cdot)$. Toward this goal, we shall provide an a priori estimate of the difference $|x_\varepsilon - w_N|$.

This is more conveniently achieved working in the $z$-coordinates. The map $s \mapsto z_\varepsilon(s)$ corresponding to $\hat{x}_\varepsilon$ provides the solution to

$$
\begin{cases}
\frac{d}{ds} z_j = f^j(z) \cdot \frac{d\gamma_0(s)}{ds} + \sum_{i=1}^m g_i^j(z) \frac{d\gamma_i(s)}{ds} & j = 1, \ldots, r, \\
\frac{d}{ds} z_j = \left( f^j(z) + \frac{1}{\varepsilon} \Phi_j(z) \right) \cdot \frac{d\gamma_0(s)}{ds} + \sum_{i=1}^m g_i^j(z) \frac{d\gamma_i(s)}{ds} & j = r+1, \ldots, n,
\end{cases}
$$
$$(4.57)$$

with initial data (4.22). On the other hand, the map $s \mapsto z_N(s)$ corresponding to $w_N$ in the $z$-coordinates satisfies the equations

$$
\begin{cases}
\frac{d}{ds} z_{N,j} = f^j(z_N) \cdot \frac{d\gamma_0(s)}{ds} + \sum_{i=1}^m g_i^j(z_N) \frac{d\gamma_i(s)}{ds} & j = 1, \ldots, r, \\
z_{N,j} = 0 & j = r+1, \ldots, n,
\end{cases}
\qquad s \notin J_N,
$$
$$(4.58)$$

$$
\frac{d}{ds} z_{N,j} = \sum_{i=1}^m g_i^j(z_N) \frac{d\gamma_i(s)}{ds} \qquad j = 1, \ldots, n, \qquad s \in J_N, \tag{4.59}
$$

while at the jump points $s_k^+$ there holds

$$
\begin{cases}
z_{N,j}(s_k^+) = \lim_{s \uparrow s_k} z_{N,j}(s) & j = 1, \ldots, r, \\
z_{N,j}(s_k^+) = 0 & j = r+1, \ldots, n.
\end{cases} \tag{4.60}
$$

**16.** Let $\varepsilon_1 > 0$ be given. Choose $\delta > 0$ such that

$$
V(z) \le 3\delta \qquad \Longrightarrow \qquad |z - \pi(z)| \le \varepsilon_1. \tag{4.61}
$$

Let $C_K \ge 1$ be the constant in Lemma 2 and choose $\varepsilon_0 \doteq \delta\, C_K^{-1}$. Consider again the countable sequence of intervals $[s_k^-, s_k^+]$ where the map $s \mapsto t(s)$ is constant, as in (4.37). Choose an integer $N$ large enough so that

$$
s_k^+ - s_k^- < \epsilon_0 \qquad \text{for all } k > N. \tag{4.62}
$$

We now estimate the distance $|z_\varepsilon(s) - w_N(s)|$, for $s \in [0, S]$.

For notational convenience we set $s_0^+ \doteq 0$, $s_{N+1}^- \doteq S$. By relabeling the intervals $[s_k^-, s_k^+]$ we can assume that

$$
0 = s_0^+ < s_1^- < s_1^+ < \cdots < s_N^- < s_N^+ \le s_{N+1}^- = S.
$$

By Lemma 1, given $\eta_0 \in\, ]0, \varepsilon_1]$, for all $\varepsilon > 0$ sufficiently small we have

$$
|z_\varepsilon(\eta_0) - \pi(\bar{z})| \le \varepsilon_1, \qquad |z_N(\eta_0) - \pi(\bar{z})| \le \varepsilon_1, \qquad V\big(z_\varepsilon(\eta_0)\big) < \delta. \tag{4.63}
$$

In particular,

$$
|z_\varepsilon(\eta_0) - z_N(\eta_0)| \le 2\varepsilon_1. \tag{4.64}
$$

Using Lemma 2 we deduce that, for $s \in [\eta_0, s_1^-]$,

$$
V\big(z_\varepsilon(s)\big) \le \delta + 2C_K \varepsilon_0 \le 3\delta. \tag{4.65}
$$

Therefore, by (4.61), for the components $j = 1, \ldots, r$ we have the estimate

$$\left| \frac{d}{ds} z^j_\varepsilon(s) - \frac{d}{ds} z^j_N(s) \right| \le C' \left( \varepsilon_1 + \sum_{i=1}^{r} \left| z^i_\varepsilon(s) - z^i_N(s) \right| \right), \qquad (4.66)$$

where the constant $C'$ depends only on the Lipschitz constants of the vector fields $f, g_i$ and of the control $\gamma$.

For $\varepsilon > 0$ sufficiently small, similar estimates can be obtained for every $k = 1, 2, \ldots, N$. Indeed, set $\eta_k \doteq 2^{-k} \eta_0$. By Lemma 2, on every interval $[s^+_k + \eta_k, s^-_{k+1}]$ we have

$$\left| z_\varepsilon(s) - \pi\big( z_\varepsilon(s) \big) \right| \le \varepsilon_1. \qquad (4.67)$$

Of course, this implies

$$\left| z^j_\varepsilon(s) \right| \le \varepsilon_1 \qquad\qquad j = r+1, \ldots, n. \qquad (4.68)$$

In turn, this implies that (4.66) remains valid for all $s \in [s^+_k + \eta_k, s^-_{k+1}]$.

In addition, when $s \in [s^-_k, s^+_k]$, the functions $z_\varepsilon$ and $z_N$ satisfy the same O.D.E. (4.59). Hence

$$\left| \frac{d}{ds} z^j_\varepsilon(s) - \frac{d}{ds} z^j_N(s) \right| \le C' \sum_{i=1}^{n} \left| z^i_\varepsilon(s) - z^i_N(s) \right|. \qquad (4.69)$$

Moreover, by Lemma 1, by possibly reducing the size of $\eta_k$ and choosing $\varepsilon > 0$ sufficiently small we obtain

$$\left| z_\varepsilon(s^+_k + \eta_k) - \pi\big( z(s^+_k) \big) \right| \le 2^{-k} \varepsilon_1, \qquad \left| z_N(s^+_k + \eta_k) - z_N(s^+_k) \right| \le 2^{-k} \varepsilon_1,$$

$$V\big( z_\varepsilon(s^+_k + \eta_k) \big) < \delta.$$

Observing that $z_N(s^+_k) = \pi\big( z_N(s^+_k-) \big)$, from the previous inequalities we deduce

$$\left| z_\varepsilon(s^+_k + \eta_k) - z_N(s^+_k + \eta_k) \right| \le 2^{1-k} \varepsilon_1 + \left| z_\varepsilon(s^+_k) - z_N(s^+_k-) \right|. \qquad (4.70)$$

In (4.70) we used the fact that $\pi : \mathbb{R}^n \mapsto \mathcal{N}$ is a perpendicular projection, when expressed in the $z$-coordinates. Hence its Lipschitz constant is 1.

**17.** We now put together all the previous estimates. For notational convenience, set

$$\alpha(s) \doteq \sum_{i=1}^{r} \left| z^i_\varepsilon(s) - z^i_N(s) \right|, \qquad\qquad \beta(s) \doteq \sum_{i=r+1}^{n} \left| z^i_\varepsilon(s) - z^i_N(s) \right|.$$

From (4.64)–(4.70) it now follows

$$\alpha(\eta_0) + \beta(\eta_0) \le 2\varepsilon_1,$$

$$\begin{cases} \dot\alpha(s) & \le C'(\alpha + \beta) \\ \beta(s) & \le \varepsilon_1 \end{cases} \qquad s \in [s^+_k + \eta_k, \; s^-_{k+1}],$$

$$\begin{cases} \dot\alpha(s) & \le C'(\alpha + \beta) \\ \dot\beta(s) & \le C'(\alpha + \beta) \end{cases} \qquad s \in [s^-_k, \; s^+_k],$$

$$\begin{cases} \alpha(s_k^+ + \eta_k) & \leq 2^{1-k}\varepsilon_1 + \alpha(s_k^-) \\ \beta(s_k^+ + \eta_k) & \leq \varepsilon_1 \,. \end{cases}$$

By induction on $k$, from the above inequalities we deduce

$$\begin{cases} \alpha(s) \leq 6\varepsilon_1 \cdot e^{2C'S}\,, \\ \beta(s) \leq \varepsilon_1\,, \end{cases} \qquad s \in [0, S] \setminus \bigcup_{k=0}^{N} [s_k^+,\, s_k^+ + \eta_k] \,. \qquad (4.71)$$

Recalling that the sequence $z_N$ converges to $z^\sharp$ as $N \to \infty$, from (4.71) we see that, as $\varepsilon \to 0$, also the sequence $z_\varepsilon$ converges to $z^\sharp$, in $\mathbf{L}^1$ and for a.e. $s \in [0, S]$. Returning to the original $x$-coordinates, we deduce the convergence $\hat{x}_\varepsilon \to \hat{x}^\sharp$, in $\mathbf{L}^1$ and pointwise almost everywhere. This achieves the proof of Theorem 2.

To prove Theorem 1, we simply observe that, if $\bar{x} \in \mathcal{N}$ and if the control function $t \mapsto u(t)$ is continuous, then $x^\flat = x^\sharp$ and the arguments used in the proof of Theorem 2 apply. Notice that the assumption (A2) was only used to study jumps in the control $u(\cdot)$. One can omit this hypothesis and prove the same result, if $u$ is continuous.

**Acknowledgement**

This material is based upon work supported by the National Science Foundation under Grant No. 0505430.

# References

[1] J. P. Aubin and A. Cellina, *Differential Inclusions*. Springer-Verlag, Berlin, 1984.

[2] A. Bressan, *On differential systems with impulsive controls*, Rend. Semin. Mat. Univ. Padova **78** (1987), 227–236.

[3] A. Bressan, *Impulsive control of Lagrangian systems and locomotion in fluids*, Discrete Cont. Dynam. Syst., to appear.

[4] A. Bressan and F. Rampazzo, *On differential systems with vector-valued impulsive controls*, Boll. Un. Mat. Ital. **2-B** (1988), 641–656.

[5] A. Bressan and F. Rampazzo, *Impulsive control systems with commutative vector fields*, J. Optim. Theory Appl. **71** (1991), 67–83.

[6] A. Bressan and F. Rampazzo, *On differential systems with quadratic impulses and their applications to Lagrangian mechanics*, SIAM J. Control Optim. **31** (1993), 1205–1220.

[7] A. Bressan and F. Rampazzo, *Impulsive control systems without commutativity assumptions*, J. Optim. Theory Appl. **81** (1994), 435–457.

[8] A. Bressan and F. Rampazzo, *Stabilization of Lagrangian systems by moving constraints*, to appear.

[9] A. Bressan, *Hyper-impulsive motions and controllizable coordinates for Lagrangean systems*, Atti Accad. Naz. Lincei, Memorie, Serie VIII, Vol. XIX (1990), 197–246.

[10] A. Bressan, *On some control problems concerning the ski or swing*, Atti Accad. Naz. Lincei, Memorie, Serie IX, Vol. I, (1991), 147–196.

[11] F. Cardin and M. Favretti, *Hyper-impulsive motion on manifolds.* Dynam. Contin. Discrete Impuls. Systems **4** (1998), 1–21.

[12] F. H. Clarke, Yu S. Ledyaev, R. J. Stern, and P. R. Wolenski, *Nonsmooth Analysis and Control Theory*, Springer-Verlag, New York, 1998.

[13] G. Dal Maso and F. Rampazzo, *On systems of ordinary differential equations with measures as controls*, Differential Integral Equat. **4** (1991), 739–765.

[14] G. Dal Maso, P. Le Floch, and F. Murat, *Definition and weak stability of nonconservative products*, J. Math. Pures Appl. (9) **74** (1995), 483–548.

[15] V. Jurdjevic, *Geometric Control Theory*, Cambridge University Press, 1997.

[16] P. LeFloch. *Hyperbolic systems of conservation laws. The theory of classical and nonclassical shock waves*, Lectures in Mathematics ETH Zürich. Birkhäuser, Basel, 2002.

[17] W. S. Liu and H. J. Sussmann, *Limits of highly oscillatory controls and the approximation of general paths by admissible trajectories*, In Proc. 30-th IEEE Conference on Decision and Control, IEEE Publications, New York, 1991, pp. 437–442.

[18] C. Marle, *Géométrie des systèmes mécaniques à liaisons actives*, In Symplectic Geometry and Mathematical Physics, 260–287, P. Donato, C. Duval, J. Elhadad, and G. M. Tuynman (Eds.), Birkhäuser, Boston, 1991.

[19] B. M. Miller, *The generalized solutions of ordinary differential equations in the impulse control problems*, J. Math. Systems Estim. Control **4** (1994).

[20] B. M. Miller, *Generalized solutions in nonlinear optimization problems with impulse controls. I. The problem of the existence of solutions*, Automat. Remote Control **56** (1995), 505–516.

[21] B. M. Miller, *Generalized solutions in nonlinear optimization problems with impulse controls. II. Representation of solutions by means of differential equations with a measure*, Automat. Remote Control **56** (1995), 657–669.

[22] M. Motta and F. Rampazzo, *Dynamic programming for nonlinear systems driven by ordinary and impulsive controls*, SIAM J. Control Optim. **34** (1996), 199–225.

[23] H. Nijmeijer and A. van der Schaft, *Nonlinear dynamical control systems*, Springer-Verlag, New York, 1990.

[24] F. Rampazzo, *On the Riemannian structure of a Lagrangian system and the problem of adding time-dependent coordinates as controls*, European J. Mechanics A/Solids **10** (1991), 405–431.

[25] B. L. Reinhart, *Foliated manifolds with bundle-like metrics*, Annals of Math. **69** (1959), 119–132.

[26] B. L. Reinhart, *Differential geometry of foliations. The fundamental integrability problem.* Springer-Verlag, Berlin, 1983.

[27] G. V. Smirnov, *Introduction to the theory of differential inclusions*, American Mathematical Society, Graduate studies in mathematics, vol. 41 (2002).

[28] E. D. Sontag, *Mathematical control theory. Deterministic finite-dimensional systems.* Second edition. Springer-Verlag, New York, 1998.

[29] H. J. Sussmann, *On the gap between deterministic and stochastic ordinary differential equations*, Ann. Prob. **6** (1978), 17–41.

Alberto Bressan
Department of Mathematics
Penn State University
US-16802 University Park, Pa.
U.S.A.
e-mail: `bressan@math.psu.edu`

Progress in Nonlinear Differential Equations
and Their Applications, Vol. 75, 105–116
© 2007 Birkhäuser Verlag Basel/Switzerland

# Lipschitz Continuity of Optimal Trajectories in Deterministic Optimal Control

Piermarco Cannarsa, Helene Frankowska, and Elsa Maria Marchini

*Dedicated to Arrigo Cellina and James Yorke*

**Abstract.** This note concerns existence, Lipschitz continuity and Du Bois-Reymond necessary optimality condition for optimal trajectories of an autonomous Bolza problem in control theory. Our results relax the usual fast growth condition on the Lagrangian.

**Mathematics Subject Classification (2000).** 49J15, 49J30, 49K15, 49K30.

**Keywords.** optimal control, Bolza problem, existence of minimizers, Lipschitz optimal trajectory, Lipschitz co-state.

## 1. Introduction

The so-called Bolza problem in optimal control concerns minimization of the functional

$$J(x, u) = \int_0^T L\big(x(t), u(t)\big)dt + \ell\big(x(T)\big) \tag{1.1}$$

over all trajectory/control pairs $(x, u)$ subject to the state equation

$$\begin{cases} x'(t) = f\big(x(t), u(t)\big) & \text{for a.e. } t \in [0, T] \\ u(t) \in U & \text{for a.e. } t \in [0, T] \\ x(0) \in C_0 . \end{cases} \tag{1.2}$$

We propose here some new results on the basic aspects of this problem: existence of solutions, optimality conditions and regularity of optimal trajectories.

In literature, many papers have been devoted to these classical questions. In particular existence of solutions was derived by several authors under very mild

Work supported in part by European Community's Human Potential Programme under contract HPRN-CT-2002-00281, Evolution Equations. This research was completed in part while the first and third authors visited the CREA, École Polytechnique, Paris and also while the second author visited the Dipartimento di Matematica, Università di Roma "Tor Vergata".

regularity assumptions on the data, provided the Lagrangian $L$ is convex with respect to $u$, and has superlinear growth at $\infty$, i.e., $L(x, u) \geq \Phi(|u|)$, for some function $\Phi : \mathbb{R} \to \mathbb{R}$ satisfying $\lim_{r \to \infty} \frac{\Phi(r)}{r} = \infty$. In such a framework, the superlinearity of $L$ can be used to deduce both existence and Lipschitz regularity results by a direct method, see for instance Cesari [4], Clarke & Vinter [5], Frankowska & Marchini [7], Sarychev & Torres [8], and Torres [9]. However, in control theory there are some relevant examples which exhibit functionals with linear growth (such as the brachistochrone problem or the area functional for minimal surfaces of revolution), or even with no growth at all. For such functionals no general existence theory is available, to our knowledge. In fact, several counterexamples to the existence of solutions are known in the literature.

For problems in the calculus of variations Clarke proposed, in [5], an approach to the existence of solutions for Bolza problems based on Du Bois-Reymond necessary optimality condition. In his paper, which also applies to state constrained problems, existence is obtained imposing a separation property on the Hamiltonian $H(x, u, p) := \langle p, u \rangle - L(x, u)$ of the form : for some constant $k > 0$

$$\sup_{|x|, |u| \leq k,\, p \in \partial_u L(x, u)} H(x, u, p) < \liminf_{c \to +\infty} \inf_{|x| \leq k,\, |u| \geq c,\, p \in \partial_u L(x, u)} H(x, u, p) \quad (1.3)$$

where $\partial_u L$ denotes the subdifferential of $L$ with respect to $u$. For other ways to relax the superlinear growth condition as well as the convexity of $L$, see also Cellina [2] and Cellina & Ferriero [3].

We generalize here condition (1.3) to cover Bolza problems of type (1.1)–(1.2) in absence of state constraints. Our method applies to control problems with slow or no growth, as well as to superlinear growth of the Lagrangian, ensuring existence and Lipschitz continuity of optimal trajectories, and the Lipschitz continuity of the corresponding co-states.

Our main results stated here are provided without proofs, that are quite lengthly and technical and will appear elsewhere. Instead we give some applications to various problems of nonlinear optimal control.

## 2. Setting and notations

Let $T > 0$ and $U \subset \mathbb{R}^m$ be closed and convex. The following notations and assumptions will be in use throughout the paper:

- $I = [0, T]$;
- $W^{1,1}(I; \mathbb{R}^N)$ denotes the space of absolutely continuous functions from $I$ to $\mathbb{R}^N$ and $W^{1,\infty}(I; \mathbb{R}^N)$ the space of Lipschitz continuous functions from $I$ to $\mathbb{R}^N$;
- we define

$$\mathcal{U} := \{u : I \longrightarrow \mathbb{R}^m \text{ measurable} : u(t) \in U, \text{ for a.e. } t \in I\}.$$

- a pair $(x, u)$ where $x \in W^{1,1}(I; \mathbb{R}^N)$ and $u \in \mathcal{U}$ is called a trajectory/control pair if $(x, u)$ satisfies (1.2);

- for $a, b \in \mathbb{R}$, we set $a \wedge b := \min\{a, b\}$;
- for any $u \in U$, $N_U(u)$ denotes the normal cone to $U$ at $u$.

Consider some functions $L : \mathbb{R}^N \times \mathbb{R}^m \longrightarrow \mathbb{R}_+$, $\ell : \mathbb{R}^N \longrightarrow \mathbb{R}_+$, $f : \mathbb{R}^N \times \mathbb{R}^m \longrightarrow \mathbb{R}^N$ and let $C_0 \subset \mathbb{R}^N$ be closed, and, for every $x \in \mathbb{R}^N$, $L(x, \cdot)$ be convex and $f(x, \cdot)$ be differentiable. We define for all $x \in \mathbb{R}^N$,

$$F(x) := \left\{ \big( f(x, u), L(x, u) + v \big) : u \in U \text{ and } v \geq 0 \right\}.$$

In what follows we set $\inf \emptyset = +\infty$.

**Assumptions (H):**

i) for some $\alpha > 1$ and every $R > 0$, $\exists C_R > 0$ such that, $\forall x, y \in B(0, R)$ and $\forall u \in U$,

i1) $|\ell(x) - \ell(y)| \leq C_R |x - y|$,

i2) $|L(x, u) - L(y, u)| \leq C_R |x - y| \big[ 1 + L(x, u) \wedge L(y, u) \big]$,

i3) $|f(x, u) - f(y, u)| \leq C_R |x - y| \big[ 1 + |f(x, u)| \wedge |f(y, u)| + \big( L(x, u) \wedge L(y, u) \big)^{1/\alpha} \big]$;

ii) $\exists \overline{u} \in \mathcal{U}$ such that, for some $\eta \in L^\alpha(I; \mathbb{R})$, for a.e. $t \in I$, and for every $x \in \mathbb{R}^N$,

$$\left| f\big(x, \overline{u}(t)\big) \right| \leq \eta(t)\big(1 + |x|\big);$$

iii) for every $R > 0$, there exists $m_R \in L^1(I; \mathbb{R})$ such that, for a.e. $t \in I$,

$$\sup_{x \in B(0,R)} L\big(x, \overline{u}(t)\big) \leq m_R(t);$$

iv) for every $x \in \mathbb{R}^N$, $F(x)$ is closed and convex.

**Assumptions (H'):** is the same as Assumptions (H) with i3) replaced by

i3') $|f(x, u) - f(y, u)| \leq C_R |x - y| \Big[ 1 + \big( L(x, u) \wedge L(y, u) \big)^{1/\alpha} \Big]$.

**Remark 2.1.** Assumptions (H') imply Assumptions (H). So, every result obtained assuming (H) still holds true under (H').

**Remark 2.2.** From (H) ii), iii) it follows that the infimum for problem (1.1)-(1.2) is finite.

## 3. The main results

The theorems contained in this section ensure existence and Lipschitzianity of solutions for the Bolza problem introduced in Section 1, the Lipschitzianity of the corresponding co-states and the validity of the Du Bois-Reymond necessary optimality condition. For $(x, u, p) \in \mathbb{R}^N \times U \times \mathbb{R}^N$, set

$$P(x, u) = \left\{ p \in \mathbb{R}^N : f_u^*(x, u)p \in \partial_u L(x, u) + N_U(u) \right\},$$

where $\partial_u L(x, u)$ denotes the subdifferential of convex analysis of $L(x, \cdot)$ at $u$, and $f_u^*(x, u)$ the adjoint of $f_u(x, u)$;. Define

$$H(x, u, p) = \big\langle p, f(x, u) \big\rangle - L(x, u).$$

**Theorem 3.1.** *Assume (H). Suppose, for some $k > 0$ and some trajectory/control pair $(x_1, u_1)$ such that $x_1' \in L^\alpha(I; \mathbb{R}^N)$ and $+\infty > J(x_1, u_1) > \inf J(x, u)$ over all trajectory/control pairs $(x, u)$ satisfying (1.2), the following holds true:*

1) *for any trajectory/control pair $(x, u)$ satisfying $J(x, u) < J(x_1, u_1)$,*

$$\|x\|_\infty \leq k \quad and \quad \operatorname{essinf}_{t \in I} |f(x(t), u(t))| < k \,;$$

2) $$\sup_{\substack{|x| \leq k, |f(x, u)| \leq k \\ p \in P(x, u)}} H(x, u, p) < \liminf_{c \to +\infty} \inf_{\substack{|x| \leq k, |f(x, u)| \geq c \\ p \in P(x, u)}} H(x, u, p) \,.$$

*Then,*

a) *problem (1.1)–(1.2) has an optimal solution $(x^*, u^*)$ such that $x^*$ is Lipschitzian and $L(x^*(\cdot), u^*(\cdot)) \in L^\infty(I; \mathbb{R}^N)$;*

b) *the Du Bois-Reymond necessary optimality condition holds true: there exist $c \in \mathbb{R}$ and $p \in W^{1,\infty}(I; \mathbb{R}^N)$ such that, for a.e. $t \in I$,*

$$c = \Big\langle p(t), f(x^*(t), u^*(t)) \Big\rangle - L(x^*(t), u^*(t)) \,.$$

The next theorem has the same conclusions but a less restrictive assumption 2). Set, for $\nu > 0$,

$$P_\nu(x, u) = P(x, u) \cap \overline{B(0, \nu)} \,.$$

**Theorem 3.2.** *Assume (H '). Under the hypotheses of Theorem 3.1, with 2) replaced by*

2') *there exists $\nu_0 > 0$ such that, for every $\nu \geq \nu_0$,*

$$\sup_{\substack{|x| \leq k, |f(x, u)| \leq k \\ p \in P_\nu(x, u)}} H(x, u, p) < \liminf_{c \to +\infty} \inf_{\substack{|x| \leq k, |f(x, u)| \geq c \\ p \in P_\nu(x, u)}} H(x, u, p) \,,$$

*the same conclusions a) and b) of Theorem 3.1 hold true.*

Proofs of the above two theorems are based on penalization and are very technical. They can be found in Cannarsa, Frankowska and Marchini [1].

**Remark 3.3.** In Theorem 3.1, replace the assumptions on $(x_1, u_1)$ and assumption 1) with the following hypotheses, already appeared in Clarke [5] in the context of the calculus of variations:

$$(x_1, u_1) \text{ satisfies } x_1' \in L^\infty(I; \mathbb{R}^N) \text{ and } J(x_1, u_1) < +\infty \,;$$

1') for any trajectory/control pair $(x, u)$ satisfying $J(x, u) \leq J(x_1, u_1)$,

$$\|x\|_\infty \leq k \quad and \quad \operatorname{essinf}_{t \in I} |f(x(t), u(t))| < k \,.$$

Then:

a') problem (1.1)–(1.2) has an optimal solution $(x^*, u^*)$ such that $x^*$ is Lipschitzian;

b') there exist $c \in \mathbb{R}$ and $p \in W^{1,1}(I; \mathbb{R}^N)$ such that, for a.e. $t \in I$,

$$c = \Big\langle p(t), f(x^*(t), u^*(t)) \Big\rangle - L(x^*(t), u^*(t)) \,.$$

Indeed, if $(x_1, u_1)$ is optimal then a') holds true taking $(x^*, u^*) = (x_1, u_1)$. The validity of b') follows from the Pontryagin Maximum Principle, see Vinter [10]. If $J(x_1, u_1) > \inf \{ J(x, u) : (x, u) \text{ solves } (1.2) \}$, then a proof similar to the one of Theorem 3.1 implies a') and b').

## 4. Examples

We provide next some examples of problems satisfying assumptions of theorems 3.1 and 3.2. So, they admit an optimal solution $(x^*, u^*)$ satisfying the Du Bois-Reymond necessary optimality condition and such that $x^*$ is Lipschitzian, and $L(x^*(\cdot), u^*(\cdot))$ is essentially bounded. Problems exhibiting functionals with linear growth, or even with no growth at all, are considered.

In this section $U$ will always be a closed convex cone in $\mathbb{R}^m$. Observe that in this case:

$$\text{for every } u \in U \text{ and every } n \in N_U(u), \quad \langle n, u \rangle = 0. \tag{4.1}$$

**Example 4.1. Superlinear growth with respect to dynamics**
Consider the problem of minimizing the functional

$$J(x, u) = \int_0^T L\big(x(t), u(t)\big) dt + \ell\big(x(T)\big) \tag{4.2}$$

over all pairs $(x, u)$ satisfying

$$\begin{cases} x'(t) = f\big(x(t)\big) + Bu(t) & \text{for a.e. } t \in I \\ u(t) \in U & \text{for a.e. } t \in I \\ x(0) = x_0, \end{cases} \tag{4.3}$$

where $I = [0, T]$, $L, \ell \geq 0$, $L(x, \cdot)$ is convex, $\forall x \in \mathbb{R}^N$, $x_0 \in \mathbb{R}^N$ is fixed, and $B$ is an $n \times m$ matrix. Suppose that $f$ is globally Lipschitz with Lipschitz constant $C_f$, and that, $\forall R > 0$, $\exists C_R > 0$ such that, for every $x, y \in B(0, R)$ and every $u \in U$,

$$|L(x, u) - L(y, u)| \leq C_R |x - y| \big[ 1 + L(x, u) \wedge L(y, u) \big], \quad |\ell(x) - \ell(y)| \leq C_R |x - y|.$$

Furthermore, assume a function $\Phi : \mathbb{R} \longrightarrow \mathbb{R}$

$$\lim_{r \to \infty} \frac{\Phi(r)}{r} = \infty \quad \text{and} \quad L(x, u) \geq \Phi\big(|f(x) + Bu|\big), \quad \forall (x, u) \in \mathbb{R}^N \times \mathbb{R}^m. \tag{4.4}$$

Consider the trajectory/control pair $(\overline{x}, \overline{u})$, with $\overline{u} \equiv 0$. If $J(\overline{x}, \overline{u}) = \inf \{ J(x, u) : (x, u) \text{ solves } (4.3) \}$, then $(\overline{x}, \overline{u})$ is the optimal solution we are looking for: by the assumptions on $L$ and $f$, $\overline{x}$ is Lipschitzian, $L(\overline{x}(\cdot), \overline{u}(\cdot))$ is essentially bounded, and the Du Bois-Reymond necessary optimality condition are satisfied, see Vinter [10].

Otherwise, we prove that the assumptions of Theorem 3.2 hold true taking $(x_1, u_1) = (\overline{x}, \overline{u})$. Indeed, to verify (H') notice that: i1), i2), i3') follow immediately from the assumptions on $L, \ell, f$; ii) follows from the Lipschitzianity of $f$; to prove iii) notice that the choice of $\overline{u}$ and the continuity of $L(\cdot, u)$ imply that, $\forall R > 0$, $L(\cdot, 0)$ is bounded on $B(0, R)$; iv) comes from the fact $L(x, \cdot)$ is convex and the dynamic is affine in the control. Using the superlinear growth and the

Lipschitzianity of $f$, it is easy to verify assumption 1) of Theorem 3.1. Indeed, let $(x_1, u_1) = (\overline{x}, \overline{u})$ and let $(x, u)$ be a trajectory/control pair of (4.3) such that $J(x, u) < J(\overline{x}, \overline{u})$ and let $\gamma > 0$ be such that,

$$|f(x) + Bu| \leq \Phi(|f(x) + Bu|), \quad \text{as } |f(x) + Bu| \geq \gamma.$$

Then, from (4.4),

$$\int_0^T |f(x(t)) + Bu(t)| dt \leq T\gamma + \int_0^T \Phi(|f(x(t)) + Bu(t)|) dt$$

$$\leq T\gamma + \int_0^T L(x(t), u(t)) dt + \ell(x(T)) < T\gamma + J(\overline{x}, 0),$$

so that, for some $k > 0$, $\|x\|_\infty \leq k$ and $\text{essinf}_{t \in I} |f(x(t)) + Bu(t)| < k$.

To show 2 '), notice that if $p \in P(x, u)$, then $B^* p \in \partial_u L(x, u) + N_U(u)$. Hence, if $|p| \leq \nu$,

$$\Phi(|f(x) + Bu|) \leq L(x, u) \leq L(x, 0) + \langle B^* p, u \rangle = L(x, 0) + \langle p, Bu \rangle$$

$$= L(x, 0) + \langle p, f(x) + Bu \rangle - \langle p, f(x) \rangle \leq L(x, 0) + \langle p, f(x) + Bu \rangle + \nu \sup_{|x| \leq k} |f(x)|.$$

It follows that, when $c$ is large enough, the set

$$\mathcal{C}_c = \{(x, u, p) : |x| \leq k, |f(x) + Bu| \geq c, p \in P(x, u), |p| \leq \nu\} \qquad (4.5)$$

is empty. So $\inf_{\mathcal{C}_c} H(x, u, p) = +\infty$ and 2 ') follows.

## Example 4.2. Superlinear growth with respect to control
Consider problem (4.2)-(4.3) introduced in the previous example. Assume that there exists a function $\Phi : \mathbb{R} \longrightarrow \mathbb{R}$ satisfying

$$\lim_{r \to +\infty} \frac{\Phi(r)}{r} = +\infty \quad \text{and} \quad L(x, u) \geq \Phi(|u|), \quad \forall (x, u) \in \mathbb{R}^N \times \mathbb{R}^m.$$

Consider the trajectory/control pair $(\overline{x}, \overline{u})$, where $\overline{u} \equiv 0$. If $J(\overline{x}, \overline{u}) = \inf \{J(x, u) : (x, u) \text{ solves (4.3)}\}$, then $(\overline{x}, \overline{u})$ is the optimal solution we are looking for.

Otherwise, we prove that the assumptions of Theorem 3.2 hold true. The validity of (H') is proved arguing as in the previous example. To prove 1) of Theorem 3.1, set $(x_1, u_1) = (\overline{x}, \overline{u})$ and consider a trajectory/control pair $(x, u)$ satisfying (4.3) and $J(x, u) < J(\overline{x}, \overline{u})$, and let $\gamma > 0$ be such that,

$$|u| \leq \Phi(|u|), \quad \text{as } |u| \geq \gamma.$$

Then, from the superlinear growth of the Lagrangian,

$$\int_0^T |u(t)| dt \leq T\gamma + \int_0^T \Phi(|u(t)|) dt$$

$$\leq T\gamma + \int_0^T L(x(t), u(t)) dt + \ell(x(T)) < T\gamma + J(\overline{x}, \overline{u}).$$

So, since, for a.e. $t \in I$, $|x'(t)| \leq C_f|x(t)| + |f(0)| + \|B\||u(t)|$, applying Gronwall's Lemma we obtain that, for some $k > 0$, $\|x\|_\infty \leq k$ and $\operatorname{essinf}_{t \in I}|f(x(t)) + Bu(t)| < k$.

To show 2'), notice that if $p \in P(x, u)$, then $B^*p \in \partial_u L(x, u) + N_U(u)$. Hence, if $|p| \leq \nu$,

$$\Phi(|u|) \leq L(x, u) \leq L(x, 0) + \langle B^*p, u \rangle \leq L(x, 0) + \nu\|B\||u| .$$

Defining $\mathcal{C}_c$ as in (4.5), we obtain that, for $c$ large, $\mathcal{C}_c = \emptyset$, and again 2') holds true.

**Example 4.3. $\alpha$-growth with respect to control**
Consider the problem of minimizing the functional

$$J(x, u) = \int_0^T L(x(t), u(t))dt + \ell(x(T))$$

over all measurable $u$ and absolutely continuous $x$ satisfying

$$\begin{cases} x'(t) = f(x(t)) + g(x(t))u(t) & \text{for a.e. } t \in I \\ u(t) \in U & \text{for a.e. } t \in I \\ x(0) = x_0 , \end{cases} \quad (4.6)$$

where $I = [0, T]$, $L, \ell \geq 0$, $L(x, \cdot)$ is convex, $\forall x \in \mathbb{R}^N$, $x_0 \in \mathbb{R}^N$ is fixed, and $f : \mathbb{R}^N \to \mathbb{R}^N, g : \mathbb{R}^N \to \mathbb{R}^m$ are globally Lipschitz with Lipschitz constants $C_f$ and $C_g$ respectively. Suppose that, $\forall R > 0$, $\exists C_R > 0$ such that

$$|L(x, u) - L(y, u)| \leq C_R|x - y|[1 + L(x, u) \wedge L(y, u)] , \quad |\ell(x) - \ell(y)| \leq C_R|x - y| ,$$

$\forall x, y \in B(0, R)$ and $\forall u \in U$. Assume that there exist $\beta > 0$ and $\alpha > 1$ such that

$$L(x, u) \geq \beta|u|^\alpha \quad \text{for all } (x, u) \in \mathbb{R}^N \times \mathbb{R}^m .$$

Arguing as in the previous examples, if the trajectory/control pair $(\overline{x}, \overline{u})$, where $\overline{u} \equiv 0$, satisfies $J(\overline{x}, \overline{u}) = \inf\{J(x, u) : (x, u) \text{ solves } (4.6)\}$, then $(\overline{x}, \overline{u})$ is the optimal solution we are looking for.

Otherwise, the assumptions of Theorem 3.2 hold true. Indeed, to verify (H') notice that: i1) and i2) hold true since $L, \ell$ are locally Lipschitz and, for every $x, y \in \mathbb{R}^N$, we have

$$|f(x) + g(x)u - f(y) - g(y)u| \leq |f(x) - f(y)| + |[g(x) - g(y)]u|$$

$$\leq C_f|x - y| + C_g|x - y||u| \leq |x - y|\left[C_f + \frac{C_g}{\beta^{1/\alpha}}(L(x, u) \wedge L(y, u))^{1/\alpha}\right]$$

$$\leq C|x - y|\left[1 + (L(x, u) \wedge L(y, u))^{1/\alpha}\right] ,$$

where $C = \max\left\{C_f, \frac{C_g}{\beta^{1/\alpha}}\right\}$. So, i3') follows. To verify ii), iii) and iv) we proceed as in the first example. Setting $(x_1, u_1) = (\overline{x}, \overline{u})$, arguing as in the previous examples, and taking into account the global Lipschitzianity of $g$, we obtain the validity of Assumption 1) of Theorem 3.2. To show 2'), notice that if $p \in P_\nu(x, u)$, then $g(x)^* p \in \partial_u L(x, u) + N_U(u)$. Hence,

$$\beta |u|^\alpha \leq L(x, u) \leq L(x, 0) + \langle g(x)^* p, u \rangle \leq L(x, 0) + \nu \|g(x)\| \|u\|.$$

It follows that, when $c$ is large enough, the set $\mathcal{C}_c$ defined as in (4.5) with $B$ replaced by $g(x)$ is empty. So, 2') follows.

**Remark 4.4.** The problems of the previous examples also satisfy the assumptions of Theorem 3.1 under the hypothesis that $f$ and $g$ are locally Lipschitz, jointly with some growth assumption as, for instance,

$$\exists\, a > 0 \text{ such that, } \forall x \in \mathbb{R}^N, \quad |f(x)| + |g(x)| \leq a(|x| + 1)$$

instead of globally Lipschitz.

**Example 4.5. Slow growth**
Consider the problem of minimizing the functional

$$J(x, y, u, v) = \int_0^1 \sqrt{1 + u(t)^2 + v(t)^2}\,dt + x(1)^2 + y(1)^2$$

over all measurable $(u, v)$ and absolutely continuous $(x, y)$ satisfying

$$\begin{cases} x'(t) = x(t)u(t) + y(t)v(t) & \text{for a.e. } t \in I \\ y'(t) = u(t) + v(t) & \text{for a.e. } t \in I \\ u(t), v(t) \in [0, +\infty) & \text{for a.e. } t \in I \\ x(0) = x_0 \in \mathbb{R}, \, y(0) = y_0 \geq 0, \end{cases} \tag{4.7}$$

where $I = [0, 1]$. We first check that the assumptions of Theorem 3.1 are satisfied. Assumptions (H) hold true. Indeed, to prove i) notice that $L$ is independent on the trajectory, and, since $u, v \geq 0$, for every $(x, y)$,

$$|\partial_{(x,y)} f(x, y, u, v)| = \sqrt{u^2 + v^2} \leq \sqrt{(xu + yv)^2 + (u + v)^2} = |f(x, y, u, v)|.$$

Taking $\bar{u} \equiv \bar{v} \equiv 0$ and arguing as in the previous examples, we deduce the validity of ii), iii) and iv). We wish to check Assumptions 1) and 2) of Theorem 3.1. Set $(u_1, v_1) \equiv (1, 0)$, so that, $\forall t \in I$, $(x_1(t), y_1(t)) = (x_0 e^t, y_0 + t)$ and $J(x_1, y_1, u_1, v_1) = \sqrt{2} + x_0^2 e^2 + (y_0 + 1)^2 > 1 + x_0^2 + y_0^2 = J(\bar{x}, \bar{y}, \bar{u}, \bar{v})$. Let $(x, y, u, v)$ be a trajectory/control pair such that

$$\int_0^1 \sqrt{1 + u(t)^2 + v(t)^2}\,dt + x(1)^2 + y(1)^2 < \sqrt{2} + x_0^2 e^2 + (y_0 + 1)^2 = k_0. \tag{4.8}$$

Since $y_0 \geq 0$ and $u, v \geq 0$, $y$ is non negative and non decreasing, so, by (4.8)

$$\|y\|_\infty \leq y(1) < \sqrt{k_0}.$$

Moreover, since, for a.e. $t \in I$, $|x'(t)| \leq |x(t)|u(t) + y(t)v(t)$, from Gronwall's Lemma, (4.8), and the estimate on $\|y\|_\infty$, we obtain that

$$\|x\|_\infty \leq e^{\int_0^1 u(t)dt} \left[ |x_0| + \int_0^1 y(t)v(t)dt \right] \leq e^{\|y\|_\infty} \left[ |x_0| + \|y\|_\infty \int_0^1 v(t)dt \right]$$

$$\leq e^{\|y\|_\infty} \left( |x_0| + \|y\|_\infty^2 \right) < e^{\sqrt{k_0}} \left( \sqrt{k_0} + k_0 \right) = k_1.$$

So,

$$\|(x,y)\|_\infty < \sqrt{k_1^2 + k_0}. \tag{4.9}$$

From (4.9),

$$\int_0^1 \left| f(x(t), y(t), u(t), v(t)) \right| dt \leq \int_0^1 \sqrt{\left[ (x(t) + y(t))^2 + 1 \right] (u(t) + v(t))^2} dt$$

$$\leq \|y\|_\infty \sqrt{2\|(x,y)\|_\infty^2 + 1} < \sqrt{k_0} \sqrt{2(k_1^2 + k_0) + 1}.$$

Taking $k = \max\{\sqrt{k_1^2 + k_0}, \sqrt{k_0}\sqrt{2(k_1^2 + k_0) + 1}\}$, 1) holds. To prove 2), observe first that, setting

$$f_0(x,y) = \begin{bmatrix} x & y \\ 1 & 1 \end{bmatrix}, \tag{4.10}$$

we have that, for every $(x, y, u, v) \in \mathbb{R}^2 \times U$, with $U = \{(u,v) \in \mathbb{R}^2 : u, v \geq 0\}$,

$$f(x, y, u, v) = f_0(x,y) \begin{bmatrix} u \\ v \end{bmatrix} \quad \text{and} \quad \partial_{(u,v)} f(x, y, u, v) = f_0(x,y). \tag{4.11}$$

So, if $p \in P(x, y, u, v)$ then

$$f_0^*(x,y)p \in \partial_{(u,v)} L(u,v) + N_U(u,v),$$

and, since $U$ is a cone, from (4.1) and (4.11) we obtain that

$$H(x, y, u, v, p) = \langle p, f(x, y, u, v) \rangle - L(u,v) = \langle f_0^*(x,y)p, (u,v) \rangle - L(u,v)$$

$$= \langle \partial_{(u,v)} L(u,v), (u,v) \rangle - L(u,v)$$

$$= \frac{1}{\sqrt{1 + u^2 + v^2}} |(u,v)|^2 - \sqrt{1 + u^2 + v^2}$$

$$= -\frac{1}{\sqrt{1 + u^2 + v^2}}.$$

Since, $\forall u, v \geq 0$, $u^2 + v^2 \leq |f(x, y, u, v)|^2$ and the function $s \mapsto -1/(1+s)$ is increasing,

$$\sup_{\substack{|(x,y)| \leq k, |f(x,y,u,v)| \leq k \\ p \in P(x,y,u,v)}} H(x, y, u, v, p) \leq \sup_{\substack{|(x,y)| \leq k \\ |f(x,y,u,v)| \leq k}} \frac{-1}{\sqrt{1 + u^2 + v^2}}$$

$$\leq \frac{-1}{1 + k^2}.$$

Moreover, since $|f(x, y, u, v)|^2 \leq \left[(x+y)^2 + 1\right](u+v)^2 \leq (2|(x,y)|^2 + 1)2(u^2 + v^2)$, $\forall c > 0$, such that

$$\inf_{\substack{|(x,y)| \leq k, |f(x,y,u,v)| \geq c \\ p \in P(x,y,u,v)}} H(x, y, u, v, p) \geq \inf_{\substack{|(x,y)| \leq k \\ |f(x,y,u,v)| \geq c}} \frac{-1}{\sqrt{1 + u^2 + v^2}}$$

$$\geq \frac{-1}{\sqrt{1 + c^2/(4k^2 + 2)}}.$$

Since $\dfrac{-1}{\sqrt{1 + c^2/(4k^2+2)}} \to 0$, as $c \to +\infty$, for $c$ sufficiently large we obtain the desired inequality. So, also 2) holds.

**Example 4.6. No growth**
Consider the problem of minimizing the functional

$$J(x, u) = \int_0^1 e^{-[u(t) + 2v(t)]} dt + x(1)^2 + y(1)^2$$

over all measurable $(u, v)$ and absolutely continuous $(x, y)$ satisfying (4.7). We prove that the assumptions of Theorem 3.1 are satisfied. Taking $\bar{u} \equiv \bar{v} \equiv 0$ and arguing as in the previous example we deduce the validity of (H). Set $(u_1, v_1) \equiv (1, 0)$, so that, $\forall t \in I$, $(x_1(t), y_1(t)) = (x_0 e^t, y_0 + t)$ and $J(x_1, y_1, u_1, v_1) = e^{-1} + x_0^2 e^2 + (y_0 + 1)^2 > 1 + x_0^2 + y_0^2 = J(\bar{x}, \bar{y}, \bar{u}, \bar{v})$. Let $(x, y, u, v)$ be a trajectory/control pair such that

$$\int_0^1 e^{-[u(t) + 2v(t)]} dt + x(1)^2 + y(1)^2 < e^{-1} + x_0^2 e^2 + (y_0 + 1)^2 = k_0. \qquad (4.12)$$

Exactly as in Example 4.5 we check that 1) of Theorem 3.1 holds true. We deduce from (4.10) and (4.11) that, if $p \in P(x, y, u, v)$, then $f_0^*(x, y)p \in \partial_{(u,v)} L(u, v) + N_U(u, v)$, and, since $U$ is a cone, from (4.1) and (4.11) we obtain that

$$H(x, y, u, v, p) = \langle p, f(x, y, u, v) \rangle - L(u, v) = \langle f_0^*(x, y)p, (u, v) \rangle - L(u, v)$$

$$= \langle \partial_{(u,v)} L(u, v), (u, v) \rangle - L(u, v) = -e^{-(u+2v)}(u + 2v) - e^{-(u+2v)}$$

$$= -e^{-(u+2v)}(1 + u + 2v).$$

Since, $\forall u, v \geq 0$, $u + 2v \leq 2|f(x, y, u, v)|$ and the function $s \mapsto \dfrac{-(1+s)}{e^s}$ is increasing,

$$\sup_{\substack{|(x,y)| \leq k, |f(x,y,u,v)| \leq k \\ p \in P(x,y,u,v)}} H(x, y, u, v, p) \leq \sup_{\substack{|(x,y)| \leq k \\ |f(x,y,u,v)| \leq k}} -\frac{(1 + u + 2v)}{e^{(u+2v)}}$$

$$\leq -\frac{(1 + 2k)}{e^{2k}}.$$

Moreover, since, $\forall u, v \geq 0$, $|f(x, y, u, v)| \leq 2(u + 2v)\sqrt{1 + 2|(x, y)|^2}$, we have $\forall c > 0$

$$\inf_{\substack{|(x,y)| \leq k, |f(x,y,u,v)| \geq c \\ p \in P(x,y,u,v)}} H(x, y, u, v, p) \geq \inf_{\substack{|(x,y)| \leq k \\ |f(x,y,u,v)| \geq c}} -\frac{(1 + u + 2v)}{e^{(u+2v)}}$$

$$\geq -\frac{1 + \frac{c}{2\sqrt{1+2k^2}}}{e^{\frac{c}{2\sqrt{1+2k^2}}}}.$$

Using that $-\frac{1 + c/(2\sqrt{1+2k^2})}{e^{c/(2\sqrt{1+2k^2})}} \to 0$, as $c \to +\infty$, for $c$ sufficiently large we obtain the desired inequality. So, also 2) holds true.

## References

[1] P. Cannarsa, H. Frankowska and E. M. Marchini, *Existence and Lipschitz regularity of solutions to Bolza problems in optimal control*, submitted 2006.

[2] A. Cellina, *The classical problem of the calculus of variations in the autonomous case: relaxation and lipschitzianity of solutions*, Trans. Am. Math. Soc., **356** (1) (2004), 415–426.

[3] A. Cellina and A. Ferriero, *Existence of lipschitzian solutions to the classical problem of the calculus of variations in the autonomous case*, Ann. Inst. H. Poincaré Anal. Non Linéaire, **20** (6) (2003), 911–919.

[4] L. Cesari (1983), *Optimization, Theory and Applications*, Springer, Berlin.

[5] F. H. Clarke, *An indirect method in the calculus of variations*, Trans. Amer. Math. Soc., **336** (2) (1993), 655–673.

[6] F. H. Clarke and R. B. Vinter, *Regularity properties of optimal controls*, SIAM J. Control Optim., **28** (4) (1990), 980–997.

[7] H. Frankowska and E. M. Marchini, *Lipschitzianity of optimal trajectories for the Bolza optimal control problem*, Calc. Var. Partial Differential Equations, **27** (2006), 4, 467–492.

[8] A. Sarychev and D. F. M. Torres, *Lipschitzian regularity of the minimizers for optimal control problems with control-affine dynamics*, Appl. Math. Optim., **41** (2) (2000), 237–254.

[9] D. F. M. Torres, *Lipschitzian regularity of the minimizing trajectories for nonlinear optimal control problems*, Math. Control Signals Systems, **16** (2-3) (2003), 158–174.

[10] R. B. Vinter, *Optimal Control*, Birkhäuser, Boston, Basel, Berlin, 2000.

Piermarco Cannarsa
Dipartimento di Matematica
Università di Roma "Tor Vergata"
Via della Ricerca Scientifica 1
I-00133, Roma
Italy
e-mail: cannarsa@axp.uniroma2.it

Helene Frankowska
CREA, École Polytechnique
1 rue Descartes
FR-75005 Paris
France
e-mail: `helene.frankowska@shs.polytechnique.fr`

Elsa Maria Marchini
Dipartimento di Matematica e Applicazioni
Università di Milano-Bicocca
Via Cozzi 53
I-20125 Milano
Italy
e-mail: `elsa.marchini@unimib.it`

Progress in Nonlinear Differential Equations
and Their Applications, Vol. 75, 117–122
© 2007 Birkhäuser Verlag Basel/Switzerland

# An Overview on Existence of Vector Minimizers for Almost Convex $1-$dim Integrals

Clara Carlota and António Ornelas

*Dedicated to Arrigo Cellina and James Yorke*

**Abstract.** We present results of existence of minimizers for the nonconvex integral $\int_a^b L(x, x')\, dt$, among the $AC$ functions $x : [a, b] \to \mathbb{R}^n$ with $x(a) = A$, $x(b) = B$. Our lagrangian $L(\cdot)$ is, e.g., lsc with coercive growth, assuming $+\infty$ values freely. We replace convexity by almost convexity, a hypothesis which in the radial case $L(s, \xi) = f(s, |\xi|)$ is automatically satisfied provided $f(s, \cdot)$ has superlinear growth and is convex at zero.

**Mathematics Subject Classification (2000).** Primary 49J05; Secondary 49K05.

**Keywords.** Calculus of variations, nonconvex nonlinear integrals, regularity properties.

## 1. Introduction

Consider the problem of minimizing the integral

$$\int_a^b L\big(x(t), x'(t)\big)\, dt \qquad \text{on } \mathcal{X}_{A,B}^n, \tag{1.1}$$

where $\mathcal{X}_{A,B}^n$ is the class of $AC$ (absolutely continuous) functions $x : [a, b] \to \mathbb{R}^n$ satisfying boundary conditions $x(a) = A$, $x(b) = B$.

This paper was presented in the meeting "Views on ODEs", in honour of Arrigo Cellina and James Yorke on their $65^{th}$ birthday, Aveiro, June 2006; and in the meeting "Variational and Differential Problems with Constraints", in honour of Arrigo Cellina on his $65^{th}$ birthday, Venezia, September 2006. The research leading to this paper was performed at Cima-ue (Centro de Investigação em Matemática e Aplicações da Universidade de Évora), financially supported by the project POCTI/Mat/56727/2004 "Cálculo das variações : problemas escalares não-convexos e não-coercivos" and by the "Financiamento Plurianual do Cima-ue" both of Fundação para a Ciência e a Tecnologia, Portugal.

Whenever the lagrangian $L : \mathbb{R}^n \times \mathbb{R}^n \to [0, +\infty]$ is *lsc* (lower semicontinuous) and has $L(s, \cdot)$ convex with superlinear growth at infinity, i.e.,

$$L(s, \xi) \geq \theta(|\xi|) \ \forall (s, \xi) \qquad \text{with} \qquad \theta(r)/r \to +\infty \ as \ r \to +\infty, \qquad (1.2)$$

Tonelli's direct method may be applied to prove existence of minimizers for the integral (1.1) .

In case $L(s, \cdot)$ is nonconvex, results of existence of minimizers have been proved, for instance, by using the following strategy. Consider the bipolar $L^{**}(s, \cdot)$ of $L(s, \cdot)$ , so that $epi\, L^{**}(s, \cdot) = \overline{co}\ epi\ L(s, \cdot)$ . One starts by proving the existence of a relaxed minimizer $y_c(\cdot)$ , i.e., a minimizer of the convexified integral

$$\int_a^b L^{**}(x(t), x'(t))dt \qquad \text{on } \mathcal{X}_{A,B}^n . \qquad (1.3)$$

The second step is to transform $y_c(\cdot)$ into a new improved relaxed minimizer $y(\cdot)$ for which $L^{**}(y(t), y'(t)) = L(y(t), y'(t))$ a.e. on $[a, b]$ ; so that $y(\cdot)$ also minimizes the original, nonconvex, integral (1.1) . This strategy has been used in the vector case $n \geq 1$, e.g., by [3]; while in the case $n = 1$ it has been refined, as follows : one already starts with an improved $y_c(\cdot)$ , i.e., satisfying convenient regularity properties, see [1, 7–12]. (In [13], Lipschitz regularity was also proved, using quite weak hypotheses. )

The last authors dedicated these efforts to the scalar case because they succeeded to obtain better results in such special case by using the above strategy in combination with the hypothesis of 0–convexity, $L(\cdot, 0) = L^{**}(\cdot, 0)$, which turned out quite useful. Indeed, the option of minimizers to take or leave the velocity zero turned out to be quite an essential feature of these minimizing problems.

Another factor leading to the successive improvements in these scalar results has been a new technique, bimonotonicity, developed there.

But the results (obtained in [2]) which we present in this overview deal with $n > 1$; and since in this vector case the hypothesis of 0-convexity does not suffice to guarantee existence of minimizers, we have used instead almost convexity. This concept was born, for multifunctions, in the paper [4], to prove existence of solutions to nonconvex differential inclusions and to time-optimal control problems, using reparametrizations.

The first result we present here assumes superlinearity to ensure existence of a relaxed minimizer, which is then changed to become a true minimizer; while in the second result existence of a relaxed minimizer, $y_c(\cdot)$ , is used as one hypothesis, and we need no growth assumption to turn $y_c(\cdot)$ into a true minimizer. We also present applications of these results to show existence of true minimizers in concrete examples, not covered by previous results.

Two techniques have been combined in [2] to prove these results. The first is the above cited reparametrizations, used by A. Cellina and collaborators during the last decade : starting in [5], to prove existence of minimizers for integrals with convex noncoercive lagrangians ; and continuing in the above stated [4] and other references (which may be seen in [2]).

The second technique is the above mentioned bimonotonicity: [2] is the first paper in which it is applied to the reparametrization.

## 2. Almost convexity

**Definition 2.1.** *For a function  $L : \mathbb{R}^m \times \mathbb{R}^n \to (-\infty, +\infty]$ ,  we call  $L(s, \cdot)$  almost convex provided*

$$\forall \xi \text{ with } L^{**}(s, \xi) < L(s, \xi) \tag{2.1}$$

$$\exists \lambda \in [0, 1] \, \exists \Lambda \in [1, +\infty) \, \exists \alpha \in [0, 1] \quad for \ which \tag{2.2}$$

$$L^{**}(s, \xi) = (1 - \alpha) L(s, \lambda \xi) + \alpha L(s, \Lambda \xi) \tag{2.3}$$

$$\xi = (1 - \alpha)(\lambda \xi) + \alpha (\Lambda \xi) . \tag{2.4}$$

*For completeness, we also set  $\lambda = 1 = \Lambda = \alpha$  at those  $\xi$  where  $L^{**}(s, \xi) = L(s, \xi)$ , in particular at  $\xi = 0$ . (The convention $0 \cdot (+\infty) = 0$  is used.)*

Clearly  $L(s, \cdot)$  convex lsc  implies $L(s, \cdot)$  almost convex. Moreover  $L(s, \cdot)$  almost convex implies $L^{**}(s, 0) = L(s, 0)$ . But the opposite implication does not hold, even for simple  $2-$dim superlinear polynomials. Indeed, e.g.,  $L(s, \xi) := h(\xi) := \left( \xi_1^2 + \xi_2^2 \right) \left( \xi_1^2 - 1 \right)^2 + \xi_2^2$  satisfies  $h^{**}(0) = h(0) = 0$  but :

$$\exists \xi = (1/2\,,1) \, \exists \lambda = 0 \ \ \exists \Lambda = 2 \ \exists \alpha = 1/2 \text{ with}$$

$$\xi = (1 - \alpha)(\lambda \xi) + \alpha (\Lambda \xi) , \ h^{**}(\lambda \xi) = h(\lambda \xi) = 0, \ h^{**}(\Lambda \xi) = h(\Lambda \xi) = 4$$

$$h^{**}(\xi) = 1 < h(\xi) < (1 - \alpha) h(\lambda \xi) + \alpha h(\Lambda \xi) = 2 \tag{2.5}$$

( and :  $\lambda$  must be zero, while  2 is the best value of  $\Lambda$ , i.e., the one yielding the smallest  *rhs* in  (2.5)) ; moreover, even though  $\ell(\cdot)$  is superlinear,

$$\exists \xi = (0, 1) : \ h^{**}(\Lambda \xi) < h(\Lambda \xi) \, \forall \Lambda \geq 1 . \tag{2.6}$$

Indeed,  $h^{**}(\xi_1, \xi_2) = \xi_2^2 \ \forall \ |\xi_1| \leq 1 \ \forall \xi_2$ .

Typical examples of almost convex functions may be obtained by increasing arbitrarily (e.g., to become  $= +\infty$ ) the values of any given  $L(\cdot)$ , satisfying  $L(\cdot) = L^{**}(\cdot)$ , as follows. Denote by  $\widehat{F}(s)$  the vertical projection into  $\mathbb{R}^n$  of any face (relatively open) face  $F(s)$  of  $epi \, L^{**}(s, \cdot)$ . Then one may change  $L(s, \xi)$  by increasing it, starting from the value  $L^{**}(s, \xi)$ , at those  $\xi \neq 0$  contained in any open bounded subset of any  $k-$dim  $\widehat{F}(s)$  which is contained in a  $k-$dim  *linear* subspace of  $\mathbb{R}^n$ ,  $1 \leq k \leq n$ .

Notice also that for $L(s, \cdot) : \mathbb{R}^n \to [0, +\infty]$  *lsc* superlinear we do have

$$L^{**}(s, 0) = L(s, 0) \qquad \Rightarrow \qquad L(s, \cdot) \ almost \ convex \tag{2.7}$$

in the scalar $n = 1$ or radial $L(s, \xi) = f(s, |\xi|)$ case. Here superlinearity is really not needed: it suffices to have boundedness of the nonconvexity faces (i.e., of the subset of each  $\widehat{F}(s)$  where  $f^{**}(s, \cdot) < f(s, \cdot)$ ).

## 3. Existence of minimizers

**Theorem 3.1.** *Let  $L : \mathbb{R}^n \times \mathbb{R}^n \to [0, +\infty]$  be a lsc superlinear function, so that for any  $A$ ,  $B$   there exists a minimizer  $y_c (\cdot)$  for the convexified integral  $(1.3)$ . Assume  $L(y_c(t), \cdot)$  to be almost convex  $\forall t \in [a, b]$  .*

*Then the nonconvex integral  $(1.1)$   has minimizers.*

*Moreover, the integrals  $(1.1)$  and  $(1.3)$   have the same minimum value ; and there exists a minimizer  $y(\cdot)$  which satisfies the following regularity:*

(i)  $y(\cdot) \equiv s'$   along some subinterval  $(a', b')$ ,   $a' \le b'$ ;
(ii)  $y'(t) \ne 0$  a.e. in  $[a, a'] \cup [b', b]$ ;
(iii)  $L(s', 0) = L^{**}(s', 0) = \min L^{**}(y([a, b]), 0)$ ;
(iv)  $L^{**}(y(t), y'(t)) = L(y(t), y'(t))$  a.e. in  $[a, b]$ .

**Corollary 3.2.** *Let  $f : \mathbb{R}^n \times [0, +\infty) \to [0, +\infty]$  be a lsc function satisfying  $f(s, r) \ge \theta(r)$  with  $\theta(r) / r \to +\infty$  as  $r \to \infty$ . Denote by  $f^{**}(s, \cdot)$  the bipolar of the function  $\mathbb{R}^n \ni \xi \mapsto f(s, |\xi|)$  and assume  $f(s, 0) = f^{**}(s, 0) \forall s$ .*

*Then for any  $A$ ,  $B$  the nonconvex integral*

$$\int_a^b f(x(t), |x'(t)|) dt \qquad on \ \mathcal{X}^n_{A,B}$$

*has minimizers.*

**Theorem 3.3.** *Let  $L : \mathbb{R}^n \times \mathbb{R}^n \to [0, +\infty]$  be a Borel measurable function, with  $L(\cdot, 0)$  lsc ; and  $L^{**}(\cdot)$  also Borel measurable. Fix  $A$ ,  $B \in \mathbb{R}^n$   and assume existence of a minimizer  $y_c(\cdot)$  for the convexified integral  $(1.3)$ . Assume moreover  $L(y_c(t), \cdot)$  to be lsc almost convex  $\forall t \in [a, b]$ .*

*Then there exists a minimizer  $y(\cdot)$  for the nonconvex integral  $(1.1)$ . Moreover, the integrals  $(1.1)$  and  $(1.3)$  have the same minimum value; and  $y(\cdot)$  satisfies the regularity properties  $(i) - (iv)$  of theorem  $3.1$ .*

Theorem 3.1 ensures the existence of a minimizer for the nonconvex integral (1.1) when  $L : \mathbb{R}^n \times \mathbb{R}^n \to [0, +\infty]$  is, e.g.,

$$L(s, \xi) = \begin{cases} |s - s_0|^2 + \left(|\xi|^2 - \gamma^2\right)^2 & for \ \xi \ne 0 \\ |s - s_0|^2 & for \ \xi = 0 . \end{cases}$$

As to Theorem 3.3, it ensures existence of a minimizer, e.g., in case  $L : \mathbb{R}^n \times \mathbb{R}^n \to [0, +\infty]$  has the form

$$L(s, \xi) = g(s) f(\xi)$$

with  $f : \mathbb{R}^n \to [0, +\infty]$ ,

$$f(\xi) = \begin{cases} \left(1 + |\xi|^2\right)^{\frac{1}{2}} & for \ |\xi| \ge 1 \\ +\infty & for \ 0 < |\xi| < 1 \\ \sqrt{2} & for \ \xi = 0, \end{cases}$$

and  $g : \mathbb{R}^n \to [1, +\infty)$    is a lsc function, locally bounded. (Indeed [6, example 2.4.2] proved existence of a relaxed minimizer in this case.)

Finally, to see a simple  2−dim example where convexity at zero does not imply existence, let  $h(\xi) = (\xi_1^2 + \xi_2^2) (\xi_1^2 - 1)^2 + \xi_2^2$, and  $L(s, \xi) = s_1^2 + h(\xi)$, $s = (s_1, s_2)$, $a = 0, A = (0, 0)$,  $b = 1$,  $B = (0, 1)$. Clearly $y_c(t) = (0, t)$ is a relaxed minimizer, giving the value  1 to the integral (1.3). However, as one easily checks, to satisfy the boundary conditions the value of the nonconvex integral (1.1) must always be  $> 1$ (while the *inf* is, clearly, $= 1$).

# References

[1] M. Amar and A. Cellina, *On passing to the limit for nonconvex variational problems* Asymp. Anal. **9** (1994), 135–148.

[2] C. Carlota and A. Ornelas, *Existence of vector minimizers for nonconvex 1-dim integrals with almost convex lagrangian* J. Diff. Equations (2007), doi:10.1016/j.jde.2007.05.019.

[3] A. Cellina and G. Colombo, *On a classical problem of the calculus of variations without convexity assumptions* Ann. Inst. Henri Poincaré **7** (1990), 97–106.

[4] A. Cellina and A. Ornelas, *Existence of solutions to differential inclusions and to time optimal control problems in the autonomous case* SIAM J. Control. Optim. **42** (2003), 260–265.

[5] A. Cellina, G. Treu and S. Zagatti, *On the minimum problem for a class of non-coercive functionals* J. Diff. Equations **127** (1996), 225–262.

[6] F. Clarke, *An indirect method in the calculus of variations* Trans. Amer. Math. Soc. **336** (1993), 655–673.

[7] N. Fusco, P. Marcellini and A. Ornelas, *Existence of minimizers for some nonconvex one-dimensional integrals* Portugaliae Mathematica **52** (2) (1998), 167–184.

[8] A. Ornelas, *Existence of scalar minimizers for nonconvex simple integrals of sum type* J. Math. Anal. Appl. **221** (1998), 559–573.

[9] A. Ornelas, *Existence and regularity for scalar minimizers of affine nonconvex simple integrals* Nonlin. Anal. **53** (2003), 441–451.

[10] A. Ornelas, *Existence of scalar minimizers for simple convex integrals with autonomous lagrangian measurable on the state variable* Nonlin. Anal. (2006), doi:10.1016/j.na.2006.08.044, preprint 2003.

[11] A. Ornelas, *Existence of scalar minimizers for autonomous simple integrals nonconvex except at zero* preprint 2003 (presented as habilitation thesis in January 2004)

[12] A. Ornelas, *Existence of scalar minimizers for nonconvex 0-convex autonomous 1-dim integrals* to appear in a refereed journal.

[13] A. Ornelas, *Lipschitz regularity for scalar minimizers of autonomous simple integrals* J. Math. Anal. Appl. **300** (2004), 285–296.

Clara Carlota
Cima-ue
rua Romão Ramalho 59
P-7000-671 Évora
Portugal
e-mail: ccarlota@uevora.pt

António Ornelas
Cima-ue
rua Romão Ramalho 59
P-7000-671 Évora
Portugal
e-mail: ornelas@uevora.pt

Progress in Nonlinear Differential Equations
and Their Applications, Vol. 75, 123–133

# Strict Convexity, Comparison Results and Existence of Solutions to Variational Problems

Arrigo Cellina

*To my twin James A. Yorke*

**Abstract.** The aim of this paper is to discuss the assumption of strict convexity in problems of the the Calculus of Variations, and to present some results that avoid introducing this assumption.

**Mathematics Subject Classification (2000).** Primary 49K20.

**Keywords.** Strict convexity, comparison theorem, uniqueness.

## 1. Introduction

We consider variational problems of the kind

$$\text{Minimize} \int_\Omega \Big[ f\big(\nabla u(x)\big) + g\big(u(x)\big) \Big] dx \text{ on } u - u_0 \in W_0^{1,1}(\Omega). \qquad (1.1)$$

Here $f$ is a convex function, possibly extended valued, whose effective domain, the set of points where $f$ takes finite values, is denoted by $Dom(f)$ and whose epigraph is denoted $epi(f)$. Since this paper is concerned with the implications of strict convexity, or of the lack of it, let us state precisely what we mean by a strictly convex function.

**Definition 1.1.** *We shall call a convex function $f$ strictly convex if for every $x \in Dom(f)$, the point $(x, f(x))$ is an extremal point of $epi(f)$.*

We shall call a function $u$ a solution to a variational problem if it yields a *finite* minimum to the problem.

The use of strict convexity appears in its most immediate form when we consider a minimization problem depending only on $\nabla u$: consider the problem of minimizing

$$\int_\Omega f\big(\nabla u(x)\big) dx \qquad (1.2)$$

with appropriate boundary conditions, and assume to have two solutions $v$ and $w$, so that

$$\int_\Omega f(\nabla v(x))dx = \int_\Omega f(\nabla w(x))dx = min.$$

Then the function $u = \frac{1}{2}v + \frac{1}{2}w$ satisfies the same boundary conditions and is such that $\nabla u(x) = \frac{1}{2}\nabla v(x) + \frac{1}{2}\nabla w(x)$ a.e. in $\Omega$. By convexity, $u$ is also a solution, since

$$\int_\Omega f(\nabla u(x))dx \le \frac{1}{2}\int_\Omega f(\nabla v(x))dx + \frac{1}{2}\int_\Omega f(\nabla w(x))dx = min.$$

Then, we must have that, a.e. in $\Omega$, $f(\nabla u(x)) = \frac{1}{2}f(\nabla v(x)) + \frac{1}{2}f(\nabla w(x))$ and the point $(\nabla(u(x)), f(\nabla u(x)))$ is not an extreme point of $epi(f)$, unless $\nabla u(x) = \nabla v(x) = \nabla w(x)$ a.e. in $\Omega$. Hence strict convexity yields *uniqueness* of the solution. The remainder of the paper will discuss other implications of strict convexity, or of the lack of it.

## 2. Existence of solutions

The main method for proving the existence of solutions is, as it is well known, the Direct Method. This method demands the assumption of convexity for $f$, to prove the weak lower semicontinuity of the integral functional, but no use is made of strict convexity. The proof is a combination of growth assumptions (superlinear growth, to have weak compactness), and of convexity, to pass to the limit with the values of the functional. Here the assumption of strict convexity not only is of no use, but would actually prevent proving the existence of solutions to those minimum problems whose Lagrangeans are generated from the convexification of Lagrangeans that were not, originally, convex, as sometimes it happens in the applications. To the other extreme of the range of the existence theorems, there is the result for the existence of solutions to the *least area problem*, in the parametric case. Here again a region $\Omega$ is given (in general, it suffices to consider $\Omega \subset \mathbb{R}^2$); on $\Omega$ a function $\phi$ is defined, the boundary condition, and one wishes to minimize

$$\int_\Omega \sqrt{1 + \|\nabla u(x)\|^2}dx \text{ on } u - \phi \in W_0^{1,1}(\Omega).$$

The map $\sqrt{1 + \|\xi\|^2}$ does not grow superlinearly and, due to the lack of weak compactness, the Direct Method cannot be applied. Still, this famous problem, a special case of Plateau's Problem, has been thoroughly investigated and a nice existence theorem has been proved for it. This theorem takes advantage of the special properties of the function $\sqrt{1 + \|\xi\|^2}$ and, above all, of the fact that it is strictly convex. The (rather delicate) existence proof can be found, e.g., in [4]. Our purpose here is simply to point out the parts of this proof that are related to the strict convexity and to comparison results.

The first step is to derive *a priori* estimates for the solution. One has

**Proposition 2.1.** *Let $f$ be strictly convex, let $\phi$ be continuous on $\overline{\Omega}$. Let $u$ be a solution to the minimization problem* (1.2) *and let $C = \sup_{\xi \in \partial\Omega} \phi(\xi)$. Then, a.e. in $\Omega$, one has*

$$u(x) \leq C\,.$$

*Proof.* In fact, assume that there exists a set $E \subset \Omega$ and a constant $\delta > 0$, such that $u(x) \geq C + \delta$ on $E$. Consider $v(x) = C$ and $w(x) = C + (u(x) - C)^+$: they both are solutions to (1.2) with boundary data $C$, and they are different on $E$. This is a contradiction to the uniqueness provided by the strict convexity.     $\square$

**Corollary 2.2.** *Let $u$ be a solution to problem* (1.2)*, where $f$ is strictly convex and assume that $u$ is enough regular. Then $\sup u$ and $\inf u$ are assumed at $\partial\Omega$.*

The next step in the existence proof is to consider two solutions, $u$ and $v$ (regular, continuous on $\overline{\Omega}$), each with its own boundary data. Again as a consequence of strict convexity, we obtain

**Proposition 2.3.** *Let $u$ and $v$ solutions, such that at $\partial\Omega$, $u(x) \leq v(x)$. Then, $u(x) \leq v(x)$ on $\Omega$.*

**Corollary 2.4.** $\sup(u - v)$ *and* $\inf(u - v)$ *are attained on $\partial\Omega$.*

*Proof.* Let $C_b = \sup_{x \in \partial\Omega} u(x) - v(x)$ so that, at $\partial\Omega$, $u(x) \leq v(x) + C_b$. Since $w(x) = v(x) + C_b$ is also a solution, the previous proposition applies.     $\square$

**Corollary 2.5.** *The supremum of the Lipschitz constant of a solution*

$$\sup\left\{ \frac{|u(x) - u(y)|}{\|x - y\|} : x, y \in \Omega; x \neq y \right\}$$

*is attained when one of the points $x$ or $y$ is at the boundary of $\Omega$.*

*Proof.* Fix a vector $k$ and consider $u(x + k)$. It is defined whenever $x \in \Omega - k$, so that the two functions $u(x)$ and $u_k(x) = u(x + k)$ have $\Omega_k = \Omega \cap (\Omega - k)$ as a common domain of definition. Both are solutions for the minimization problem on $\Omega_k$, each with its own boundary data at $\partial\Omega_k$. So, $\sup(u - u_k)$ and $\inf(u - u_k)$ are attained on $\partial(\Omega - k)$. One of the two points is on $\partial\Omega$.     $\square$

Having established this fundamental fact, then the existence proof proceeds by showing that one can build a "barrier" for the solution. Our purpose here was to point out that:

**Remark 2.6.** The previous steps, needed to obtain the property of a solution stated in the proposition above, all depend only on $f$ being strictly convex, and not on other properties of $f$.

However, consider the following example

**Example 2.7.** Let $\Omega$ be the interval $[-1, 1]$; let

$$f(\xi) = \left\{ \begin{array}{ll} 0 & |\xi| \leq 1 \\ (|\xi| - 1)^2 & |\xi| > 1 \end{array} \right.$$

and consider the problem

$$\text{minimize} \int_{[-1,1]} f\big(u'(x)\big)\, dx\,; \qquad u \in W_0^{1,1}\big([-1, 1]\big)\,.$$

The function $v(x) \equiv 0$ and $w(x) = -|x| + 1$ are solutions, $w \leq v$ at $\partial\Omega$ but it is not true that $w \leq v$ on $\Omega$.

Moreover, it is not true that $\sup u$ and $\inf u$ are assumed at $\partial\Omega$. In addition, again considering the two solutions $v$ and $w$, it is not true that the $\sup(w - v)$ is attained at the boundary.

Hence, without the assumption of strict convexity for $f$, the whole approach to this existence ought to be carefully checked.

**Remark 2.8.** Sometimes, one finds the suggestion: in case $f$ is not strictly convex, simply add $\varepsilon\|\xi\|^2$ to it. This will make the new $f$ strictly convex; you can use the results for strict convexity and pass these results to the limit.

Let us add $\varepsilon\|\xi\|^2$ to the Lagrangean of the previous Example: now we wish to minimize

$$\int_\Omega f_\varepsilon\big(\nabla u(x)\big)\,dx \tag{2.1}$$

where $f_\varepsilon(\xi) = \varepsilon\|\xi\|^2$ when $\|\xi\| \leq 1$, $= +\infty$ otherwise. When the boundary condition is such that $u_0 \leq 0$ at $\partial\Omega$, it is true that *every* solution $v$ to (2.1) is such that $v \leq 0$ a.e. in $\Omega$, but it is *not* true that *every* solution $v$ to the limit problem is such that $v \leq 0$ a.e. in $\Omega$. At most, one can say that, for the limit problem, *there exists a solution* $v$ satisfying $v \leq 0$ a.e. in $\Omega$. However, for the purpose of obtaining *a priori* estimates, this statement is not strong enough. In fact, consider the least area problem of minimizing

$$\int_\Omega \sqrt{1 + \|\nabla u(x)\|^2}\,dx\,. \tag{2.2}$$

By adding $\varepsilon\|\nabla u(x)\|^2$ to the Lagrangean, we obtain a strictly convex coercive functional, so that solutions exist no matter what the boundary data and $\Omega$ are (sufficiently regular). However, it is not true that for the limiting problem (the least area problem), one has existence of solutions no matter what the boundary data and $\Omega$ are.

## 3. Comparison results without strict convexity

If one carefully examines the line of thought followed in the existence proof in the previous paragraph, one finds that strict convexity has been used to ensure a comparison result between solutions; in some cases, while one of the solutions was unknown, the other was simply a constant. It might be possible to have existence theorems based on the construction of barriers without the assumption of strict convexity; one should carefully reconsider the comparison results. For some purposes, it would be enough to obtain results when one of the two belongs to a special class of solutions, for instance, the affine functions, and one might aim at results for this more restricted class of problems. In [1] a comparison result was proved. This result concerns a special class of solutions to the minimum problem 1.2).

The class of solutions to be considered is described below. It is a generalization, of the class of affine maps. Before introducing it, let us revisit the affine maps. Consider

$$\langle a, x - x^0 \rangle + r :$$

we have

$$\langle a, x - x^0 \rangle + r = \langle x - x^0, a \rangle + r = \sup_z \left\{ \langle x - x^0, z \rangle - I_{\{a\}}(x) \right\} + r = \left( I_{\{a\}} \right)^* (x - x^0) + r .$$

When a function $f$ is not strictly convex, its graph contains, so to say, some "flat" parts; in turn, this is reflected on the polar of $f$, whose graph contains angles, i.e., the subdifferential of the polar is not single-valued whenever $f$ is not strictly convex. This is essential in the definition of the class of solutions we want to consider.

**Definition 3.1.** *For $z \in Dom(f^*)$ , $x^0 \in \mathbb{R}^N$ and $r \in \mathbb{R}$, set*

$$h^+_{z,x^0,r}(x) = \left( I_{f^*(z)} \right)^* (x - x^0) + r \ and \ h^-_{z,x^0,r}(x) = - \left( I_{f^*(z)} \right)^* (x - x^0) + r .$$

Next Theorem shows that the maps just introduced are solutions to the minimum problem; we recall that by saying that a function $v$ is a solution, we mean that it is a solution among all those functions that have the same boundary data as $v$. In it, we assume the following growth assumption: $Dom(f^*)$ is open. This assumption is very general; in particular we have

**Proposition 3.2.** *Let $f$ be an extended valued, convex, lower semicontinuous function with superlinear growth; then $Dom(f^*) = \mathbb{R}^N$.*

However, a Lagrangean as $f(t) = |t| - \sqrt{|t|}$, whose polar is $f^*(p) = \frac{1}{4} \frac{1}{1-|p|}$ for $p \in (-1, 1)$, satisfies the condition $Dom(f^*)$ open, without being of superlinear growth.

The only case of Lagrangean (among those usually encounterd in Variational Problems!), that does not satisfy the assumption that $Dom(f^*)$ is open, is the least area Lagrangean $f(t) = \sqrt{1 + t^2}$, whose polar is $f^*(p) = -\sqrt{1 - p^2}$, so that $Dom(f^*) = \overline{B(0,1)}$.

**Theorem 3.3.** *Let $Dom(f^*)$ be open. For $z \in Dom(f^*)$ , $x^0 \in \mathbb{R}^N$ and $r \in \mathbb{R}$, the maps $h^+_{z,x^0,r}(x)$ and $h^-_{z,x^0,r}(x)$ are solutions to the minimum problem 1.2), among those $u$ in*

$$\mathcal{S}^+_{z,x^0,r} = \left\{ u \in W^{1,1}(\Omega) : u - h^+_{z,x^0,r} \in W^{1,1}_0(\Omega) \right\}$$

*and*

$$\mathcal{S}^-_{z,x^0,r} = \left\{ u \in W^{1,1}\left(\Omega\right) : u - h^-_{z,x^0,r} \in W^{1,1}_0\left(\Omega\right) \right\}$$

*respectively.*

Notice that, in the case where $f$ is strictly convex, $\partial f^*(z)$ is single-valued and the maps $h^+_z$ and $h^-_z$ are affine maps. Hence the above class of functions can be seen as the natural generalization of the affine maps when $f$ is not strictly convex.

Let us consider the following examples.

**Example 3.4.**

$$f(\xi) = \left\{ \begin{array}{ll} 0 & \|\xi\| \le 1 \\ \frac{1}{2}\left(\|\xi\| - 1\right)^2 & \|\xi\| > 1 \end{array} \right.$$

whose polar is

$$f^*(p) = \frac{1}{2}\|p\|^2 + \|p\|$$

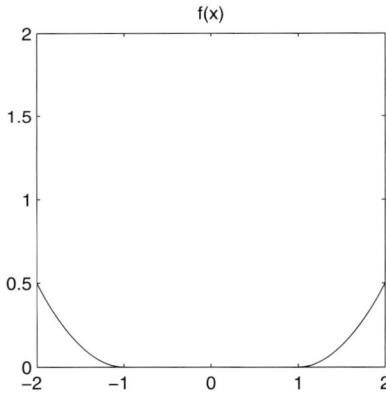
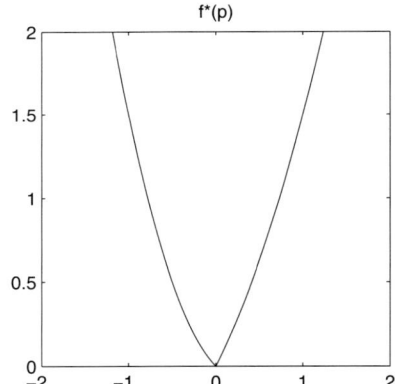

we have

$$\partial f^*(p) = \left\{ \begin{array}{ll} B(0,1) & p = 0 \\ p + \frac{p}{\|p\|} & p \ne 0 \end{array} \right. \tag{3.1}$$

and we obtain that $\left(I_{\partial f^*(z)}\right)^*(x)$ is the family of maps

$$\left(I_{\partial f^*(z)}\right)^*(x) = \left\{ \begin{array}{ll} \|x\| & z = 0 \\ \left\langle z + \frac{z}{\|z\|}, x \right\rangle & z \ne 0 \end{array} \right.$$

**Example 3.5.** for the Lagrangean (see [5])

$$f(\xi) = \left\{ \begin{array}{ll} \sqrt{2}\|\xi\| & \|\xi\| \le \sqrt{2} \\ 1 + \frac{1}{2}\|\xi\|^2 & \|\xi\| \ge \sqrt{2} \end{array} \right. \tag{3.2}$$

whose polar is

$$f^*(p) = \left\{ \begin{array}{ll} 0 & \|p\| \le \sqrt{2} \\ \frac{1}{2}\|p\|^2 - 1 & \|p\| \ge \sqrt{2} \end{array} \right. \tag{3.3}$$

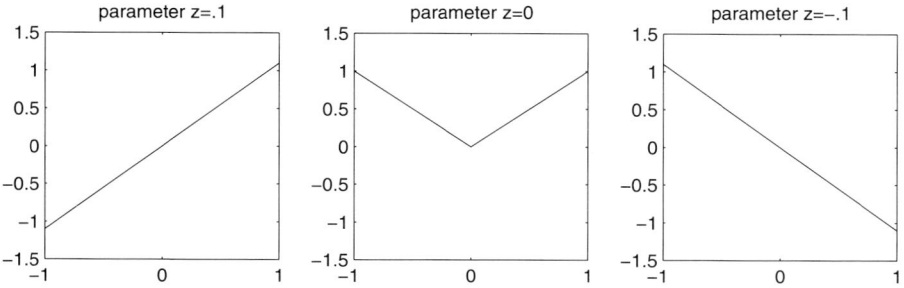

we have

$$\partial f^*(p) = \begin{cases} 0 & \|p\| \leq \sqrt{2} \\ \{\alpha p\} & 0 \leq \alpha \leq 1, \|p\| = \sqrt{2} \\ \frac{1}{2}\|p\|^2 - 1 & \|p\| \geq \sqrt{2} \end{cases} \tag{3.4}$$

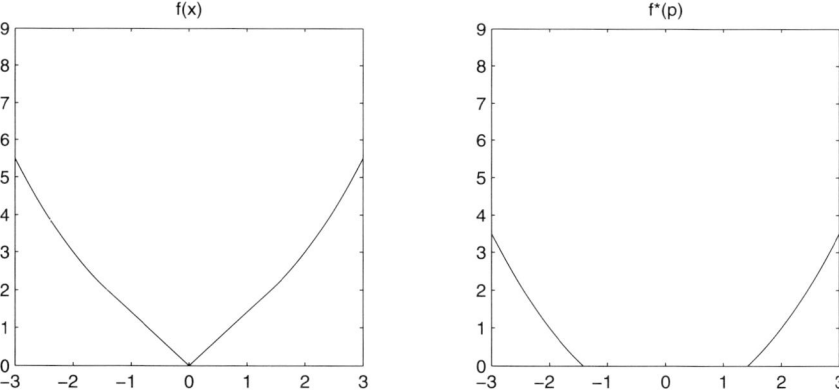

and we obtain that $\left(I_{\partial f^*(z)}\right)^*(x)$ is the family of maps

$$\left(I_{\partial f^*(z)}\right)^*(x) = \begin{cases} 0 & \|z\| \leq \sqrt{2} \\ \sqrt{2}\left\langle \frac{z}{\|z\|}, x\right\rangle \chi_{\{x:\langle z,x\rangle \geq 0\}} & \|z\| = \sqrt{2} \\ \langle z, x\rangle & \|z\| > \sqrt{2} \end{cases}$$

Besides being solutions, the maps introduced above enjoy some kind of "maximality" condition, that yields comparison results, as in the theorem that follows. In it, the assumption of convexity of the domain $\Omega$ appears: this assumption seems to be needed for technical reasons only.

**Theorem 3.6.** *Let $\Omega$ be convex, let $f$ be a (possibly extended valued) lower semicontinuous, convex function such that $Dom(f^*)$ is open. Let $w$ be a solution to the*

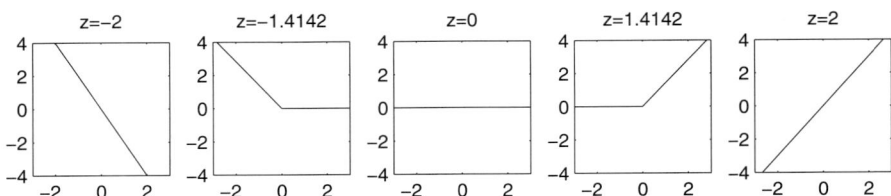

*problem of minimizing the functional*

$$\mathcal{J}(u) = \int_\Omega f\big(\nabla u(x)\big)\,dx$$

on $\{u : u - u^0 \in W_0^{1,1}(\Omega)\}$.

*Assume that, for $z \in Dom(f^*)$, $x^0 \in (\Omega)^c$ and $r \in \mathbb{R}$, we have $h_{z,x^0,r}^+ \geq w$ on $\partial\Omega$. Then, $h_{z,x^0,r}^+ \geq w$ on $\Omega$.*

**Remark 3.7.** Notice that any affine function $\ell(x) = \langle a, x\rangle + b$ can be written as $\ell(x) = \langle a, x - x^0\rangle + r$ with $x^0 \notin \Omega$ and $r = b + \langle a, x^0\rangle$.

Notice also that Example 1 shows that the analogous theorem, where we had an affine function $\ell$ (in particular, $\ell(x) \equiv 0$) instead of the convex function $h_{z,x^0,r}^+$, would be false.

Finally notice that the functions $u(x) \equiv 0$ and $h_{0,0,-1}^+(x) = -1 + |x|$ are solutions to the problem of Example 1; still, $h_{0,0,-1}^+ \geq u$ on $\partial\Omega$, but it is not true that $h_{0,0,-1}^+ \geq u$ on $\Omega$. Here, the point $x^0 = 0 \in \Omega$, opposite to our assumptions.

## 4. Another class of minimization problems

Another class of minimization problems, connected to the class previously considered, are the problems of the kind

$$\text{Minimize} \int_\Omega \big[ f(\nabla u(x)) + \alpha u(x)\big]\,dx \quad \text{on} \quad u - u_0 \in W_0^{1,1}(\Omega) \qquad (4.1)$$

where $\alpha$ is a given parameter. We emphasize this point, since some similar problems occur where the parameter is a Lagrange multiplier (expressing, for instance, a problem under volume constraints), and this multiplier is to be determined by the solution to the problem itself. A case where this arises is the problem of capillary surfaces without gravity, as in [3]: in this case $f(\xi) = \sqrt{1 + |\xi|^2}$.

In the previous section we have introduced a class of solutions to problem (1.2) that is explicit, i.e., such that it can be computed directly from the Lagrangean; in this section we will show that the same can be done for problem (4.1). At the end of it we shall examine the behaviour of these solutions as the parameter $\alpha$ tends to zero, obtaining a result that, to this author, was rather surprising. In [2] the following result is proved.

**Theorem 4.1.** *Let $\Omega$ be an open bounded set, enough regular so that the Divergence Theorem holds, and let $f : \mathbb{R}^N \to \mathbb{R} \cup \{+\infty\}$ be an extended valued, convex, lower semicontinuous function. For $x_0$ and $z$ in $\mathbb{R}^N$ and $c \in \mathbb{R}$, consider the function*

$$\omega_\alpha(x) = \frac{N}{\alpha} f^* \left( z + \frac{x - x_0}{N} \alpha \right) + c \,.$$

*If $\omega_\alpha$ is defined on $\Omega$ and belongs to $W^{1,1}(\Omega)$, then it is the only minimum of the functional*

$$\mathcal{J}(u) = \int_\Omega \Big[ f\big(\nabla u(x)\big) + \alpha u(x) \Big] dx \,,$$

*in the class of functions*

$$\mathcal{S} = \left\{ u \in W^{1,1}(\Omega) , u - \omega_\alpha \in W_0^{1,1}(\Omega) \right\} .$$

For instance, for $N = 1$, a solution to the problem of minimizing

$$\int_\Omega \Big[ f\big(u'(x)\big) + u(x) \Big] dx$$

where

$$f(\xi) = \begin{cases} 0 & \text{if } \|\xi\| \leq 1 \\ \|\xi\| - 1 & \text{if } 1 \leq \|\xi\| \leq 2 \\ +\infty & \text{elsewhere} \end{cases}$$

is the map $\omega_1$ defined by

$$\omega_1 = f^*(x) = \begin{cases} |x| & \text{if } |x| \leq 1 \\ 2|x| - 1 & \text{if } 1 \leq |x| \leq 2 \end{cases}$$

So the previous result yields at once an explicit formula for a solution to a varia-

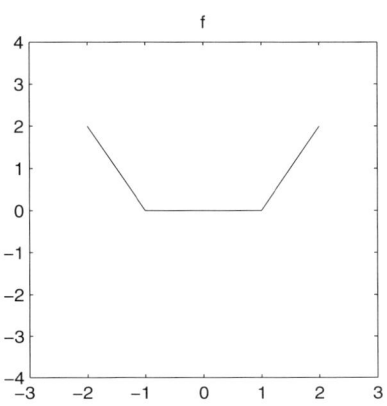

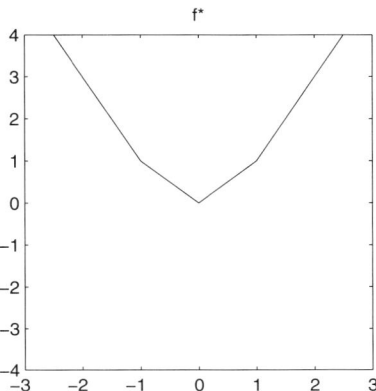

tional problem and an uniqueness result, under very few assumptions on $f$, mainly that it is a convex, lower semicontinuous function. There is *some* uniqueness even though the map $f$, as in the example above, can be far from being strictly convex.

Moreover, assume that the domain of $f^*$ is open. Then, being a convex function Lipschitzian in the interior of its domain, there exists some open ball $B(x^0, r)$ strictly contained in the domain of $f^*$, where $f^*$ is lipschitzian; hence, taking $\Omega = \{x : \|Nz + (x - x^0)\| < r\}$, we have that $\omega_\alpha$ belongs to $W^{1,\infty}(\Omega)$. This remark proves the following Corollary.

**Corollary 4.2.** *Let $Dom(f^*)$ contain an open set. Then there exists a region $\Omega$ and suitable boundary conditions, such that the problem* (4.1) *admits a solution.*

The condition that the effective domain of $f$ contains an open set is very weak, and is satisfied by any reasonable problem; for instance, when $f(\xi) = \sqrt{1 + |\xi|^2}$, as in the capillary problem, $f^*(p) = -\sqrt{1 - p^2}$, so that $Dom(f^*) = \overline{B}(0, 1)$. In this case, whenever $\overline{\Omega} \subset B(0, 1)$, problem (4.1) admits some solution. This might suggest that in general, problem (4.1) admits solutions when $\Omega$ is sufficiently small (and, may be, the mean curvature of $\partial\Omega$ is positive). The author does not know of any such result.

A class of problems clearly excluded from the application of Theorem (4.1) is when $f(x) = \langle a, x \rangle + b$: in this case, the domain of the polar to $f$ reduces to one point.

# 5. Passing to the limit

The formula for a solution presented in Theorem 4.1 is not defined for $\alpha = 0$, and does not hold for this case. Still, as the parameter $\alpha$ tends to 0, a rather surprising connection, among the different classes of solutions we have presented, arises. It is enough to write the solution $\omega_\alpha$ in the form

$$\frac{N}{\alpha} f^* \left( z + \frac{x - x_0}{N} \alpha \right) - \frac{N}{\alpha} f^*(z)$$

to realize that, in fact, the solution $\omega_\alpha$ is a differential quotient. It not surprising, then, that the following result holds, as proved in [2].

**Theorem 5.1.** *Let $f$ be an extended valued, convex, lower semicontinuous function with superlinear growth. Then:*

a) *when $f^*$ is differentiable at $z$, as $\alpha$ tends to 0, the function $\omega_{(\alpha, z, \beta)}$ converges to the affine map $\langle \nabla f^*(z), x - x_0 \rangle + \beta$, a solution to problem 1.2).*

b) *in general, as $\alpha$ tends to $0^+$, the function $\omega_{(\alpha, z, \beta)}$ converges to $h^+_{z, x_0, \beta}$, the solution to problem 1.2, presented in Theorem 1 of [1]; as $\alpha$ tends to $0^-$, $\omega_{(\alpha, z, \beta)}$ converges to $h^-_{z, x_0, \beta}$.*

**Acknowledgment**

I wish to warmly thank Prof. Vasile Staicu for the excellent organization of the meeting and for inviting me to Aveiro, where this paper was written under the research project POCI MAT 55524/2004 from FCT of Portugal.

# References

[1] A. Cellina, *Comparison results and estimates on the gradient without strict convexity*, SIAM J. Contrrol Optim., 46(2007), 738–749.

[2] A. Cellina, *Uniqueness and comparison results for functionals depending on $\nabla u$ and on $u$*, SIAM J. Optim., (to appear).

[3] R. Finn, *Equilibrium Capillary Surfaces*. Grundlehren der Mathematischen Wissenschaften, 284. Springer-Verlag, New York, 1986.

[4] E. Giusti, *Minimal Surfaces and Functions of Bounded Variation*. Monographs in Mathematics, 80. Birkhäuser Verlag, Basel, 1984.

[5] B. Kawohl, J. Stará, G. Wittum, *Analysis and numerical studies of a problem of shape design*. Arch. Rational Mech. Anal. **114** (1991), no. 4, 349–363.

Arrigo Cellina
Department of Mathematics
University of Milan "Bicocca"
Via Cozzi 53
I-20122 Milano
Italy
e-mail: `arrigo.cellina@unimib.it`

Progress in Nonlinear Differential Equations
and Their Applications, Vol. 75, 135–142
© 2007 Birkhäuser Verlag Basel/Switzerland

# Necessary Optimality Conditions for Discrete Delay Inclusions

Aurelian Cernea

*Dedicated to Arrigo Cellina and James Yorke*

**Abstract.** The aim of this paper is to present a short survey of several new results concerning optimization of discrete delay inclusions. We study an optimization problem given by a discrete delay inclusion with end point constraints and we present several approaches concerning first and second-order necessary optimality conditions for this problem.

**Mathematics Subject Classification (2000).** Primary 93C30; Secondary 49J30.

**Keywords.** Tangent cone, discrete delay inclusion, maximum principle.

## 1. Introduction

Consider the problem

$$\text{minimize} \quad g\big(x(N)\big) \tag{1.1}$$

over the solutions of the discrete inclusion

$$x(t+1) \in F_t\big(x(t), x(t-k)\big); \quad t = 0, 1, \ldots, N-1, \quad x(l) = \overline{x}(l), \quad l = -k, \ldots, 0, \tag{1.2}$$

with end point constraints of the form

$$x(N) \in X_N, \tag{1.3}$$

where $F_t(.,.) : \mathbf{R}^n \times \mathbf{R}^n \to \mathcal{P}(\mathbf{R}^n)$, $t = 0, \ldots, N-1$, $\overline{x}(l) \in \mathbf{R}^n$, $l = -k, \ldots, 0$, $X_N \subset \mathbf{R}^n$ and $g : \mathbf{R}^n \to \mathbf{R}$ are given.

The aim of this paper is to present a short survey of several new results concerning first and second-order necessary optimality conditions for problem (1.1)–(1.3).

At the beginning we obtain necessary optimality conditions for a solution $\overline{x} = (\overline{x}(-k), \ldots, \overline{x}(0), \ldots, \overline{x}(N))$ to the problem (1.1)–(1.3) in terms of a variational inclusion associated to the problem (1.2) and in terms of the cone of interior directions (Dubovitskij-Miljutin tangent cone) to the set $X_N$ at $\overline{x}(N)$. Afterwards this result is improved by replacing the cone of interior directions with the concept of derived cone introduced by Hestenes ( [5]) and using a remarkable "intersection property" of derived cones obtained by Mirica ( [8]). Finally, we present an approach concerning second-order necessary optimality conditions for the problem (1.1)–(1.3).

Optimal control problems for systems described by discrete inclusions have been studied by many authors ( [1,6,9–11,13] etc.). In the framework of multivalued problems sufficient conditions for local controllability along a reference trajectory of a discrete delay inclusion are obtained in [7]. At the same time, necessary optimality conditions for an optimization problem without end point constraints and given by a discrete inclusion (i.e., without delay) may be found in [12]. The idea in [12] is to use a special (Warga's) open mapping theorem to obtain a sufficient condition for local controllability of the discrete inclusion around a given trajectory and as a consequence, via a separation result, to obtain the maximum principle.

In contrast with the above mentioned approach, the general idea present in our approach seems to be conceptually very simple, relying only 2-3 clear-cut steps and using a minimum of auxiliary results from finite dimensional analysis. Moreover it can be adapted in order to obtain second-order necessary conditions for the optimization problem.

The paper is organized as follows: in Section 2 we present the notations and preliminary results to be used in the sequel and in Section 3 we state our results.

## 2. Preliminaries

Denote by $S_F$ the solution set of inclusion (1.2), i.e.,

$$S_F := \Big\{ x(.) = \big(x(-k), \ldots, x(N)\big); \quad x(.) \text{ is a trajectory of } (1.2) \Big\}.$$

and by $R_F^N := \{x(N); \quad x \in S_F\}$ the reachable set of inclusion (1.2).

Let $\overline{x} = (\overline{x}(-k), \ldots, \overline{x}(N)) \in S_F$ be a trajectory of (1.2).

Since the reachable set $R_F^N$ is, generally, neither a differentiable manifold, nor a convex set, its infinitesimal properties may be characterized only by tangent cones in a generalized sense, extending the classical concepts of tangent cones in Differential Geometry and Convex Analysis, respectively.

From the multitude of the intrinsic tangent cones in the literature, the contingent, the quasitangent and Clarke's tangent cones, defined, respectively, by

$$K_x X = \big\{ v \in \mathbf{R}^n; \quad \exists s_m \to 0+, \; x_m \in X : \tfrac{x_m - x}{s_m} \to v \big\}$$
$$Q_x X = \big\{ v \in \mathbf{R}^n; \quad \exists c(.) : [0, s_0) \to X, \; c(0) = x, \; c'(0) = v \big\}$$
$$C_x X = \big\{ v \in \mathbf{R}^n; \quad \forall (x_m, s_m) \to (x, 0+), \; x_m \in X, \; \exists y_m \in X : \tfrac{y_m - x_m}{s_m} \to v \big\}$$

seem to be among the most oftenly used in the study of different problems involving nonsmooth sets and mappings.

The *second-order quasitangent set* to $X$ at $x$ relative to $v \in Q_x X$ is defined by

$$Q^2_{(x,v)} X = \{ w \in \mathbf{R}^n; \quad \forall s_m \to 0+, \ \exists w_m \to w : \ x + s_m v + s_m^2 w_m \in X \}$$

We recall that, in contrast with $K_x X, Q_x X$, the cone $C_x X$ is convex and one has $C_x X \subset Q_x X \subset K_x X$.

Another important tangent cone is the *cone of interior directions* (Dubovitskij-Miljutin tangent cone) defined by

$$I_x X := \left\{ v \in \mathbf{R}^n; \quad \exists s_0, \ r > 0 : \ x + s B(v,r) \subset X \ \forall s \in [0, s_0) \right\},$$

$$B(v,r) := \left\{ w \in \mathbf{R}^n; \quad ||w - v|| < r \right\}, \quad \overline{B}(v,r) := \mathrm{cl} B(v,r).$$

From the properties of the quasitangent cones we recall only the following

$$Q_x X_1 \cap I_x X_2 \subset Q_x (X_1 \cap X_2). \tag{2.1}$$

The concept of derived set was introduced for the first time by Hestenes ( [5]).

**Definition 2.1.** *A subset $M \subset \mathbf{R}^n$ is said to be a derived set to $X \subset \mathbf{R}^n$ at $x \in X$ if for any finite subset $\{v_1, \ldots, v_k\} \subset M$, there exist $s_0 > 0$ and a continuous mapping $a(.) : [0, s_0]^k \to X$ such that $a(0) = x$ and $a(.)$ is (conically) differentiable at $s = 0$ with the derivative $\mathrm{col}[v_1, \ldots, v_k]$ in the sense that*

$$\lim_{\mathbf{R}^k_+ \ni \theta \to 0} \frac{||a(\theta) - a(0) - \sum_{i=1}^k \theta_i v_i||}{||\theta||} = 0.$$

We shall write in this case that the derivative of $a(.)$ at $s = 0$ is given by

$$Da(0)\theta = \sum_{i=1}^k \theta_j v_j, \quad \forall \theta = (\theta_1, \ldots, \theta_k) \in \mathbf{R}^k_+ := [0, \infty)^k.$$

A subset $C \subset \mathbf{R}^n$ is said to be a *derived cone* of $X$ at $x$ if it is a derived set and also a convex cone.

For the basic properties of derived sets and cones we refer to Hestenes ( [5]); we recall that if $M$ is a derived set then $M \cup \{0\}$ as well as the convex cone generated by $M$, defined by

$$cco(M) = \left\{ \sum_{i=1}^k \lambda_j v_j ; \quad \lambda_j \geq 0, \ k \in \mathbf{N}, \ v_j \in M, \ j = 1, \ldots, k \right\}$$

is also a derived set, hence a derived cone.

The fact that the derived cone is a proper generalization of the classical concepts in Differential Geometry and Convex Analysis is illustrated by the following results ( [5]): if $X \subset \mathbf{R}^n$ is a differentiable manifold and $T_x X$ is the tangent space in the sense of Differential Geometry to $X$ at $x$

$$T_x X = \{ v \in \mathbf{R}^n; \quad \exists c : (-s, s) \to X, \ \text{of class} \ C^1, c(0) = x, c'(0) = v \}$$

then $T_x X$ is a derived cone; also, if $X \subset \mathbf{R}^n$ is a convex subset then the tangent cone in the sense of Convex Analysis defined by

$$TC_x X = cl\{t(y - x); \quad t \geq 0, \, y \in X\}$$

is also a derived cone. By $cl\,A$ we denote the closure of the set $A \subset \mathbf{R}^n$.

Since any convex subcone of a derived cone is also a derived cone, such an object may not be uniquely associated to a point $x \in X$; moreover, simple examples show that even a maximal with respect to set-inclusion derived cone may not be uniquely defined: if the set $X \subset \mathbf{R}^2$ is defined by

$$X = C_1 \bigcup C_2, \quad C_1 = \{(x, x); x \geq 0\}, \quad C_2 = \{(x, -x), x \leq 0\}$$

then $C_1$ and $C_2$ are both maximal derived cones of $X$ at the point $(0,0) \in X$.

We recall that two cones $C_1, C_2 \subset \mathbf{R}^n$ are said to be *separable* if there exists $q \in \mathbf{R}^n \backslash \{0\}$ such that:

$$\langle q, v \rangle \leq 0 \leq \langle q, w \rangle \quad \forall v \in C_1, \, w \in C_2.$$

We denote by $C^+$ the positive dual cone of $C \subset \mathbf{R}^n$

$$C^+ = \{q \in \mathbf{R}^n; \quad \langle q, v \rangle \geq 0, \quad \forall v \in C\}$$

The negative dual cone of $C \subset \mathbf{R}^n$ is $C^- = -C^+$.

The following "intersection property" of derived cones, obtained by Mirică ( [8]), is a key tool in the proof of necessary optimality conditions.

**Lemma 2.2.** *Let $X_1, X_2 \subset \mathbf{R}^n$ be given sets, $x \in X_1 \cap X_2$, and let $C_1, C_2$ be derived cones to $X_1$, resp. to $X_2$ at $x$. If $C_1$ and $C_2$ are not separable, then*

$$Cl(C_1 \cap C_2) = \big(Cl(C_1)\big) \cap \big(Cl(C_2)\big) \subset Q_x(X_1 \cap X_2).$$

For a mapping $g(.) : X \subset \mathbf{R}^n \to \mathbf{R}$ which is not differentiable, the classical (Fréchet) derivative is replaced by some generalized directional derivatives. We recall, in the case when $g(.)$ is locally-Lipschitz at $x \in int(X)$, Clarke's generalized directional derivative, defined by:

$$D_C g(x; v) = \limsup_{(y,\theta) \to (x,0+)} \frac{g(y + \theta v) - g(y)}{\theta}, \quad v \in \mathbf{R}^n.$$

The first and second order uniform lower Dini derivative are defined as follows

$$D_\uparrow g(x; v) = \liminf_{(v',\theta) \to (v,0+)} \frac{g(x + \theta v') - g(x)}{\theta},$$

$$D_\uparrow^2 g(x, v; w) = \liminf_{(w',\theta) \to (w,0+)} \frac{g(x + \theta v + \theta^2 w') - g(x) - \theta D_\uparrow g(x; v)}{\theta^2}.$$

When $g(.)$ is of class $C^2$ one has

$$D_\uparrow g(x, v) = g'(x)^T v, \quad D_\uparrow^2 g(x, v; w) = g'(x)^T z + \frac{1}{2} v^T g''(x) v.$$

The results in the sequel will be expressed, in the case where $g(.)$ is locally-Lipschitz at $x$, in terms of the Clarke generalized gradient, defined by:

$$\partial_C g(x) = \left\{ q \in \mathbf{R}^n ; \quad \langle q, v \rangle \leq D_C g(x; v) \quad \forall v \in \mathbf{R}^n \right\}.$$

By $\mathcal{P}(\mathbf{R}^n)$ we denote the family of all subsets of $\mathbf{R}^n$.

Corresponding to each type of tangent cone, say $\tau_x X$ one may introduce a *set-valued directional derivative* of a multifunction $G(.) : X \subset \mathbf{R}^n \to \mathcal{P}(\mathbf{R}^n)$ (in particular of a single-valued mapping) at a point $(x, y) \in Graph(G)$ as follows

$$\tau_y G(x; v) = \left\{ w \in \mathbf{R}^n; (v, w) \in \tau_{(x,y)} Graph(G) \right\}, \quad v \in \tau_x X.$$

Similarly one may define second-order directional derivatives of the set-valued map $G(.)$. For example the second-order quasitangent derivative of $G$ at $(x, u)$ relative to $(y, v) \in Q_{(x,u)}(graph(G(.))$ is the set-valued map $Q^2_{(u,v)} G(x, y, .)$ defined by

$$graph Q^2_{(u,v)} G(x, y; .) = Q^2_{((x,u),(y,v))}\big(graph G(.)\big).$$

We recall that a set-valued map, $A(.) : \mathbf{R}^n \to \mathcal{P}(\mathbf{R}^m)$ is said to be a *convex* (respectively, *closed convex*) *process* if $Graph(A(.)) \subset \mathbf{R}^n \times \mathbf{R}^m$ is a convex (respectively, closed convex) cone.

## 3. Results

In what follows, we shall assume the following hypothesis. **Hypothesis.**

i) *The values of $F_t(.,.)$ are compact convex, $\forall t \in \{0, \ldots, N-1\}$.*

ii) *There exists $l(t) > 0$ such that $F_t(.,.)$ is Lipschitz with the Lipschitz constant $l(t)$, $\forall t \in \{0, \ldots, N-1\}$.*

iii) *There exists $A_t : \mathbf{R}^n \times \mathbf{R}^n \to \mathcal{P}(\mathbf{R}^n)$, $t = 0, 1, \ldots, N-1$ a family of closed convex processes such that*

$$A_t(u, v) \subset Q_{\overline{x}(t+1)} F_t((\overline{x}(t), \overline{x}(t-k)); (u, v)) \quad \forall (u, v) \in \mathbf{R}^n \times \mathbf{R}^n,$$

$$\forall t \in \{0, 1, \ldots, N-1\}.$$

To the problem (1.2) we associate the linearized problem

$$w(t+1) \in A_t(w(t), w(t-k)), \ t = 0, 1, \ldots, N-1, \quad w(l) = 0, \ l = -k, \ldots, 0, \quad (3.1)$$

Denote by $S_A$ the solution set of inclusion (3.1) and by $R_A^N$ the reachable set of inclusion (3.1).

We recall that if $A : \mathbf{R}^n \to \mathcal{P}(\mathbf{R}^m)$ is a set-valued map then the adjoint of $A$ is the multifunction $A^* : \mathbf{R}^m \to \mathcal{P}(\mathbf{R}^n)$ defined by

$$A^*(p) = \left\{ q \in \mathbf{R}^m ; \langle q, v \rangle \leq \langle p, v' \rangle \quad \forall (v, v') \in graph A(.) \right\}.$$

The next result, due to Minchenko and Sirotko ( [7]) characterizes the positive dual of the reachable set $R_A^N$ of problem (3.1).

**Lemma 3.1.** *Assume that Hypothesis is satisfied. Then, one has $(R_A^N)^+ =$*

$$\Big\{\eta \in \mathbf{R}^n; \; \exists \; q(t), p(t) \quad t = 0, \ldots, N \quad \text{such that} \quad \eta = p(N), \quad (p(t), q(t))$$

$$\in A_t^* \big(p(t+1)\big) + \big(q(t+k), 0\big), \quad t = N-1, \ldots, 0 \quad \text{and} \quad q(t) = 0 \text{ for } t \geq N\Big\}.$$

Using the property in (2.1), the fact that $R_A^N \subset Q_{\overline{x}(N)} R_F^N$ and Lemma 3.1 we obtain a Maximum Principle for problem (1.1)–(1.3).

**Theorem 3.2.** *Let* $X_N \subset \mathbf{R}^n$ *be a closed set, let* $\overline{x} \in S_F$ *be an optimal solution for problem* (1.1)–(1.3) *such that Hypothesis is satisfied and let* $g(.) : \mathbf{R}^n \to \mathbf{R}$ *be a locally Lipschitz function.*

*Then for any convex cone* $C_N \subset I_{\overline{x}(N)} X_N$ *there exist* $\lambda \in \{0, 1\}$ *and* $(p(0), p(1), \ldots, p(N)) \in \mathbf{R}^{(N+1)n}$, $(q(0), q(1), \ldots, q(N)) \in \mathbf{R}^{(N+1)n}$, *such that*

$$\big(p(t) - q(t+k), q(t)\big) \in A_t^* \big(p(t+1)\big), \quad t = N-1, \ldots, 0, \quad q(t) = 0 \text{ for } t \geq N, \quad (3.2)$$

$$p(N) \in \lambda \partial_C g(\overline{x}(N)) - C_N^+, \tag{3.3}$$

$$\big\langle -p(t+1), \overline{x}(t+1) \big\rangle = \max\big\{ < -p(t+1), v >; \quad v \in F_t(\overline{x}(t), \overline{x}(t-k)) \big\}, \quad (3.4)$$
$$t = 0, \ldots, N-1,$$

$$\lambda + ||p(0)|| + \ldots + ||p(N)|| > 0. \tag{3.5}$$

For the details of the proof see [4].

In Theorem 3.2 an important hypothesis is that the terminal set $X_N$ is assumed to have a nonempty cone of interior directions. Such type of assumptions may be overcome using the concept of derived cone.

Using the fact that the reachable set $R_A^N$ is a derived cone to $R_F^N$ at $\overline{x}(N)$ ( [2]), Lemma 2.2 and Lemma 3.1 we have the next version of the Maximum Principle for problem (1.1)–(1.3).

**Theorem 3.3.** *Let* $X_N \subset \mathbf{R}^n$ *be a closed set, let* $\overline{x} \in S_F$ *be an optimal solution for problem* (1.1)–(1.3) *such that Hypothesis is satisfied and let* $g(.) : \mathbf{R}^n \to \mathbf{R}$ *be a locally Lipschitz function.*

*Then for any derived cone* $C_N$ *of* $X_N$ *at* $\overline{x}(N)$ *there exist* $\lambda \in \{0, 1\}$ *and* $(p(0), p(1), \ldots, p(N)) \in \mathbf{R}^{(N+1)n}$, $(q(0), q(1), \ldots, q(N)) \in \mathbf{R}^{(N+1)n}$, *such that* (3.2)–(3.5) *hold true.*

The proof can be found in [2].

Denote by $R_Q^N$ the reachable set of the discrete delay inclusion

$$w(t+1) \in Q_{\overline{x}(t+1)} F_t\Big((\overline{x}(t), \overline{x}(t-k)) \, ; \, \big(w(t), w(t-k)\big)\Big) \quad t = 0, 1, \ldots, N-1,$$
$$w(l) = 0, \; l = -k, \ldots, 0.$$
$$\tag{3.6}$$

Let $\overline{y} = (\overline{y}(-k), \ldots, \overline{y}(N))$ satisfy (3.6) and let $R_Q^2$ denote the reachable set of the discrete delay inclusion

$$v(t+1) \in Q^2_{(\overline{x}(t+1), \overline{y}(t+1))} F_t\big(\overline{x}(t), \overline{x}(t-k)\big), \big(\overline{y}(t), \overline{y}(t-k)\big); \big(v(t), v(t-k)\big)$$
$$t = 0, 1, \ldots, N-1, \quad w(l) = 0, \quad l = -k, \ldots, 0.$$

In the next result we obtain second-order necessary optimality conditions for problem (1.1)–(1.3).

**Theorem 3.4.** *Assume that Hypothesis is satisfied, let $g(.) : \mathbf{R}^n \to \mathbf{R}$ be a locally Lipschitz function, let $\overline{x} = (\overline{x}(-k), \ldots, \overline{x}(N)) \in S_F$ be an optimal solution for problem (1.1)–(1.3) and assume that the following constraint qualification is satisfied*

$$\left\{ \begin{array}{l} \eta \in \mathbf{R}^n; \; \exists \;\; q(t), p(t) \quad t = 0, \ldots, N \quad such \; that \quad \eta = p(N), \\[2mm] \big(p(t), q(t)\big) \in \Big( C_{\overline{x}(t+1)} F_t\Big(\big(\overline{x}(t), \overline{x}(t-k)\big), (.,.)\Big)\Big)^* \big(p(t+1)\big) + \big(q(t+k), 0\big), \\[2mm] t = N-1, \ldots, 0 \quad and \quad q(t) = 0 \; for \; t \geq N \end{array} \right\} \cap (C_{\overline{x}(N)} X_N)^+ = \{0\}. \quad (3.7)$$

*Then we have the first-order necessary condition*

$$D_{\uparrow} g\big(\overline{x}(N); y(N)\big) \geq 0 \quad \forall y_N \in R_Q^N \cap Q_{\overline{x}(N)} X_N.$$

*Furthermore, if equality holds for some $\overline{y}(N)$, then we have the second-order necessary condition*

$$D_{\uparrow}^2 g\big(\overline{x}(N), \overline{y}(N); w(N)\big) \geq 0 \quad \forall w_N \in R_Q^2 \cap Q^2_{(\overline{x}(N), \overline{y}(N))} X_N.$$

The proof, that can be found in [3], is based on a general (abstract) optimality condition formulated by Zheng ( [14]) and use also several first and second-order approximations of the reachable set $R_F^N$ at $\overline{x}(N)$ ( [3]).

# References

[1] V. G. Boltjanskii, *Optimal Control for Discrete Systems*, Nauka, Moskow (in Russian), 1973.

[2] A. Cernea, *Controllability and maximum principle for discrete delay inclusions using derived cones*, Revue Roumaine Math. Pures Appl. **50** (2005), 19–29.

[3] A. Cernea, *On some second-order necessary conditions for discrete delay inclusion problems*, Math. Reports, vol. 8, **58** (2006), 259–265.

[4] A. Cernea and C. Georgescu, *The maximum principle for discrete delay inclusions with end point constraints*, Bull. Math. Soc. Sci. Math. Roum. **48(96)** (2005), 277–284.

[5] M. R. Hestenes, *Calculus of Variations and Optimal Control Theory*, Wiley, New York, 1966.

[6] A. G. Kusraev, *Discrete maximum principle*, Math. Notes. **34** (1984), 617–619.

[7] L. I. Minchenko and S. I. Sirotko, *On local controllability conditions, for discrete delay inclusions*, Diff. Equations. **38** (2002), 1058–1060.

[8] Ş. Mirică, *New proof and some generalizations of the Minimum Principle* In Optimal Control. J. Optim. Theory Appl., **74** (1992), 487–508.

[9] V. N. Phat, *Controllability of nonlinear discrete systems without differentiability assumption* Optimization **19** (1988), 133–142.

[10] A. I. Propoi, *The maximum principle for discrete control systems*, Automat. Remote Control. **26** (1965), 451–461.

[11] R. Pytlak, *A variational approach to discrete maximum principle*, IMA J. Math. Control and Information. **9** (1992), 197–220.

[12] H. D. Tuan and I. Ishizuka, *On controllability and maximum principle for discrete inclusions*. Optimization **34** (1995), 293–316.

[13] N.D. Yen and T.C. Dien, *On local controllability of nondifferentiable discrete time systems with nonconvex constraints on control*, Optimization **20** (1989), 889–899.

[14] H. Zheng, *Second-order necessary conditions for differential inclusion problems*, Appl. Math. Opt. **30** (1994), 1–14.

Aurelian Cernea
Faculty of Mathematics and Informatics
University of Bucharest
Academiei 14
RO-010014 Bucharest
Romania
e-mail: `acernea@math.math.unibuc.ro`

Progress in Nonlinear Differential Equations
and Their Applications, Vol. 75, 143–156
© 2007 Birkhäuser Verlag Basel/Switzerland

# Necessary Conditions in Optimal Control and in the Calculus of Variations

Francis Clarke

*Dedicated to Arrigo Cellina and James Yorke*

**Abstract.** The goal of this article is to find, for the two most standard paradigms in dynamic optimization, the *simplest* proofs that can be based on the techniques invented and refined over the last thirty years in connection with the nonsmooth analysis approach. Specifically, we present a proof of Theorem 2.1 below, which asserts all the first-order necessary conditions for the basic problem in the calculus of variations, and a proof of Theorem 3.1, which is the Pontryagin maximum principle in a classical context.

## 1. Introduction

The theory of necessary conditions in the calculus of variations is a classical subject whose birth can be traced back to the famous monograph published by Euler in 1744. Within the more general framework of dynamic optimization (which includes optimal control), the subject has remained active ever since. One modern approach (among others) to the issues involved has been based on the methods of *nonsmooth analysis*, a branch of the subject that began in 1973 with the author's thesis [2]. A number of people have contributed in the past decades to the substantial progress that has been made along these lines; we refer to the recent monograph [7] for details, comments, and references.

It is natural that the dominant theme in this work has been an ongoing effort to make the results as general, powerful, and unifying as possible. It is our view that the results of [7] are a culmination of these efforts in many ways.[1] Furthermore, it turns out that the nonsmooth analysis approach has given rise to the current state of the art in the subject.

---

[1] This is not to say, however, that we are announcing the end of history in this regard.

The goal of this article, however, lies in a different direction. We attempt here to find, for the two most standard paradigms in dynamic optimization, the *simplest* proofs that can be based on the techniques invented and refined over the last thirty years in connection with the nonsmooth analysis approach. Specifically, we present a proof of Theorem 2.1 below, which asserts all the first-order necessary conditions for the basic problem in the calculus of variations, and a proof of Theorem 3.1, which is the Pontryagin maximum principle in a classical context.

The devices used below (such as decoupling, penalization, use of an approximate minimization principle) are now familiar in the subject; they were introduced in the given references for much the same purposes as here. It is the elementary, self-contained, and economical nature of the proofs which is new. Of course, a concept such as "self-contained" is relative (and subjective), so let us now identify the pre-requisites in detail.

## Background

The proofs of the theorems below mostly call upon standard tools of measure and integration. Frequent use is made of measurable set-valued mappings and their selections, a theory which is perhaps not as well known as it should be. Our needs are elementary, and the short overview given in [6, pp. 149-151] is adequate. We also require a sequential compactness result that is a consequence of the Dunford-Pettis criterion for weak compactness in $L^1$, together with Ascoli's theorem. We state it now for convenience (a proof is given in [5, p. 119]).

An absolutely continuous function $x : [a, b] \to \mathbb{R}^n$ (where $[a, b]$ is a fixed interval in $\mathbb{R}$) is called an *arc*. Let a sequences of arcs $x_i$ be given which satisfy the inclusion

$$x_i'(t) \in \Gamma\big(t, x_i(t) + y_i(t)\big) + r_i(t)\overline{B}, \quad t \in \Omega_i \text{ a.e.},$$

where the set-valued mapping $\Gamma(t, x)$ from $[a, b] \times \mathbb{R}^n$ to the closed convex subsets of $\mathbb{R}^n$ is measurable in $t$ and has closed graph relative to $x$. The functions $y_i$ and $r_i$ are assumed to go to 0 in $L^1$, and the measure of the sets $\Omega_i$ converges to $b - a$. It is assumed that the sequence $x_i(a)$ is bounded, and that for some summable function $k$ we have $\Gamma(t, x) \subset k(t)\overline{B}$ and, for each $i$, $|x_i'(t)| \leq k(t)$ a.e. Then there is an arc $\bar{x}$ satisfying $\bar{x}'(t) \in \Gamma(t, \bar{x}(t))$ a.e. which is the uniform limit of a subsequence $x_{n_i}$ having the additional property that $x_{n_i}'$ converges weakly in $L^1$ to $\bar{x}'$.

Also used in the proofs is the variational principle of Ekeland (see for example [5]), which at this point can be viewed as a familiar property of complete metric spaces. Finally, a few simple facts from nonsmooth calculus will be invoked. However, it is a striking feature of the proofs given here that (in contrast to other work) they require very little beyond the actual definitions of the objects in question; we give these now (see [6] for details).

Given a lower semicontinuous function $f : X \to \mathbb{R} \cup \{+\infty\}$ and a point $x$ at which $f$ is finite, we say that $\zeta$ is a *proximal subgradient* of $f$ at $x$ if there exists $\sigma \geq 0$ such that

$$f(x') - f(x) + \sigma \, \|x' - x\|^2 \geq \langle \zeta, x' - x \rangle$$

for all $x'$ in a neighborhood of $x$. The set of such $\zeta$, which may be empty, is denoted $\partial_P f(x)$ and referred to as the *proximal subdifferential*. A closure operation then defines the *limiting subdifferential*:

$$\partial_L f(x) := \big\{ \lim \zeta_i : \zeta_i \in \partial_P f(x_i), x_i \to x, f(x_i) \to f(x) \big\}.$$

It can be shown that the same limiting subdifferential is obtained if this closure operation is applied to the subgradients usually employed in the theory of viscosity solutions.

The *indicator* of a set $S$ is the function denoted $I_S$ which takes the value $0$ on $S$ and $+\infty$ elsewhere. If $S$ is a closed subset of $\mathbb{R}^n$, and if $x$ is a point in $S$, then the *limiting normal cone* $N_S^L(x)$ to $S$ at $x$ can be defined as $\partial_L I_S(x)$. This normal cone reduces to the familiar normal vectors when $S$ is convex, or when $S$ is a manifold or a manifold with boundary.

## 2. The Lipschitz problem of Bolza

We study in this section a version of the basic problem in the calculus of variations. The problem (P) consists of minimizing the (Bolza) functional

$$J(x) := \ell_0\big(x(a)\big) + \ell_1\big(x(b)\big) + \int_a^b L\big(t, x(t), x'(t)\big)\, dt \tag{2.1}$$

over all arcs $x : [a, b] \to \mathbb{R}^n$ satisfying the constraints

$$x(a) \in C_0, \;\; x(b) \in C_1, \;\; x'(t) \in V(t) \text{ a.e.} \tag{2.2}$$

where $[a, b]$ is a given fixed interval in $\mathbb{R}$, $C_0, C_1$ are closed subsets of $\mathbb{R}^n$, $\ell_0, \ell_1 : \mathbb{R}^n \to \mathbb{R}$ are locally Lipschitz functions, and $V$ is a measurable mapping from $[a, b]$ to the closed convex subsets of $\mathbb{R}^n$.

An arc $x$ is said to be *admissible* for (P) if $x$ satisfies the constraints (2.2), and the integral in (2.1) is well-defined and finite. We are given an admissible arc $x_*$ which is a local minimum in the sense that, for some $\epsilon_* > 0$, for any admissible arc $x$ satisfying $\|x - x_*\|_\infty \le \epsilon_*$, we have $J(x_*) \le J(x)$.

The Lagrangian $L(t, x, v)$ is a mapping from $[a, b] \times \mathbb{R}^n \times \mathbb{R}^n$ to $\mathbb{R}$; we assume that it is measurable with respect to $t$ and Lipschitz with respect to $(x, v)$ near $x_*$ in the following sense: for a summable function $k : [a, b] \to \mathbb{R}$, we have, for almost all $t$, for all $x, y \in \overline{B}(x_*(t), \epsilon_*)$ and $v, w \in V(t)$,

$$|L(t, x, v) - L(t, y, w)| \le k(t)\big\{|x - y| + |v - w|\big\}. \tag{2.3}$$

The theorem below requires the following *interiority hypothesis* which distinguishes the problem from one in optimal control: There is a positive $\delta$ such that $B(x'_*(t), \delta) \subset V(t)$ a.e.

**Theorem 2.1.** *There exists an arc $p$ which satisfies the* **Euler inclusion**

$$p'(t) \in \overline{\mathrm{co}}\big\{\omega : \big(\omega, p(t)\big) \in \partial_L L\big(t, x_*(t), x'_*(t)\big)\big\} \text{ a.e.} \quad t \in [a, b] \tag{2.4}$$

*together with the* **Weierstrass condition**

$$\langle p(t), v\rangle - L\big(t, x_*(t), v\big)$$
$$\leq \langle p(t), x_*'(t)\rangle - L\big(t, x_*(t), x_*'(t)\big) \ \forall v \in V(t), \quad \text{a.e. } t \in [a, b] \tag{2.5}$$

*and the* **transversality condition**

$$p(a) \in \partial_L \ell_0\big(x_*(a)\big) + N_{C_0}^L\big(x_*(a)\big), \quad -p(b) \in \partial_L \ell_1\big(x_*(b)\big) + N_{C_1}^L\big(x_*(b)\big). \tag{2.6}$$

**Remark 2.2.** The limiting subdifferential $\partial_L L$ in the Euler inclusion is taken with respect to the $(x, v)$ variables for each fixed $t$. It reduces to a singleton when $L$ is of class $C^1$ in $(x, v)$ (or just strictly differentiable); in that case, (2.4) becomes

$$p'(t) = \nabla_x L\big(t, x_*(t), x_*'(t)\big), \quad p(t) = \nabla_v L\big(t, x_*(t), x_*'(t)\big),$$

the familiar integral (or duBois-Reymond) form of the Euler equation. This subsumes the classical first Erdmann condition: that is, the essential continuity of the function $t \mapsto \nabla_v L(t, x_*(t), x_*'(t))$, a property which serves as the gateway to higher order regularity.

*Proof.* We begin by identifying certain additional hypotheses that can be made without any loss of generality, by simple reformulations.

Note first that the theorem's hypotheses and conclusions are unaffected if we redefine $L(t, x, v)$ to be $L(t, \pi_t(x), \pi_t'(v))$, where $\pi_t(x)$ denotes the projection of $x$ onto the set $\overline{B}(x_*(t), \epsilon_*)$ and $\pi_t'(v)$ is the projection of $v$ onto $V(t)$. Since $\pi_t$ and $\pi_t'$ are globally Lipschitz, this convention allows us to suppose that the Lipschitz condition (2.3) holds globally. By similar arguments, we may suppose that $\ell_0$ and $\ell_1$ are bounded below and globally Lipschitz, and that $C_0$ is compact. We suppose as well that $k(t) \geq 1$. Finally, by reformulating we may take $x_* \equiv 0$ and $[a, b] = [0, 1]$.

We proceed now to prove the theorem under two additional hypotheses whose removal will constitute the last step in the proof.

**Temporary hypotheses:** (**TH1**) $C_1 = \mathbb{R}^n$ (so that there is no explicit constraint on $x(b)$).

(**TH2**) There exists $R > 0$ such that $V(t) \subset B(x_*'(t), R)$ a.e.

**A.** We proceed to define via penalization a sequence of *decoupled* problems converging in an appropriate sense to (P). We introduce, for a given sequence of positive numbers $n_i$ tending to $+\infty$,

$$\ell_i^1(y) := \min_{\beta \in \mathbb{R}^n} \left\{ \ell_1(\beta) + n_i |y - \beta|^2 \right\}, \tag{2.7}$$

a type of expression known as a *quadratic inf-convolution*, and which figures in the Moreau-Yosida approximation to $\ell_1$. Since $\ell_1$ is globally Lipschitz and bounded below, there is a constant $c$ such that

$$\ell_i^1 \leq \ell \leq \ell_i^1 + c/\sqrt{n_i}.$$

We set

$$L_i(t, x, v) := \min_{u \in \mathbb{R}^n} \left\{ L(t, u, v) + n_i k(t) |u - x|^2 \right\}$$

$$J_i(x) := \ell_0\big(x(0)\big) + \ell_i^1\big(x(1)\big) + \int_0^1 L_i\big(t, x(t), x'(t)\big) \, dt$$

and we define $I_i$ to be the infimum of $J_i(x)$ over all arcs $x$ satisfying

$$x(0) \in C_0, \quad x'(t) \in V(t) \text{ a.e.}, \quad |x(0)| \leq \epsilon_*/2, \quad \int_0^1 |x'(t)| \, dt \leq \epsilon_*/2. \tag{2.8}$$

Note that these constraints imply $\|x\|_\infty \leq \epsilon_*$. Because of (TH2) we have (for some constant $c_0$)

$$c_0 \leq I_i \leq J_i(0) \leq \ell_0(0) + \ell_1(0) + \int_0^1 L(t, 0, 0) \, dt = J(0).$$

**Lemma 2.3.** $\lim_{i \to +\infty} I_i = J(0)$.

To see this, let $x_i$ satisfy (2.8) together with $J_i(x_i) \leq I_i + n_i^{-1}$, and let $u_i$ be a measurable function such that $u_i(t)$ is (almost everywhere) a point at which the minimum defining $L_i(t, x_i(t), x_i'(t))$ is achieved:

$$L_i\big(t, x_i(t), x_i'(t)\big) = L\big(t, u_i(t), x_i'(t)\big) + n_i k(t) |u_i(t) - x_i(t)|^2 \text{ a.e.}$$

(This is the first of several times that the existence of a measurable selection is left as an exercise.) This equality together with (2.3) leads to $|u_i(t) - x_i(t)| \leq n_i^{-1}$ a.e. We now observe

$$I_i + n_i^{-1} \geq J_i(x_i)$$

$$= \ell_0\big(x_i(0)\big) + \ell_i^1\big(x_i(1)\big) + \int_0^1 \left\{ L(t, u_i, x_i') + n_i k(t) |u_i - x_i|^2 \right\} dt$$

$$\geq \ell_0\big(x_i(0)\big) + \ell^1\big(x_i(1)\big) - c/\sqrt{n_i} + \int_0^1 \left\{ L(t, x_i, x_i') - k(t) |u_i - x_i| \right\} dt$$

$$\geq J(0) - c/\sqrt{n_i} - \|k\|_1/n_i,$$

and the assertion of the lemma follows.

We may view the problem defining $I_i$ as one that is defined relative to the couples $(x(0), x')$ in the complete metric space $\mathbb{R}^n \times L^1$ lying in the closed set $S$ defined by (2.8). The lemma implies that the arc $x_* = 0$ is $\epsilon_i^2$-optimal for the problem, where $\epsilon_i$ is a positive sequence tending to 0. We apply Ekeland's theorem (see [5]) to deduce the existence of an arc $x_i \in S$ satisfying

$$|x_i(0)| + \int_0^1 |x_i'(t)| \, dt \leq \epsilon_i, \tag{2.9}$$

and which minimizes over $S$ the functional $J'_i$ defined by

$$J'_i\big(x(0), x'\big) := \ell_0\big(x(0)\big) + \epsilon_i|x(0) - x_i(0)| + \ell^1_i\big(x(1)\big)$$

$$+ \int_0^1 \min_u \Big\{ L\big(t, u, x'(t)\big) + n_i k(t)|u - x(t)|^2 + \epsilon_i|x'(t) - x'_i(t)| \Big\} \, dt \, .$$

We may pass to a subsequence (without relabeling) to assure $x'_i(t) \to 0$ a.e.
**B.** We now fix $i$ and reformulate the optimality of $x_i$ for $J'_i$ in a more useful manner, one that will allow us to identify an arc $p_i$ that is "close" to satisfying the necessary conditions. Let $u_i$ be a measurable function such that almost everywhere the minimum

$$\min_u \Big\{ L\big(t, u, x'_i(t)\big) + n_i k(t)|u - x_i(t)|^2 \Big\}$$

is achieved at $u_i(t)$; it follows as in the proof of the lemma that $|u_i(t) - x_i(t)| \leq n_i^{-1}$ a.e. Now let $\beta_i$ be a point achieving the minimum in (2.7) when $y = x_i(1)$. We proceed to define an arc $p_i$ via

$$p'_i(t) = 2n_i k(t)\big(x_i(t) - u_i(t)\big), \quad p_i(1) = -2n_i\big(x_i(1) - \beta_i\big) \, .$$

Then we have (by choice of $\beta_i$)

$$-p_i(1) \in \partial_P \ell_1(\beta_i) \, . \tag{2.10}$$

Using the observation

$$\ell^1_i(y) \leq \ell_1(\beta_i) + n_i|y - \beta_i|^2 \; \forall \, y \, ,$$

(with equality for $y = x_i(1)$) together with the identity

$$n_i k|u - x|^2 = n_i k|u_i - x_i|^2 - \langle p'_i, u - u_i \rangle + \langle p'_i, x - x_i \rangle + n_i k|(x - x_i) - (u - u_i)|^2 \, ,$$

and integration by parts, we see that the cost functional $\Phi(u, \alpha, v)$ defined on $L^\infty \times \mathbb{R}^n \times L^1$ by

$$\ell_0(\alpha) + \epsilon_i|\alpha - x_i(0)| - \langle p_i(0), \alpha \rangle + n_i |x(1) - \beta_i|^2 + \langle p_i(1), x(1) \rangle$$

$$+ \int_0^1 \Big\{ L\big(t, u(t), v(t)\big) - \langle p_i(t), v(t) \rangle - \langle p'_i(t), u(t) \rangle + \epsilon_i|v(t) - x'_i(t)| \Big\} \, dt$$

$$+ 2n_i \int_0^1 k(t)\Big\{ |u(t) - u_i(t)|^2 + |x(t) - x_i(t)|^2 \Big\} \, dt$$

satisfies (letting $x(t)$ stand for $\alpha + \int_0^t v(s) \, ds$, and for a certain constant $c_i$):

$$\Phi(u, \alpha, v) \geq J'_i\big(x(0), x'\big) + c_i \, ,$$

with equality when $(u, \alpha, v) = (u_i, x_i(0), x'_i)$. It follows that $\Phi(u, \alpha, v)$ is minimized relative to the constraints

$$\alpha \in C_0 \, , \quad |\alpha| \leq \epsilon_*/2 \, , \quad v(t) \in V(t) \text{ a.e.} \, , \quad \int_0^1 |v(t)| dt \leq \epsilon_*/2 \tag{2.11}$$

at $(u_i, x_i(0), x'_i)$.

It is easy to see (by substituting for $x$ and $p$) that the last two boundary terms in the expression for $\Phi$ may be rewritten in the form $n_i\{|x(1) - x_i(1)|^2 + |\beta_i|^2 - |x_i(1)|^2\}$. It follows then that the functional $\Psi(u, \alpha, v)$ defined by

$$\ell_0(\alpha) + \epsilon_i|\alpha - x_i(0)| - \langle p_i(0), \alpha\rangle + n_i|x(1) - x_i(1)|^2$$

$$+ \int_0^1 \left\{L\big(t, u(t), v(t)\big) - \langle p_i(t), v(t)\rangle - \langle p_i'(t), u(t)\rangle + \epsilon_i|v(t) - x_i'(t)|\right\} dt$$

$$+ 2n_i \int_0^1 k(t)\left\{|u(t) - u_i(t)|^2 + |x(t) - x_i(t)|^2\right\} dt$$

is minimized relative to the constraints (2.11) at $(u_i, x_i(0), x_i')$.

**C.** The next step consists of a variational analysis (for $i$ still fixed) of the minimum of $\Psi$ just mentioned. Let us first fix $u = u_i$ and $v = x_i'$. Then the function $\alpha \mapsto \Psi(u_i, \alpha, x_i')$ attains a local minimum (for $i$ sufficiently large, since $x_i(0) \to 0$) relative to $\alpha \in C_0$ at $x_i(0)$. The corresponding necessary condition is

$$p_i(0) \in \partial_L\{\ell_0 + I_{C_0}\}\big(x_i(0)\big) + \epsilon_i\overline{B}. \tag{2.12}$$

This, together with (2.10), is the precursor of the transversality condition (2.6).

We now exploit the minimum in $v$ of $\Psi(u_i, x_i(0), v)$ to derive a forerunner of the Weierstrass condition. The constraint on $v$ in (2.11) is slack for $i$ sufficiently large, and a simple argument by contradiction shows that we must then have, for almost every $t$,

$$\langle_i(t), v\rangle - L(t, u_i(t), v) - \epsilon_i|v - x_i'(t)| \le$$

$$\langle p_i(t), x_i'(t)\rangle - L\big(t, u_i(t), x_i'(t)\big) \ \forall\, v \in V(t). \tag{2.13}$$

Let us give this argument. If (2.13) does not hold, then there exists $r > 0$ and a subset $S$ of $[a, b]$ of positive measure $m$ such that, for some measurable function $w$ defined on $S$ and taking values in $V(t)$, we have

$$L\big(t, u_i(t), w(t)\big) + \epsilon_i|w(t) - x_i'(t)| - \langle p_i(t), w(t)\rangle \le$$

$$L\big(t, u_i(t), x_i'(t)\big) - \langle p_i(t), x_i'(t)\rangle - r, \ t \in S \text{ a.e.}$$

Of course, $m$ can be taken arbitrarily small. If we let $v$ be the function equal to $w$ on $S$ and equal to $x_i'$ elsewhere, and if $x(t)$ signifies $x_i(0) + \int_0^t v(s)\, ds$, then we have $\|x - x_i\|_\infty \le Km$ for a constant $K$ independent of $m$ (the hypothesis (TH2) is used for this). It follows that

$$\Psi\big(u_i, x_i(0), v\big) - \Psi\big(u_i, x_i(0), x_i'\big) \le -rm + K^2 n_i\big(1 + 2\|k\|_1\big)m^2 < 0$$

for $m$ sufficiently small. Further, $v$ satisfies the constraint in (2.11) if $m$ is small enough. This contradicts the optimality of $(u_i, x_i(0), x_i')$ and concludes the argument.

Making use of such evident estimates as

$$|x(t) - x_i(t)|^2 \le 2|x(0) - x_i(0)|^2 + 2\int_0^1 |x'(s) - x_i'(s)|^2 ds$$

and rearranging, we deduce that the cost functional $\Psi^+(u, \alpha, v)$ defined by

$$\ell_0(\alpha) + \epsilon_i |\alpha - x_i(0)| - \langle p_i(0), \alpha \rangle + 6n_i |\alpha - x_i(0)|^2$$

$$+ \int_0^1 \left\{ L\big(t, u(t), v(t)\big) - \langle p_i(t), v(t) \rangle - \langle p'_i(t), u(t) \rangle + \epsilon_i |v(t) - x'_i(t)| \right\} dt$$

$$+ 2n_i \int_0^1 k(t) |u(t) - u_i(t)|^2 \, dt + 6n_i \int_0^1 k(t) |v(t) - x'_i(t)|^2 \, dt$$

also attains a minimum relative to the constraints (2.11) at $(u_i, x_i(0), x'_i)$.

Setting $\alpha = x_i(0), v = x'_i$ in $\Psi^+$, the attainment of the minimum relative to $u \in L^\infty$ implies by measurable selection theory that for $t$ a.e., it is the case that $u_i(t)$ minimizes freely the integrand in $\Psi^+$. This fact yields

$$p'_i(t) \in \partial_P \Big\{ L\big(t, \cdot, x'_i(t)\big) \Big\} \big( u_i(t) \big) \quad \text{a.e.},$$

which in turn gives

$$|p'_i(t)| \le k(t) \quad \text{a.e.} \tag{2.14}$$

When the constraint on $v$ in (2.11) is slack, it follows that for almost every $t$, the minimum with respect to $(u, v) \in \mathbb{R}^n \times V(t)$ of the integrand in $\Psi^+$ is attained at $(u_i(t), x'_i(t))$; this implies an intermediate version of the Euler inclusion:

$$\big(p'_i(t), p_i(t)\big) \in \partial_L L\big(t, u_i(t), x'_i(t)\big) + \{0\} \times \epsilon_i \overline{B}, \quad t \in \Omega_i \ \text{a.e.} \tag{2.15}$$

where $\Omega_i := \{t \in [0, 1] : x'_i(t) \in \operatorname{int} V(t)\}$. Note that the measure of $\Omega_i$ tends to 1 as $i \to \infty$, in light of the interiority hypothesis.

**D.** The next step is to let $i$ tend to infinity. The conditions (2.14) and (2.10) allow us to deduce (for a subsequence, without relabeling) that $p_i$ converges uniformly to an arc $p$ and $p'_i$ converges weakly to $p'$. Passing to the limit in (2.10) and (2.12) (note (2.9), and that $\beta_i \to x_*(1)$), we see that $p$ satisfies the transversality condition (2.6) (for $C_1 = \mathbb{R}^n$). The Euler inclusion (2.4) follows from (2.15) by the sequential compactness result stated in the Introduction. From (2.13) we conclude that almost everywhere we have

$$\langle p(t), v \rangle - L(t, 0, v) \le \langle p(t), 0 \rangle - L(t, 0, 0) \ \forall v \in V(t),$$

which is the desired Weierstrass condition. The theorem is therefore proven, in the presence of the Temporary Hypotheses (TH1)(TH2).

**E.** The final step in the proof is the removal of the Temporary Hypotheses. It is clear at this point (given our expertise at passing to the limit in the necessary conditions along subsequences) that it suffices to deduce, for any $\eta > 0$ sufficiently small, the necessary conditions for $x_*$ in which the Weierstrass condition holds for the following subset of $V(t)$:

$$V_\eta(t) := \Big\{ v \in V(t) \cap \overline{B}\big(x'_*(t), 1/\eta\big) : v + \eta \overline{B} \subset V(t) \Big\}$$

(note the role of the interiority hypothesis in this reduction.) In this setting, the case of an arbitrary $C_1$ is reduced to the one in which $C_1 = \mathbb{R}^n$ by an exact penalization device, as follows. A simple argument by contradiction (as in [3]) shows that for some $K > 0$ sufficiently large, $x_*$ solves the problem of minimizing

$$J_K(x) := \ell_0\big(x(0)\big) + \ell_1\big(x(1)\big) + K d_{C_1}\big(x(1)\big) + \int_0^1 L\big(t, x(t), x'(t)\big) \, dt$$

over the arcs $x$ satisfying

$$x(0) \in C_0, \; \|x - x_*\|_\infty \le \epsilon_*/2, \; x'(t) \in V_\eta(t) \text{ a.e.}$$

The argument goes as follows. If the assertion is false, then there exists for each positive integer $j$ an arc $x_j$ admissible for this problem with $J_j(x_j) < J_j(x_*) = J(x_*)$. Since $J_j(x_j)$ is bounded below, it follows that $d_{C_1}(x_j(1)) \to 0$. Let $\sigma_j$ be a closest point in $C_1$ to $x_j(1)$. Then for $j$ sufficiently large, the arc $y_j(t) := x_j(t) + t(\sigma_j - x_j(1))$ satisfies (2.2) and we have, for a certain constant $K_0$ depending only upon $k(\cdot)$,

$$\begin{aligned} J(y_j) &\le J(x_j) + K_0 d_{C_1}\big(x_j(1)\big) \\ &\le J_j(x_j) < J(x_*), \end{aligned}$$

contradicting the optimality of $x_*$.

This new penalized problem satisfies (TH1) and (TH2), so we may apply the theorem already proven to deduce the existence of an arc $p$ satisfying the Euler inclusion and the Weierstrass condition (for $V_\eta$, as agreed). It remains to see that transversality holds. But we have (invoking two simple facts from nonsmooth calculus)

$$\begin{aligned} -p(1) \in \partial_L(\ell_1 + K d_{C_1})\big(x_*(1)\big) &\subset \partial_L \ell_1\big(x_*(1)\big) + K \partial_L d_{C_1}\big(x_*(1)\big) \\ &\subset \partial_L \ell_1\big(x_*(1)\big) + N^L_{C_1}\big(x_*(1)\big), \end{aligned}$$

and the theorem is proved. $\qquad\square$

**Remark 2.4.** Until this final step, no Lipschitz behavior of $L$ with respect to $v$ is used in the proof. It can in fact be dispensed with, but then the necessary conditions may hold only in *abnormal* form. This is but one of several ways in which the theorem can be extended (at considerable technical expense, however). As regards necessary conditions for one-dimensional problems in the calculus of variations with finite-valued Lagrangians, the state of the art is currently given by [7, Theorem 4.4.1].

It is also possible to consider the problem (P) with extended-valued Lagrangians, as in Theorem 4.1.1 of [7]. From this result, the classical *multiplier rule* for mixed constraints such as $\psi(t, x(t), x'(t)) = 0$ a.e. (for example) follows as a special case. In addition, the nature of the local minimum is more general; it is linked to the hypotheses and conclusions in a *stratified* manner. The Lipschitz hypothesis is replaced in such a context by a much weaker *pseudo-Lipschitz* one. Finally, one can develop structural criteria on the Lagrangian (the *generalized Tonelli-Morrey conditions*) that have the important property of automatically yielding the required pseudo-Lipschitz behavior near the given arc (whatever it may be); see [7, Theorem 4.3.2].

The advantage of Theorem 2.1 in comparison to these more general results stems solely from the directness and relative simplicity of the proof given above. As we have said, the hypotheses can be considerably weakened. Nonetheless, as we shall now see, Theorem 2.1 can play a very useful role in obtaining necessary conditions for optimal control problems. The fact that nonsmooth Lagrangians are admitted is crucial in this regard.

## 3. The maximum principle

We consider now the standard control system

$$x'(t) = f\big(t, x(t), u(t)\big), \ t \in [a, b] \text{ a.e.} \tag{3.1}$$

where $u$ is a measurable function on $[a, b]$ whose values are constrained as follows:

$$u(t) \in U(t) \text{ a.e.},$$

where $U$ is a measurable mapping from $[a, b]$ to the subsets of $\mathbb{R}^m$. Such a function $u$ is called a *control*, and the corresponding function $x(t)$ is termed a *trajectory*. Our interest centers around a given control-trajectory pair $(u_*, x_*)$ satisfying $x_*(a) \in C_0$, where $C_0$ is a given set in $\mathbb{R}^n$.

For $\epsilon_* > 0$ given, let us define the (local) *attainable set from* $C_0$, denoted $\mathcal{A}[C_0]$, as the set of all points $x(b)$, where $x(t)$ satisfies (3.1) for some control $u(t)$, as well as $x(a) \in C_0$ and $\|x - x_*\|_\infty \leq \epsilon_*$. Now let a function $\Phi : \mathbb{R}^n \to \mathbb{R}^N$ be given. Then $x_*$ is called a local $\Phi$-*boundary trajectory* if $\Phi(x_*(b)) \in \text{bdry } \Phi(\mathcal{A}[C_0])$.

Our purpose is to give necessary conditions for such a trajectory and its associated control. The hypotheses on the data are resolutely simple: $f$ and $\Phi$ are continuously differentiable and $C_0$ compact; as for $U$, we assume that $U(t)$ is a closed subset of a compact set $U_0$ for each $t$.[2]

**Theorem 3.1.** *Let* $x_*$ *be a local* $\Phi$-*boundary trajectory corresponding to the control* $u_*$. *Then there exists an arc* $p$ *satisfying the* **adjoint equation**

$$-p'(t) = D_x f\big(t, x_*(t), u_*(t)\big)^T p(t) \text{ a.e.} \tag{3.2}$$

*as well as the* **maximum condition**

$$\max_{u \in U(t)} \Big\langle p(t), f\big(t, x_*(t), u\big) \Big\rangle = \Big\langle p(t), f\big(t, x_*(t), u_*(t)\big) \Big\rangle \text{ a.e.} \tag{3.3}$$

*and the* **transversality condition:** *for some unit vector* $\nu \in \mathbb{R}^N$,

$$p(a) \in N^L_{C_0}\big(x_*(a)\big), \quad p(b) = D\Phi\big(x_*(b)\big)^T \nu. \tag{3.4}$$

---

[2]Thus we do not pursue the greatest generality as regards the regularity of the data. However, there are structural hypotheses *not* made here that would significantly reduce the complexity of the problem. Notable among these are: linearity of $f$ in $x$, one state endpoint being free, and the convexity of the sets $f(t, x, U(t))$ (this last condition is connected to the *relaxation* of the problem).

*Proof.* **A.** We may take $[a, b] = [0, 1]$ and assume (without loss of generality) that $f$ and $\Phi$ are globally Lipschitz. We proceed under the following
**Temporary Hypothesis:** There exists $\eta > 0$ with the following property: for almost each $t$, for every $u \in U(t)$ different from $u_*(t)$, one has

$$\left| f\big(t, x_*(t), u\big) - f\big(t, x_*(t), u_*(t)\big) \right| \geq \eta. \tag{3.5}$$

By the definition of boundary, for any $\epsilon \in (0, 1)$, there is a point $\gamma \notin \Phi(\mathcal{A}[C_0])$ such that $|\Phi(x_*(1)) - \gamma| < \epsilon^2$. Thus $x_*$ is $\epsilon^2$-optimal for the problem of minimizing $|\Phi(x(1)) - \gamma|$ over all trajectories $x$ satisfying $x(0) \in C_0$ and $\|x - x_*\|_\infty \leq \epsilon_*$. We take the distance between two such trajectories $x_1, x_2$ to be

$$|x_1(0) - x_2(0)| + \int_0^1 |x_1'(t) - x_2'(t)|\, dt.$$

It follows from Ekeland's theorem that some trajectory $\bar{x}$ satisfying

$$|\bar{x}(0) - x_*(0)| + \int_0^1 |\bar{x}'(t) - x_*'(t)|\, dt \leq \epsilon \tag{3.6}$$

minimizes the functional $J_0(x)$ defined by

$$\left| \Phi\big(x(1)\big) - \gamma \right| + \epsilon |x(0) - \bar{x}(0)| + \epsilon \int_0^1 |x'(t) - \bar{x}'(t)|\, dt \tag{3.7}$$

over all the trajectories $x$ in question. Observe that $|\Phi(\bar{x}(1)) - \gamma| \neq 0$ (since $\gamma \notin \Phi(\mathcal{A}[C_0])$) and that for $\epsilon$ sufficiently small we have $\|\bar{x} - x_*\|_\infty < \epsilon_*/2$.
**B.** The next step is to reinterpret the problem above in such a way that Theorem 2.1 can be applied to it. We employ an exact penalization technique that hinges on the following approximation fact.

**Lemma 3.2.** *There is a constant $K$ with the following property: if $z$ is any arc on $[0, 1]$ emanating from $C_0$, then there is a trajectory $y$ with $y(0) = z(0)$ such that*

$$\int_0^1 |y'(t) - z'(t)|\, dt \leq K \int_0^1 \min_{u \in U(t)} \left| z'(t) - f\big(t, z(t), u\big) \right| dt. \tag{3.8}$$

Proof: Let $u$ have values in $U(t)$ a.e. and satisfy

$$\left| z'(t) - f\big(t, z(t), u(t)\big) \right| = \min_{u \in U(t)} \left| z'(t) - f\big(t, z(t), u\big) \right| \text{ a.e.},$$

and let $y$ be the trajectory generated by the control $u$, with initial condition $y(0) = z(0)$. Then, letting $K_f$ denote a Lipschitz constant for $f$, we have almost everywhere

$$\begin{aligned}
|y'(t) - z'(t)| &= \left| f\big(t, y(t), u(t)\big) - z'(t) \right| \\
&\leq \left| f\big(t, z(t), u(t)\big) - z'(t) \right| + K_f |y(t) - z(t)| \\
&= \min_{u \in U(t)} \left| z'(t) - f\big(t, z(t), u\big) \right| + K_f |y(t) - z(t)|,
\end{aligned}$$

and the estimate (3.8) follows from Gronwall's Lemma.

Let us set

$$L_0(t, x, v) := \min_{u \in U(t)} |v - f(t, x, u)|.$$

Let $K_\Phi$ be a Lipschitz constant for $\Phi$ and set $K_0 := K(K_\Phi + 2)$. We claim that the arc $\bar{x}$ minimizes the functional

$$J(x) := J_0(x) + K_0 \int_0^1 L_0(t, x(t), x'(t)) \, dt$$

over the arcs $x$ satisfying $\|x - \bar{x}\|_\infty \leq \epsilon_*/2$ and $x(0) \in C_0$. If this were false, there would be an admissible arc $z$ such that $J(z) < J(\bar{x})$. Apply the lemma to obtain a trajectory $y$ as indicated. Then

$$J_0(y) \leq J_0(z) + (K_\Phi + 2) \int_0^1 |y'(t) - z'(t)| \, dt$$

$$\leq J_0(z) + (K_\Phi + 2)K \int_0^1 \min_{u \in U(t)} |z'(t) - f(t, z(t), u)| \, dt$$

$$= J(z) < J(\bar{x}) = J_0(\bar{x}),$$

contradicting the optimality of $\bar{x}$ relative to (3.7).

**C.** We now apply the necessary conditions of Theorem 2.1 relative to the minimization of $J$ by $\bar{x}$. We deduce the existence of an arc $\bar{p}$ satisfying

$$\bar{p}(0) \in N_{C_0}^L(\bar{x}(0)) + \epsilon \bar{B}, \quad -\bar{p}(1) = D\Phi(\bar{x}(1))^T \nu, \tag{3.9}$$

where $\nu$ is the unit vector $(\Phi(\bar{x}(1)) - \gamma)/|\Phi(\bar{x}(1)) - \gamma|$, together with the Euler equation (2.4) and the Weierstrass condition (2.5). The latter evidently implies (take $v$ of the form $f(t, \bar{x}(t), u)$ for $u \in U(t)$) that we have almost everywhere

$$\left\langle \bar{p}(t), f(t, \bar{x}(t), u) \right\rangle - \epsilon |f(t, \bar{x}(t), u) - \bar{x}'(t)| \leq \langle \bar{p}(t), \bar{x}'(t) \rangle, \quad u \in U(t). \tag{3.10}$$

We now examine the Euler inclusion. To begin with, it yields the estimate

$$|\bar{p}'(t)| \leq k(t), \quad t \in [a, b] \text{ a.e.} \tag{3.11}$$

Next, let us define

$$\Omega := \left\{ t : K_f |\bar{x}(t) - x_*(t)| + |\bar{x}'(t) - x_*'(t)| < \eta \right\}. \tag{3.12}$$

Consider a value of $t \in \Omega$ for which $\bar{x}'(t) = f(t, \bar{x}(t), \bar{u}(t))$, and let $(x, v)$ and $L_0(t, x, v)$ satisfy

$$\left|(x, v) - (\bar{x}(t), \bar{x}'(t))\right| + L_0(t, x, v) < \delta$$

for some $\delta > 0$ (note that $L_0(t, \bar{x}(t), \bar{x}'(t)) = 0$). If $u \in U(t)$ provides the minimum in the definition of $L_0(t, x, v)$, then it follows that for $\delta$ sufficiently small we have $|f(t, x_*(t), u) - f(t, x_*(t), u_*(t))| < \eta$, so that (by the Temporary Hypothesis (3.5)), $u = u_*(t)$. Locally therefore, $L_0(t, x, v)$ is given by $|v - f(t, x, u_*(t))|$.

A simple exercise is in order: let $(q, p) \in \partial_P h(x, v)$, where $h(x, v) = |v - g(x)|$; then $q = -Dg(x)^T p$. Invoking this, the Euler equation is seen to imply

$$\bar{p}'(t) + D_x f(t, \bar{x}(t), u_*(t))^T \bar{p}(t) \in \epsilon \bar{B}, \quad t \in \Omega \text{ a.e.} \tag{3.13}$$

It is now time for $\epsilon$ to experience its usual fate and go to 0 (at least along a sequence $\epsilon_i$). The arc $\bar{x}$ depends of course on this parameter: $\bar{x} = \bar{x}_{\epsilon_i}$, as do the arc $\bar{p}$, the unit vector $\nu$, and the set $\Omega$ defined by (3.12). In light of (3.11), (3.9) and (3.6), for an appropriate subsequence we have $\bar{p}_{\epsilon_i}$ converging to an arc $p$, $\bar{x}_{\epsilon_i}$ converging uniformly to $x_*$, $\bar{x}'_{\epsilon_i}(t)$ converging almost everywhere to $x'_*(t)$, and $\nu_{\epsilon_i}$ converging to a unit vector $\nu$. The measure of $\Omega_{\epsilon_i}$ goes therefore to 1, and it follows from (3.13), (3.10) and (3.9) that the limiting arc $p$ satisfies all the requirements of Theorem 3.1.

**D.** There remains the Temporary Hypothesis to deal with. We define

$$U_\eta(t) := \{u_*(t)\} \cup \left\{u \in U(t) : |f(t, x_*(t), u) - f(t, x_*(t), u_*(t))| \geq \eta\right\}.$$

When we replace $U$ by $U_\eta$, $x_*$ is still a local $\Phi$-boundary trajectory for the new system, and all the hypotheses of the theorem continue to hold, as well as the Temporary Hypothesis (3.5). We therefore deduce the existence of an arc $p_\eta$ satisfying all the required conditions, except that the maximum condition holds only for the control values in $U_\eta$. An appeal to a subsequence of $\eta_i$ converging to 0 gives the required arc $p$, in a now familiar fashion.                                   $\square$

**Remark 3.3.** It is well-known that the necessary conditions for a boundary trajectory subsume those that correspond to optimality. To be specific, consider now the minimization of a cost functional

$$\ell_1(x(b)) + \int_a^b L(t, x(t), u(t))\, dt$$

over the same trajectory-control pairs $(x, u)$ as above satisfying $x(a) \in C_0$, $x(b) \in C_1$, where we take $L$ smooth, $\ell_1$ Lipschitz, $C_0, C_1$ closed. Then if $x_*$ is a local solution corresponding to the control $u_*$ there is an arc $p$ satisfying the following versions of the three necessary conditions: the adjoint equation

$$-p'(t) = D_x f(t, x_*(t), u_*(t))^T p(t) - \lambda L_x(t, x_*(t), u_*(t)) \quad \text{a.e.},$$

the maximum condition

$$\max_{u \in U(t)} \left\{ \big\langle p(t), f(t, x_*(t), u) \big\rangle - \lambda L(t, x_*(t), u) \right\}$$
$$= \big\langle p(t), f(t, x_*(t), u_*(t)) \big\rangle - \lambda L(t, x_*(t), u_*(t)) \quad \text{a.e.}$$

and the transversality condition

$$p(a) \in N_{C_0}(x_*(a)), \quad -p(b) \in \lambda \partial_L \ell_1(x_*(b)) + N_{C_1}(x_*(b)).$$

Here, $\lambda$ is a scalar equal to 0 or 1 (the case $\lambda = 1$ being called "normal") and it is also asserted that $(\lambda, p(t))$ is nonvanishing. These necessary conditions follow from Theorem 3.1 by simple reformulation; see for example [5, pp. 212].

Once again, Theorem 3.1 is a long way from being the best that can be obtained. In [7, Theorem 5.1.1], under greatly reduced regularity hypotheses, necessary conditions are given in the setting of *differential inclusions* (see [1]) and *generalized control systems*. It is also possible to treat a new hybrid problem [7, Theorem 5.3.1] that goes well beyond the standard formulation of optimal control, allowing for example the consideration of mixed state-control constraints $\psi(t, x(t), u(t)) \leq 0$ (see also [8]). To obtain these more general results requires considerably greater investment, however.

## References

[1] J. P. Aubin and A. Cellina. *Differential Inclusions*. Springer-Verlag, 1984.

[2] F. H. Clarke. *Necessary Conditions for Nonsmooth Problems in Optimal Control and the Calculus of Variations*. Doctoral thesis, University of Washington, 1973. (Thesis director: R. T. Rockafellar).

[3] F. H. Clarke. *The Euler-Lagrange differential inclusion*. J. Differential Equations, **19** (1975), 80–90.

[4] F. H. Clarke. *Maximum principles without differentiability*. Bulletin Amer. Math. Soc., **81** (1975), 219–2221.

[5] F. H. Clarke. *Optimization and Nonsmooth Analysis*. Wiley-Interscience, New York, 1983. Republished as Vol. 5 of *Classics in Applied Mathematics*, SIAM, 1990.

[6] F. H. Clarke, Yu. S. Ledyaev, R. J. Stern, and P. R. Wolenski. *Nonsmooth Analysis and Control Theory*. Graduate Texts in Mathematics, Vol. 178. Springer-Verlag, New York, 1998.

[7] F. Clarke. *Necessary Conditions in Dynamic Optimization*. Memoirs of the Amer. Math. Soc., No. 816, Vol. 173. 2005.

[8] F. Clarke. *The maximum principle in optimal control*. J. Cybernetics and Control, **34** (2005), 709–722.

[9] R. B. Vinter. *Optimal Control*. Birkhäuser, Boston, 2000.

Francis Clarke
Institut universitaire de France et Institut Camille Jordan
Université Claude Bernard Lyon 1
FR-69622 Villeurbanne
France
e-mail: `clarke@math.univ-lyon1.fr`

Progress in Nonlinear Differential Equations
and Their Applications, Vol. 75, 157–163

# Almost Periodicity in Functional Equations

C. Corduneanu

*Dedicated to Arrigo Cellina and James Yorke*

**Abstract.** The paper is aimed at providing some results on the almost periodicity of solutions to some general functional or functional differential equations. The term "general" is meant in the sense that the equations involve operators of general form, acting on the space of almost periodic functions. First order and second order equations are dealt with.

**Mathematics Subject Classification (2000).** Primary 99Z99; Secondary 00A00.

**Keywords.** Functional equations, almost periodicity.

## 1.

First, we shall consider the functional differential equation of the form

$$\dot{x}(t) = (Lx)(t) + (Fx)(t), \quad t \in \mathbb{R} \tag{1}$$

where $x : \mathbb{R} \to \mathcal{C}^n$ is the unknown map, while $L$ is a linear continuous operator on the space $AP(\mathbb{R}, \mathcal{C}^n)$. The operator $F : AP(\mathbb{R}, \mathcal{C}^n) \to AP(\mathbb{R}, \mathcal{C}^n)$ is assumed continuous, and will be subject to some restrictions. In (1), the operator $F$ plays the role of the nonlinearity.

The problem of almost periodicity of bounded solutions to (1), i.e., $x \in BC(\mathbb{R}, \mathcal{C}^n)$, which stands for the space (Banach) of continuous and bounded maps from $\mathbb{R}$ into $\mathcal{C}^n$, with the supremum norm, has been considered under various hypotheses. In our recent paper [4], we have provided a rather general result, which can be stated as follows:

**Theorem 1.1.** *Consider the system* (1), *under the following hypotheses:*

1) $L : AP(\mathbb{R}, \mathcal{C}^n) \to AP(\mathbb{R}, \mathcal{C}^n)$ *is a linear continuous operator.*
2) *The linear equation*

$$\dot{x}(t) = (Lx)(t) + f(t), \quad t \in \mathbb{R} \tag{2}$$

*has a unique solution in* $AP(\mathbb{R}, \mathcal{C}^n)$, *for each* $f \in AP(\mathbb{R}, \mathcal{C}^n)$.

3) $F : AP(\mathbb{R}, \mathcal{C}^n) \to AP(\mathbb{R}, \mathcal{C}^n)$ *is an operator, generally nonlinear, such that*

$$|Fx - Fy|_{AP} \leq \lambda |x - y|_{AP}, \tag{3}$$

*with* $\lambda > 0$ *a constant, for each* $x, y \in AP(\mathbb{R}, \mathcal{C}^n)$.

*Then* (1) *has a unique almost periodic solution* $x \in AP(\mathbb{R}, \mathcal{C}^n)$, *provided* $\lambda$ *is small enough.*

The proof of Theorem 1.1 is given in detail in [4], where the main auxiliary tool is the fact that the map $f \to x$, with $x$ the unique solution in $AP(\mathbb{R}, \mathcal{C}^n)$ of (2), is a continuous map from $AP(\mathbb{R}, \mathcal{C}^n)$ into itself.

It is also proved in [4], that the Lipschitz condition (3) can be substituted by another growth condition on $F$. But in this case, even with the assumption of compactness of $F$ on $AP(\mathbb{R}, \mathcal{C}^n)$, uniqueness is lost, in general.

**Remark 1.2.** The condition (2) of Theorem 1.1 is verified, for instance, in the following situations:

a) When $(Lx)(t) = Ax(t)$, with $A$ a constant square matrix of order $n$, whose characteristic numbers have nonzero real part.

b) When $(Lx)(t) = A(t)x(t)$, with $A(t)$ periodic, and such that its characteristic exponents (see, for instance, [3, 6]) have nonzero real parts. Other cases are also considered.

c) When $(Lx)(t) = A(t)x(t)$, with $A(t)$ almost periodic and of triangular form (i.e., $a_{ij} \equiv 0$ for $i < j$), such that the mean values of $a_{ii}(t), 1 \leq i \leq n$, have the real part different from zero. This case is due, basically, to A. Calderon and J. L. Massera [7].

## 2.

In this second part we shall deal with systems involving monotone operators. Normally, we shall consider systems of the form

$$\dot{x}(t) = (fx)(t), t \in \mathbb{R} \tag{4}$$

or

$$\ddot{x}(t) = (fx)(t), t \in \mathbb{R} \tag{5}$$

where $f$ is a monotone operator on the space $BC(\mathbb{R}, \mathbb{R}^n)$. The case when $\mathbb{R}^n$ is substituted by $\mathcal{C}^n$ is not much different.

Actually, one can consider as underlying space the space $BC(\mathbb{R}, \mathbb{H})$, with $\mathbb{H}$ a Hilbert space.

The following qualitative lemma turns out to be a very useful auxiliary result, in investigating the almost periodicity of solutions to equations with monotone operators.

**Lemma 2.1.** [5] *Consider the differential inequalities:*

$$\dot{u}(t) \geq \omega\big(u(t)\big), \quad t \in \mathbb{R} \tag{6}$$

$$\ddot{u}(t) \geq \omega\big(u(t)\big), \quad t \in \mathbb{R} \tag{7}$$

and assume $\omega : \mathbb{R}_+ \to \mathbb{R}$ is continuous, while $u \in BC(\mathbb{R}, \mathbb{R})$.

(1) If $u : \mathbb{R} \to \mathbb{R}$ is differentiable, satisfies (6), $u \in BC(\mathbb{R}, \mathbb{R})$ and $\omega$ is such that

$$\omega(u) > 0 \text{ for } u > \alpha, \tag{8}$$

then

$$u(t) \leq \alpha, t \in \mathbb{R}. \tag{9}$$

(2) If $u : \mathbb{R} \to \mathbb{R}$ is twice differentiable, satisfies (7), with $u \in BC(\mathbb{R}, \mathbb{R})$, while $\omega$ is satisfying (8), then $u$ satisfies (9).

We can now state and prove the following results, related to (4) and (5):

**Theorem 2.2.** *Let us consider* (4) *and* (5), *under the following assumptions:*

1) $f : BC(\mathbb{R}, \mathbb{R}^n) \to BC(\mathbb{R}, \mathbb{R}^n)$ *is a monotone operator, i.e., there exists* $m > 0$, *such that*

$$\langle (fx)(t) - (fy)(t), x(t) - y(t) \rangle \geq m|x(t) - y(t)|^2, \quad t \in \mathbb{R} \tag{10}$$

*for any* $x, y \in BC(\mathbb{R}, \mathbb{R}^n)$

2) *There exists* $z : \mathbb{R} \to \mathbb{R}^n, z \in BC(\mathbb{R}, \mathbb{R}^n)$, *such that* $fz \in AP(\mathbb{R}, \mathbb{R}^n)$.

*If* $x \in BC(\mathbb{R}, \mathbb{R}^n)$ *is a solution of* (4) *or* (5), *then necessarily* $x \in AP(\mathbb{R}, \mathbb{R}^n)$. *When* $x$ *satisfies* (4), $\dot{x} \in AP(\mathbb{R}, \mathbb{R}^n)$, *and when* $x$ *satisfies* (5), $\dot{x}, \ddot{x} \in AP(\mathbb{R}, \mathbb{R}^n)$.

Before providing the proof of Theorem 2.2, we shall make the following remark.

Condition (2) requires that $f$ takes at least one element of $BC(\mathbb{R}, \mathbb{R}^n)$ into $AP(\mathbb{R}, \mathbb{R}^n)$. If we keep in mind that $AP \subset BC$, and assume, for instance, that $f$ leaves invariant the subspace $AP$ of $BC$, then condition (2) is superfluous. As it is stated, condition (2) requires the least necessary for obtaining the almost periodicity. Obviously, without loss of generality we can assume that $z(t) = \theta =$ the null element of $\mathbb{R}^n$ (and of $BC(\mathbb{R}, \mathbb{R}^n)$).

*Proof of Theorem* 2.2. We shall distinguish two different situations, in accordance with the choice of (4) or (5) for proof.

Since the case of (4) is the simplest, we shall dwell only with the proof of Theorem 2.2, in case the second order (5) is considered.

We shall rewrite (5) in the following equivalent form:

$$\ddot{x}(t) = (gx)(t) + h(t), \quad t \in \mathbb{R} \tag{11}$$

where

$$(gx)(t) = (fx)(t) - (f\theta)(t), \quad t \in \mathbb{R}$$
$$h(t) = (f\theta)(t), \quad t \in \mathbb{R} \tag{12}$$

Let $x \in BC(\mathbb{R}, \mathbb{R}^n)$ be a solution of (11), and take an arbitrary $\tau \in \mathbb{R}$, which will remain fixed in the proof. Then (12) implies for each $t \in \mathbb{R}$

$$\ddot{x}(t + \tau) - \ddot{x}(t) = (gx)(t + \tau) - (gx)(t) + h(t + \tau) - h(t).$$

Multiplying scalarly (in $\mathbb{R}^n$) both sides by $x(t+\tau) - x(t)$, and taking (10) into account, one obtains the inequality

$$\langle \ddot{x}(t+\tau) - \ddot{x}(t)\, x(t+\tau) - x(t)\rangle \geq m|x(t+\tau) - x(t)|^2 - |h(t+\tau) - h(t)||x(t+\tau) - x(t)| \, .$$

But $\langle \ddot{u}, u \rangle = \frac{d}{dt}\langle \dot{u}, u \rangle - |\dot{u}|^2 = \frac{1}{2}\frac{d^2}{dt^2}|u|^2 - |\dot{u}|^2.$

Therefore, applying this formula with $u(t) = x(t+\tau) - x(t)$, one obtains from above the second order inequality

$$\frac{1}{2}\frac{d^2}{dt^2}|x(t+\tau) - x(t)|^2 \geq m|x(t+\tau) - x(t)|^2 - |h(t+\tau) - h(t)||x(t+\tau) - x(t)| \, ,$$

which means, for $v(t) = |x(t+\tau) - x(t)|^2, t \in \mathbb{R}$,

$$\frac{1}{2}\frac{d^2 v}{dt^2} \geq mv - \sqrt{v}|h(t+\tau) - h(t)| \, . \tag{13}$$

Now, we shall choose $\tau \in \mathbb{R}$ in such a way that it is an almost period of $h(t)$, corresponding to $m\epsilon$, with $\epsilon > 0$ arbitrary. Then (13) becomes

$$\frac{1}{2}\frac{d^2 v}{dt^2} \geq mv - m\sqrt{v}\epsilon \, , \quad t \in \mathbb{R} \, , \tag{14}$$

which is an inequality of the form (7) in the above Lemma. Since the right hand side of (14) satisfies

$$mv - m\sqrt{v}\epsilon > 0 \quad \text{for} \quad v > \epsilon^2 \, , \tag{15}$$

there results by applying the lemma

$$v(t) = |x(t+\tau) - x(t)|^2 \leq \epsilon^2 \, , \quad t \in \mathbb{R} \, , \tag{16}$$

which implies $|x(t+\tau) - x(t)| \leq \epsilon$ for $t \in \mathbb{R}$, proving the almost periodicity of $x(t)$.

The case of (4) can be treated in the same way, relying on the first part of the lemma.

This ends the proof of Theorem 2.2.                                      □

**Remark 2.3.** Existence of solutions in $BC(\mathbb{R}, \mathbb{R}^n)$ can be obtained under some supplementary condition. One such example is presented in our book [5]. Let's point out that the literature is very rich in regard to monotone operators and related topics [8].

**Remark 2.4.** It would be interesting to see whether the monotonicity condition (10), which is in $\mathbb{R}^n$, could be replaced by a monotonicity condition in a function space. For instance, taking into account the fact that $AP(\mathbb{R}, \mathbb{R}^n) \subset B^2(\mathbb{R}, \mathbb{R}^n)$ (see [5]), and $B^2$ can be regarded as a Hilbert space, to impose the condition in $B^2(\mathbb{R}, \mathbb{R}^n)$.

## 3.

In this section we shall investigate differential systems of the form

$$\dot{x}(t) = f\big(x(t)\big) + g\big(t, x(t)\big), \quad t \in \mathbb{R} \tag{17}$$

or

$$\ddot{x}(t) = f\big(x(t)\big) + g\big(t, x(t)\big), \quad t \in \mathbb{R}. \tag{18}$$

It is possible to consider the more general case when $f$ stands for an operator acting on a certain function space, say $BC(\mathbb{R}, \mathbb{R}^n)$. The difference appearing in the discussion is only cosmetic.

We shall assume that, in case we consider (17), the following conditions are satisfied:

1) $f : \mathbb{R}^n \to \mathbb{R}^n$ is continuous and monotone, which means that for any $x, y \in \mathbb{R}^n$, one has:

$$\big\langle f(x) - f(y), x - y \big\rangle \geq \lambda(|x - y|^2), \tag{19}$$

with $\lambda(r), r \in \mathbb{R}_+$ continuous, $\lambda(0) = 0$ and $\lambda(r)$ increasing for $r > 0$.

2) $\lambda(r)$ satisfies the conditions:

$$\int_{0+} \frac{dr}{\lambda(r)} = \infty \tag{20}$$

and

$$\nu(r) = \frac{\lambda(r)}{\sqrt{r}}, \quad r > 0, \quad \nu(0) = 0 \tag{21}$$

is increasing for $r > 0$ (hence, invertible).

3) $g : \mathbb{R} \times \mathbb{R}^n \to \mathbb{R}$ is continuous, Stepanov almost periodic in $t$, uniformly with respect to $x \in \mathbb{R}^n$

4) $g$ is convex in respect to the second argument, for each $t \in \mathbb{R}$, i.e.,

$$\big\langle g(t, x) - g(t, y), x - y \big\rangle \geq 0, \quad x, y \in \mathbb{R}^n. \tag{22}$$

Before we state Theorem 3.1, let us point out the fact that condition (2) on $\lambda(r)$ is satisfied for $\lambda(r) = mr, m > 0$, which covers the classical case of monotonicity. Also, $\lambda(r) = mr^\alpha$ satisfies condition (2), for $\alpha \geq 1, m > 0$.

It is obvious that (17) and (18) constitute perturbed variants of (4) and (5), respectively, and (22) shows that we admit only convex perturbations.

We can now state the following results regarding (17) and (18).

**Theorem 3.1.** *Let us consider* (17) *and* (18), *under assumptions* 1)–4). *Then any solution* $x \in BC(\mathbb{R}, \mathbb{R}^n)$, *of either* (17) *or* (18), *is also in* $AP(\mathbb{R}, \mathbb{R}^n)$.

*Proof.* Consider the system (17), and let $x \in BC(\mathbb{R}, \mathbb{R}^n)$ be a solution of (17). The function $x_\tau(t) = x(t + \tau), t \in \mathbb{R}$, for a fixed $\tau \in \mathbb{R}$, is a solution of the equation

$$\dot{x}(t + \tau) = f\big(x(t + \tau)\big) + g\big(t + \tau, x(t + \tau)\big), \quad t \in \mathbb{R}. \tag{23}$$

From (23) and (17) we obtain, by subtraction, the equation

$$\dot{x}(t + \tau) - \dot{x}(t) = f\big(x(t + \tau)\big) - f\big(x(t)\big) + g\big(t + \tau, x(t + \tau)\big) - g\big(t, x(t)\big). \tag{24}$$

We add and then subtract $g(t, x(t+\tau))$ from the right hand side of (24) and then multiply scalarly both sides by $x(t+\tau) - x(t)$. One obtains the following equation:

$$\frac{1}{2}\frac{d}{dt}|x(t+\tau) - x(t)|^2 = \Big\langle f\big(x(t+\tau)\big) - f\big(x(t)\big), x(t+\tau) - x(t)\Big\rangle$$

$$+ \Big\langle g\big(t+\tau, x(t+\tau)\big) - g\big(t, x(t+\tau)\big), x(t+\tau) - x(t)\Big\rangle$$

$$+ \Big\langle g\big(t, x(t+\tau)\big) - g\big(t, x(t)\big), x(t+\tau) - x(t)\Big\rangle.$$

Taking into account our assumptions 1) and 2), we obtain from the above equation the inequality

$$\frac{1}{2}\frac{d}{dt}|x(t+\tau) - x(t)|^2 \geq \lambda\big(|x(t+\tau) - x(t)|^2\big) - |g(t+\tau, \xi) - g(t, \xi)||x(t+\tau) - x(t)|, \ t \in \mathbb{R} \tag{25}$$

where $\xi = x(t+\tau) \in \mathbb{R}^n$. We notice that in (25) we have neglected the term

$$\Big\langle g\big(t, x(t+\tau)\big) - g\big(t, x(t)\big), x(t+\tau) - x(t)\Big\rangle,$$

on behalf of condition (22) for $g$.

Denoting $u(t) = |x(t+\tau) - x(t)|^2, t \in \mathbb{R}$, the inequality (25) leads to

$$\frac{1}{2}\dot{u}(t) \geq \lambda\big(u(t)\big) - |g(t+\tau, \xi) - g(t, \xi)|\sqrt{u(t)}, \tag{26}$$

which must be verified for any $t \in \mathbb{R}$ and only for $\xi = x(t+\tau)$.

The inequality (26) represents a qualitative inequality of the form

$$\dot{u}(t) \geq \lambda\big(u(t)\big) - f(t)\mu\big(u(t)\big), \quad t \in \mathbb{R}, \tag{27}$$

with $f(t) = |g(t+\tau, \xi) - g(t, \xi)| \in S(\mathbb{R}, \mathbb{R}) \subset M(\mathbb{R}, \mathbb{R})$, which is treated in detail in our book [5], as well as in our paper [2].

Such an inequality implies

$$\sup_{t \in \mathbb{R}} u(t) \leq \nu^{-1}\left(K \sup_{\xi \in \mathbb{R}} |g(t+\tau, \xi) - g(t, \xi)|_S\right) \tag{28}$$

in which $\nu(r)$ is given by (21), while $K > 0$ is a constant determined by our conditions.

Due to our hypothesis 2), one can make

$$|g(t+\tau, \xi) - g(t, \xi)|_S < \epsilon, \quad t \in \mathbb{R}, \tag{29}$$

if $\xi$ is arbitrary in $\mathbb{R}^n$ (Stepanov almost periodicity in $t$, uniformly with respect to $\xi \in \mathbb{R}^n$), provided one chooses $\tau \in \mathbb{R}$ to be an $\epsilon$-almost period of $g$, regarded as an element of Stepanov's space of almost periodic functions.

Hence, we can write, on behalf of (28), the inequality

$$|x(t+\tau) - x(t)|^2 \leq \nu^{-1}\left(K \sup_{\xi \in \mathbb{R}^n} |g(t+\tau, \xi) - g(t, \xi)|_S\right), \tag{30}$$

and relying on what we've discussed above, there results that $x \in AP(\mathbb{R}, \mathbb{R}^n)$.

The second case appearing in Theorem 3.1, for second order equations of the form (18), can be treated in a completely similar manner, leading to the same result.

This ends the proof of Theorem 3.1                                  $\square$

**Remark 3.2.** The assumption on the convexity of $g(t, x)$ does not appear to be essential for the validity of almost periodicity in the perturbed system. It is likely due to our method of proof, and it remains as an open problem the possibility of eliminating this assumption, or replacing it.

In concluding the paper, we want to call attention on the qualitative inequalities discussed in [2, 5], by means of which it is certainly possible to obtain new results on the almost periodicity of solutions of functional or functional-differential equations.

# References

[1] C. Corduneanu, *Almost Periodic Functions,* $2^{nd}$ ed. Chelsea Publ. Co, New York, 1989 (Distributed by AMS and Oxford Univ. Press).

[2] C. Corduneanu, *Two qualitative inequalities*, J. Diff. Equations, **168** (1986), pp. 16–25.

[3] C. Corduneanu, *Integral Equations and Applications.* Cambridge University Press, 1991.

[4] C. Corduneanu, *Almost Periodic Solutions for a Class of Functional-Differential Equations.* Functional Differential Equations (Israel) in print.

[5] C. Corduneanu, *Almost Periodic Oscillations and Waves*, vol. 14, 2007, pp. 223–229.

[6] A. Halanay, *Differential Equations: Stability, Oscillations, Time Lag.* Academic Press, 1966.

[7] J. L. Massera, *Un criterio de existencia de soluciones casiperiodicas de ecuaciones diferenciales.* Publ. Inst. Mat. y Estad., Fac. Ing. Agrim, 3 (1958), pp. 89–102 (Montevideo).

[8] Yu. V. Trubnikov and A. I. Perov, *Differential Equations with Monotone Nonlinearities* (Russian). Nauka i Tehnika, Minsk, 1986.

C. Corduneanu
The University of Texas at Arlington
and
Romanian Academy
e-mail: concord@uta.edu

Progress in Nonlinear Differential Equations
and Their Applications, Vol. 75, 165–174
© 2007 Birkhäuser Verlag Basel/Switzerland

# Age-dependent Population Dynamics with the Delayed Argument

Antoni Leon Dawidowicz and Anna Poskrobko

*Dedicated to Arrigo Cellina and James Yorke*

**Abstract.** The model of age-dependent population dynamics was for the first time described by von Foerster (1959) This model is based on the first-order partial differential equation with the standard initial condition and the non-local boundary condition in integral form. Gurtin and MacCamy in their paper (1974) analyzed the more general model, where the progress of the population depends on its number. They established the existence of the unique solution of this model for all time. In our presentation the results of Gurtin and MacCamy will be generalized on the case, when the dependence on number of population is delayed.

## 1. Introduction

The model proposed by Gurtin and MacCamy [4] was based on the assumption that the progress of the population depends on its number. However, it is common knowledge that other factors can have an influence on the reproduction. Natural generalization of a population dynamics is taking into consideration, for example two sex, a period of gestation or a period of response of a system to a stimulus (see Busoni and Palczewski [1]). Two last examples suggest to consider descriptions with a delayed parameter. Apparently, the delay is rather natural assumption in every biological models. We should draw attention to fact that one of the most important equations with the delayed parameter was proposed by Polish scientists (Ważewska-Czyżewska and Lasota [9]) in the description of the growth of a population of red blood cells. The delay appears also in the papers concerning mathematical modeling in epidemiology and immunology, in particulary see Marchuk [8], Foryś [3], Leszczyński and Zwierkowski [6]. The analysis of the equations with the delayed parameter can be found also in Łoskot [7], Haribash [5] and Forystek [2]. The inspiration of this paper was the results of Gurtin and Mac-Camy [4]. We consider similar model of age-dependent population dynamics but with the delayed parameter. Our intention is researching local and global existence of a unique solution.

## 2. General description of model

Denote by $z(t) = \int_0^\infty u(a,t)da$ the total population at time $t$. Here $u(a,t)$ is the population of age $a$ at time $t$. This description is more exact as the unit of time is shorter. Consider the group of individuals who are of age $a$ at time $t$. The age of these individuals at the moment $t + h$ equals $a + h$. The rate

$$Du(a,t) = \lim_{h \to 0} \frac{u(a+h, t+h) - u(a,t)}{h} \tag{2.1}$$

denotes the intensity of changing of the population of this group in time. The sum this rate and the number $d(a,t)$ of individuals (per unit age and time) of age $a$ who die at time $t$ equals zero

$$Du(a,t) + d(a,t) = 0 \,.$$

The quotient

$$\lambda(a) = \frac{d(a,t)}{u(a,t)}$$

denotes a probability that an individual, who is of age $a$ at time $t$ will die to time $t + 1$. Let call the rate $\lambda$ the *death-modulus*. The birth process is described by the "renewal equation"

$$u(0,t) = \int_0^\infty \beta(a)u(a,t)da \,.$$

The quantity $\beta(a)$ is called the *birth-modulus*, and it is the average number of offsprings produced (per unit time) by an individual of age $a$. Here the birth and death moduli are independent of the population $z$. We rectify this and introduce the dependence of $\beta$ and $\lambda$ on $z_t$, i.e., the population considered according to time. Summarizing, we shall build the model of age-dependent population dynamics with delayed parameter. Our theory is based on the following system of equations:

$$Du(a,t) = -\lambda(a, z_t)u(a,t) \,, \tag{2.2}$$

$$u(0,t) = \int_0^\infty \beta(a, z_t)u(a,t)da \,, \tag{2.3}$$

$$z(t) = \int_0^\infty u(a,t)da \,, \tag{2.4}$$

where

$$z_t : [-r, 0] \to \mathbb{R}_+ \,, \quad r \geqslant 0 \,, \quad \mathbb{R}_+ = [0, \infty) \tag{2.5}$$

and

$$z_t = z(t + s) \tag{2.6}$$

with initial condition

$$u(a,0) = \varphi(a). \tag{2.7}$$

When $u$ is differentiable everywhere there is

$$Du = \frac{\partial u}{\partial a} + \frac{\partial u}{\partial t}.$$

In our model, in different manner as in Gurtin-MacCamy theory, the dependence of $\lambda$ and $\beta$ on $z$ is functional one. We assume that both, the reproduction as the death depend on the population in any preceding period of time.

## 3. Local existence

From now on we make the following assumptions:

$(H_1)$ $\varphi \in L^1(\mathbb{R}_+)$ is piecewise continuous;

$(H_2)$ $\lambda, \beta \in C\left(\mathbb{R}_+ \times C([-r,0])\right)$; the Fréchet derivatives $D\lambda$ of $\lambda(a,\psi)$ and $D\beta$ of $\beta(a,\psi)$ with respect to $\psi$ exist for all $a \geqslant 0$ and $\psi \geqslant 0$;

$(H_3)$ The functions $\lambda(\cdot,\psi)$, $\beta(\cdot,\psi)$ belong to $C\left(C([-r,0]); L^\infty(\mathbb{R}_+)\right)$;

$(H_4)$ The Fréchet derivatives $D_{\psi_0}\lambda$ and $D_{\psi_0}\beta$ in the point $\psi_0$ as a function of $\psi_0$ belong to $C\left(C([-r,0]); \mathcal{L}\left(C([-r,0]); L^\infty(\mathbb{R}_+)\right)\right)$, where $\mathcal{L}(X,Y)$ denotes the Banach space of all bounded linear operators from $X$ to $Y$;

$(H_5)$ $\varphi \geqslant 0$, $\lambda \geqslant 0$, $\beta \geqslant 0$.

**Theorem 3.1.** *Let $u$ be a solution of the age-dependent population problem up to time $T > 0$. Then the population $z_t$ and the birth-rate $B$ satisfy on $[0,T]$ the operator equations*

$$z_t(s) = \int_0^{t+s} B(a)e^{-\int_a^{t+s}\lambda(\tau-a,z_\tau)d\tau}\,da + \int_0^\infty \varphi(a)e^{-\int_0^{t+s}\lambda(a,z_\tau)d\tau}\,da \tag{3.1}$$

*and*

$$B(t) = \int_0^t \beta(t-a,z_t)B(a)e^{-\int_a^t \lambda(\tau-a,z_\tau)d\tau}\,da \tag{3.2}$$
$$+ \int_0^\infty \beta(a+t,z_t)\varphi(a)e^{-\int_0^t \lambda(a+\tau,z_\tau)d\tau}\,da.$$

*Conversely, if $z_t$ and $B$ are non-negative continuous functions satisfying (3.1) and (3.2) on $[0,T]$, and if $u$ is defined on $\mathbb{R}_+ \times [0,T]$ by the formula*

$$u(a,t) = \begin{cases} \varphi(a-t)e^{-\int_0^t \lambda(a-t+\tau,z_\tau)d\tau} & \text{for} \quad a \geqslant t \\ B(t-a)e^{-\int_0^a \lambda(\alpha,z_{t-a+\alpha})d\alpha} & \text{for} \quad t > a \end{cases}, \tag{3.3}$$

*then $u$ is a solution of the age-dependent population problem up to time $T$.*

*Proof.* Let $u$ be a solution up to time $T$. It means that $u$ is a non-negative function on $\mathbb{R}_+ \times [0,T]$ with the following properties: $Du$ exists on $\mathbb{R}_+ \times [0,T]$, $u(\cdot,t) \in L^1(\mathbb{R}_+)$ and $u$ fulfils the system of equations (2.2)–(2.7).

Let $(a_0, t_0) \in \mathbb{R}_+ \times [0, T]$, and let

$$\overline{u}(h) = u(a_0 + h, t_0 + h), \quad \overline{\lambda}(h) = \lambda(a_0 + h, z_{t_0 + h})$$

then (2.1) and (2.2) imply

$$\frac{d\overline{u}}{dh} + \overline{\lambda}(h)\overline{u} = 0.$$

This equation has the unique solution

$$u(a_0 + h, t_0 + h) = u(a_0, t_0)e^{-\int_0^h \overline{\lambda}(\eta)d\eta}. \tag{3.4}$$

In particular, if we take $(a_0, t_0) = (a - t, 0)$ and $h = t$ in (3.4) we get

$$u(a, t) = \varphi(a - t)e^{-\int_0^t \lambda(a - t + \tau, z_\tau)d\tau} \quad \text{for } a \geqslant t.$$

On the other hand, writing $(a_0, t_0) = (0, t - a)$ and $h = a$ in (3.4) we recover

$$u(a, t) = B(t - a)e^{-\int_0^a \lambda(\alpha, z_{t-a+\alpha})d\alpha} \quad \text{for } t > a,$$

where

$$B(t) = u(0, t).$$

Substituting the above two conditions into (2.3), (2.4) and (2.6) yields integral equations for age dependent population $z_t$ and birth-rate $B$

$$
\begin{aligned}
z_t(s) &= \int_0^\infty u(a, t + s)da = \int_0^{t+s} B(t + s - a)e^{-\int_0^a \lambda(\alpha, z_{t+s-a+\alpha})d\alpha}da \\
&\quad + \int_{t+s}^\infty \varphi(a - t - s)e^{-\int_0^{t+s} \lambda(a-t-s+\tau, z_\tau)d\tau}da \\
&= \int_0^{t+s} B(a)e^{-\int_0^{t+s-a} \lambda(\alpha, z_{a+\alpha})d\alpha}da + \int_0^\infty \varphi(a)e^{-\int_0^{t+s} \lambda(a, z_\tau)d\tau}da \\
&= \int_0^{t+s} B(a)e^{-\int_a^{t+s} \lambda(\tau - a, z_\tau)d\tau}da + \int_0^\infty \varphi(a)e^{-\int_0^{t+s} \lambda(a, z_\tau)d\tau}da,
\end{aligned}
$$

$$
\begin{aligned}
B(t) &= \int_0^t \beta(t - a, z_t)B(a)e^{-\int_a^t \lambda(\tau - a, z_\tau)d\tau}da \\
&\quad + \int_0^\infty \beta(a + t, z_t)\varphi(a)e^{-\int_0^t \lambda(a+\tau, z_\tau)d\tau}da.
\end{aligned}
$$

To prove the second part of the theorem we assume that $z_t \geqslant 0$ and $B \geqslant 0$ are continuous functions on the interval $[0, T]$ fulfilling conditions (3.1) and (3.2). Let $u$ be defined on $\mathbb{R}_+ \times [0, T]$ by the formula (3.3). Since $\varphi$ and $\beta$ are non-negative so is function $u$. A trivial verification shows that (2.7) is hold, and $u(0, t) = B(t)$ for $t > 0$, and $u \in L^1(\mathbb{R}_+)$ because $\lambda$, $\beta$ and $z_t$ are continuous and $\varphi \in L^1(\mathbb{R})$. It follows from (3.1), (3.2) and (3.3) that (2.3), (2.4) and (2.6) are satisfied. To complete the proof let notice that (2.1) and (3.3) imply existing $Du$ on $\mathbb{R}_+ \times [0, T]$ and that equality (2.2) is hold. $\qquad \square$

Let $C^+[a, b]$ denote the space of non-negative continuous functions on $[a, b]$. To solve the population problem up to time $T$, $(T > 0)$ it is sufficient to find functions $z \in C^+[-r, T]$ and $B \in C^+[0, T]$ satisfying the operator equations (3.1) and (3.2). We should begin with equation (3.2) which is a linear Volterra equation of $B$ for fixed $z \in C^+[-r, T]$. It has a unique solution on $[0, T]$ which one we denote by

$$B(t) = \mathcal{B}_T(z)(t) \,.$$

It permits us to define an operator $\mathcal{Z}_T$ on $C^+[-r, T]$.

$$\mathcal{Z}_T(z)(s) = \int_0^{t+s} \mathcal{B}_T(z)(a) e^{-\int_a^{t+s} \lambda(\tau-a, z_\tau) d\tau} da \qquad (3.5)$$

$$+ \int_0^\infty \varphi(a) e^{-\int_0^{t+s} \lambda(a, z_\tau) d\tau} da \,.$$

Defining the operator $\mathcal{Z}_T$ we used the right-hand side of equation (3.1) with $B$ replaced by $\mathcal{B}_T(z)$. Hypotheses $(H_1) - (H_5)$ imply that

$$\mathcal{B}_T : C^+[-r, T] \to C^+[0, T]$$
$$\mathcal{Z}_T : C^+[-r, T] \to C^+[-r, T] \,.$$

**Lemma 3.2.** *There exists $T > 0$ such that the operator $\mathcal{Z}_T : C^+[-r, T] \to C^+[-r, T]$ defined by (3.5) has a unique fixed point.*

*Proof.* Let us consider the Banach space $C[-r, T]$ with supremum norm $\|\cdot\|_T$. Let

$$\Phi = \int_0^\infty \varphi(a) da$$

and

$$\Sigma_T = \left\{ f : f \in C^+[-r, T], \|f - \Phi\|_T \leqslant m \right\}$$

for fixed $m > 0$. The set $\Sigma_T$ is closed. It is sufficient to show that $\mathcal{Z}_T : \Sigma_T \to \Sigma_T$ and is contracting, then the proof of Lemma will be an easy consequence of the Banach fixed-point theorem. Let

$$\Omega = \left\{ (a, z) : a \geqslant 0, z \geqslant 0, |z - \Phi| \leqslant m \right\} \,.$$

Assumptions $(H_2)$-$(H_4)$ imply that the quantities

$$\begin{array}{ll}
\lambda_0 = \sup_{(a,z) \in \Omega} \lambda(a, z) & \lambda_1 = \max_{\psi_0 \in C([-r,0])} \|D_{\psi_0} \lambda\| \\
\beta_0 = \sup_{(a,z) \in \Omega} \beta(a, z) & \beta_1 = \max_{\psi_0 \in C([-r,0])} \|D_{\psi_0} \beta\|
\end{array} \qquad (3.6)$$

are finite. Here the norm $\|.\|$ is considered in the Banach space $\mathcal{L}(C([-r, 0]); L^\infty(\mathbb{R}_+))$. For $z \in \Sigma_T$ we have from (3.2)

$$\mathcal{B}_T(z)(t) \leqslant \beta_0 \int_0^t \mathcal{B}_T(z)(a) da + \beta_0 \Phi$$

and by Gronwall's inequality

$$\mathcal{B}_T(z)(t) \leqslant \beta_0 \Phi e^{\beta_0 t} \,. \qquad (3.7)$$

Next, we shall estimate $|\mathcal{Z}_T(z)(s) - \Phi|$. We use (3.5), (3.6) and (3.7). For $0 \leqslant t \leqslant T$, $-r \leqslant s \leqslant 0$ and $z \in \Sigma_T$,

$$
\begin{aligned}
|\mathcal{Z}_T(z)(s) - \Phi| &\leqslant \beta_0 \Phi \int_0^{t+s} e^{\beta_0 a} da + \int_0^\infty \left| e^{-\int_0^{t+s} \lambda(a, z_\tau) d\tau} - 1 \right| \varphi(a) da \\
&\leqslant \Phi \left( e^{\beta_0 T} - 1 \right) + \Phi \sup_{\substack{a \geqslant 0 \\ -r \leqslant t \leqslant T}} \left| e^{-\int_0^{t+s} \lambda(a, z_\tau) d\tau} - 1 \right|.
\end{aligned}
$$

Since

$$
|e^k - 1| \leqslant |k| e^{|k|} \tag{3.8}
$$

we have

$$
|\mathcal{Z}_T(z)(s) - \Phi| \leqslant \Phi \left( e^{\beta_0 T} - 1 \right) + \Phi \lambda_0 T e^{\lambda_0 T}.
$$

For sufficiently small $T$ we have $\mathcal{Z}_T(z) \in \Sigma_T$. Next we show that for small $T$, the map $\mathcal{Z}_T$ is contracting. Let choose $z, \hat{z} \in \Sigma_T$. By (3.5) we get

$$
\begin{aligned}
\|\mathcal{Z}_T(z) \quad - \quad & \mathcal{Z}_T(\hat{z})\|_T \\
&\leqslant \sup_{\substack{t \in [0,T] \\ s \in [-r,0]}} \int_0^{t+s} \left| e^{-\int_a^{t+s} \lambda(\tau-a, z_\tau) d\tau} \right. \tag{3.9} \\
& \hspace{4cm} \left. - e^{-\int_a^{t+s} \lambda(\tau-a, \hat{z}_\tau) d\tau} \right| \mathcal{B}_T(z)(a) da \\
&+ \sup_{\substack{t \in [0,T] \\ s \in [-r,0]}} \int_0^{t+s} e^{-\int_a^{t+s} \lambda(\tau-a, \hat{z}_\tau) d\tau} |\mathcal{B}_T(z)(a) - \mathcal{B}_T(\hat{z})(a)| da \tag{3.10} \\
&+ \sup_{\substack{t \in [0,T] \\ s \in [-r,0]}} \int_0^\infty \left| e^{-\int_0^{t+s} \lambda(a, z_\tau) d\tau} - e^{-\int_0^{t+s} \lambda(a, \hat{z}_\tau) d\tau} \right| \varphi(a) da \tag{3.11}
\end{aligned}
$$

We will estimate each of the above elements. We begin with (3.9) using the relation (3.6) and (3.8) we get

$$
\begin{aligned}
\left| e^{-\int_a^{t+s} \lambda(\tau-a, z_\tau) d\tau} - e^{-\int_a^{t+s} \lambda(\tau-a, \hat{z}_\tau) d\tau} \right| &\leqslant \left| 1 - e^{-\int_a^{t+s} [\lambda(\tau-a, z_\tau) - \lambda(\tau-a, \hat{z}_\tau)] d\tau} \right| \\
&\leqslant e^{-\int_a^{t+s} |\lambda(\tau-a, z_\tau) - \lambda(\tau-a, \hat{z}_\tau)| d\tau} \int_a^{t+s} |\lambda(\tau - a, z_\tau) - \lambda(\tau - a, \hat{z}_\tau)| \, d\tau \\
&\leqslant \lambda_1 T e^{2\lambda_0 T} \|z - \hat{z}\|_T.
\end{aligned}
$$

In similar way we can estimate

$$
\left| e^{-\int_0^{t+s} \lambda(a, z_\tau) d\tau} - e^{-\int_0^{t+s} \lambda(a, \hat{z}_\tau) d\tau} \right| \leqslant \lambda_1 T e^{2\lambda_0 T} \|z - \hat{z}\|_T.
$$

Summarizing, for (3.9) and (3.11) we get

$$
\begin{aligned}
\sup_{\substack{t \in [0,T] \\ s \in [-r,0]}} \int_0^{t+s} & \left| e^{-\int_a^{t+s} \lambda(\tau-a, z_\tau) d\tau} - e^{-\int_a^{t+s} \lambda(\tau-a, \hat{z}_\tau) d\tau} \right| \mathcal{B}_T(z)(a) da \\
& \hspace{3cm} \leqslant \lambda_1 T \Phi e^{2\lambda_0 T} \left( e^{\beta_0 T} - 1 \right) \|z - \hat{z}\|_T
\end{aligned}
$$

and

$$\sup_{\substack{t \in [0,T] \\ s \in [-r,0]}} \int_0^\infty \left| e^{-\int_0^{t+s} \lambda(a, z_\tau) d\tau} - e^{-\int_0^{t+s} \lambda(a, \hat{z}_\tau) d\tau} \right| \varphi(a) da$$

$$\leqslant \lambda_1 T \Phi e^{2\lambda_0 T} \|z - \hat{z}\|_T .$$

Therefore, for $T$ sufficiently small (3.9) and (3.11) are each less then $C\|z - \hat{z}\|_T$ with the constant $C \in [0, \frac{1}{3})$ (independent of $z$ and $\hat{z}$). Thus to complete proof we should show that (3.10) fulfils the identical inequality with a constant from the interval $[0, \frac{1}{3})$. From (3.2) and the definition of $\mathcal{B}_T$ we have

$$\mathcal{B}_T(z)(t) \quad - \quad \mathcal{B}_T(\hat{z})(t)$$

$$= \int_0^t \beta(t - a, z) e^{-\int_a^t \lambda(\tau - a, z_\tau) d\tau} \left( \mathcal{B}_T(z)(a) - \mathcal{B}_T(\hat{z})(a) \right) da \quad (3.12)$$

$$+ \int_0^t \mathcal{B}_T(\hat{z})(a) \left( \beta(t - a, z) e^{-\int_a^t \lambda(\tau - a, z_\tau) d\tau} \right. \quad (3.13)$$

$$\left. - \beta(t - a, \hat{z}) e^{-\int_a^t \lambda(\tau - a, \hat{z}_\tau) d\tau} \right) da$$

$$+ \int_0^\infty \varphi(a) \left( \beta(a + t, z) e^{-\int_0^t \lambda(a + \tau, z_\tau) d\tau} \right. \quad (3.14)$$

$$\left. - \beta(a + t, \hat{z}) e^{-\int_0^t \lambda(a + \tau, \hat{z}_\tau) d\tau} \right) da .$$

Let denote (3.13) and (3.14) by $f(t)$, then

$$\mathcal{B}_T(z)(t) - \mathcal{B}_T(\hat{z})(t) \leqslant \beta_0 \int_0^t \left( \mathcal{B}_T(z)(a) - \mathcal{B}_T(\hat{z})(a) \right) da + |f(t)| \quad (3.15)$$

and hence, by Gronwall's inequality

$$\mathcal{B}_T(z)(t) - \mathcal{B}_T(\hat{z})(t) \leqslant |f(t)| + \beta_0 \int_0^t |f(a)| e^{\beta_0(t-a)} da . \quad (3.16)$$

Proceeding as before, we can verify that

$$\|f\|_T \leqslant C_0 \|z - \hat{z}\|_T ,$$

where $C_0$ is a constant depending only on $\beta_0$, $\beta_1$, $\lambda_0$, $\lambda_1$ and $T$. In view of the definitions of (3.10), (3.15) and (3.16) we have the conclusion that for sufficiently small $T$ the quantity (3.10) is less than $C\|z - \hat{z}\|_T$ when $C \in [0, \frac{1}{3})$. This shows the contraction of $\mathcal{Z}_T$ and completes the proof.    $\square$

According to the above Lemma we can draw a conclusion about the existence and uniqueness of the solution of the population problem. So we can formulate the following theorem:

**Theorem 3.3.** *There exists $T > 0$ such that the population problem has a unique solution up to time $T$.*

## 4. Global existence

Let assume that the average number of offsprings (per unit time) $\beta(a, z_t)$ is uniformly bounded for all $a$ and $z_t$, that is

$$\overline{\beta} = \sup_{\substack{a \geqslant 0 \\ z_t \geqslant 0}} \beta(a, z_t) < \infty. \tag{4.1}$$

Under the hypothesis $(H_5)$

$$\underline{\lambda} = \inf_{\substack{a \geqslant 0 \\ z_t \geqslant 0}} \lambda(a, z_t) \tag{4.2}$$

is finite and non-negative. When $\overline{\beta} < \infty$ we can call the bounding *growth rate*

$$\delta = \overline{\beta} - \underline{\lambda}.$$

**Theorem 4.1.** *If condition* (4.1) *holds and if $u$ is a solution of the population problem up to time $T$, then for $0 \leqslant t \leqslant T$*

$$z_t(s) \leqslant \Phi e^{\delta(t+s)}, \tag{4.3}$$

$$B(t) \leqslant \overline{\beta} \Phi e^{\delta t} \tag{4.4}$$

*and*

$$u(a, t) \leqslant \overline{\beta} \Phi e^{\delta t} e^{-\lambda a} \quad (a < t), \quad u(a, t) \leqslant \|\varphi\|_t e^{-\lambda t} \quad (a \geqslant t), \tag{4.5}$$

*where $\Phi$ is the initial total population $\Phi = \int_0^\infty \varphi(a)da$, $\Phi = z_0(0)$ and $\|\varphi\|_t = \sup_{[0,t]} \varphi$.*

*Proof.* Using (4.1) we can write

$$e^{-\int_a^t \lambda(\tau-a, z_\tau)d\tau} \leqslant e^{-\underline{\lambda}(t-a)}$$

$$e^{-\int_0^t \lambda(a+\tau, z_\tau)d\tau} \leqslant e^{-\underline{\lambda}t}.$$

According to (3.2) and (4.1) we have

$$B(t) \leqslant \overline{\beta} \int_0^t B(a)e^{-\underline{\lambda}(t-a)}da + \overline{\beta}\Phi e^{-\underline{\lambda}t}.$$

The desired conclusion (4.4) is an easy consequence of Gronwall's inequality. If we substitute the result (4.4) into (3.1), we get immediately (4.3). Finally, conditions (3.3), (4.3) and (4.4) lead to (4.5). $\square$

**Theorem 4.2.** *Assume that* (4.1) *holds. Then the age-dependent population problem has a unique solution for all time.*

*Proof.* Let $\mathcal{U}$ be the set of all functions $u_t$, $t > 0$, solutions of the age-dependent population problem up to time $t$. The set $\mathcal{U}$ is not empty by Theorem 3.3. We define the relation in $\mathcal{U}$

$$u_{t_1} \leqslant u_{t_2} \quad \text{when} \quad [0, t_1] \subset [0, t_2].$$

This is the relation of partial order. Let $\mathcal{W}$ be totally ordered set in $\mathcal{U}$. Let $u \in \mathcal{W}$ and $u = u_T$ where $[0, T] = \bigcup_{i \geqslant 1}[0, t_i]$. $u$ is the solution of the age-dependent population problem and it is the majorant of the totally ordered set $\mathcal{W}$. So from the Kuratowski–Zorn lemma it follows that there exists the maximal element in the set $\mathcal{U}$. It is the solution of the population problem for $T = \infty$.

To prove uniqueness assume that $u_1$ and $u_2$ are the solutions for all time. Then $(z_{t,1}, B_1)$ and $(z_{t,2}, B_2)$ are solutions of (3.1) and (3.2) for all time. From Theorem 3.3 we have local existence. So there exists $T > 0$ that $(z_{t,1}, B_1) = (z_{t,2}, B_2)$ on $[0, T]$. From the continuity of functions $z_t$ and $B$ we have $T = \infty$, and finally uniqueness $u_1 = u_2$.    □

**Theorem 4.3.** *Assume that*

($A_1$) $\varphi \in C^1(\mathbb{R}_+)$ *with* $\dot{\varphi} \in L^1(\mathbb{R}_+)$;

($A_2$) $\lambda, \beta \in C^1(\mathbb{R}_+ \times C([-r, 0]))$;

($A_3$) *the mappings carrying* $(t, z_t)$ *into the functions* $a \mapsto \beta_a(a + t, z_t)$ *and* $a \mapsto D\beta(a + t, z_t)$ *belong to* $C(\mathbb{R}_+ \times C([-r, 0]); L^\infty(\mathbb{R}_+))$;

($A_4$) $u$ *is a solution of the population problem up to time* $T$;

($A_5$) $\dot{z}_0(0) = B(0) - \int_0^\infty \lambda(a, z_0)\varphi(a)da$

*then* $u \in C^1(\mathbb{R}_+ \times [0, T])$ *if and only if* $\varphi$ *satisfies conditions*

$$\varphi(0) = \int_0^\infty \beta(a, z_0)\varphi(a)da \tag{4.6}$$

*and*

$$\dot{\varphi}(0) = [-\lambda(0, z_0) - \beta(0, z_0)]\,\varphi(0) \tag{4.7}$$
$$- \int_0^\infty [\beta_a(a, z_0) + D\beta(a, z_0)\dot{z}_0 - \beta(a, z_0)\lambda(a, z_0)]\varphi(a)da\,.$$

*Proof.* The function $u$ is continuous when $\varphi \in C^1(\mathbb{R}_+)$ and $\varphi(0) = B(0^+)$ so when $u$ is continuous across the characteristic $t = a$. By (3.2) $B(0^+) = \int_0^\infty \beta(a, z_0)\varphi(a)da$, so it equals the right-hand side of (4.6). We should show that (4.7) is a necessary and sufficient condition that $u$ is of class $C^1$ on $\mathbb{R}_+ \times [0, T]$. From hypothesis $(H_1)$–$(H_5)$ and conditions (3.1) and (3.2) we know that $z_t, B(t) \in C(\mathbb{R}_+)$. It follows from (3.3) that $u$ is of class $C^1$ when its derivative is continuous across the characteristic $t = a$, so when

$$\dot{B}(0) = -\dot{\varphi}(0) - \varphi(0)\lambda(0, z_0)\,. \tag{4.8}$$

We can calculate the quantity $\dot{B}(0)$ using (3.2)

$$\dot{B}(0) = \beta(0, z_0)B(0) + \int_0^\infty [\beta_a(a, z_0) + D\beta(a, z_0)\dot{z}_0 - \beta(a, z_0)\lambda(a, z_0)]\varphi(a)da\,.$$

Using the formula (4.8) we obtain (4.7).    □

**Acknowledgment**

The second author acknowledges the support from Białystok Technical University (Grant No. S/WI/1/07).

# References

[1] G. Busoni and A. Palczewski, *Dynamics of a two sex population with gestation period.* Appl. Math., **27** (2000), no. 1, 21–34.

[2] E. Forystek, *On the system of nonlinear von Foerster equations.* Univ. Iagel. Acta Math., **38** (2000), 199–204.

[3] U. Foryś, *Biological delay systems and the Mikhailov criterion of stability.* J. Biol. Sys., **12** (1) (2001), 1–16.

[4] M. E. Gurtin and R. C. MacCamy, *Non-linear age-dependent population dynamics.* Arch. Rat. Mech. Anal., **54** (1974), 281–300.

[5] N. Haribash, *Delayed von Foerster equation.* Univ. Iagel. Acta Math., **39** (2001), 239–248.

[6] H. Leszczyński and P. Zwierkowski, *Existence of solutions to generlized von Foerster equations with functional dependence.* Ann. Polon. Math., **83** (2004), no. 3, 201–210.

[7] K. Łoskot, *Bounded solutions of a system of partial differential equations describing interacting cell populations.* Bull. Polish Acad. Sci. Math., **42** (1994), no. 4, 315–343.

[8] G. I. Marchuk, *Mathematical Models in Immunology.* Springer, New York, 1983.

[9] M. Ważewska-Czyżewska and A. Lasota, *Matematyczne problemy dynamiki układu krwinek czerwonych.* Roczniki PTM, Matematyka Stosowana, **VI** (1976), 23–40.

Antoni Leon Dawidowicz
Institute of Mathematics
Jagiellonian University
ul. Reymonta 4
PL-30-059 Kraków,
Poland
e-mail: Antoni.Leon.Dawidowicz@im.uj.edu.pl

Anna Poskrobko
Faculty of Computer Science
Bialystok Technical University
ul. Wiejska 45A
PL-15-351 Białystok
Poland
e-mail: aposkrobko@wp.pl

Progress in Nonlinear Differential Equations
and Their Applications, Vol. 75, 175–194
© 2007 Birkhäuser Verlag Basel/Switzerland

# Chaos in the Störmer Problem

## Rui Dilão and Rui Alves-Pires

*Dedicated to Arrigo Cellina and James Yorke*

**Abstract.** We survey the few exact results on the Störmer problem describing the dynamics of charged particles in the Earth magnetosphere. The analysis of this system leads to the the conclusion that charged particles are trapped in the Earth magnetosphere or escape to infinity, and the trapping region is bounded by a torus-like surface, the Van Allen inner radiation belt. In the trapping region, the motion of the charged particles can be periodic, quasi-period or chaotic. The three main effects observed in the Earth magnetosphere, radiation belts, radiation aurorae and South Atlantic anomaly, are described in the framework described here. We discuss some new mathematical problems suggested by the analysis of the Störmer problem.

**Mathematics Subject Classification (2000).** Primary 34D23; Secondary 45D45.

**Keywords.** Störmer problem; chaos; Van Allen inner radiation belt; quasi-periodic motion.

## 1. Introduction

Stellar and planetary magnetic environments or magnetospheres are generated by the motion of charged particles inside the core of stars and planets. In the magnetosphere of the Earth, incoming charged particles have intricate trajectories and are in the origin of observable radiation phenomena as is the case of radiation aurorae [16], the Van Allen inner radiation belts [19], and the South Atlantic anomaly [18].

The magnetic field of the Earth has a strong dipolar component, [14], and it is believed that, for low altitudes ($\leq 3000\,\mathrm{km}$), the radiation phenomena occurring in the Earth magnetosphere can be understood by studying the motion of nonrelativistic charged particles in a pure dipole field. If the dipolar component of the Earth magnetic field is considered aligned with the rotation axis of the Earth, the equations of motion of a charged particle in the dipolar field of the Earth reduce to a non-linear autonomous Hamiltonian dynamical system. This is the Störmer problem, [16].

The analysis of the Störmer problem presents big challenges from the theoretical, applied and computational points of view.

From the computational point of view, the determination of the trajectories of high energy charged particles in the Earth magnetosphere for long periods of time is inaccurate and time consuming, being difficult to extract information about the several aspects of the radiation phenomena observed in the Earth magnetosphere. For example, in order to explain some of the dynamic aspects associated with aurorae and magnetic mirrors, adiabatic *ad-hoc* arguments have been introduced into the theory, and the motion of charged particles in the magnetosphere has been assumed similar to the cyclotron type motion in constant magnetic fields, [5,11,19].

The long lived and transient radiation belts observed in the Earth magnetosphere have adverse effects on the electronics of spacecrafts, and affect communications, [5, 15]. Therefore, a qualitative and quantitative understanding of the Störmer problem has important applications.

The first theoretical studies about the properties of the trajectories of charged particle in a dipolar field where done by DeVogelaere [6], and Dragt [7]. In the work of these authors, the existence of a trapping region for charged particles in the dipole field of the Earth was implicitly established.

Based on the qualitative theory of conservative maps of the plane, Dragt and Finn [8] carried a comparative study between the phase space topology of the orbits of a generic area-preserving map of the plane and the numerically computed Poincaré sections of the Störmer problem. They have presented numerical evidence about the existence of homoclinic points in the Poincaré sections. According to these authors, this shows that the Störmer problem is insoluble, implying that the adiabatic magnetic moment series diverges, a basic theoretical argument used by Van Allen to explain some features of the radiation phenomena in planetary magnetospheres. The KAM approach to the Störmer problem has been developed by Braun in a sequence of papers, [1–3]. Within this approach, it has been shown that trapped particles can have quasi-periodic motion, and can penetrate arbitrarily close to the dipole axis.

Due to its intrinsic difficulty, the analysis of the Störmer problem has been done using a mixture of analytical and numerical techniques. Here, we are interested in surveying the exact results on the Störmer problem, separating the results that are numeric from the exact ones. All the exact results are summarized in Propositions 3.1 and 4.1. We have made extensive simulations of trajectories of charged particles in the Earth magnetosphere, and we have obtained the shape of the trapping regions for charged particles – Van Allen inner radiation belts. This contrast with the usual approach used in radiation environment studies, where radiation belt boundaries are correlated with the dipole field lines, [5].

This paper is organized as follows. In the next section, we derive the equations of motion for the Störmer problem and we obtain its conservation laws. In Section 3, we study the motion of charged particles in the equatorial plane of the Earth. In Section 4, we analyse the general case for the motion on the three-dimensional configuration space. In the final Section 5, we summarize the main results from the theoretical and applied points of views, and we discuss some of the mathematical problems suggested by the analysis done previously.

## 2. Equations of motion and conservation laws

The equation of motion of a nonrelativistic charged particle of mass $m$ and charge $q$ in a magnetic field $\vec{B}$ has the Lorentz form,

$$m\ddot{\vec{r}} = q\left(\dot{\vec{r}} \times \vec{B}\right). \tag{2.1}$$

We use the international system of units and $B$ is measured in Tesla. At the surface of the Earth, $B$ is in the range $(0.5 \times 10^{-4} - 1.0 \times 10^{-4})$ Tesla $= (0.5 - 1.0)$ Gauss.

Magnetic fields are produced by moving charges and currents. A magnetic dipole field can be produced by a current loop on a planar surface, and the resulting field is proportional to the current intensity times the area delimited by the current loop, ( [9], pp. 7–14). For a current loop in the horizontal $xy$-plane, flowing counterclockwise with current intensity $I$, the dipole momentum is $\vec{\mu}_z = \mu \vec{e}_z$, where $\mu = I \times area\ of\ the\ loop$, and $\vec{e}_z$ is the unit vector of the $z$ axis. This current loop produces a dipole field with a dipole momentum pointing in the positive direction of the $z$-axis. This dipole field derives from the vector potential,

$$\vec{A} = \frac{1}{4\pi\varepsilon_0 c^2} \frac{1}{r^2} \vec{\mu}_z \times \vec{e}_r = \frac{1}{4\pi\varepsilon_0 c^2} \frac{1}{r^3} \vec{\mu}_z \times \vec{r} = M_z \frac{1}{r^3}\left(-y\vec{e}_x + x\vec{e}_y\right) \tag{2.2}$$

where $r = \sqrt{x^2 + y^2 + z^2}$, $M_z$ is the (scalar) dipole momentum, for short, and the vector potential is independent of the $z$ coordinate. As $\vec{B} = \mathrm{rot}\vec{A}$, by (2.2), the Lorentz equation (2.1) describing the motion of a charge particle in a dipole field is,

$$\begin{cases} \ddot{x} = 3\alpha\dfrac{z}{r^5}(\dot{y}z - \dot{z}y) - \alpha\dot{y}\dfrac{1}{r^3} \\[2mm] \ddot{y} = -3\alpha\dfrac{z}{r^5}(\dot{x}z - \dot{z}x) + \alpha\dot{x}\dfrac{1}{r^3} \\[2mm] \ddot{z} = 3\alpha\dfrac{z}{r^5}(\dot{x}y - \dot{y}x) \end{cases} \tag{2.3}$$

where $\alpha = qM_z/m$. For the Earth dipolar field, the dipole momentum is $M_z = 7.9 \times 10^{25}$ G cm$^3$ $= 7.9 \times 10^{15}$ T m$^3$ ( [14], 1975 IGRF value), and, for electrons and protons, we have,

$$\alpha = \begin{cases} -1.45 \times 10^{27}\ \mathrm{m}^3/\mathrm{s}\ \text{(electrons)} \\ 7.88 \times 10^{23}\ \mathrm{m}^3/\mathrm{s}\ \text{(protons)}. \end{cases} \tag{2.4}$$

We now rescale the system of equations (2.3). With $X = x/r_0$, $Y = y/r_0$, and $Z = z/r_0$, where $r_0 = 6378136$ m is the radius of the Earth, we rewrite the system of equations (2.3) in the form,

$$
\begin{cases}
\ddot{X} = 3\alpha_1 \dfrac{Z}{R^5}(\dot{Y}Z - \dot{Z}Y) - \alpha_1 \dot{Y} \dfrac{1}{R^3} \\[2ex]
\ddot{Y} = -3\alpha_1 \dfrac{Z}{R^5}(\dot{X}Z - \dot{Z}X) + \alpha_1 \dot{X} \dfrac{1}{R^3} \\[2ex]
\ddot{Z} = 3\alpha_1 \dfrac{Z}{R^5}(\dot{X}Y - \dot{Y}X)
\end{cases}
\tag{2.5}
$$

where $R = \sqrt{X^2 + Y^2 + Z^2}$,

$$
\alpha_1 = \frac{\alpha}{r_0^3} =
\begin{cases}
-5.588 \times 10^6 \ \text{s}^{-1} \ (\text{electrons}) \\[1ex]
3.037 \times 10^3 \ \text{s}^{-1} \ (\text{protons}).
\end{cases}
\tag{2.6}
$$

In this rescaled coordinate system, if $R(t) = 1$, the charged particle hits the surface of the Earth.

The Lorentz equations (2.5) can be derived from a Lagrangian. By standard Lagrangian mechanics techniques, and by (2.2), we have,

$$
\begin{aligned}
L &= \frac{1}{2}m(\dot{x}^2 + \dot{y}^2 + \dot{z}^2) + q\vec{r}.\vec{A} = \frac{1}{2}m(\dot{x}^2 + \dot{y}^2 + \dot{z}^2) - \frac{m\alpha}{r^3}(\dot{x}y - \dot{y}x) \\[2ex]
&= \frac{1}{2}m(\dot{X}^2 + \dot{Y}^2 + \dot{Z}^2) - \frac{m\alpha_1}{R^3}(\dot{X}Y - \dot{Y}X).
\end{aligned}
\tag{2.7}
$$

The conjugate momenta to the coordinates $X$, $Y$, and $Z$ are,

$$
\begin{aligned}
p_X &= \frac{\partial L}{\partial \dot{X}} = m\dot{X} - m\alpha_1 \frac{Y}{R^3} \\[2ex]
p_Y &= \frac{\partial L}{\partial \dot{Y}} = m\dot{Y} + m\alpha_1 \frac{X}{R^3} \\[2ex]
p_Z &= \frac{\partial L}{\partial \dot{Z}} = m\dot{Z}
\end{aligned}
\tag{2.8}
$$

and the Hamiltonian is,

$$
\begin{aligned}
H &= \vec{p}.\dot{\vec{R}} - L = \frac{1}{2}m(\dot{X}^2 + \dot{Y}^2 + \dot{Z}^2) \\[2ex]
&= \frac{1}{2m}(p_X{}^2 + p_Y{}^2 + p_Z{}^2) - \alpha_1 p_Y \frac{X}{R^3} + \alpha_1 p_X \frac{Y}{R^3} + \frac{m\alpha_1^2}{2} \frac{X^2 + Y^2}{R^6}.
\end{aligned}
\tag{2.9}
$$

Hence, the system of equations (2.5) has the conservation law,

$$
H = \frac{1}{2}m(\dot{X}^2 + \dot{Y}^2 + \dot{Z}^2).
\tag{2.10}
$$

We show now that the system of equations (2.5) has a second constant of motion.

To determine the second constant of motion, we introduce cylindrical coordinates. With, $X = \rho\cos\phi$, and $Y = \rho\sin\phi$, the Lagrangian (2.7) becomes,

$$
L' = \frac{1}{2}m(\dot{\rho}^2 + \rho^2\dot{\phi}^2 + \dot{Z}^2) + \frac{m\alpha_1}{R^3}\rho^2\dot{\phi}
\tag{2.11}
$$

where $R = \sqrt{\rho^2 + Z^2}$. In this coordinate system, the conjugate momenta become,

$$p_\rho = \frac{\partial L}{\partial \dot\rho} = m\dot\rho$$

$$p_\phi = \frac{\partial L}{\partial \dot\phi} = m\rho^2\dot\phi + m\alpha_1\rho^2\frac{1}{R^3} \qquad (2.12)$$

$$p_Z = \frac{\partial L}{\partial \dot Z} = m\dot Z$$

and the new Hamiltonian is now,

$$H = \frac{1}{2}m(\dot X^2 + \dot Y^2 + \dot Z^2) = \frac{1}{2m}\left(p_\rho{}^2 + p_Z{}^2 + \left(\frac{p_\phi}{\rho} - m\alpha_1\frac{\rho}{R^3}\right)^2\right). \qquad (2.13)$$

As the Hamiltonian (2.13) is independent of $\phi$, we have $\dot p_\phi = -\partial H/\partial\phi = 0$, and by (2.12), the second conservation law is,

$$p_\phi = \frac{\partial L}{\partial \dot\phi} = m\rho^2\dot\phi + m\alpha_1\rho^2\frac{1}{R^3} = c_2 = constant. \qquad (2.14)$$

As $p_\phi = c_2$ is the second equation of motion, we can write the Hamiltonian (2.13) in the form,

$$H = \frac{1}{2m}\left(p_\rho{}^2 + p_Z{}^2 + \left(\frac{c_2}{\rho} - m\alpha_1\frac{\rho}{R^3}\right)^2\right). \qquad (2.15)$$

Then, the equations of motion of a nonrelativistic charged particle in a dipole field reduce to,

$$\begin{cases} \dot\rho &= \frac{\partial H}{\partial p_\rho} = \frac{1}{m}p_\rho \\[4pt] \dot p_\rho &= -\frac{\partial H}{\partial\rho} = \frac{1}{m\rho^3}c_2^2 + 3m\alpha_1^2\frac{\rho^3}{R^8} - m\alpha_1^2\frac{\rho}{R^6} - 3\alpha_1 c_2\frac{\rho}{R^5} \\[4pt] \dot Z &= \frac{\partial H}{\partial p_Z} = \frac{1}{m}p_Z \\[4pt] \dot p_Z &= -\frac{\partial H}{\partial Z} = 3m\alpha_1^2\frac{\rho^2 Z}{R^8} - 3\alpha_1 c_2\frac{Z}{R^5} \end{cases}$$

or,

$$\begin{cases} \ddot\rho &= \frac{1}{m^2\rho^3}c_2^2 + 3\alpha_1^2\frac{\rho^3}{R^8} - \alpha_1^2\frac{\rho}{R^6} - 3\frac{\alpha_1 c_2}{m}\frac{\rho}{R^5} \\[4pt] \ddot Z &= x13\alpha_1^2\frac{\rho^2 Z}{R^8} - 3\frac{\alpha_1 c_2}{m}\frac{Z}{R^5} \end{cases} \qquad (2.16)$$

together with the conservation laws,

$$\begin{cases} \frac{1}{2m^2}\left(p_\rho{}^2 + p_Z{}^2 + \left(\frac{c_2}{\rho} - m\alpha_1\frac{\rho}{R^3}\right)^2\right) = c_1 \\[4pt] m\rho^2\dot\phi + m\alpha_1\rho^2\frac{1}{R^3} = c_2 \end{cases} \qquad (2.17)$$

where $R = \sqrt{\rho^2 + Z^2}$. Note that, we have rescaled the Hamiltonian of the Störmer problem to,

$$H_{eff} = \frac{1}{2}\left(\dot{\rho}^2 + \dot{Z}^2\right) + \frac{1}{2m^2}\left(\frac{c_2}{\rho} - m\alpha_1\frac{\rho}{R^3}\right)^2 = T + V_{eff}(\rho, Z) \qquad (2.18)$$

where $T$ is a scaled kinetic energy.

Hence, the motion of charged particle in a dipole field is described by a two-degree of freedom Hamiltonian system. The effective Hamiltonian is parameterized by the second conservation law in (2.17), which determines the time dependence of the angular cylindrical coordinate.

Before analysing the general topology of the orbits of the phase space flow of equations (2.16), we first consider the case where the motion restricted to the plane $Z = 0$.

## 3. Motion in the equatorial plane of the dipole field

Here, we consider the case where charged particles are constrained to the equatorial plane of the Earth, the plane $Z = 0$, for every $t \geq 0$. In this case, by (2.16), the equations of motion reduce to,

$$\ddot{\rho} = \frac{c_{20}^2}{m^2}\frac{1}{\rho^3} - 3\frac{\alpha_1 c_{20}}{m}\frac{1}{\rho^4} + 2\alpha_1^2\frac{1}{\rho^5} = -\frac{d\bar{V}_{eff}}{d\rho} \qquad (3.1)$$

where $c_{20}$ is the value of the constant $c_2$ evaluated at $Z = 0$, and, by (2.18), the potential function $V_{eff}(\rho)$ is,

$$\bar{V}_{eff}(\rho) = \frac{c_{20}^2}{2m^2}\frac{1}{\rho^2} - \frac{\alpha_1 c_{20}}{m}\frac{1}{\rho^3} + \frac{\alpha_1^2}{2}\frac{1}{\rho^4}. \qquad (3.2)$$

Introducing $p_Z = 0$ and $Z = 0$ into (2.17), the two conservation laws reduce to,

$$\frac{1}{2m^2}\left(p_\rho{}^2 + \left(\frac{c_{20}}{\rho} - m\alpha_1\frac{1}{\rho^2}\right)^2\right) = c_{10}$$

$$m\rho^2\dot{\phi} + m\alpha_1\frac{1}{\rho} = c_{20} \qquad (3.3)$$

where,

$$\begin{cases} c_{10} = \frac{1}{2}\dot{\rho}(0)^2 + \frac{1}{2m^2}\left(\frac{c_{20}}{\rho(0)} - m\alpha_1\frac{1}{\rho^2(0)}\right)^2 \\ c_{20} = m\rho(0)^2\dot{\phi}(0) + m\alpha_1\frac{1}{\rho(0)} \end{cases} \qquad (3.4)$$

and $c_{10}$ is the value of the effective total energy evaluated at $Z = 0$.

Integrating the second conservation law in (3.3) by quadratures, we obtain, for the angular coordinate $\phi(t)$,

$$\phi(t) = \phi(0) + \int_0^t \left(\frac{c_{20}}{m\rho(s)^2} - \frac{\alpha_1}{\rho(s)^3}\right) dt. \qquad (3.5)$$

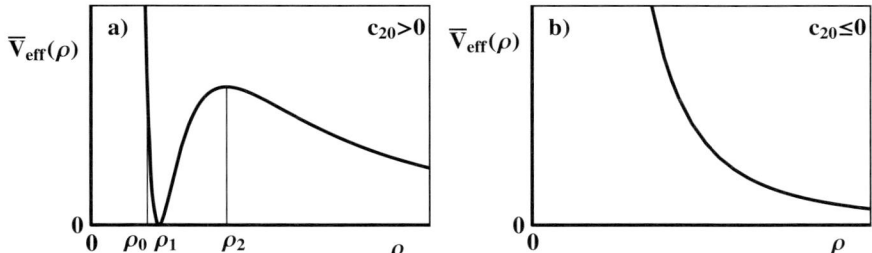

FIGURE 1. Effective potential associated with the motion of a charged particle (protons, $\alpha_1 > 0$) in the equatorial plane of the Earth ($Z = 0$). a) If $c_{20} > 0$ and $\rho \geq 0$, the effective potential has a maximum for $\rho = \rho_2$, a minimum for $\rho = \rho_1$, $\bar{V}_{eff}(\rho_1) = 0$, and $\bar{V}_{eff}(\rho_2) = c_{20}^4/(32m^4\alpha_1^2)$. At $\rho = \rho_0 = (\sqrt{2} - 1)\rho_2$, $\bar{V}_{eff}(\rho_0) = \bar{V}_{eff}(\rho_2)$. b) If $c_{20} \leq 0$ and $\rho \geq 0$, $\bar{V}_{eff}(\rho)$ is a monotonically decreasing function of the argument, and the effective energy surfaces have no compact components.

So, if the solution $\rho(t)$ of (3.1) is known, the temporal dependency of angular coordinate is obtained from (3.5).

In these conditions, the motion of the charged particle in the equatorial plane of the Earth is completely determined by (3.1), derived from the effective Hamiltonian,

$$\bar{H}_{eff}(\rho, \dot{\rho}) = \bar{T} + \bar{V}_{eff} = \frac{1}{2}\dot{\rho}^2 + \frac{c_{20}^2}{2m^2}\frac{1}{\rho^2} - \frac{\alpha_1 c_{20}}{m}\frac{1}{\rho^3} + \frac{\alpha_1^2}{2}\frac{1}{\rho^4} \qquad (3.6)$$

where $\bar{T}$ is a scaled kinetic energy. If $c_{20} > 0$ and for $\rho > 0$, the potential function $\bar{V}_{eff}(\rho)$ has one local minimum and one local maximum at,

$$\rho_1 = \frac{m\alpha_1}{c_{20}} \quad \text{and} \quad \rho_2 = 2\frac{m\alpha_1}{c_{20}} \qquad (3.7)$$

respectively, Figure 1a. By direct calculation, we have, $\bar{V}_{eff}(\rho_2) = c_{20}^4/(32m^4\alpha_1^2)$, and $\bar{V}_{eff}(\rho_1) = 0$. Therefore, for $0 \leq \bar{H}_{eff}(\rho, \dot{\rho}) \leq \bar{V}_{eff}(\rho_2)$, the constant effective energy surface contains a compact component.

If $c_{20} \leq 0$ and for positive values of $\rho$, the potential function $\bar{V}_{eff}(\rho)$ is a monotonically decreasing function of the argument, Figure 1b. Then, we have:

**Proposition 3.1.** *In the equatorial plane of a dipole field, a nonrelativistic (positively) charged particle precesses around the dipole axis, along a circle with radius $\rho_1$, provided:*

(i) *$\bar{H}_{eff}(\rho(0), \dot{\rho}(0)) < \bar{V}_{eff}(\rho_2)$, $c_{20} > 0$, $\rho(0) \neq \rho_1$ and $\rho(0) < \rho_2$, where $\bar{H}_{eff}$ is defined in (3.6), the constants $\rho_1$ and $\rho_2$ are given in (3.7), and the constant $c_{20}$ is defined in (3.4). For initial conditions near the circumference of radius $\rho = \rho_1$, the Larmor or precession period is, $T_L = 2\pi m^3 \alpha_1^2/c_{20}^3$. The* *phase advance per Larmor period is,*

$$\Delta\phi = 2\pi \frac{\rho_1^2}{\rho(0)^2}\left(1 - \frac{\rho_1}{\rho(0)}\right) - 4\pi^2 \frac{\rho_1^3}{\rho(0)^3}\frac{m}{c_{20}}\rho_1\dot\rho(0)\left(1 - \frac{3}{2}\frac{\rho_1}{\rho(0)}\right) + \cdots$$

*and the period of rotation around the Earth is* $T_r = 2\pi T_L/\Delta\phi.$

(ii) *If* $\bar{H}_{eff}(\rho(0), \dot\rho(0)) = \bar{V}_{eff}(\rho_2)$, $c_{20} > 0$, *and* $\rho(0) < \rho_2$, *or,* $\rho(0) > \rho_2$ *and* $\dot\rho(0) < 0$, *then,* $\rho(t) \to \rho_2$, *as* $t \to \infty$.

(iii) *If* $\rho(0) = \rho_2[resp.,\ \rho(0) = \rho_1]$, $\dot\rho(0) = 0$, $\dot\phi(0) \neq 0$, *and* $c_{20} > 0$, *then, the charged particle has a circular trajectory with radius* $\rho = \rho_2[resp.,\ \rho = \rho_1]$, *and the period of rotation around the Earth is* $T_r = 2\pi/\dot\phi(0)$. *If* $\dot\phi(0) = 0$, *the charged particle is at rest.*

(iv) *If* $\bar{H}_{eff}(\rho(0), \dot\rho(0)) > \bar{V}_{eff}(\rho_2)$ *and* $c_{20} > 0$, *or,* $c_{20} \leq 0$, *then,* $\rho(t) \to \infty$, *as* $t \to \infty$, *and we have escape trajectories.*

*Proof.* To prove the proposition, we must ensure first that the solution of the differential equation (3.1) exists, and is defined for every $t \geq 0$. In the cases (i)–(iii), the existence of solutions for every $t \geq 0$ follows because the initial conditions on phase space are on the compact components of the level sets of the Hamiltonian function (3.6), ( [4], pp. 187; [13], pp. 8).

A simple phase space analysis shows that the vector field associated with (3.1) has a centre type fixed point with phase space (cylindrical) coordinates $(\rho_1, \dot\rho = 0)$. This centre type fixed point is inside the homoclinic loop of a saddle point with coordinates $(\rho_2, \dot\rho = 0)$. The conditions in (i) correspond to initial conditions inside the homoclinic loop, and away from the centre fixed point, $\rho(0) \neq \rho_1$. To calculate the Larmor frequency and the phase advance per Larmor period, we linearize (3.1) around $\rho_1$, and we obtain,

$$\ddot{x} + \frac{c_{20}^6}{m^6\alpha_1^4}x = 0$$

where $x = \rho - \rho_1$, and $c_{20}^6/(m^6\alpha_1^4) = \omega_L^2$. The Larmor period is $T_L = 2\pi/\omega_L$. Note that, by a direct calculation, $\omega_L^2 = \dfrac{d^2\bar{V}_{eff}}{d\rho^2}(\rho = \rho_1)$.

By (3.5), the phase advance per Larmor period is,

$$\Delta\phi = \int_0^{T_L}\left(\frac{c_{20}}{m\rho(s)^2} - \frac{\alpha_1}{\rho(s)^3}\right)dt.$$

From the above linearized differential equation, in the vicinity of the circumference of radius $\rho = \rho_1$, $\rho(t) = \rho_1 + (\rho - \rho_1)\cos(\omega_L t) + \dot\rho(0)\sin(\omega_L t)/\omega_L$ and after substitution into the expression for $\Delta\phi$, we obtain the result. The rotation period around the Earth is obtained from the conditions, $n\Delta\phi = 2\pi$ and $T_r = nT_L$.

The conditions in (ii) correspond to initial conditions on the stable branches of the homoclinic orbit of the saddle point with coordinates $(\rho_2, \dot\rho = 0)$.

For case (iii), as $(\rho_2, \dot\rho = 0)$ is a fixed point of the differential equation (3.1), if $\dot\phi(0) \neq 0$, the charged particle has a circular trajectory around the origin of coordinates. To calculate the period of the trajectory, by (3.5), we have,

$$2\pi = \left( \frac{c_{20}}{m\rho_2^2} - \frac{\alpha_1}{\rho_2^3} \right) T_r \,.$$

Introducing the value of the constant $c_{20}$ into the above expression, by (3.4), we obtain $T_r = 2\pi/\dot{\phi}(0)$. For $\rho(0) = \rho_1$, the proof is similar.

In case (iv), the effective potential function monotonically decreases as $\rho$ increases from $\rho = 0$, and we are in the conditions of escape trajectories. The prolongation of solutions for every $t \geq 0$, follows from the condition that $f_i(\rho)/\|\rho\| \rightarrow$ *constant* as $\rho \rightarrow \infty$ ( [13], pp. 9). $\qquad\square$

By Proposition 3.1, the trajectory of a charged particle on the equatorial plane of an axially symmetric magnetosphere can be unbounded, can be trapped in an annular region around the Earth, or can collide with the surface of the Earth. As $\rho_0$, $\rho_1$ and $\rho_2$ depend on $c_{20}$, there are charged particles with different energies trapped in the annular region $[\rho_0, \rho_2]$. This annular region is the equatorial cross section of the Van Allen inner radiation belt, [19].

As $\rho_0$, $\rho_1$ and $\rho_2$ depend on the initial conditions $\rho(0)$ and $\dot{\phi}(0)$ through $c_{20}$, for the same belt parameter, there are particles that escape from the trapping region, escaping to infinity or hitting the surface of the Earth. This is due to the fact that the belt parameters are independent of $\dot{\rho}(0)$. By Proposition 3.1-(iv), and by the conservation of energy, a charged particle escapes from the inner Van Allen radiation belt if,

$$|\dot{\rho}(0)| > \sqrt{2|\bar{V}_{eff}(\rho_2) - \bar{V}_{eff}(\rho(0))|} \qquad (3.8)$$

By the same argument, if $|\dot{\rho}(0)| \geq |c_{20}/m - \alpha_1|$, a particle with initial velocity $\dot{\rho}(0) < 0$ at infinity, and velocity vector within the equatorial plane of the Earth, after an infinite time, hits the surface of the Earth.

In Figure 2, we show the limits of the equatorial cross section of a Van Allen inner radiation belt, the trajectories of trapped protons, and an escape trajectory. In the three cases shown, the Van Allen parameters are the same. The trajectories shown in Figure 2 have been calculated by the numerical integration of (3.1), (see the Appendix). The angular variable $\phi$ has been obtained from the discretization of (3.5),

$$\phi_{n+1} = \phi_n + \Delta t \left( \frac{c_{20}}{m\rho_n^2} - \frac{\alpha_1}{\rho_n^3} \right) \qquad (3.9)$$

where $\Delta t$ is the discretization time step, $\phi_n = \phi(n\Delta t)$, $\rho_n = \rho(n\Delta t)$, and $n = 0, 1, \dots$.

As the radiation belt parameters $\rho_0$, $\rho_1$ and $\rho_2$ depend on the angular velocity $\dot{\phi}(0)$, we can have several Van Allen radiation belts at different altitudes, and trapped particles with different energies.

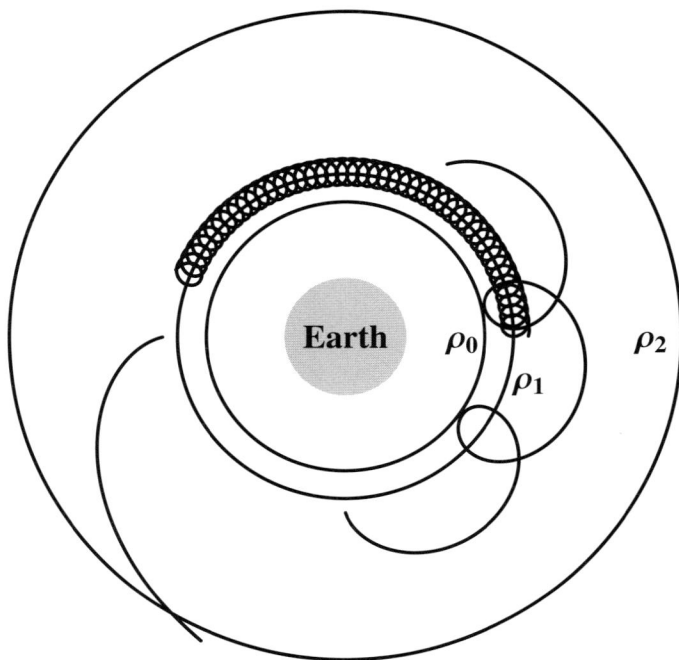

FIGURE 2. Equatorial cross section of the Van Allen inner radiation belt, for protons in the equatorial plane of the Earth.
We show the trajectories of three protons with initial conditions
(cylindrical coordinates): a) $\rho(0) = 3.0$, $\dot{\rho}(0) = 10.0$, $\phi(0) = 0.0$,
and $\dot{\phi}(0) = 10.0$; b) $\rho(0) = 3.0$, $\dot{\rho}(0) = 80.0$, $\phi(0) = 3\pi/2$,
and $\dot{\phi}(0) = 10.0$; c) $\rho(0) = 3.0$, $\dot{\rho}(0) = 100.0$, $\phi(0) = \pi$, and
$\dot{\phi}(0) = 10.0$. The first and the second trajectories, a) and b),
correspond to the precession of protons around the Earth. The
third case is an escape trajectory. By (3.8), the escape condition
is $|\dot{\rho}(0)| \geq 95.42$. For the three trajectories shown, the radiation
belt parameters are $\rho_0 = 2.282$, $\rho_1 = 2.755$, $\rho_2 = 5.510$, and
$c_{20} = 1.844 \times 10^{-24}$. The total effective energies are: a) $c_{10} = 500$;
b) $c_{10} = 3650$, and, c) $c_{10} = 5450$. The trajectories have been
calculate with the Störmer–Verlet numerical method (Appendix)
with the time step $\Delta t = 0.0001$, and the angular coordinate has
been calculated by (3.9). For protons with precessing trajectories
near the circumference of radius $\rho_1$, by Proposition 3.1, the Larmor period is $T_L = 0.043$ s. The first two particle trajectories have
been calculated from $t = 0$ up to $t = 2$ s.

## 4. Motion in the three-dimensional space

The equations of motion of a charged particle in a dipole field – (2.16), are derived from the effective potential function,

$$V_{eff}(\rho, Z) = \frac{1}{2m^2} \left( \frac{c_2}{\rho} - m\alpha_1 \frac{\rho}{R^3} \right)^2 \tag{4.1}$$

where $R = \sqrt{\rho^2 + Z^2}$. In Figure 3a), we show the graph of the potential function $V_{eff}(\rho, Z)$, and its level lines, for $c_2 > 0$.

As we have seen in the previous section, on the equatorial plane $Z = 0$, the exterior boundary of the trapping region is the circumference of radius $\rho_2$, which corresponds to the unique local maximum of the potential function $\bar{V}_{eff}(\rho)$. The equation $V_{eff}(\rho, Z) = \bar{V}_{eff}(\rho_2)$ defines a bounded region in the $(\rho, Z)$ plane whose closure is a compact set of maximal area[1], Figure 3b). The compact components of the level sets of the effective Hamiltonian function (2.18) are obtained as the topological product of a compact plane set with the compact sets in the interior of the region defined by the equation $V_{eff}(\rho, Z) = \bar{V}_{eff}(\rho_2)$. Therefore, any charged particle with an effective energy on a compact component of the effective Hamiltonian is trapped in a torus-like region around the Earth. This trapping region is the Van Allen inner radiation belt of the Earth.

In the rescaled coordinate system introduced in Section 2, the level lines $V_{eff}(\rho, Z) = \bar{V}_{eff}(\rho_2)$ hit the surface of the Earth for $\sqrt{\rho^2 + Z^2} = 1$. So, we define the contact points of a specific radiation belt as the points at the surface of the Earth where $V_{eff}(\rho, Z) = 0$. If $c_2 > 0$, this potential function takes its minimum value ($V_{eff} = 0$) along the line $c_2/\rho = m\alpha_1 \frac{\rho}{R^3}$, Figure 3b), and, for $R = 1$, we have,

$$\rho_c = \sqrt{\frac{c_2}{m\alpha_1}} = \frac{1}{\sqrt{\rho_1}} \tag{4.2}$$

which corresponds to the latitudes,

$$\theta_c = \arccos \rho_c = \arccos \sqrt{\frac{c_2}{m\alpha_1}} . \tag{4.3}$$

We now summarize the main features of the dynamics associated with the two degrees of freedom Hamiltonian (2.18).

**Proposition 4.1.** *We consider the motion of a nonrelativistic (positively) charged particle in a dipole field. If one of the coordinates $Z(0)$ or $\dot{Z}(0)$ of the initial condition is different from zero, and $(\rho(0), Z(0)) \neq (0,0)$, then the charged particle has bounded motion, being trapped in a torus-like region around the Earth, provided:*

---

[1]The compact set is obtained by adding the point $(0,0)$ to the open set defined by $V_{eff}(\rho, Z) = \bar{V}_{eff}(\rho_2)$. In the following, compact level sets are always obtained by adding this exceptional point to the open and bounded level sets of the effective potential function $V_{eff}(\rho, Z)$.

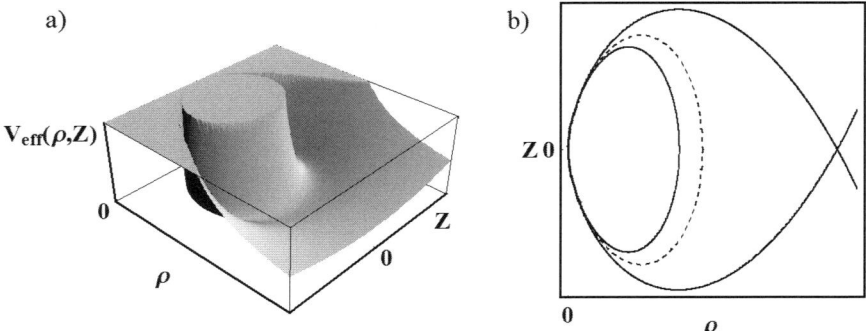

FIGURE 3. a) Graph of the effective potential function (4.1), with
$c_2 > 0$, for the three dimensional Störmer problem for protons.
b) Effective potential energy level line (heavy lines), $V_{eff}(\rho, Z) =$
$\bar{V}_{eff}(\rho_2)$. For particles characterized by the constant of motion
$c_2 > 0$, the Van Allen inner radiation belt is the toric-like surface
obtained by rotating the level lines shown in b) around the axis
of the dipole field of the Earth. The annular region shown in
Figure 2, delimited by the circumferences of radius $\rho_0$ and $\rho_2$, is
the $Z = 0$ cross section of the Van Allen inner radiation belt. The
dotted line is the local minimum of the effective potential energy
function $V_{eff}(\rho, Z)$.

(i) $H_{eff}(\rho(0), \dot{\rho}(0), Z(0), \dot{Z}(0)) \leq \bar{V}_{eff}(\rho_2)$, $c_2 > 0$, and $\rho(0) < \rho_2$, where $\rho_2$
is defined in (3.7), the Hamiltonian function $H_{eff}$ is defined in (2.18), and
the constant of motion $c_2$ is defined in (2.17). In the particular case where,
$\dot{Z}(0) = 0$, $\dot{\rho}(0) = 0$, and $c_2/\rho(0) = m\alpha_1\rho(0)/\sqrt{\rho(0)^2 + Z(0)^2}^3$, the charged
particle remains, for every $t \geq 0$, in the plane $Z = Z(0)$, and $Z(0) \in$
$[-2m\alpha_1/(3\sqrt{3}c_2), 2m\alpha_1/(3\sqrt{3}c_2)]$. If $\dot{\phi}(0) \neq 0$, it rotates around the dipole
axis. If $\dot{\phi}(0) = 0$, the particle is at rest in the plane $Z = Z(0)$.
(ii) If one of the coordinates $Z(0)$ or $\dot{Z}(0)$ of the initial condition is different
from zero, and one of the above conditions is not verified, then $\rho(t) \to \infty$, as
$t \to \infty$, and we have an escape trajectory.

*Proof.* If $Z(0) = \dot{Z}(0) = 0$, the motion is restricted to the plane perpendicular to the dipole axis, and we obtain the case of Proposition 3.1 of the previous
section. If $(\rho(0), Z(0)) = (0, 0)$, the equation of motion (2.16) is not defined. In
case (i), the initial condition is in a compact component of the effective Hamiltonian, and the motion is bounded for every $t \geq 0$. If, $\dot{Z}(0) = 0$, $\dot{\rho}(0) = 0$,
and $c_2/\rho(0) = m\alpha_1\rho(0)/\sqrt{\rho(0)^2 + Z(0)^2}^3$, where $c_2$ is defined in (2.17), and
$Z(0) \in [-2m\alpha_1/(3\sqrt{3}c_2), 2m\alpha_1/(3\sqrt{3}c_2)]$, the differential equation (2.16) has one
or two fixed points in the plane $Z = Z(0)$.

In case (ii), the level sets of the Hamiltonian function are not bounded, and the prolongation of solutions as $t \to \infty$ follows as in case (iv) of Proposition 3.1. $\square$

In Figure 4, we show the trajectories of several particles trapped in a Van Allen inner radiation belt characterized by the constant $c_2 > 0$. The trajectories have been calculate with the Störmer–Verlet numerical method (Appendix) with integration time step $\Delta t = 0.00002$, and total integration time $t = 2\,\mathrm{s}$. From (2.17), it follows that the angular coordinate $\phi(t)$ is given by,

$$\phi(t) = \phi(0) + \int_0^t \left( \frac{c_2}{m\rho(s)^2} - \frac{\alpha_1}{(\sqrt{\rho(s)^2 + Z(t)^2})^3} \right) dt$$

and by discretization, we obtain,

$$\phi_{n+1} = \phi_n + \Delta t \left( \frac{c_2}{m\rho_n^2} - \frac{\alpha_1}{\left( \sqrt{\rho_n^2 + Z_n^2} \right)^3} \right) \tag{4.4}$$

where $\Delta t$ is the discretization time step, $\phi_n = \phi(n\Delta t)$, $\rho_n = \rho(n\Delta t)$, $Z_n = Z(n\Delta t)$, and $n = 0, 1, \ldots$. In the trajectories of Figure 4, the angular coordinate $\phi(t)$ has been calculated with (4.4).

In the plots on the right-hand side of Figure 4, we show the projection of the boundary of the level set of the effective Hamiltonian on the $(\rho, Z)$ plane. The trajectories of trapped charged particles are in the interior of these boundary curves. Particles with low effective energy have trajectories concentrated in the vicinity of the equatorial plane of the Earth, Figure 4a. Increasing the effective energy of the particles, their trajectories approach the polar regions of the Earth and eventually hit the surface of the Earth. By (4.3), and for the parameter values of Figure 4, the contact points of the Van Allen inner radiation belt with the Earth are located at the latitudes $\theta_c = \pm 53.76^o$. These simulations suggest that the constant effective energy surface is not filled density by a unique trajectory.

To characterize more precisely the dynamics of the charged particles trapped in the dipole field of the Earth, we can eventually use KAM techniques, [12], or construct a Poincaré map for the equations of motion (2.16).

To pursue a KAM approach, [12], the first step is to find an integrable Hamiltonian system leaving invariant a two-dimensional torus in the four-dimensional phase space of the differential equations (2.16). For that, we develop in Taylor series around the points $(\rho = \rho_1, Z = 0)$ and $(\rho = \rho_2, Z = 0)$ the second members of the differential equations in (2.16). By Proposition 3.1, at these points the motion is integrable and periodic in the equatorial plane of the Earth. In the first case, the differential equations (2.16) become,

$$\begin{cases} \ddot{\rho} = -\dfrac{c_2^6}{m^6 \alpha_1^4} (\rho - \rho_1) \\ \ddot{Z} = 0 \end{cases} \tag{4.5}$$

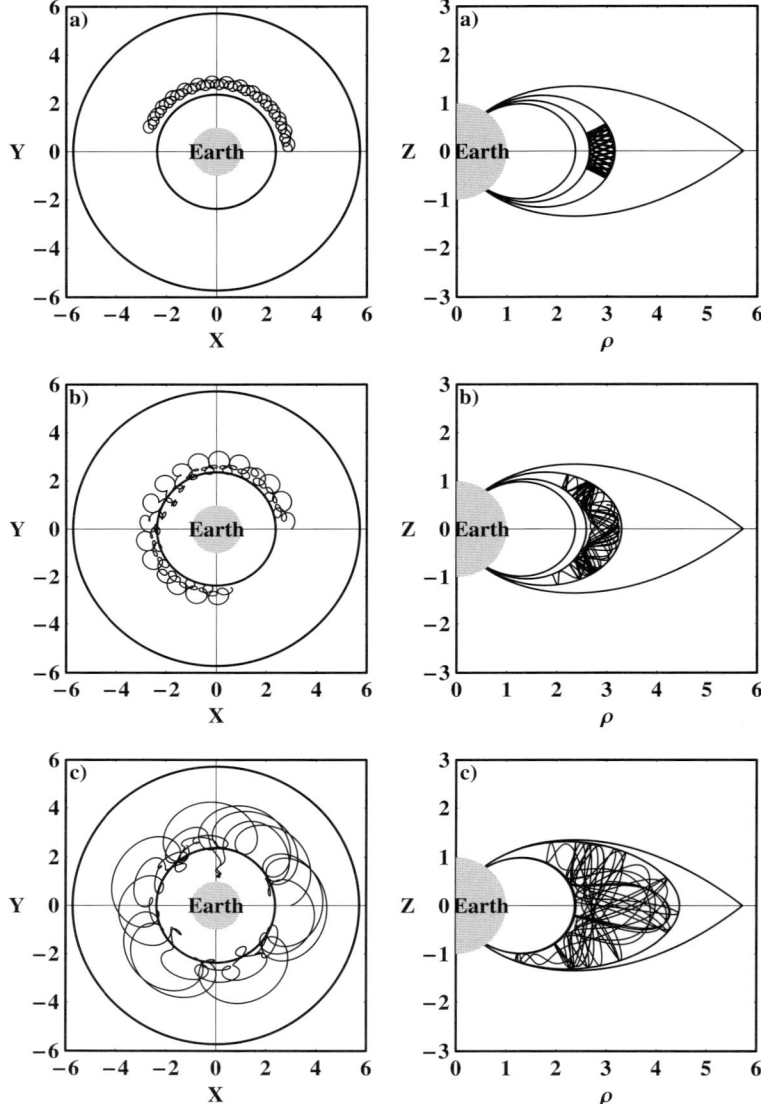

FIGURE 4. Trajectories of three protons in the Van Allen inner radiation belt, with initial conditions: a) $\rho(0) = 3.0$, $\dot{\rho}(0) = 10.0$, $\phi(0) = 0.0$, $\dot{\phi}(0) = 10.0$, $Z(0) = 0.5$, and $\dot{Z}(0) = 0.0$; b) $\rho(0) = 3.0$, $\dot{\rho}(0) = 30.0$, $\phi(0) = 0.0$, $\dot{\phi}(0) = 10.0$, $Z(0) = 0.5$, and $\dot{Z}(0) = 0.0$; c) $\rho(0) = 3.0$, $\dot{\rho}(0) = 80.0$, $\phi(0) = 0.0$, $\dot{\phi}(0) = 10.0$, $Z(0) = 0.5$, and $\dot{Z}(0) = 0.0$. The radiation belt parameters are $\rho_0 = 2.370$, $\rho_1 = 2.861$, $\rho_2 = 5.722$, and $c_2 = 1.776 \times 10^{-24}$. The total effective energies are: a) $c_1 = 500$; b) $c_1 = 900$, and c) $c_1 = 3650$.

and, in the second case, we obtain,

$$\begin{cases} \ddot{\rho} = \dfrac{c_2^6}{32m^6\alpha_1^4}(\rho - \rho_2) \\[2mm] \ddot{Z} = -\dfrac{3c_2^6}{64m^6\alpha_1^4}Z \,. \end{cases} \qquad (4.6)$$

As the solutions of the linear equations (4.5) and (4.6) are unbounded, we loose the property of boundness already contained in Proposition 4.1- (i) and also the possibility of having a family of invariant two-dimensional torus in the four-dimensional phase space of the unperturbed systems (4.5) and (4.6).

The other way of characterizing the dynamics of the charged particles in the Van Allen inner belt is to find a Poincaré map for the equations of motion (2.16).

We consider that at most one of the coordinates $Z(0)$ and $\dot{Z}(0)$ of the initial condition of the differential equation (2.16) is different from zero, and the particle has bounded motion ($c_2 > 0$). To avoid degenerate situations, we also consider that, $(\rho(0), Z(0)) \neq (0,0)$ and the effective energy function is positive, $c_1 > 0$. Under these conditions, by Proposition 4.1-(i), the initial condition is on a compact component of the level sets of the Hamiltonian function (2.18). So, the solution of the differential equation (2.16) exists, and is defined for all $t \geq 0$ ( [4], pp. 187; [13], pp. 8).

The compact components of the level sets of the Hamiltonian function $H_{eff}$ are obtained as the topological product of a compact plane set with the compact level set of the potential function $V_{eff}(\rho, Z)$. These level sets are compact, provided $c_2 > 0$. The compact level sets of the Hamiltonian function are three-dimensional compact manifolds embedded in the four dimensional phase space, and their intersection with the three-dimensional hyperplane $Z = 0$ is a two-dimensional compact manifold $\Sigma_{c_2}$. This two-dimensional manifold is a compact set in the four dimensional phase space of the differential equation (2.16) and, by (2.18), has local coordinates $\rho$ and $\dot{\rho}$.

In the conditions of Proposition 4.1-(i), any orbit initiated in an initial condition on the effective energy level sets can cross the hyperplane $Z = 0$, intersecting transversally the two-dimensional compact manifold $\Sigma_{c_2}$. Therefore, the compact two-dimensional manifold $\Sigma_{c_2}$ is a good candidate for a Poincaré section of the differential equation (2.16). This construction is used to justify the existence of two-dimensional conservative Poincaré maps for two-degrees of freedom Hamiltonian systems, ( [12], pp. 17–20). However, there are two main difficulties in proving that $\Sigma_{c_2}$ is the domain of a Poincaré map. The first case, it is difficult to prove that any trajectory that crosses transversally the section $\Sigma_{c_2}$ will return to it after a finite time. The second difficult questions concerns the unicity of the trajectories crossing the plane $Z = 0$.

From Proposition 4.1- (i), it follows that there are orbits of trapped particles that never cross $\Sigma_{c_2}$. However, due to the invariance of the differential equation (2.16) for transformations of $t$ into $-t$, any initial condition on the plane

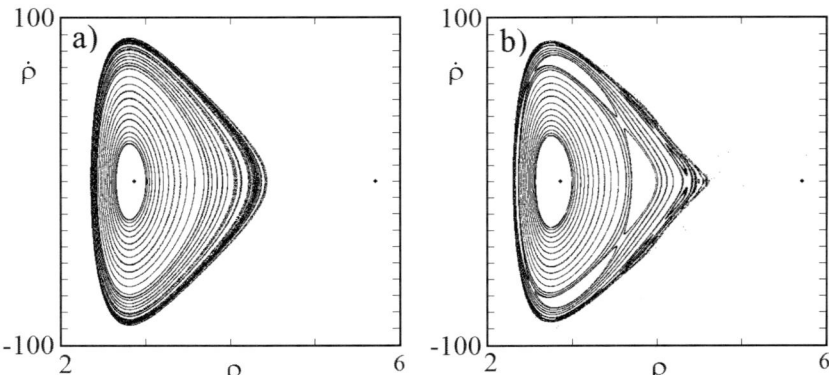

FIGURE 5. Poincaré maps for the constant of motion $c_2 = 1.776 \times 10^{-24}$. The Poincaré map have been calculated with the initial conditions: a) $\rho(0) = 3.0$, $\phi(0) = 0.0$, $\dot{\phi}(0) = 5.47089$, $Z(0) = 0.0$, and $\dot{Z}(0) = 0.5$; b) $\rho(0) = 3.0$, $\phi(0) = 0.0$, $\dot{\phi}(0) = 10.0$, $Z(0) = 0.5$, and $\dot{Z}(0) = 0.5$. In both figures, the initial conditions $\dot{\rho}(0)$ have been changed in the interval $[0, 85]$. The dots have co-ordinates $(\rho_1, 0)$ and $(\rho_2, 0)$, with $\rho_1 < \rho_2$. Both Poincaré sections have been generated with different families of initial conditions. This suggests that there are different particle trajectories with the same effective energy that cross the two-dimensional section $\Sigma_{c_2}$ at the same point.

$Z = 0$ with $\dot{Z} \neq 0$, has a prolongation for positive and negative values of $t$, cross-ing transversally the plane $Z = 0$ at most once. With this property, we can test numerically the structure of the orbits that cross the set $\Sigma_{c_2}$. In Figure 5, we show the computed orbits on the Poincaré section $\Sigma_{c_2}$, for two families of initial conditions and the same value of the constant of motion $c_2$.

In Figure 5a, we have chosen initial conditions on the plane $Z(0) = 0$, with $\dot{Z}(0) \neq 0$. For all the analysed initial conditions, the orbits crossed several times the surface of section $\Sigma_{c_2}$. Due to the structure of orbits, the motion appears to be quasi-periodic. However, if the initial conditions are away from the plane $Z(0) = 0$ and for the same value of $c_2$, the topology of the orbits on the section $\Sigma_{c_2}$ changes drastically, Figure 5b. Comparing Figure 5a with 5b, the structure of the orbits are incompatible, in the sense that different orbits cross $\Sigma_{c_2}$ at the same point. This shows that $\Sigma_{c_2}$ can not be the domain of a Poincaré map for the differential equations(2.16). On the other hand, the structure of orbits in Figure 5b suggests the existence of transversal homoclinic intersections and therefore chaotic motion inside the trapping region of the Störmer problem.

## 5. Conclusions

The Hamiltonian dynamical system describing the motion of a charged particle in a dipole field (Störmer problem) has been reduced to a two-degrees of freedom system. This reduced system has two constants of motion. It has been shown that the trajectories of charged particles can be periodic and quasi-periodic, and that, for a suitable choice of the initial conditions in between a minimal and maximal height from the equatorial plane, the planes perpendicular to the dipole axis are invariant for the motion of charged particles.

From a more global point of view, the trajectories of the charged particles in the Earth dipole field can be trapped in a torus-like region surrounding the Earth, or can be scattered and escape to infinity. The torus-like trapping region around the Earth can be interpreted as the Van Allen inner radiation belt, as measured by particle detectors in spacecrafts. The physical effects associated with radiation phenomena intrinsic to accelerated charged particles (*Bremsstrahlung*) account for the phenomena of radiation aurorae. The numerically computed trajectories of trapped charged particles suggests the existence of chaotic motion inside the Van Allen inner belts around the dipole axis of the Earth. The properties of the Störmer dynamical systems are summarized in Propositions 3.1 and 4.1.

From the applied physics point of view, the Störmer problem deviates from the real situation in essentially three ways. In the first case, the magnetic field of the Earth has a strong quadrupolar component that was not considered. The second drastic simplification has to do with the fact that the dipole axis is not coincident with the rotation axis of the Earth. The rotation of the Earth introduces a time periodic forcing into the equations of motion. In the third simplification, we have considered implicitly that charged particles do not radiate (*Bremsstrahlung*) when subject to accelerating forces. However, the three main effects observed in the Earth magnetosphere, radiation belts, radiation aurorae and South Atlantic anomaly, are described by the simplified model.

The existence of periodic, quasi-periodic and chaotic trajectories in the Störmer problem together with the *Bremsstrahlung* effect, shows that radiation belts make a shield protection for the high energy charged particles arriving at Earth. These particles loose kinetic energy by *Bremsstrahlung* when they are scattered and when they are trapped in the Van Allen inner radiation belts.

From the mathematical point of view, the Störmer problem poses some open problems for the dynamics of two-degrees of freedom Hamiltonian systems. The Störmer dynamical systems is not generated from the perturbation of a two-dimensional torus in a four-dimensional phase space, suggesting the existence of non KAM mechanism for the generation of bounded motion in two-degrees of freedom Hamiltonian systems. On the other hand, the arguments used in the construction of Poincaré sections in two-degrees of freedom Hamiltonian systems with two conservation laws lead to the non-unicity of the orbits on these surfaces of section.

# Appendix

To integrate numerically the equations of motion (2.16) and (3.1), we have used the explicit Störmer–Verlet method of order 2, ( [10], pp. 14 and 177),

$$\begin{cases} p_{n+1/2} = p_n - \Delta t \dfrac{\partial H}{\partial q}(q_i) \\[2mm] q_{n+1} = q_n + \Delta t \dfrac{\partial H}{\partial p}(p_{n+1/2}) \\[2mm] p_{n+1} = p_{n+1/2} - \Delta t \dfrac{\partial H}{\partial q}(q_{i+1}) \end{cases}$$

where $H(q_1, \ldots, q_n, p_1, \ldots, p_n)$ is the Hamiltonian function, and $\Delta t$ is the integration time step. For example, for (3.1), the Störmer–Verlet method reduces to,

$$\begin{cases} \rho_{n+1} = \rho_n + \Delta t \dot\rho_n + \dfrac{\Delta t^2}{2} f(\rho_n) \\[2mm] \dot\rho_{n+1} = \dot\rho_n + \dfrac{\Delta t}{2} f(\rho_n) + \dfrac{\Delta t}{2} f(\rho_{n+1}) \end{cases}$$

where $\rho_n = \rho(n\Delta t)$, $\dot\rho_n = \dot\rho(n\Delta t)$, $f(\rho) = -\dfrac{dV_{eff}}{d\rho}$. This method is symplectic, being area preserving.

The Störmer–Verlet method has the advantage of being explicit, and the integration accuracy is obtained by decreasing the time step $\Delta t$. However, it does not conserve the energy function. We have tested other higher order numerical integrators but, with this Störmer–Verlet method, the overall behaviour of the solutions is closer to the exact results. For a detailed theoretical discussion about the Störmer–Verlet method see [17].

## Acknowledgments

This work has been partially supported by the POCTI Project /FIS/10117/2001 (Portugal) and by a pluriannual funding grant to GDNL.

# References

[1] M. Braun, *Particle motions in a magnetic field*, J. Diff. Equ., **8** (1970) 294–332.

[2] M. Braun, *On the stability of the Van Allen radiation belt*, SIAM J. Appl. Math., **37** (1979) 664–668.

[3] M. Braun, *Mathematical remarks on the Van Allen radiation belt: a survey of old and new results*, SIAM Review, **23** (1981) 61–93.

[4] D. R. J. Chillingworth, *Differential Topology with a View to Applications*, Pitman Publishing, London, 1976.

[5] E. J. Daly, *The Evaluation of Space Radiation Environments for ESA Projects*, ESA Jopurnal, **12** (1988) 229–247.

[6] R. DeVogelaere, On the structure of symmetric periodic solutions of conservative systems, with applications. In: S. Lefschetz (Ed.) *Contributions to the Theory of Nonlinear Oscillations*, pp. 53–84. Princeton University Press, Princeton, 1958.

[7] A. J. Dragt, *Trapped Orbits in a Magnetic Dipole Field*, Rev. Geophysics, **3** (1965) 255–298.

[8] A. J. Dragt and J. M. Finn, *Insolubility of trapped particle motion in a magnetic dipole field*, J. Geophys. Res., **81** (1976) 2327–2340.

[9] R. P. Feynman, R. B. Leighton, and M. Sands, *Lectures on Physics*, vol. II, Addison-Wesley, Reading, 1964.

[10] E. Hairer, C. Lubich, and G. Wanner, *Geometric Numerical Integration, Structure-Preserving Algorithms for Ordinary Differential Equations*. Springer-Verlag, Berlin, 2002.

[11] W. N. Hess, *The Radiation Belt and Magnetosphere*. Blaisdel, Waltham, MA, 1968.

[12] A. J. Lichtenberg and M. A. Lieberman, *Regular and Stochastic Motion*. Springer-Verlag, Berlin, 1983.

[13] V. V. Nemitskii and V. V. Stepanov, *Qualitative Theory of Differential Equations*, Princeton University Press, Princeton, 1960.

[14] T. Rikitake and Y. Honkura, *Solid Earth Geomagnetism*, Terra Scientific Publishing Co., Tokyo, 1985.

[15] E. G. Stassinopoulos and J. P. Raymond, *The Space Radiation Environment for Electronics*, Proceedings IEEE, **76** n$^o$11 (1988) 1423–1442.

[16] C. Störmer, *The Polar Aurora*, Oxford at the Clarendon Press, Oxford, 1955.

[17] P. F. Tupper, *Ergodicity and the Numerical Simulation of Hamiltonian Systems*, SIAM J. Applied Dynamical Systems, **4** n$^o$3 (2005) 563–587.

[18] C. Underwood, D. Brock, P. Williams, S. Kim, R. Dilão, P. Ribeiro Santos, M. Brito, C. Dyer, and A. Sims, *Radiation Environment Measurements with the Cosmic Ray Experiments On-Board the KITSAT-1 and PoSAT-1 Micro-Satellites*, IEEE Transactions on Nuclear Sciences, **41** (1994) 2353–2360.

[19] J. A. Van Allen and L A. Frank, *Radiation Around the Earth to a Radial Distance of* 107, 400 *KM*, Nature, **183** (1959) 430–434.

Rui Dilão
Nonlinear Dynamics Group
Instituto Superior Técnico
Av. Rovisco Pais
P-1049-001 Lisbon
Portugal
e-mail: `rui@sd.ist.utl.pt`; `ruidilao@gmail.com`

Rui Alves-Pires
Faculdade de Engenharia
Universidade Católica Portuguesa
Estrada de Talaíde
P-2635-631 Rio de Mouro
Portugal
e-mail: pires@fe.ucp.pt; pires@sd.ist.utl.pt

Progress in Nonlinear Differential Equations
and Their Applications, Vol. 75, 195–209
© 2007 Birkhäuser Verlag Basel/Switzerland

# Hausdorff Dimension versus Smoothness

Flávio Ferreira, Alberto A. Pinto, and David A. Rand

*Dedicated to Arrigo Cellina and James Yorke*

**Abstract.** There is a one-to-one correspondence between $C^{1+H}$ Cantor exchange systems that are $C^{1+H}$ fixed points of renormalization and $C^{1+H}$ diffeomorphisms $f$ on surfaces with a codimension 1 hyperbolic attractor $\Lambda$ that admit an invariant measure absolutely continuous with respect to the Hausdorff measure on $\Lambda$. However, there is no such $C^{1+\alpha}$ Cantor exchange system with bounded geometry that is a $C^{1+\alpha}$ fixed point of renormalization with regularity $\alpha$ greater than the Hausdorff dimension of its invariant Cantor set. The proof of the last result uses that the stable holonomies of a codimension 1 hyperbolic attractor $\Lambda$ are not $C^{1+\theta}$ for $\theta$ greater than the Hausdorff dimension of the stable leaves of $f$ intersected with $\Lambda$.

**Mathematics Subject Classification (2000).** Primary 37D20; Secondary 37E30.

**Keywords.** Hyperbolic systems, attractors, Hausdorff dimension.

## 1. Introduction

The works of Veech, Penner, Thurston and Masur show a strong link between affine interval exchange maps and Anosov and pseudo-Anosov maps. We develop a smooth version of the above link proving that every $C^{1+H}$ diffeomorphism $f$ on a surface, with a codimension 1 hyperbolic attractor, induces a $C^{1+H}$ Cantor exchange system $\Phi_f$. E. Ghys and D. Sullivan observed that Anosov diffeomorphisms on the torus determine circle diffeomorphisms that have an associated renormalization operator. We explain that every $C^{1+H}$ diffeomorphism $f$ on a surface, with a codimension 1 hyperbolic attractor, determines a renormalization operator acting on the topological conjugacy class $[\Phi_f]_{C^0}$ of $\Phi_f$. We show that, every $C^{1+H}$ diffeomorphism $g$, topologically conjugate to $f$, determines a $C^{1+H}$ Cantor exchange system $\Phi_g \in [\Phi_f]_{C^0}$ with bounded geometry that is a $C^{1+H}$ fixed point of renormalization $[R\Phi_g]_{C^{1+H}} = [\Phi_g]_{C^{1+H}}$ and vice-versa. Denjoy has shown the existence of upper bounds for the smoothness of Denjoy maps. We prove that there is no $C^{1+\alpha}$ Cantor exchange system $\Psi \in [\Phi_f]_{C^0}$, with bounded geometry, that is a $C^{1+\alpha}$ fixed point of renormalization with regularity $\alpha$ greater than the Hausdorff dimension of the Cantor invariant set of $\Psi$. As an example, for $C^{1+H}$ derived Anosov maps we show that the induced $C^{1+H}$ Cantor exchange systems correspond to $C^{1+H}$ Denjoy maps. In this case, the renormalization operator is a simple generalization to Denjoy maps of the usual renormalization operator for rotations of the circle and for comuting pairs as introduced by O. Lanford and D. Rand. We get that there is an infinite dimensional space of $C^{1+H}$ Denjoy maps semi-conjugate to the golden rotation and with bounded geometry that are $C^{1+H}$ fixed points of renormalization. However, there is no such $C^{1+\alpha}$ Denjoy map that is a $C^{1+\alpha}$ fixed point of renormalization with regularity $\alpha$ greater than the Hausdorff dimension of its nonwandering set. This result partially confirms the conjecture of J. Harrison [11]. Another result that partially proves the conjecture of J. Harrison has been done by A. Norton [18] using box dimension instead of Hausdorff dimension. To prove these last results, we use that codimension 1 hyperbolic attractors $\Lambda$ do not have affine models as we explain in the paper.

## 2. Cantor exchange systems

A $C^{1+\alpha}$ *exchange system* $\Phi = \{\phi_i; i = 1, \dots, n\}$ is a finite set of $C^{1+\alpha}$ diffeomorphisms $\phi_i : I_i \to J_i$ with the following properties:

(i) The domains $I_i$ of $\phi_i$ are closed intervals in the reals;
(ii) If $\phi_i \in \Phi$ then $\phi_i^{-1} \in \Phi$;
(iii) Every $x \in \mathbf{R}$ has at most two distinct images by the maps in $\Phi$.

A $C^{1+H}$ *exchange system* $\Phi$ is a $C^{1+\alpha}$ exchange system, for some $\alpha > 0$.

We note that condition (i) implies that the intervals $J_i$ are also closed intervals. We say that a finite sequence $\{\phi_{i_n} \in \Phi\}_{n=1}^m$ or an infinite sequence $\{\phi_{i_n} \in \Phi\}_{n \geq 1}$ is *admissible* with respect to $x$, if $\phi_{i_n} \circ \dots \circ \phi_{i_1}(x) \in I_{i_{n+1}}$ and $\phi_{i_n} \neq \phi_{i_{n-1}}^{-1}$ for all $n > 1$. We define the invariant set $\Omega_\Phi$ of $\Phi$ as being the set of all points $x \in \mathbf{R}$ for which there are two distinct infinite admissible sequences with respect to $x$. We denote the Hausdorff dimension of $\Omega_\Phi$ by $HD(\Omega_\Phi)$. If $0 < HD(\Omega_\Phi) < 1$, we call $\Phi$ a $C^{1+H}$ *Cantor exchange system*.

We say that a Cantor exchange system $\Phi$ is determined by a map $\phi : I \to J$ if all the maps $\phi_i : I_i \to J_i$ contained in $\Phi$ are the restriction of the map $\phi$ or its inverse $\phi^{-1}$ to $I_i$. In this case, we call $\phi$ a *Cantor exchange map*. We note that not all Cantor exchange systems are determined by a Cantor exchange map.

We say that two $C^{1+\alpha}$ Cantor exchange systems $\Phi = \{\phi_i : I_{\phi_i} \to J_{\phi_i}; i = 1, \dots, n\}$ and $\Psi = \{\psi_i : I_{\psi_i} \to J_{\psi_i}; i = 1, \dots, n\}$, with $\alpha > 0$, are $C^0$ *conjugate* if there is a $C^0$ homeomorphism $h : \cup I_{\phi_i} \to \cup I_{\psi_i}$ such that $h \circ \phi_i(x) = \psi_i \circ h(x)$ for all $x \in \Omega_\Phi$. We denote by $[\Phi]_{C^0}$ the set of all $C^{1+H}$ Cantor exchange systems that are $C^0$ conjugate to $\Phi$. If $h$ is a $C^{1+\alpha}$ diffeomorphism, then we say that $\Phi$ and $\Psi$ are $C^{1+\alpha}$ *conjugate*. We denote by $[\Phi]_{C^{1+H}}$ the set of all $C^{1+H}$ Cantor exchange systems $\Psi$ that are $C^{1+\alpha(\Psi)}$ conjugate to $\Phi$, for some $\alpha(\Psi) > 0$ depending on $\Psi$.

## 2.1. Renormalization

Let $\Phi = \{\phi_i : I_{\phi_i} \to J_{\phi_i} : i = 1, \ldots, n\}$ and $\Psi = \{\psi_i : I_{\psi_i} \to J_{\psi_i} : i = 1, \ldots, m\}$ be $C^{1+H}$ Cantor exchange systems. We say that $\Psi$ is a *renormalization of* $\Phi$ if there is a *renormalization sequence set* $S = S(\Phi, \Psi) = \{\underline{s}^1, \ldots, \underline{s}^m\}$ with the following properties:

(i) For every $i \in \{1, \ldots, n\}$, we have that

$$\psi_i = \phi_{s^i_{k(\underline{s}^i)}} \circ \cdots \circ \phi_{s^i_1} | I_{\psi_i} \,,$$

where $k(\underline{s}^i)$ is the length of the sequence $\underline{s}^i \in S$. In particular, $\Omega_\Psi \subset \Omega_\Phi$ and $I_{\psi_i} \subset I_{\phi_{s^i_1}}$.

(ii) For every $x \in \Omega_\Phi \setminus \Omega_\Psi$, there are exactly two distinct sequences $\underline{s}^i, \underline{s}^j \in S$ with the property that there are points $y_i \in I_{\psi_i}$, $y_j \in I_{\psi_j}$ such that

$$x = \phi_{s^i_{k(x,i)}} \circ \cdots \circ \phi_{s^i_1}(y_i) \quad \text{and} \quad x = \phi_{s^j_{k(x,j)}} \circ \cdots \circ \phi_{s^j_1}(y_j) \,,$$

for some $0 < k(x, i) < k(\underline{s}^i)$ and $0 < k(x, j) < k(\underline{s}^j)$.

We say that the $C^{1+H}$ Cantor exchange system $\Gamma = \{\gamma_i : I_{\gamma_i} \to J_{\gamma_i} : i = 1, \ldots, m\}$ is an *affine renormalization* of the $C^{1+H}$ Cantor exchange system $\Phi = \{\phi_i : I_{\phi_i} \to J_{\phi_i} : i = 1, \ldots, n\}$ if there are affine maps $A_i : \mathbf{R} \to \mathbf{R}$ and $B_i : \mathbf{R} \to \mathbf{R}$ such that

$$\Psi = \left\{ \psi_i = B_i^{-1} \gamma_i A_i : A_i^{-1}(I_{\gamma_i}) \to B_i(I_{\gamma_j}); i = 1, \ldots, m \right\}$$

is a renormalization of $\Phi$. If $\Psi$ is a renormalization of $\Phi$, with renormalization sequence set $S(\Phi, \Psi)$, then there is a unique renormalization operator $R = R_{S(\Phi, \Psi)} : [\Phi]_{C^0} \to [\Psi]_{C^0}$ defined as follows: Let $\underline{\Phi}$ be a $C^{1+H}$ Cantor exchange system topologically conjugate to $\Phi$ by $\xi$, and set

$$\underline{\Psi} = \left\{ \underline{\psi}_i = \underline{\phi}_{s^i_{k(\underline{s}^i)}} \circ \cdots \circ \underline{\phi}_{s^i_1} : \xi(I_{\psi_i}) \to \xi(J_{\psi_i}), \text{ for every } \underline{s}^i \in S(\Phi, \Psi) \right\}.$$

By construction, $\underline{\Psi}$ is topologically conjugate to $\Psi$, and so $\underline{\Psi}$ is a $C^{1+H}$ Cantor exchange system that is a renormalization of $\underline{\Phi}$ with respect to the renormalization sequence set $S(\underline{\Phi}, \underline{\Psi}) = S(\Phi, \Psi)$. Hence, the renormalization operator $R$ is well-defined by $R\underline{\Phi} = \underline{\Psi}$.

Let $R : [\Phi]_{C^0} \to [\Psi]_{C^0}$ be a renormalization operator. We say that a $C^{1+\alpha}$ Cantor exchange system $\Gamma \in [\Phi]_{C^0}$ is a $C^{1+\alpha}$ *fixed point* of the renormalization operator $R$, if $R\Gamma$ is $C^{1+\alpha}$ conjugated to $\Gamma$, i.e., $[R\Gamma]_{C^{1+\alpha}} = [\Gamma]_{C^{1+\alpha}}$. We say that a $C^{1+H}$ Cantor exchange system $\Gamma \in [\Phi]_{C^0}$ is a $C^{1+H}$ *fixed point* of the renormalization operator $R$, if $\Gamma$ is a $C^{1+\alpha}$ fixed point of the renormalization operator $R$, for some $\alpha > 0$ depending upon $\Gamma$.

## 2.2. Bounded geometry

Let $\Phi$ and $\Psi$ be $C^{1+H}$ Cantor exchange systems such that $\Psi$ is a renormalization of $\Phi$ with renormalization sequence set $S = S(\Phi, \Psi)$. Let us suppose that $\Psi$ is topologically conjugate to $\Phi$, i.e., $\Phi$ is a $C^0$ fixed point of renormalization $[R\Phi]_{C^0} =$

$[\Phi]_{C^0}$. In this case $\Phi$ is an infinitely renormalizable $C^{1+H}$ Cantor exchange system, i.e., there is an infinite sequence

$$\left( R^m \Phi = \{ \phi_i^{(m)} : I_{\phi_i}^{(m)} \to J_{\phi_i}^{(m)}; i = 1, \ldots, n(m) \} \right)_{m \geq 1}$$

of Cantor exchange systems inductively determined, for every $m \geq 1$, by $R^m \Phi = R(R^{m-1}\Phi)$ with $S(R^m\Phi, R^{m-1}\Phi) = S(\Phi, \Psi)$.

Set

$$L_m^{(1)} = \left\{ \phi_{s_k^i}^{(m)} \circ \cdots \circ \phi_{s_1^i}^{(m)} \left( I_{\phi_i}^{(m+1)} \right) : I_{\phi_i}^{(m+1)} \subset I_{\phi_{s_1^i}}^{(m)}, 0 \leq k \leq k(\underline{s}_i), \underline{s}_i \in S \right\}.$$

Set, inductively on $j \geq 1$, the sets

$$L_m^{(j)} = \left\{ \phi_{s_k^i}^{(m)} \circ \cdots \circ \phi_{s_1^i}^{(m)}(I) : I \in L_{m+1}^{(j-1)}, I \subset I_{\phi_{s_1^i}}^{(m)}, 0 \leq k \leq k(\underline{s}_i), \underline{s}_i \in S \right\}.$$

By construction, $L_m^{(j+1)} \subset L_m^{(j)}$ and $\Omega_{R^m\Phi} = \cap_{j \geq 1} L_m^{(j)}$. We call $L_m^{(j)}$ the *j-th level of the partition* of $R^m\Phi$. Let the *j-gap set* $G_m^{(j)}$ of $R^m\Phi$ be the set of all maximal closed intervals $I$ such that $I \subset J$ for some $J \in L_m^{(j-1)}$ and $\text{int}I \cap K = \emptyset$, for every $K \in L_m^{(j)}$. We say that the $C^{1+H}$ Cantor exchange system $\Phi$ has *bounded geometry*, if there is $c > 1$ such that, for all $j \geq 1$ and all intervals $I \in L_0^{(j)} \cup G_0^{(j)}$ contained in a same interval $K \in L_0^{(j-1)}$, we have $c^{-1} < |I|/|K| < c$.

## 3. Codimension 1 hyperbolic attractors

Throughout this paper, $(f, \Lambda, \mathbf{M})$ is a $C^{1+H}$ diffeomorphism $f$ with a codimension 1 hyperbolic attractor $\Lambda$ and with a Markov partition $\mathbf{M}$ on $\Lambda$ satisfying the disjointness property that we pass to define.

We say that $(f, \Lambda)$ is a $C^{1+H}$ *diffeomorphism $f$ with a codimension 1 hyperbolic attractor* $\Lambda$, if $(f, \Lambda)$ has the following properties:

(i) $f : S \to S$ is a $C^{1+H}$ diffeomorphism of a compact surface $S$ with respect to a $C^{1+H}$ structure on $S$.

(ii) $\Lambda$ is a hyperbolic invariant subset of $S$ such that $f|\Lambda$ is topologically transitive and $\Lambda$ has a local product structure.

(iii) There is an open set $O \subset S$ such that $\Lambda = \cap_{n \geq 0} f^n O$.

A $C^{1+H}$ diffeomorphism $(f, \Lambda)$ with codimension 1 hyperbolic attractor has the property that the local stable leaves intersected with $\Lambda$ are Cantor sets and the local unstable leaves are 1 dimensional manifolds. Let $HD(\Lambda^s)$ be the Hausdorff dimension of the stable leaves intersected with the basic set. Furthermore, $(f, \Lambda)$ has a Markov partition on $\Lambda$ (see [29]) with the following disjointness property (see [2, 17] and [30]): The unstable leaf boundaries of any two Markov rectangles do not intersect except, possibly, at their endpoints.

Suppose that $M$ and $N$ are Markov rectangles, and $x \in M$ and $y \in N$. We say that $x$ and $y$ are *s-holonomically related* if (i) there is an $u$-leaf segment $\ell^u(x, y)$ such that $\partial \ell^u(x, y) = \{x, y\}$, and (ii) $\ell^u(x, y) \subset \ell^u(x, M) \cup \ell^u(y, N)$. Let $P = P_{\mathbf{M}}$

be the set of all pairs $(M, N)$ such that there are points $x \in M$ and $y \in N$ stable holonomically related.

For every Markov rectangle $M \in \mathbf{M}$, choose a spanning leaf segment $\ell_M$ in $M$. Let $\mathbf{I} = \{\ell_M : M \in \mathbf{M}\}$. For every pair $(M, N) \in P$, there are maximal leaf segments $\ell_M^D \subset \ell_M$, $\ell_N^C \subset \ell_N$ such that the holonomy $h_{(M,N)} : \ell_M^D \to \ell_N^C$ is well-defined. We call such holonomies $h_{(M,N)} : \ell_M^D \to \ell_N^C$ the (stable) primitive holonomies associated to the Markov partition $\mathbf{M}$.

**Definition 3.1.** *The set $\mathbf{H} = \{h_{(M,N)} : \ell_M^D \to \ell_N^C; (M, N) \in P\}$ is a complete set of stable holonomies.*

For every leaf segment $\ell_M \in \mathbf{I}$, let $\hat{\ell}_M$ be the smallest full leaf segment containing $\ell_M$ (see definition in Section 5). By the Stable Manifold Theorem, there are $C^{1+H}$ diffeomorphisms $k_M : \hat{\ell}_M \to K_M \subset \mathbf{R}$. We choose the $C^{1+H}$ diffeomorphisms $k_M : \hat{\ell}_M \to K_M \subset \mathbf{R}$ with the extra property that their images are pairwise disjoint, i.e., $K_M \cap K_N = \emptyset$ for all $M, N \in \mathbf{M}$ such that $M \neq N$. Set

$$K_{\mathbf{M}} = \bigcup_{i=1}^{n} K_{M_i}, \quad \hat{L}_{\mathbf{M}} = \bigcup_{i=1}^{n} \hat{\ell}_{M_i} \quad \text{and} \quad L_{\mathbf{M}} = \hat{L}_{\mathbf{M}} \bigcap \Lambda_f. \tag{3.1}$$

Let $k : \hat{L}_{\mathbf{M}} \to K_{\mathbf{M}}$ be the map defined by $k|\hat{\ell}_M = k_M$, for every $M \in \mathbf{M}$. Let

$$\pi : \bigcup_{i=1}^{n} M_i \to L_{\mathbf{M}} \tag{3.2}$$

be the projection defined by $\pi(x_i) = y_i$, where $y_i \in \ell_{M_i}^u(x_i) \cap L_{\mathbf{M}}$ for every $x_i \in M_i$.

**Lemma 3.2.** *The triple $(f, \Lambda, \mathbf{M})$ induces a $C^{1+H}$ Cantor exchange system*

$$\Phi = \Phi_{f,\mathbf{M}} = \left\{ e_{(M,N)} : k_M \left( \hat{\ell}_{(M,N)}^D \right) \to k_N \left( \hat{\ell}_{(M,N)}^C \right) | (M, N) \in \mathbf{P} \right\},$$

*with bounded geometry. Furthermore, for every $(M, N) \in \mathbf{P}$:*

(i) $e_{(M,N)}|k_M \left( \ell_{(M,N)}^D \right) = k_M \circ h_{(M,N)} \circ k_N^{-1}$;
(ii) $\Omega_\Phi = k_{\mathbf{M}}(L_{\mathbf{M}})$.

*Proof.* Define $e_{(M,N)}|k_M \left( \ell_{(M,N)}^D \right) = k_M \circ h_{(M,N)} \circ k_N^{-1}$. By Theorem 2.1 in [21], the map $k_M \circ h_{(M,N)} \circ k_N^{-1}|k_M \left( \ell_{(M,N)}^D \right)$ extends (not uniquely) to a $C^{1+\alpha}$ diffeomorphism $e_{(M,N)} : k(\hat{\ell}_{(M,N)}^D) \to k(\hat{\ell}_{(M,N)}^C)$ for some $\alpha > 0$. By construction, the set $\{e_{(M,N)} : (M, N) \in \mathbf{P}\}$ satisfies properties (i),(ii) and (iii) of the definition of a $C^{1+H}$ Cantor exchange system, and $\Omega_\Phi = k_{\mathbf{M}}(L_{\mathbf{M}})$. $\qquad \square$

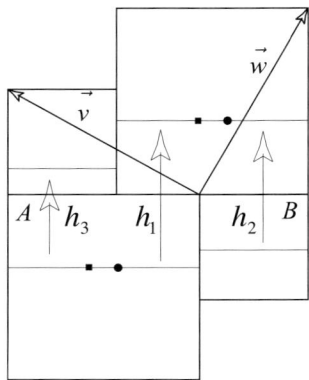

FIGURE 1. The complete set of holonomies $\mathbf{H} = \{h_1, h_2, h_3,$ $h_1^{-1}, h_2^{-1}, h_3^{-1}\}$ for the Anosov map $f : \mathbf{R}^2 \setminus (\mathbf{Z}\vec{v} \times \mathbf{Z}\vec{w}) \rightarrow$ $\mathbf{R}^2 \setminus (\mathbf{Z}\vec{v} \times \mathbf{Z}\vec{w})$ defined by $f(x, y) = (x + y, y)$ and with Markov partition $\mathbf{M} = \{A, B\}$.

### 3.1. Renormalization

In this section, given a $C^{1+H}$ diffeomorphism $f$ with codimension 1 hyperbolic attractor $\Lambda$ and with a Markov partition $\mathbf{M}$ satisfying the disjointness property, we present an explicit construction of a renormalization operator $R = R_{f,\mathbf{M}}$ acting on the topological conjugacy class of the $C^{1+H}$ Cantor exchange system $\Phi_{f,\mathbf{M}}$ induced by $(f, \Lambda, \mathbf{M})$. Let the Markov partition $\mathbf{N} = f_*\mathbf{M}$ be the pushforword of the Markov partition $\mathbf{M}$, i.e., for every $M \in \mathbf{M}$, $N = f(M) \in \mathbf{N}$.

**Lemma 3.3.** *Let $\Phi_{f,\mathbf{M}}$ and $\Phi_{f,\mathbf{N}}$ be the $C^{1+H}$ Cantor exchange systems induced (as in Lemma 3.2), respectively, by $(f, \Lambda, \mathbf{M})$ and $(f, \Lambda, \mathbf{N})$. There is a well-defined renormalization operator*

$$R = R_{f,\mathbf{M}} : [\Phi_{f,\mathbf{M}}]_{C^0} \rightarrow [\Phi_{f,\mathbf{N}}]_{C^0} .$$

*Proof.* For simplicity of notation, let us denote $k_{\mathbf{M}}$ by $k$ (see (3.1)). We choose a map

$$\sigma : \{1, \dots, n\} \rightarrow \{1, \dots, n\} \tag{3.3}$$

with the property that $N_i \cap M_{\sigma(i)} \neq \emptyset$, where $N_i \in \mathbf{N}$ and $M_{\sigma(i)} \in \mathbf{M}$. For each $N_i \in \mathbf{N}$, let $\ell_{N_i}$ be the stable spanning leaf segment $\ell_{M_{\sigma(i)}} \cap \pi(N_i)$, and let $\hat{\ell}_{N_i}$ be the corresponding full stable spanning leaf (i.e., $\hat{\ell}_{N_i} \cap \Lambda = \ell_{N_i}$), where $\pi : \bigcup_{i=1}^n M_i \rightarrow L_{\mathbf{M}}$ is the natural projection as defined in (3.1). Set

$$\mathbf{L_N} = \bigcup_{i=1}^n \ell_{N_i} \quad \text{and} \quad \hat{\mathbf{L}}_{\mathbf{N}} = \bigcup_{i=1}^n \hat{\ell}_{N_i} .$$

Let $\mathbf{H_N} = \{h_{(N_i,N_j)} : \ell^D_{(N_i,N_j)} \rightarrow \ell^C_{(N_i,N_j)} | (N_i, N_j) \in \mathbf{P_N}\}$ be the (stable) complete set of holonomies associated to the Markov partition $\mathbf{N}$. By construction, for every

$(N_i, N_j) \in \mathbf{P_N}$ there is a sequence $h_{\alpha_1}, \ldots, h_{\alpha_n}$ of holonomies in $\mathbf{H_M}$ such that

$$h_{(N_i, N_j)} = h_{\alpha_n} \circ \cdots \circ h_{\alpha_1} | \tilde{\ell}^D_{N_i} .$$

Let

$$e_{(N_i, N_j)} : k_{M_{\sigma(i)}} \left( \hat{\ell}^D_{(N_i, N_j)} \right) \to k_{M_{\sigma(j)}} \left( \hat{\ell}^D_{(N_i, N_j)} \right)$$

be given by $e_{(N_i, N_j)} = e_{\alpha_n} \circ \cdots \circ e_{\alpha_1}$, where $e_{\alpha_i} \in \Phi_{f,\mathbf{M}}$ and $e_{\alpha_i} | k \left( \ell^D_{(N_i, N_j)} \right) = k \circ h_{\alpha_i} \circ k^{-1} | k \left( \ell^D_{(N_i, N_j)} \right)$. Set

$$\Psi = \left\{ e_{(N_i, N_j)} : k \left( \hat{\ell}^D_{(N_i, N_j)} \right) \to k \left( \hat{\ell}^C_{(N_i, N_j)} \right) | (N_i, N_j) \in \mathbf{P_N} \right\} .$$

By construction, $\Psi = \Phi_{f,\mathbf{N}}$ (as constructed in Lemma 3.2), and so $\Psi$ is a $C^{1+H}$ Cantor exchange system. Since the set $S(\Phi_{f,\mathbf{M}}, \Phi_{f,\mathbf{N}})$ of all sequences $\alpha_1 \ldots \alpha_n$ such that $e_{(N_i, N_j)} = e_{\alpha_n} \circ \cdots \circ e_{\alpha_1}$, for some $(N_i, N_j) \in P_{\mathbf{N}}$, form a renormalizable sequence set, the $C^{1+H}$ Cantor exchange system $\Phi_{f,\mathbf{N}}$ is a renormalization of $\Phi_{f,\mathbf{M}}$. Therefore, by §2.1, there is a well-defined renormalization operator $R = R_{f,\mathbf{M}} : [\Phi_{f,\mathbf{M}}]_{C^0} \to [\Phi_{f,\mathbf{N}}]_{C^0}$. □

**Theorem 3.4.** *The $C^{1+H}$ Cantor exchange system $\Phi_{f,\mathbf{M}}$ is a $C^{1+H}$ fixed point of renormalization, i.e., $[R\Phi_{f,\mathbf{M}}]_{C^0} = [\Phi_{f,\mathbf{M}}]_{C^0}$, where $R = R_{f,\mathbf{M}} : [\Phi_{f,\mathbf{M}}]_{C^0} \to [\Phi_{f,\mathbf{N}}]_{C^0}$ is the renormalization operator (as constructed in Lemma 3.3).*

See proof of the above theorem in [26].

### 3.2. Hausdorff dimension versus smoothness

Before proceeding, we present the following notion of $C^{1,HD}$ regularity of a function.

**Definition 3.5.** *Let $\phi : I \to J$ be a homeomorphism between open sets $I \subset \mathbf{R}$ and $J \subset \mathbf{R}$. If $0 < \alpha < 1$, then $\phi$ is said to be $C^{1,\alpha}$ if it is differentiable and for all points $x, y \in I$*

$$|\phi'(y) - \phi'(x)| \leq \chi_\phi(|y - x|) \qquad (3.4)$$

*where the positive function $\chi_\phi(t)$ satisfies $\lim_{t \to 0} \chi_\phi(t)/t^\alpha = 0$.*

In particular, a $C^{1+\beta}$ diffeomorphism is $C^{1,\alpha}$ for all $0 < \alpha < \beta$. Furthermore, a $C^{1,\alpha}$ homeomorphism is $C^{1+\alpha}$.

**Theorem 3.6.** *Let $\mathbf{C}_{f,\mathbf{M}}$ be the topological conjugacy class of $C^{1+H}$ Cantor exchange systems determined by a $C^{1+H}$ diffeomorphism $f$ with codimension 1 hyperbolic attractor $\Lambda$ and with a Markov partition $\mathbf{M}$ satisfying the disjointness property. There is no $C^{1,HD(\Omega_\Phi)}$ Cantor exchange system $\Phi \in \mathbf{C}_{f,\mathbf{M}}$, with bounded geometry, that is a $C^{1,HD(\Omega_\Phi)}$ fixed point of renormalization operator, i.e., $[R_{f,\mathbf{M}}\Phi]_{C^{1,HD(\Omega_\Phi)}} = [\Phi]_{C^{1,HD(\Omega_\Phi)}}$.*

See proof of the above theorem in [26].

### 3.3. An example: Denjoy maps

Let $r_\gamma : \mathbf{S}^1 \to \mathbf{S}^1$ be the rigid rotation of the circle $\mathbf{S}^1 = \mathbf{R}/\mathbf{Z}$, with rotation number $w = 1/(1+\gamma)$, where $\gamma$ satisfies $\gamma^{-1} = \gamma + a$ for some integer $a \geq 1$, i.e., $\gamma = 1/(a+1/(a+\ldots))$. Let $q_1, q_2, \ldots$ be the sequence defined inductively as follows: $q_1 = 1$, and $q_{n+1}$ is the smallest positive integer with the property that the distance between $r^{q_{n+1}}(0)$ and $0$ is smaller than the distance between $r^{q_n}(0)$ and $0$. Let $\tilde{S}$ be the circle obtained from the circle $\mathbf{S}^1$ as follows: At $0$, we cut $\mathbf{S}^1$ and join an arc with length $1$, and, for every $n \geq 1$ and every integer $m$ such that $q_n \leq |m| < q_{n+1}$, we cut $\mathbf{S}^1$ at each point $r_\gamma^m(0)$ and we join an arc $a_m$ with length $\gamma^{-n}$. This construction of $\tilde{S}$ determines a natural projection map $\pi : \tilde{S} \to \mathbf{S}^1$ by collapsing all the arcs $a_m$. Let $\tilde{d}_\gamma : \tilde{S} \to \tilde{S}$ be a homeomorphism such that $\pi \circ \tilde{d}_\gamma(x) = r_\gamma \circ \pi(x)$ for all $x \in \tilde{S}$. Fix the points $\tilde{0}_l$ and $\tilde{0}_r$ as being the left and right endpoints of the interval $\pi^{-1}(0)$. Let $\mathrm{cl}O^+(\tilde{0}_l)$ and $\mathrm{cl}O^+(\tilde{0}_r)$ be the closures of the forward orbits of $\tilde{0}_l$ and $\tilde{0}_r$, respectively. By construction of the map $\tilde{d}_\gamma$, the sets $\mathrm{cl}O^+(\tilde{0}_l)$, $\mathrm{cl}O^+(\tilde{0}_r)$ and the nonwandering set $\tilde{\Omega}$ of $\tilde{d}_\gamma$ are equal.

Before proceeding, we present the following notion of quasisymmetric regularity of a function: A homeomorphism $h : L \subset \mathbf{R} \to \mathbf{R}$ is *quasisymmetric* (*qs*) if for every constant $c > 1$ there exists $k(c) > 1$ such that

$$k(c)^{-1} < \big(h(z) - h(y)\big)/\big(h(y) - h(x)\big) < k(c),$$

for every points $x, y, z \in L$ with the property that $x < y < z$ and that $c^{-1} < (z-y)/(y-x) < c$.

**Definition 3.7.** *A $C^{1+\alpha}$ diffeomorphism $d : \mathbf{S}^1 \to \mathbf{S}^1$ is a $C^{1+\alpha}$ Denjoy map with bounded geometry and rotation number $\gamma$, if there is a homeomorphism $h : \mathbf{S}^1 \to \mathbf{S}^1$ which conjugates $\tilde{d}_\gamma$ with $d$, and the map $h|\tilde{\Omega}$ is quasisymmetric. Let us denote the point $h(\tilde{0}_l)$ by $0_l$, the point $h(\tilde{0}_r)$ by $0_r$, and the set $h(\tilde{\Omega})$ by $\Omega_d$. Let $HD(\Omega_d)$ be the Hausdorff dimension of the nonwandering set $\Omega_d$ of $d$.*

Let $d : \mathbf{S}^1 \to \mathbf{S}^1$ be a Denjoy map with rotation number $\gamma$ satisfying $\gamma^{-1} = \gamma + a$, for some integer $a \geq 1$. Let $I_R$ be the arc with endpoints $d^a(0_r)$ and $d^{a+1}(0_r)$ and containing $0_r$. Set $I_1^D = [0_r, d^{a+1}(0_l)] \subset I_R$ and $I_2^D = [d^a(0_r), 0_l] \subset I_R$. Let $S_R$ be the circle obtained by identifying the endpoints of $I_R$. We say that a $C^{1+\alpha}$ diffeomorphism $d_R : S_R \to S_R$ is a *renormalization of $d$*, if

$$d_R|_{I_1^D}(x) = d^a(x) \quad \text{and} \quad d_R|_{I_2^D}(x) = d^{a+1}(x).$$

Since $\gamma^{-1} = \gamma + a$, for some integer $a \geq 1$, we obtain that there exists a map $h : \Omega_d \to \Omega_{d_R}$ which conjugates $d$ on its nonwandering set $\Omega_d$ with $d_R$ on its nonwandering set $\Omega_{d_R}$, and $h$ extends to a homeomorphism $\tilde{h} : \mathbf{S}^1 \to S_R$. The proof follows from using that the rigid rotation $r_\gamma$ is a fixed point of the usual renormalization operator for diffeomorphisms of the circle and for comuting pairs as introduced by O. Lanford and D. Rand. A $C^{1+\alpha}$ diffeomorphism $d : \mathbf{S}^1 \to \mathbf{S}^1$ is a $C^{1+\alpha}$ *fixed point of renormalization*, if there exists a $C^{1+\alpha}$ diffeomorphism $d_R : S_R \to S_R$ that is a renormalization of $d$, and the conjugacy $h : \Omega_d \to \Omega_{d_R}$

between $d|\Omega_d$ and $d_R|\Omega_{d_R}$ extends to a $C^{1+\alpha}$ diffeomorphism of the circle, with $\alpha > 0$. Let $\mathbf{D}_\gamma$ be the set of all $C^{1+H}$ conjugacy classes of $C^{1+H}$ Denjoy maps $d : \mathbf{S}^1 \to \mathbf{S}^1$, with bounded geometry and rotation number $\gamma$, that are $C^{1+H}$ fixed points of renormalization.

**Theorem 3.8.** *The set $\mathbf{D}_\gamma$ is an infinite dimensional space characterized by a Teich-müller space consisting of solenoid functions. However, there is no $C^{1,HD(\Omega_d)}$ Den-joy map $d : \mathbf{S}^1 \to \mathbf{S}^1$, with bounded geometry and rotation number $\gamma$, that is a $C^{1,HD(\Omega_d)}$ fixed point of renormalization.*

See proof of the above theorem in [26].

# 4. No affine models

The proof of Theorems 3.6 and 3.8 rely in the non-existence of affine models for codimension 1 hyperbolic attractors as we pass to explain.

The *stable basic holonomies* are maps defined between $s$-leaf segments defined by travelling along the unstable manifolds (see Section 5.4). In [21], it is proved that these maps have $C^{1+\alpha}$ extensions to the full $s$-leaf segments for some $\alpha > 0$ with respect to the full $s$-leaf charts. Therefore, all $s$-leaf segments have the same Hausdorff dimension which we denote by $HD^s$. Unstable basic holonomies, $u$-leaf segments and $HD^u$ are similarly defined.

**Definition 4.1.** *An* hyperbolic affine model *for $f$ on $\Lambda$ is an atlas $\mathbf{A}$ with the following properties:*

(i) *the union of the domains $U$ of the charts $i : U \to \mathbf{R}^2$ of $\mathbf{A}$ (which are open in $M$) cover $\Lambda$;*

(ii) *any two charts $i : U \to \mathbf{R}^2$ and $j : V \to \mathbf{R}^2$ in $\mathbf{A}$ have overlap maps $j \circ i^{-1} : i(U \cap V) \to \mathbf{R}^2$ with affine extensions to $\mathbf{R}^2$;*

(iii) *$f$ is affine with respect to the charts in $\mathbf{A}$;*

(iv) *$L$ is a basic hyperbolic set;*

(v) *the images of the stable and unstable local leaves under the charts in $\mathbf{A}$ are contained in horizontal and vertical lines; and*

(vi) *the basic holonomies have affine extensions to the stable and unstable leaves with respect to the charts in $\mathbf{A}$.*

We will use, from now on, $\iota$ to denote an element of the set $\{s, u\}$ of the stable and unstable superscripts and $\iota'$ to denote the element of $\{s, u\}$ that is not $\iota$. For $\iota \in \{s, u\}$, a *full $\iota$-leaf segment $I$* (or, equivalently, a *local $\iota$-leaf*) is defined as a connected subset of $W^\iota(x)$, and an *$\iota$-leaf segment* is the intersection with $\Lambda$ of a full $\iota$-leaf segment. The *endpoints* of an $\iota$-leaf segment $I$ are the endpoints of the minimal closed full $\iota$-leaf segment containing $I$ (see Section 5.1).

**Definition 4.2.** *A twinned pair of $u$-leaves $(I, J)$ in a basic set $\Lambda$ consists of a pair of $u$-leaf segments $I$ and $J$ with the following properties (see Figure 2):*

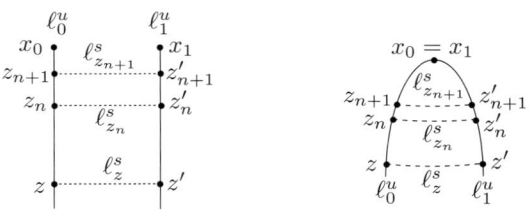

FIGURE 2. An illustration of twinned pair of $u$-leaves..

(i) *an endpoint $p$ of $I$ and an endpoint $q$ of $J$ are periodic points under $f$;*
(ii) *$(I \setminus \{p\}) \cap (J \setminus \{q\}) = \emptyset$;*
(iii) *for all $z \in I \setminus \{p\}$ there is a full $s$-leaf segment $\gamma_z$ in the stable manifold through $z$ which has endpoints $z$ and $z'$ such that $z' \in J \setminus \{q\}$ and $\gamma_z \cap \Lambda = \{z, z'\}$.*

It follows from this that if a sequence $z_n \in I \setminus \{p\}$ converges to $p$ then the corresponding sequence $z'_n \in J \cap \gamma_{z_n}$ converges to $q$. Also, it follows that the periodic points $p$ and $q$ must have the same period. A twinned pair of $s$-leaves in a basic set $\Lambda$ is similarly defined.

**Remark 4.3.** In the previous definition we allow the points $p$ and $q$ to coincide. However, if $p$ is different from $q$ then there is no stable leaf containing both $p$ and $q$ (otherwise they would converge under iteration by $f$).

If $\Lambda$ is a codimension 1 hyperbolic attractor then $\Lambda$ contains a twinned pair of $u$-leaves. Furthermore, if $\Lambda$ contains a twinned pair of $\iota$-leaves then there are no affine models for $f$ on $\Lambda$ (see [25]).

A complete set of holonomies $\mathbf{H}^\iota$ is $C^{1, HD^\iota}$ if for every holonomy $h_\alpha : I \to J$ in $\mathbf{H}^\iota$ with $I \subset I_i$ and $J \subset I_j$, the map $u_j \circ h_\alpha \circ u_i^{-1}$ and its inverse have a $C^{1, HD^\iota}$ diffeomorphic extension to $\mathbf{R}$ such that the modulus of continuity does not depend upon $h_\alpha \in \mathbf{H}^\iota$.

**Theorem 4.4.** *If $\Lambda$ be a codimension 1 hyperbolic attractor, for a $C^{1+\gamma}$ diffeomorphism $f$ on a surface with $\gamma > HD^s$, then the complete set of $s$-holonomies $\mathbf{H}^s$ is not $C^{1, HD^s}$.*

The proof of the above theorem follows from the non-existence of affine models for codimension 1 hyperbolic attractors (see [25]).

## 5. Appendix

In this Appendix, we present some basic facts on hyperbolic dynamics, that we include for clarity of the exposition.

## 5.1. Leaf segments

Let $d$ be a metric on $M$, and define the map $f_\iota = f$ if $\iota = u$, or $f_\iota = f^{-1}$ if $\iota = s$. For $\iota \in \{s, u\}$, if $x \in \Lambda$ we denote the local $\iota$-manifolds through $x$ by

$$W^\iota(x, \varepsilon) = \left\{ y \in M : d\big(f_\iota^{-n}(x), f_\iota^{-n}(y)\big) \leq \varepsilon, \text{ for all } n \geq 0 \right\}.$$

By the Stable Manifold Theorem [29], these sets are respectively contained in the stable and unstable immersed manifolds

$$W^\iota(x) = \bigcup_{n \geq 0} f_\iota^n \left( W^\iota\big(f_\iota^{-n}(x), \varepsilon_0\big) \right)$$

which are the image of a $C^{1+\gamma}$ immersion $\kappa_{\iota, x} : \mathbf{R} \to M$. An *open* (resp. *closed*) *full $\iota$-leaf segment $I$* is defined as a subset of $W^\iota(x)$ of the form $\kappa_{\iota, x}(I_1)$ where $I_1$ is an open (resp. closed) subinterval (non-empty) in $\mathbf{R}$. An *$\iota$-open* (resp. *closed*) *leaf segment* is the intersection with $\Lambda$ of a full open (resp. closed) $\iota$-leaf segment such that the intersection contains at least two distinct points. If the intersection is exactly two points we call this $\iota$-closed leaf segment an *$\iota$-leaf gap*. An *$\iota$-full leaf segment* is either an open or closed $\iota$-full leaf segment. An *$\iota$-leaf segment* is either an open or closed $\iota$-leaf segment. The *endpoints* of a full $\iota$-leaf segment are the points $\kappa_{\iota, x}(u)$ and $\kappa_{\iota, x}(v)$ where $u$ and $v$ are the endpoints of $I_1$. The *endpoints* of an $\iota$-leaf segment $I$ are the points of the minimal closed full $\iota$-leaf segment containing $I$. The *interior* of a $\iota$-leaf segment $I$ is the complement of its boundary. In particular, a $\iota$-leaf segment $I$ has empty interior if, and only if, it is an $\iota$-leaf gap. A map $c : I \to \mathbf{R}$ is an *$\iota$-leaf chart* of an $\iota$-leaf segment $I$ if has an extension $c_E : I_E \to \mathbf{R}$ to a full $\iota$-leaf segment $I_E$ with the following properties: $I \subset I_E$ and $c_E$ is a homeomorphism onto its image. An *$\iota$-full leaf segment* is either an open or close full leaf segment.

## 5.2. Rectangles

Since L is a hyperbolic invariant set of a diffeomorphism $f : M \to M$, for $0 < \varepsilon < \varepsilon_0$ there is $\delta = \delta(\varepsilon) > 0$, such that for all points $w, z \in$ L with $d(w, z) < \delta$, $W^u(w, \varepsilon)$ and $W^s(z, \varepsilon)$ intersect in an unique point that we denote by $[w, z]$. Since we assume that the hyperbolic set has a *local product structure*, we have that $[w, z] \in \Lambda$. Furthermore, the following properties are satisfied: (i) $[w, z]$ varies continuously with $w, z \in$ L; (ii) the bracket map is continuous on a $\delta$-uniform neighbourhood of the diagonal in L × L; and (iii) whenever both sides are defined $f([z, w]) = [f(z), f(w)]$. Note that the bracket map does not really depend on $\delta$ provided it is sufficiently small.

Let us underline that it is a standing hypothesis that all the hyperbolic sets considered here have such a local product structure.

A *rectangle $R$* is a subset of L which is (i) closed under the bracket, i.e., $x, y \in R \Rightarrow [x, y] \in R$, and (ii) proper, i.e., is the closure of its interior in L. This definition imposes that a rectangle has always to be proper which is more restrictive than the usual one which only insists on the closure condition.

If $\ell^s$ and $\ell^u$ are respectively stable and unstable leaf segments intersecting in a single point then we denote by $[\ell^s, \ell^u]$ the set consisting of all points of the form $[w, z]$ with $w \in \ell^s$ and $z \in \ell^u$. We note that if the stable and unstable leaf segments $\ell$ and $\ell'$ are closed then the set $[\ell, \ell']$ is a rectangle. Conversely in this 2-dimensional situations, any rectangle $R$ has a product structure in the following sense: for each $x \in R$ there are closed stable and unstable leaf segments of L, $\ell^s(x, R) \subset W^s(x)$ and $\ell^u(x, R) \subset W^u(x)$ such that $R = [\ell^s(x, R), \ell^u(x, R)]$. The leaf segments $\ell^s(x, R)$ and $\ell^u(x, R)$ are called *stable and unstable spanning leaf segments* for $R$. For $\iota \in \{s, u\}$, we denote by $\partial\ell^\iota(x, R)$ the set consisting of the endpoints of $\ell^\iota(x, R)$, and we denote by $\text{int}\ell^\iota(x, R)$ the set $\ell^\iota(x, R) \setminus \partial\ell^\iota(x, R)$. The *interior of R* is given by $\text{int}R = [\text{int}\ell^s(x, R), \text{int}\ell^u(x, R)]$, and the *boundary of R* is given by $\partial R = [\partial\ell^s(x, R), \ell^u(x, R)] \bigcup [\ell^s(x, R), \partial\ell^u(x, R)]$.

## 5.3. Markov partitions

A *Markov partition of f* is a collection $\mathbf{M} = \{R_1, \ldots, R_k\}$ of such rectangles such that (i) $\Lambda \subset \bigcup_{i=1}^{k} R_i$; (ii) $R_i \bigcap R_j = \partial R_i \bigcap \partial R_j$ for all $i$ and $j$; (iii) if $x \in \text{int }R_i$ and $fx \in \text{int }R_j$ then

(a) $f(\ell^s(x, R_i)) \subset \ell^s(fx, R_j)$ and $f^{-1}(\ell^u(fx, R_j)) \subset \ell^u(x, R_i)$
(b) $f(\ell^u(x, R_i)) \bigcap R_j = \ell^u(fx, R_j)$ and $f^{-1}(\ell^s(fx, R_j)) \bigcap R_i = \ell^s(x, R_i)$.

The last condition means that $f(R_i)$ goes across $R_j$ just once. In fact, it follows from condition (a) providing the rectangles $R_j$ are chosen sufficiently small (see Mañé [14]). The rectangles which make up the Markov partition are called *Markov rectangles*.

We note that there is a Markov partition $\mathbf{M}$ of $f$ with the following *disjointness property* (see Bowen [2], Newhouse and Palis [17] and Sinai [30]):

(i) if $0 < \delta_{f,s} < 1$ and $0 < \delta_{f,u} < 1$ then the stable and unstable leaf boundaries of any two Markov rectangles do not intesect.
(ii) if $0 < \delta_{f,\iota} < 1$ and $0 < \delta_{f,\iota'} = 1$ then the $\iota'$-leaf boundaries of any two Markov rectangles do not intersect except, possibly, at their endpoints.

If $\delta_{f,s} = \delta_{f,u} = 1$, the disjointness property does not apply and so we consider that it is trivially satisfied for every Markov partition. For simplicity of our exposition, we consider Markov partitions that satisfy the disjointness property.

## 5.4. Basic holonomies

Suppose that $x$ and $z$ are two points inside any rectangle $R$ of L. Let $I$ and $J$ be two stable leaf segments respectively containing $x$ and $z$ and inside $R$. Then we define $h : I \to J$ by $h(w) = [w, z]$. Such maps are called the *basic stable holonomies*. They generate the pseudo-group of all stable holonomies. Similarly we define the basic unstable holonomies.

## Acknowledgments

We are very grateful to Welington de Melo and Stefano Luzzatto for very useful discussions on this work. We thank IHES, CUNY, IMPA, Stony Brook and University of Warwick for their hospitality. We thank Calouste Gulbenkian Foundation,

PRODYN-ESF, POCTI and POCI by FCT and Ministério da Ciência, Tecnologia e do Ensino Superior, and Centro de Matemática da Universidade do Porto for their financial support of A. A. Pinto and F. Ferreira.

# References

[1] C. Bonatti and R. Langevin, *Difféomorphismes de Smale des surfaces.* Asterisque **250**, 1998.

[2] R. Bowen, *Equilibrium States and the Ergodic Theory of Anosov Diffeomorphisms.* Lecture Notes in Mathematics **470**, Springer-Verlag, New York, 1975.

[3] A. Denjoy, *Sur les courbes définies par les équations différentielles á la surface du tore.* J. Math. Pure et Appl. **11**, série 9 (1932), 333–375.

[4] A. Fahti, F. Laudenbach, and V. Poenaru, *Travaux de Thurston sur les surfaces.* Astérisque **66-67** (1991), 1–286.

[5] E. de Faria, W. de Melo, and A. A. Pinto, *Global hyperbolicity of renormalization for $C^r$ unimodal mappings.* Annals of Mathematics. To appear. 1–96.

[6] F. Ferreira and A. A. Pinto, *Explosion of smoothness from a point to everywhere for conjugacies between diffeomorphisms on surfaces.* Ergod. Th. & Dynam. Sys. **23** (2003), 509–517.

[7] F. Ferreira and A. A. Pinto, *Explosion of smoothness from a point to everywhere for conjugacies between Markov families.* Dyn. Sys. **16**, no. 2 (2001), 193–212.

[8] J. Franks, *Anosov Diffeomorphisms.* Global Analysis (ed. S Smale) AMS Providence (1970), 61–93.

[9] J. Franks, *Anosov diffeomorphisms on tori.* Trans. Amer. Math. Soc. **145** (1969), 117–124.

[10] A. A. Gaspar Ruas, *Atratores hiperbólicos de codimensão um e classes de isotopia em superfícies.* Tese de Doutoramento. IMPA, Rio de Janeiro, 1982.

[11] J. Harrison, *An introduction to fractals.* Proceedings of Symposia in Applied Mathematics **39** American Mathematical Society (1989), 107–126.

[12] M. Hirsch and C. Pugh, *Stable manifolds and hyperbolic sets.* Proc. Symp. Pure Math., Amer. Math. Soc. **14** (1970), 133–164.

[13] H. Kollmer, *On the structure of axiom-A attractors of codimension one.* XI Colóquio Brasileiro de Matemática, vol. II (1997), 609–619.

[14] R. Mañé, *Ergodic Theory and Differentiable Dynamics.* Springer-Verlag Berlin, 1987.

[15] A. Manning, *There are no new Anosov diffeomorphisms on tori.* Amer. J. Math. **96** (1974), 422.

[16] S. Newhouse, *On codimension one Anosov diffeomorphisms.* Amer. J. Math. **92** (1970), 671–762.

[17] S. Newhouse and J. Palis, *Hyperbolic nonwandering sets on two-dimension manifolds.* Dynamical systems (ed. M Peixoto), 1973.

[18] A. Norton, *Denjoy's Theorem with exponents.* Proceedings of the American Mathematical Society **127**, no. 10 (1999), 3111–3118.

[19] A. A. Pinto and D. A. Rand, *Solenoid Functions for Hyperbolic sets on Surfaces*. A survey paper to be published by *MSRI book series* (*published by Cambridge University Press*) (2006), 1–29.

[20] A. A. Pinto and D. A. Rand, *Rigidity of hyperbolic sets surfaces*. Journal London Math. Soc. **71**, 2 (2004), 481–502.

[21] A. A. Pinto and D. A. Rand, *Smoothness of holonomies for codimension 1 hyperbolic dynamics*. Bull. London Math. Soc. **34** (2002), 341–352.

[22] A. A. Pinto and D. A. Rand, *Teichmüller spaces and HR structures for hyperbolic surface dynamics*. Ergod. Th. Dynam. Sys. **22** (2002), 1905–1931.

[23] A. A. Pinto and D. A. Rand, *Classifying $C^{1+}$ structures on hyperbolical fractals: 1 The moduli space of solenoid functions for Markov maps on train-tracks*. Ergod. Th. Dynam. Sys. **15** (1995), 697–734.

[24] A. A. Pinto and D. A. Rand, *Classifying $C^{1+}$ structures on hyperbolical fractals: 2 Embedded trees*. Ergod. Th. Dynam. Sys. **15** (1995), 969–992.

[25] A. A. Pinto, D. A. Rand, and F. Ferreira, *Hausdorff dimension bounds for smoothness of holonomies for codimension 1 hyperbolic dynamics,* J. Differential Equations. Special issue dedicated to A. Cellina and J. A. Yorke. Accepted for publication (2007).

[26] A. A. Pinto, D. A. Rand, and F. Ferreira, *Cantor exchange systems and renormalization*. Submitted (2006).

[27] A. A. Pinto and D. Sullivan, *The circle and the solenoid*. Dedicated to Anatole Katok On the Occasion of his 60th Birthday. DCDS-A **16**, no. 2 (2006), 463–504.

[28] T. Sauer, J. A. Yorke, and M. Casdagli, *Embedology*, Journal of Statistical Physics **65**, no. 3–4 (1991), 579–616.

[29] M. Schub, *Global Stability of Dynamical Systems*. Springer-Verlag, 1987.

[30] Ya Sinai, *Markov partitions and C-diffeomorphisms*. Anal. and Appl. **2** (1968), 70–80.

[31] D. Sullivan, *Differentiable structures on fractal-like sets determined by intrinsic scaling functions on dual Cantor sets*. Proceedings of Symposia in Pure Mathematics **48** American Mathematical Society, 1988.

[32] W. Thurston, *On the geometry and dynamics of diffeomorphisms of surfaces*. Bull. Amer. Math. Soc. **19** (1988), 417–431.

[33] R. F. Williams, *Expanding attractors*. Publ. I.H.E.S. **43** (1974), 169–203.

Flávio Ferreira
E.S.E.I.G.
Instituto Politécnico do Porto
R. D. Sancho I, 981
P-4480-876 Vila do Conde
Portugal
e-mail: flavioferreira@eseig.ipp.pt

Alberto A. Pinto
Departamento de Matemática
Universidade do Minho
4710-057 Braga
Portugal
e-mail: `aapinto1@gmail.com`

David A. Rand
Mathematics Institute
University of Warwick
Coventry CV4 7AL
UK
e-mail: `david_rand@mac.com`

Progress in Nonlinear Differential Equations
and Their Applications, Vol. 75, 211–222
© 2007 Birkhäuser Verlag Basel/Switzerland

# On Bounded Trajectories for Some Non-Autonomous Systems

Andrea Gavioli and Luis Sanchez

*Dedicated to Arrigo Cellina and James Yorke*

**Abstract.** We study conditions for the existence of heteroclinics connecting $\pm 1$ for a nonautonomous equation of the form

$$\ddot{u} = a(t)f(u) \tag{0.1}$$

where $a(t)$ is a bounded positive function and $f(\pm 1) = 0$. In addition, we consider the existence of a solution to the boundary value problem in the half line

$$\begin{cases} \ddot{x} + c\dot{x} = a(t)V'(x) \\ x(0) = 0, \quad x(+\infty) = 1. \end{cases} \tag{0.2}$$

where $c \geq 0$ and $V$ is a $C^1$, non-negative function, such that $V(0) = V(1) = 0$. If $c = 0$ and $a$ and $V$ are even, it turns out that these solutions yield heteroclinics for a special class of symmetric systems which connect the two non-consecutive equilibria $\pm 1$ at the same minimum level of the potential $V$. Therefore, the existence of such a solution in the case $c = 0$ means that the system (0.2) behaves in significantly different way from its autonomous counterpart.

**Mathematics Subject Classification (2000).** 34B40, 34C37.

**Keywords.** Bounded solution, heteroclinic, non-autonomous equation.

## 1. Introduction and main results

Let us consider the simple general autonomous scalar equation

$$\ddot{u} = F'(u), \tag{1.1}$$

where

A. Gavioli is supported by CNR, Italy and L. Sanchez is supported by GRICES and Fundação para a Ciência e Tecnologia, program POCI (Portugal/FEDER-EU).

$(H_F)$ $F \in C^1(\mathbb{R}, \mathbb{R})$ is a non negative function, $F(-1) = F(1) = 0$ and $F > 0$ in $]-1, 1[$.

Hence (1.1) has two equilibria, $u = \pm 1$, at the (same) zero level of the potential. As for solutions of (1.1) energy is conserved, that is

$$\frac{\dot{u}^2}{2} - F(u) = K \tag{1.2}$$

for some constant $K$, it makes sense to look for heteroclinic solutions connecting $-1$ and $1$, i.e., solutions such that

$$u(\pm\infty) := \lim_{x \to \pm\infty} u(x) = \pm 1 \quad \text{and} \quad \dot{u}(\pm\infty) := \lim_{x \to \pm\infty} \dot{u}(x) = 0$$

or the same properties with the roles of the $+\infty$ and $-\infty$ reversed. In fact, for such solutions we must have $K = 0$ in (1.2) and they can be easily found by separation of variables. It is easily seen that they do not reach the equilibria $\pm 1$ in finite time whenever there exists $c > 0$ so that

$$F(u) \leq c(u \pm 1)^2$$

in a neighborhood of $-1$ and $+1$ respectively.

However a heteroclinic of (1.1) may be seen from another angle. Instead of using elementary integration techniques, it can be characterized by a variational property. Formally, (1.1) is the Euler–Lagrange equation of the functional

$$\mathbf{I}(u) := \int_{-\infty}^{+\infty} \left( \frac{\dot{u}^2}{2} + F(u) \right) dt , \tag{1.3}$$

We look for the heteroclinics of (1.1) as minimizers of $\mathbf{I}$ in the functional space

$$\mathcal{E} := \left\{ u \in H^1_{loc}(\mathbb{R}, \mathbb{R}) \mid u(\pm\infty) = \pm 1 \right\} .$$

In fact it is not difficult to see that we may confine ourselves to functions taking values in $[-1, 1]$ by simply assuming that $F$ is extended by 0 on $]-\infty, -1[ \cup ]+1, +\infty[$. It can be shown that (see [2])

**Theorem 1.1.** *Let $F \in C^1([-1, 1], \mathbb{R})$, extended by 0 outside the interval $]-1, 1[$, satisfy the assumption $(H_F)$. Then the functional $\mathbf{I}$ defined by (1.3) attains a minimum in $\mathcal{E}$. A minimizer is a heteroclinic solution of (1.1) connecting $-1$ and $1$.*

Let us move to a less trivial situation: consider the second order non-autonomous differential equation

$$\ddot{u} = a(t)f(u) , \tag{1.4}$$

where a primitive $F$ of $f \in C(\mathbb{R}, \mathbb{R})$ satisfies the assumption $(F)$ and $a \in L^\infty(\mathbb{R}, \mathbb{R})$ is such that

$(H_A)$ there exist $a_1, a_2 \in \mathbb{R}$ so that $0 < a_1 \leq a(t) \leq a_2$ for all $t \in \mathbb{R}$.

We look for a heteroclinic connection between the equilibria $-1$ and $+1$. In the absence of a conservation law, the variational argument appears as a natural device. So we now consider the functional

$$\mathbf{J}(u) := \int_{-\infty}^{+\infty} \left( \frac{\dot{u}^2}{2} + a(t)F(u) \right) dt \tag{1.5}$$

and seek conditions that allow to minimize it in $\mathcal{E}$. The fact that $\mathbf{I}$, defined by (1.3), is translation invariant, is a powerful argument to obtain compactness of modified minimizing sequences. As long as $\mathbf{J}$ is concerned, we have to face a possible loss of compactness if $a$ does not possess any symmetry or periodicity property. A simple setting where this is overcome is the following.

**Theorem 1.2.** *Assume that $f \in C(\mathbb{R}, \mathbb{R})$, $F' = f$ in $\mathbb{R}$, so that $F$ and $a \in L^{\infty}(\mathbb{R}, \mathbb{R})$ satisfy $(H_F)$–$(H_A)$. If in addition*

$$\lim_{|t| \to \infty} a(t) = a_2$$

*and $a(t) < a_2$ in some subset with nonzero measure, then (1.4) has a heteroclinic solution from $-1$ to $1$. This solution takes values in $[-1, 1]$.*

A proof can be found in [2]. See [4] for related results.

In order to exploit a simple symmetry, we shall consider the boundary value problem

$$\ddot{x} = a(t)V'(x) \tag{1.6}$$
$$x(0) = 0, \quad x(+\infty) = 1. \tag{1.7}$$

Here we shall assume

$(S_1)$ $V \in C^1(\mathbb{R}, \mathbb{R})$ is a non negative function, $V(1) = 0$ and $V > 0$ in $[0, 1[$.
$(S_2)$ The function $a \in L^{\infty}(0, +\infty)$ is such that there exist $t_0 \geq 0$, $0 < a_1 \leq a_2$ with the property that $a_1 \leq a(t) \leq a_2 \ \forall t \in [t_0, +\infty[$.

**Theorem 1.3.** *Assume that $(S_1)$ and $(S_2)$ hold. Then the boundary value problem (1.6), (1.7) has at least one solution that takes values in $[0, 1]$.*

**Corollary 1.4.** *Let $a : \mathbb{R} \to \mathbb{R}$ and $V \in C^1(\mathbb{R})$ be even functions satisfying $(S_1)$ and $(S_2)$. Moreover assume that $1$ is an isolated local minimizer of $V$ in $[0, 1]$. Then (1.6) has an odd heteroclinic solution connecting the equilibria $-1$ and $1$.*

Theorem 1.2 and Corollary 1.4 illustrate the fact that some basic characteristics of autonomous systems are naturally inherited by their non-autonomous counterparts. However one may encounter striking differences in the behaviour of each type of system. As an example let us mention that in [5] Coti Zelati and Rabinowitz have obtained, for some non-autonomous $n$-dimensional systems, heteroclinics that connect two equilibria at different levels of the potential. The problem we introduce next, dealing with a trickier situation than the preceeding one, leeds to another example of a similar phenomenon.

Consider a smooth scalar potential $V(x)$, $x \in \mathbb{R}$, which is positive in $]0,1[$ and such that $V(0) = V(1) = 0$, a scalar positive function $a$, and $c \geq 0$. Moreover, if $c = 0$, let $a(t)$ have a positive infimum in $\mathbb{R}$. We shall deal with the existence of solutions to the following boundary value problem on the interval $]0,\infty[$:

$$\ddot{x} + c\dot{x} = a(t)V'(x) \tag{1.8}$$

$$x(0) = 0, \quad x(+\infty) = 1. \tag{1.9}$$

Before stating precise assumptions, let us make some comments on our interest in this problem.

Suppose that $c = 0$, $a$ is a positive constant and $V$ is extended as an even function in $\mathbb{R}$. According to our previous remarks, there exists a heteroclinic connection between 0 and 1 for (1.8), but the conservation of energy prevents the existence of a heteroclinic connection between the equilibria $-1$ and 1, since the equilibrium 0 would have to be crossed in finite time. In particular, the problem (1.8), (1.9) has no solution in the autonomous case, since otherwise by anti-symmetry we would obtain the mentioned heteroclinic between $\pm 1$.

Anyway, if $c = 0$ and $a$ and $V$ are even functions, our problem is equivalent to that of finding a heteroclinic connection between non-consecutive equilibria $\pm 1$ of a potential $V$ having three minima at the same level; we shall give simple conditions on the time dependence of the coefficient $a(t)$ that ensure the existence of such kind of trajectory. Such connections do not exist in the autonomous case, again by the conservation principle. We note nevertheless that for higher order conservative (autonomous) equations where more degrees of freedom are available, heteroclinic connections between non-consecutive equilibria may exist: we refer to the paper by Bonheure, Sanchez, Tarallo, and Terracini [3].

Our results show that with respect to problem (1.8), (1.9) the differences do exist for $c = 0$ but not for $c > 0$. In fact, in the case $c > 0$ a solution of our boundary value problem exists in the autonomous case as well. Incidentally, we shall see that the case $c > 0$ may be dealt with under simpler assumptions.

We shall state two existence theorems for (1.8), (1.9) under distinct sets of conditions on the data. It turns out that the way the (increasing) function $a(t)$ approaches its limit plays an important role in the sufficient conditions. More precisely, Theorem 1.5 deals with problem (0.2) in a situation where weak regularity assumptions on $V$ and its minima are assumed; while with respect to $a(t)$ it is required that $a(t)$ tends to its limit $l$ in such a way that, if $l < \infty$, $l - a(t)$ is slower than $1/t$. On the other hand, in Theorem 1.7, dealing with the case $c = 0$ only, we prove the existence of solutions for a wider class of functions $a(t)$, while confining ourselves to the class of $C^2$ potentials $V$.

The description of our assumptions follows.

$(H_1)$ $V \in C^1(\mathbb{R})$ is a non negative function, $V(0) = V(1) = 0$ and $V > 0$ in $]0,1[$.

$(H_2)$ There exist $\delta > 0$ and $A_1$, $A_2 > 0$ such that $A_1 x^2 \leq V(x) \leq A_2 x^2$ for $|x| < \delta$.

$(H_3)$ The function $a : [0, +\infty[ \to ]0, +\infty[$ is such that there exists $t_0 \geq 0$ with the property that $a$ is increasing in $[t_0, +\infty[$.

**Theorem 1.5.** *Assume that $(H_1)$ holds and $a \in L^\infty(0, +\infty)$ is non-negative. If $c = 0$ assume that $(H_2)$ and $(H_3)$ hold as well and in addition that $\eta := \inf_{t \geq 0} a(t) > 0$ and $l := \lim_{t \to +\infty} a(t)$ has the property*

$$\lim_{t \to +\infty} t(l - a(t)) = +\infty \qquad (1.10)$$

*then the boundary value problem (1.8), (1.9) has at least one solution that takes values in $[0, 1]$.*

**Remark 1.6.** *If $l \in \mathbb{R}^+$ it is easy to check that (1.10) holds if $a(t)$ is of the form $a(t) = l - \frac{\gamma}{(1+t)^\beta}$, $0 < \gamma < l$, $\beta < 1$.*

**Theorem 1.7.** *Assume $c = 0$. Let $V \in C^2(\mathbb{R})$ satisfy $(H_1)$ and $V''(0) > 0$. Let $a : \mathbb{R} \to ]0, +\infty[$ be a function of bounded variation with $\eta := \inf_{t \geq 0} a(t) > 0$ satisfying $(H_3)$ and the property*

$$\lim_{t \to +\infty} (l - a(t)) e^{2\mu t} = +\infty, \qquad (1.11)$$

*where $l := \lim_{t \to +\infty} a(t)$ and $\mu = \sqrt{\eta V''(0)}$. Then the boundary value problem (1.8), (1.9) has at least one solution taking values in $[0, 1]$.*

**Corollary 1.8.** *Let $c = 0$, $a$ and $V$ be even functions satisfying the assumptions of Theorem 1.5 or Theorem 1.7. Moreover assume that 1 is an isolated minimizer of $V$. Then (1.8) has a heteroclinic solution connecting the equilibria $-1$ and $1$.*

Theorems 1.2 and 1.3 may be proved by a minimization procedure, as is the case for Theorem 1.1. Theorem 1.7 combines shooting and variational arguments. Some proofs will be given in the last section. For further details see [2] and [6].

## 2. A comparison between autonomous and non automous problems

First, note that if $c = 0$ and $a(t)$ is constant, then problem (0.2), which corresponds now to an autonomous equation, has no solution. Then as a next step and still maintaining $c = 0$, the simplest non autonomous system one can discuss is one where $a(t)$ is a 'bang-bang' function with only one switch. More precisely, if $0 < a < b$ and $T > 0$, we define

$$a(t) := \begin{cases} a, & 0 \leq t \leq T \\ b, & t > T. \end{cases} \qquad (2.1)$$

Then, consider a $C^1$ potential $V(x)$ as above and which is locally bounded from above by $Ax^2$ and $A(x-1)^2$, $A > 0$, respectively around $x = 0$ and $x = 1$.

We note that problem (0.2) has a solution if and only if there exists a solution of $\ddot{x} = aV'(x)$ on $[0, T]$, with $x(0) = 0$ such that the corresponding solution curve $(x(t), \dot{x}(t))$ in the phase plane $(x, \dot{x})$ intersects at time $T$ the heteroclinic orbit between $(0, 0)$ and $(0, 1)$ corresponding to the equation $\ddot{x} = bV'(x)$.

Set $\xi = x(T) \in ]0, 1[$. By the conservation of energy, the heteroclinic solution of the second equation satisfies $\dot{x} = \sqrt{2bV(x)}$, whereas the solution of the first equation with $x(0) = 0$ satisfies $\dot{x} = \sqrt{2aV(x) + C}$ for some constant $C$. Then, imposing that the two solution curves intersect at time $T$ in the phase plane, we get

$$C = 2(b - a)V(\xi) \tag{2.2}$$

and, if our problem admits a solution, then the following representation holds for $T = T(\xi)$:

$$T(\xi) = \int_0^\xi \frac{dx}{\sqrt{2aV(x) + 2(b - a)V(\xi)}} \tag{2.3}$$

By the quadratic growth of $V(x)$ in a neighbourhood of $x = 0$, there exists a constant $\underline{c}$ such that

$$\underline{c} \int_0^\xi \frac{dx}{\sqrt{2ax^2 + 2(b - a)\xi^2}} \leq T(\xi) \tag{2.4}$$

for any sufficiently small $\xi > 0$. Since

$$\int_0^\xi \frac{dx}{\sqrt{2ax^2 + 2(b - a)\xi^2}} = \frac{1}{\sqrt{2a}} \log\left(\frac{\sqrt{a} + \sqrt{b}}{\sqrt{b}}\right)$$

we infer that $T(\xi)$ is bounded away from zero in a right neighbourhood of $\xi = 0$. In a similar way it can be shown that $T(\xi) \to +\infty$ as $\xi \to 1^-$. Then, since $T(\xi)$ is a continuous function of $\xi$, we conclude that there exists $T_0 > 0$ such that $T(]0, 1[) = [T_0, +\infty[$ or $T(]0, 1[) = ]T_0, +\infty[$ and therefore problem (0.2) has no solution if $T < T_0$ (and admits a solution for any $T > T_0$).

In the simple example above the switch time $T_0$ for the function $a(t)$ actually depends on $V(x)$ through $\underline{c}$. This suggests that, generally speaking, despite the fact that the variables $x$ and $t$ are separate in the right-hand side of our equation, the conditions given on $a(t)$ to solve problem (0.2) when $c = 0$ may naturally involve the potential $V$. This is apparent in Theorem 1.7 and unlike the heteroclinic problem $x(-\infty) = 0$, $x(+\infty) = 1$ associated to the same equation for the class of potentials considered above.

We now turn to consider the features of the case $c > 0$. If we again take $a$ to be a constant, the problem (1) may be solved by an elementary reduction of order technique. Indeed, if one looks for strictly monotone solutions then (1) is easily transformed into a first order problem for the new unknown function $\psi = \Phi^2$ where $\Phi$ describes the graph of the curve $\dot{x} = \Phi(x)$ in the phase plane. See [2, 7] for similar examples. The new formulation may be written as

$$\begin{cases} \psi' = 2(aV'(x) - c\sqrt{\psi}) \\ \psi(1) = 0, \quad \psi(x) > 0 \ \forall x \in [0, 1[. \end{cases} \tag{2.5}$$

It is not difficult to conclude, directly from phase-plane analysis or by studying (2.5), that for any $c > 0$ and $a > 0$ constant, the problem (0.2) has a solution. In fact if we consider, for $\epsilon > 0$, the Cauchy problem

$$\begin{cases} \psi' = 2(aV'(x) - c\sqrt{\psi_+ + \epsilon}) \\ \psi(1) = 0, \end{cases} \tag{2.6}$$

it turns out that it has a solution in $[0, 1]$ that stays above $2aV(x)$. Then, by taking the limit as $\epsilon \to 0$, our claim follows. Theorem 1.5 shows that the solution still exists when $a$ depends on $t$.

## 3. Some proofs

*Proof of Theorem* 1.3. We need the following auxiliary fact which is a slight modification of Lemma 3.6 in [8]. A similar lemma is used also in the proof of Theorem 1.2, cf. [2]. We are going to use the functional space

$$X := \left\{ x \in C\big([0, +\infty[\big) \cap H^1_{loc}\big([0, +\infty[\big) \; : \; x(+\infty) = 1 \right\}. \qquad \square$$

**Lemma 3.1.** *Let $V$ be as in $(S_1)$ or $(H_1)$, $a$ be as in $(S_2)$. Fix $\varepsilon \in \,]0, 1[$ and*

$$\beta_\varepsilon := \min \left\{ V(z) \mid 1 - \varepsilon \leq z \leq 1 - \frac{\varepsilon}{2} \right\}.$$

*If $u \in X$ has the property that there exist $t_1, t_2 \geq t_0$ such that $u(t_1) = 1 - \frac{\varepsilon}{2}$ and $u(t_2) = 1 - \varepsilon$, then we have*

$$\left| \int_{t_1}^{t_2} \left( \frac{\dot{u}^2}{2} + a(t)V(u) \right) dt \right| \geq \frac{\varepsilon \sqrt{a_1 \beta_\varepsilon}}{\sqrt{2}}.$$

**Remark 3.2.** *If instead $V$ satisfies $(H_1)$ there is a corresponding statement as follows:* Let $\varepsilon \in \,[0, 1/2[$ and

$$\beta_\varepsilon := \min \left\{ V(z) \mid 1 - \varepsilon \leq z \leq 1 - \frac{\varepsilon}{2} \text{ or } \frac{\varepsilon}{2} \leq z \leq \varepsilon \right\}.$$

*If $u \in X$ has the property that there exist $t_1, t_2 \in \mathbb{R}$ such that $u(t_1) = 1 - \frac{\varepsilon}{2}$ and $u(t_2) = 1 - \varepsilon$ (or $u(t_1) = \varepsilon/2$ and $u(t_2) = \varepsilon$), then we have*

$$\left| \int_{t_1}^{t_2} \left( \frac{\dot{u}^2}{2} + V(u) \right) dt \right| \geq \frac{\varepsilon \sqrt{\beta_\varepsilon}}{\sqrt{2}}.$$

We extend $V$ to $\mathbb{R}$ so that it is constant in $]-\infty, 0]$ and $[0, +\infty[$ and consider the functional

$$\mathcal{F}(x) := \int_0^{+\infty} \left( \frac{\dot{x}(t)^2}{2} + a(t)V\big(x(t)\big) \right) dt$$

defined in

$$X_0 = \left\{ x \in X \; : \; x(0) = 0 \right\}.$$

Then $\inf_{X_0} \mathcal{F}$ is finite, since $\int_0^{t_0} a(t)V(x(t))\, dt \geq -t_0 \|a\|_\infty \|V\|_\infty \; \forall x \in X_0$. Take a minimizing sequence $x_n \in X_0$:

$$\mathcal{F}(x_n) \to \inf_{X_0} \mathcal{F}.$$

We may assume $0 \leq x_n(t) \leq 1$, since otherwise we could replace $x_n$ with $\min(\max(x_n, 0), 1)$, still obtaining a minimizing sequence. By virtue of our preceeding remark it is clear that, for any $M > 0$, $(x_n)_n$ is bounded in $H^1(0, M)$ and $(\dot{x}_n)_n$ is bounded in $L^2(0, +\infty)$. Then we can take a subsequence, still denoted by $(x_n)_n$, which converges to some absolutely continuous function $x$ uniformly on compact sets; $x$ takes values in $[0, 1]$ and

$$\dot{x}_n \rightharpoonup \dot{x} \quad \text{in} \quad L^2(0, +\infty).$$

Using the compact imbedding $H^1(0, t_0) \hookrightarrow C[0, t_0]$, the weak lower semicontinuity of the norm and Fatou's lemma we find

$$\mathcal{F}(x) \leq \liminf \mathcal{F}(x_n).$$

In particular, $\int_0^{+\infty} a(t)V(x(t))\, dt$ converges. We now assert that $x(+\infty) = 1$. Otherwise, we would be able to find $\varepsilon > 0$ and construct an infinite sequence of disjoint intervals $[t_{1n}, t_{2n}]$ with $t_{1n}, t_{2n} \geq t_0$ in the conditions of the preceeding lemma, implying

$$\int_0^{+\infty} a(t)V(x(t))\, dt = +\infty.$$

We conclude that $u \in X_0$, so that $u$ minimizes $\mathcal{F}$. By well known arguments, $u$ is a solution of (1.6).

*Proof of Corollary* 1.4. Under the assumptions of Corollary 1.4, (1.6)–(1.9) has a solution $x(t)$ taking values in $[0, 1]$. Then the function

$$w(t) = \begin{cases} x(t) & t \geq 0 \\ -x(-t) & t < 0. \end{cases}$$

is a solution of (1.6) such that

$$\lim_{t \to \pm\infty} w(t) = \pm 1.$$

This is indeed a heteroclinic solution because $\lim_{t \to \pm\infty} \dot{w}(t) = 0$. In fact, integrating (1.8) between 0 and $t > 0$ we have

$$\dot{x}(t) - \dot{x}(0) = \int_0^t a(s)V'(x(s))\, ds.$$

Since the integrand in the right-hand side does not change sign in a neighbourhood of $+\infty$, we conclude that $\int_0^{+\infty} a(s)V'(x(s))\, ds$ converges. Therefore $\lim_{t \to +\infty} \dot{x}(t)$ exists, and the boundedness of $x(t)$ implies that $\lim_{t \to +\infty} \dot{x}(t) = 0$. $\qquad\square$

*Proof of Theorem* 1.7. For $I \subset [0, +\infty[$ let

$$\mathcal{F}(x, I) := \int_I \left( \frac{\dot{x}(t)^2}{2} + a(t)V(x(t)) \right) dt.$$

Consider the equation

$$\ddot{x} = a(t)V'(x) \qquad (3.1)$$

with the boundary condition (1.9).

Consider the space $X$ as before and put:

$$X(\xi) = \{x \in X \,|\, x(0) = \xi,\ x(+\infty) = 1\}, \quad \xi \in \mathbb{R}. \qquad (3.2)$$

In $X$ we will make use of the norm $x \mapsto (|x(0)|^2 + \|\dot{x}\|_2^2)^{1/2}$. $\qquad\square$

We need the following three Lemmas.

**Lemma 3.3.** $(l - a(t))\gamma(t)^2 \to +\infty$ as $t \to +\infty$, where $\gamma$ is the solution of the Cauchy problem

$$\begin{cases} \gamma''(t) = a(t)V''(0)\gamma(t) \\ \gamma(t_0) = 0, \quad \gamma'(t_0) = 1. \end{cases} \tag{3.3}$$

*Proof.* From (1.11) we obviously get, as $t \to +\infty$,

$$(l - a(t))\rho(t)^2 \to +\infty, \tag{3.4}$$

where $\rho(t) = (e^{\mu(t-t_0)} - e^{-\mu(t-t_0)})/4\mu$. On the other hand, since $\rho''(t) = \mu^2\rho(t)$, $\rho(t_0) = 0$, $\rho'(t_0) = 1/2$, it is easy to see that $\gamma(t) > \rho(t)$ for $t > t_0$. Indeed, from the initial conditions we obtain the assertion in a right neighbourhood of $t = t_0$. By contradiction, let $\tau$ the first point after $t_0$ at which $\gamma(\tau) = \rho(\tau)$. Then we must have $\gamma''(\sigma) < rho''(\sigma)$ at some point $\sigma \in ]t_0, \tau[$, while, on this interval,

$$\gamma''(t) \geq \eta V''(0)\gamma(t) = \mu^2\gamma(t) \geq \mu^2\rho(t) = \rho''(t).$$

Then $\gamma > \rho > 0$ on $]t_0, +\infty[$, so that $\gamma^2 > \rho^2$, and our claim follows from (3.4). □

**Lemma 3.4.** For any $\xi \in ]0,1]$ $\mathcal{F}$ attains its minimum on the class $X(\xi)$.

*Proof.* Let $(y_k)_k$ be a minimizing sequence for $\mathcal{F}$ on $X(\xi)$, and let $t_1 > t_0$ be fixed. For any $k \in \mathbf{Z}^+$ the following properties may be assumed to hold:

(a) $0 \leq y_k(t) \leq 1$,
(b) $y_k$ solves (3.1) on $J := [0, t_1]$.

Indeed, if these conditions are not satisfied it is enough to replace $y_k$, respectively:

(a) by $\min(\max(y_k, 0), 1)$
(b) on $J$ by a function which minimizes $\mathcal{F}(\cdot, J)$ on $y_k + H_0^1(J)$.

It is easy to check that $\mathcal{F}$ does not increase after these procedures. Furthermore, $(y_k)_k$ is bounded in $H^1(J)$: then we can suppose, up to a subsequence, that $(y_k)_k$ converges uniformly on $J$ to some absolutely continuous function $y$ and that $\dot{y}_k \to \dot{y}$ weakly in $L^2(J)$. Then, on the interval $J$ where $y_k$ solves (3.1), $\ddot{y}_k$ is uniformly bounded: since the sequence $(\dot{y}_k)_k$ is bounded in $L^2$, we conclude that it is actually bounded in $H^1(J)$ and, even more so, in $L^\infty(J)$. Then we may suppose that $(\dot{y}_k(0))_k$ converges, so that the continuous dependence on initial data of the solutions of a differential equation ensures that the limit function $y$ solves (3.1) on $J$ (and also the $C^1$-convergence on that interval). Furthermore, $0 \leq y(t) \leq 1$. On the other hand, if $y$ vanishes at $t_0$, we should also get $\dot{y}(t_0) = 0$, since $t_0$ is in the interior of $J$ and $y$ cannot take negative values. But the conditions $y(t_0) = \dot{y}(t_0) = 0$, together with (3.1), would imply $y(t) \equiv 0$ on $J$, in contrast with $y(0) = \xi > 0$. Hence $y(t_0) > 0$, and we can find $\delta > 0$ such that $y_k(t_0) \geq \delta$ for large $k$'s. We then redefine $y_k$ by replacing its restriction to $S := [t_0, +\infty[$ by the shifted function $t \mapsto y_k(t + \tau_k)$, where $\tau_k \geq t_0$ is the last point such that $y_k(\tau) = y_k(t_0)$. Since $a$ is increasing in $S$, this operation does not increase the value of $\mathcal{F}$. If we still denote by $y_k$ the modified functions, we can actually suppose that $y_k(t) \geq y_k(t_0)$ for any

$t \in S := [t_0, +\infty[$. Now we can take a subsequence, still denoted by $(y_k)_k$, which converges to some absolutely continuous function $x$ uniformly on compact sets and in such a way that $\dot{y}_k \rightharpoonup \dot{x}$ in $L^2(0, +\infty)$. Of course, $x \equiv y$ in $J$. By pointwise convergence,

$$x(t) \geq \delta \quad \text{for all} \quad t \in S. \tag{3.5}$$

Furthermore, $\mathcal{F}(x) < +\infty$. Using the analogue of Lemma 3.1 adapted to our $V$ (see Remark 2), this entails that $x(+\infty) \in \{0, 1\}$. But (3.5) excludes the case $x(+\infty) = 0$, and actually $x \in X(\xi)$. Now, thanks again to the weak lower semicontinuity of $\mathcal{F}$, $x$ minimizes $\mathcal{F}$ on $X(\xi)$. Of course, $x$ takes values in $[0, 1]$. $\square$

Now, let $\xi_i \to 0^+$ as $i \to +\infty$, and apply the previous Lemma on the class $X(\xi_i)$ for any $i \in \mathbf{Z}^+$, so as to get functions $x_i$ such that

$$\mathcal{F}(x_i) \leq \mathcal{F}(y) \quad \text{for any} \quad y \in X(\xi_i).$$

As before we can suppose, up to a subsequence, that $(x_i)_i$ converges uniformly on compact sets to some function $x \in X$. Furthermore, the same arguments as in the proof of the previous Lemma allow to suppose that

$$\dot{x}_i(t_0) \to \dot{x}(t_0). \tag{3.6}$$

Then $x$ solves (3.1), like $x_i$, and the following properties hold: $0 \leq x(t) \leq 1$, $x(t) \geq x(t_0)$ on $S$, $x(0) = 0$, $x(+\infty) \in \{0, 1\}$.

**Lemma 3.5.** $x(t_0) > 0$

*Proof.* Let us suppose, by contradiction, $x(t_0) = 0$: since $x \geq 0$, we have $\dot{x}(t_0) = 0$ as well, and from (3.1) we actually get $x(t) \equiv 0$. In particular, $x_i(t_0) < r$ for large $i's$, where $r > 0$ is such that $V'' > 0$ in $[0, r]$. Furthermore $t_i \to +\infty$ as $i \to +\infty$, where $t_i$ is the first time at which $x_i$ reaches the value $r$. We put $\rho_i = x_i(t_0)$, $\eta_i = \dot{x}_i(t_0)$ and recall that each $x_i$ solves (3.1). Then, for any $\tau \geq t_0$:

$$\frac{1}{2}\dot{x}_i(\tau)^2 - \frac{1}{2}\dot{x}_i(t_0)^2 = \int_{t_0}^{\tau} \dot{x}_i(s)\ddot{x}_i(s)\, ds = \int_{t_0}^{\tau} a(s)V\big(x_i(s)\big)\dot{x}_i(s)\, ds$$

$$= \Big[a(s)V\big(x_i(s)\big)\Big]_{t_0}^{\tau} - \int_{t_0}^{\tau} V\big(x_i(s)\big)\, da(s).$$

Since $\dot{x}_i(\tau) \to 0$ and $V(x_i(\tau)) \to V(1) = 0$ as $\tau \to +\infty$, we get

$$\frac{1}{2}\eta_i^2 = a(t_0)V\big(x_i(t_0)\big) + \int_{t_0}^{+\infty} V\big(x_i(s)\big)\, da(s) \geq \int_{t}^{t_i} V\big(x_i(s)\big)\, da(s),$$

for any $t \in [t_0, t_i]$. Now, let us denote by $t \mapsto \phi(t; \xi, \eta)$ the solution of (3.1) which fulfils the conditions $x(t_0) = \xi$, $\dot{x}(t_0) = \eta$, so as to write $x_i(t) = \phi(t; \rho_i, \eta_i)$. For $t \geq t_0$, and as long as $\phi(t; \xi, \eta) \leq r$, it is easy to check that the function $V(\phi(t; \xi, \eta))$ is increasing with respect to all its arguments, so that $V(x_i(s)) = V(\phi(s; \rho_i, \eta_i)) \geq W(t, \eta_i)$ for any $s \in [t, t_i]$, where we put $W(s, \eta) = V(\phi(s; 0, \eta))$. Then

$$\frac{1}{2} \geq \frac{W(t, \eta_i)}{\eta_i^2}\big(a(t_i) - a(t)\big), \tag{3.7}$$

and we can let $i \to +\infty$. Since, by virtue of (3.6), $\eta_i \to \dot{x}(t_0) = 0$, we look for the behaviour of $W(t, \eta)/\eta^2$ as $\eta \to 0^+$, which depends on the partial derivatives of $W$ (hence of $\phi$) with respect to $\eta$. To this end we apply well-known results on differentiability with respect to initial data of the solution of a differential equation, which hold because the differential of the map $(x, y) \mapsto f(t, x, y) = (y, a(t)V'(x))$ is bounded uniformly with respect to $t$. Since $\phi(t; 0, 0) \equiv 0$, the evolution of $\gamma(t) = \phi'_\eta(t; 0, 0)$ is ruled by (3.3). Hence

$$\frac{\partial W}{\partial \eta}(t, 0) = V'(0)\gamma(t) = 0 \,, \quad \frac{\partial^2 W}{\partial \eta^2}(t, 0) = V''(0)\gamma(t)^2 \,,$$

so that

$$\lim_{\eta \to 0^+} \frac{W(t, \eta)}{\eta^2} = \frac{1}{2}V''(0)\gamma(t)^2 \,.$$

Now (3.7) entails $1 \geq V''(0)\gamma(t)^2(l - a(t))$, in contrast with Lemma 3.2. Then $x(t_0) > 0$, as claimed.  $\square$

We can now conclude the proof of Theorem 1.4.

Since $x(t) \geq x(t_0) > 0$ for $t \geq t_0$, the previous arguments show that $x(+\infty) = 1$, so that $x \in X$. Now, let $y \in X$, $i \in \mathbf{Z}^+$: we can modify $y$ by putting $y_i = y + u_i$, where $u_i(0) = \xi_i$, $u \equiv 0$ in $[1, +\infty[$, $\dot{u} \equiv -\xi_i$ in $[0, 1]$ , so that $y_i \in X(\xi_i)$ and $\varepsilon_i := |\mathcal{F}(y_i) - \mathcal{F}(y)| \to 0$ as $i \to +\infty$. Then $\mathcal{F}(x_i) \leq \mathcal{F}(y_i) \leq \mathcal{F}(y) + \varepsilon_i$, and the lower limit as $i \to +\infty$ yields $\mathcal{F}(x) \leq \mathcal{F}(y)$. Hence $x$ minimizes $\mathcal{F}$ on $X$. Of course, $x$ takes values in $[0, 1]$.

# References

[1] R. P. Agarwal and D. O'Regan, *Infinite interval problems for differential, difference and integral equations*, Kluwer Ac. Publ. Dordrecht, 2001.

[2] D. Bonheure, L. Sanchez, *Heteroclinic orbits for some classes of second and fourth order differential equations*, Handbook of Differential Equations: Ordinary Differential Equations, Vol. 3, A. Cañada, P. Drabek, A. Fonda, editors, Elsevier, 2006.

[3] D. Bonheure, L. Sanchez, M. Tarallo, S. Terracini, *Heteroclinic connections between nonconsecutive equilibria of a fourth order differential equations*, Calculus of Variations and Partial Differential Equations, **17**, 341–356 (2003).

[4] C.-N. Chen and S.-Y. Tzeng, *Existence and multiplicity results for heteroclinic orbits of second order Hamiltonian systems*, J. Differential Equations, **158** (1999), no. 2, 211–250.

[5] V. Coti Zelati and P. H. Rabinowitz, *Heteroclinic solutions between stationary points at different energy levels*, Top. Meth. Nonlinear Analysis, **17** (2001) 1–21.

[6] A. Gavioli and L. Sanchez, *On a class of bounded trajectories for some non-autonomous systems* , to appear in Math. Nachr.

[7] L. Malaguti and C. Marcelli, *Travelling wavefronts in reaction-diffusion equations with convection effects and non-regular terms*, Math. Nachr., **242** (2002) 148–164.

[8] P. H. Rabinowitz, *Periodic and heteroclinic orbits for a periodic hamiltonian system*, Ann. Inst. Henri Poincaré, **6–5** (1989), 331–346.

Andrea Gavioli
Dipartimento di Matematica Pura ed Applicata
Via Campi, 213b
I–41100 Modena
Italy


Luis Sanchez
Faculdade de Ciências da Universidade de Lisboa
CMAF, Avenida Professor Gama Pinto, 2
P–1649-003 Lisboa
Portugal

Progress in Nonlinear Differential Equations
and Their Applications, Vol. 75, 223–229
© 2007 Birkhäuser Verlag Basel/Switzerland

# On Generalized Differential Quotients and Viability

Ewa Girejko and Zbigniew Bartosiewicz

*Dedicated to Arrigo Cellina and James Yorke*

**Abstract.** Differential inclusions with constraints of the form
$$\dot{y}(t) \in F\big(t, y(t)\big), \quad y(t) \in K(t).$$
and the viability problem for such inclusions are studied. It is assumed that
$t \twoheadrightarrow K(t)$ is a set-valued map that has a GDQ-regular multiselection and
$(t, y) \twoheadrightarrow F(t, y)$ is a set-valued map measurable with respect to $t$ and upper
semi-continuous with respect to $y$. Some auxiliary results on Cellina continuously approximable multifunctions and Generalized Differential Quotients are
given.

**Mathematics Subject Classification (2000).** Primary 34A60; Secondary 49J55.

**Keywords.** Viability, Generalized Differential Quotients.

## 1. Introduction

A multifuction (a set-valued map) $F$ from a set $X$ to a set $Y$ assigns to every
$x \in X$ a subset, maybe empty, of $Y$. We denote this by $F : X \twoheadrightarrow Y$ or $X \ni x \mapsto\joinrel\twoheadrightarrow$
$F(x) \subseteq Y$. The domain of $F$ is $Do(F) = \{x \in X : F(x) \neq \emptyset\}$ and the graph
of $F$ is the set $Gr(F) = \{(x, y) \in X \times Y : y \in F(x)\}$. In applications one often
wants to control the infinitesimal behavior of the multifunction around a point
$(x, y)$ belonging to its graph. For this some generalization of ordinary derivative is
needed. We recall here two concepts: more classical contingent derivative (see [1–4])
and newer Generalized Differential Quotient (GDQ) introduced by H. Sussmann
(see [15, 16]). Both can be also used for single-valued functions that do not have
ordinary derivatives. We show relations between these concepts for multifunctions
from $\mathbb{R}$ to $\mathbb{R}^n$. Such multifunctions appear as constraints in viability problems.

This work was supported by Bialystok Technical University grant S/WI/1/07.

Let $K : [0,1] \twoheadrightarrow \mathbb{R}^n$ and $F : Gr(K) \twoheadrightarrow \mathbb{R}^n$. Then $F$ is called a time dependent orientor field with the restriction multifunction $K$. Consider the initial value problem for $F$:

$$\dot{x}(t) \in F(t, x(t)), \quad x(t_0) = x_0, \tag{1.1}$$

where $(t_0, x_0) \in Gr(K)$. We say that $F$ is *viable* if for every $(t_0, x_0) \in Gr(K)$ (1.1) has an absolutely continuous solution $x : [t_0, 1] \to \mathbb{R}^n$ such that $x(t) \in K(t)$ for all $t \in [t_0, 1]$ and the inclusion in (1.1) is satisfied almost everywhere on $[t_0, 1]$.

Many different criteria of viability can be found in the literature. The reader can consult, e.g., [1,3,5–7,10] for the results and properties used in the statements (like upper semi-continuity or left upper semi-continuity). The authors impose various conditions on $K$ and $F$ and propose some tangency requirements, which generalize the original condition given by Nagumo for time independent vector fields.

In the tangency condition usually the contingent derivative of $K$ is used. In [10] and in this paper, we prefer to use GDQ of $K$. We feel that it is more natural than the contingent derivative and GDQ-differentiability implies other important properties of $K$ that are used in the viability theorem – the main result of this paper. It extends the result of [10] by relaxing some restrictions imposed on $F$.

## 2. Motivation and basic definitions

In the viability problem one often assumes that the orientor field satisfies the following growth condition:

$$\|F(t, x)\| \leq c(t)(1 + \|x\|) \tag{2.1}$$

where $c$ is an $L_1$ function. This is a natural generalization of a similar growth condition assumed for differential equations to achieve existence of global solutions. The following example shows that for differential inclusions and viability problems this assumption is no longer needed.

**Example 2.1.** Consider a set-valued map $K : [-1, 0] \twoheadrightarrow \mathbb{R}$ such that

$$K(t) = \begin{cases} \{|t \sin \frac{1}{t}|\} \cup \{0\} & \text{if } t \neq 0 \\ 0 & \text{if } t = 0. \end{cases}$$

Let $\dot{x} \in F(t, x)$ and $x(t_0) = x_0$ where $F(t, x) = [-1 - \frac{1}{t}, 1 + \frac{1}{t}]$ and $F(0, 0) = 0$ is defined on $Gr(K)$. Observe that $F$ does not satisfy (2.1). But for every $t_0 \in [-1, 0)$ and $x_0 \in K(t_0)$ there exists a global solution $x : [t_0, 0] \to \mathbb{R}$ defined as follows

$$x(t) = \begin{cases} t \sin \frac{1}{t} & \text{if } t \in [t_0, \tau_0] \\ 0 & \text{if } t \in (\tau_0, 0], \end{cases}$$

where $\sin \frac{1}{\tau_0} = 0$.

We shall see later that it is enough to assume (2.1) on some subset of $Gr(K)$ containing $(t_0, x_0)$ to achieve viability.

Let us recall now some basic definitions.

Let $X$ and $Y$ be metric spaces and $A, B \subseteq X$. Then

$$\triangle(A, B) = \sup \{dist(q, B) : q \in A\}$$

is called the *semi-distance* between sets $A$ and $B$ where $dist(a, B) := \inf_{b \in B} d(a, b)$

**Definition 2.2.** *We say that a sequence of set-valued maps $F_n : X \twoheadrightarrow Y$, $n \in \mathbb{N}$, graph-converges to a set-valued map $F : X \twoheadrightarrow Y$, and write $F_n \xrightarrow{gr} F$, if*

$$\lim_{n \to \infty} \triangle\big(Gr(F_n), Gr(F)\big) = 0 \,.$$

**Definition 2.3.** *A set-valued map $F : X \twoheadrightarrow Y$ is called* Cellina continuously approximable *(abbr. CCA) if for every compact subset $K$ of $X$*

(1) *$Gr(F \mid_K)$ is compact;*
(2) *there exists a sequence $\{f_j\}_{j=1}^{\infty}$ of single-valued continuous maps $f_j : K \to Y$ that graph-converges to $F \mid_K$.*

**Example 2.4.** Consider $F : \mathbb{R} \twoheadrightarrow \mathbb{R}$ such that

$$F(x) = \begin{cases} -1 & \text{if } x < 0 \\ [-1, 1] & \text{if } x = 0 \\ 1 & \text{if } x > 0 \end{cases}$$

The $F$ is CCA. The following figure shows how to approximate $F$ by continuous functions.

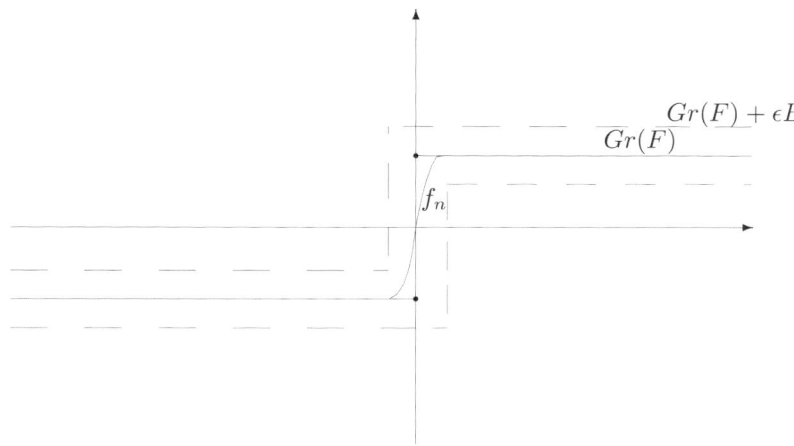

**Definition 2.5.** *Let $F : \mathbb{R}^m \twoheadrightarrow \mathbb{R}^n$ be a set-valued map, $\overline{x} \in \mathbb{R}^m$, $\overline{y} \in \mathbb{R}^n$, $\overline{y} \in F(\overline{x})$ and let $\Lambda$ be a nonempty compact subset of $\mathbb{R}^{n \times m}$. Let $S$ be a subset of $\mathbb{R}^m$. We say that $\Lambda$ is a* generalized differential quotient *(GDQ) of $F$ at $(\overline{x}, \overline{y})$ in the direction $S$, and write $\Lambda \in GDQ(F; \overline{x}, \overline{y}; S)$ if for every positive real number $\delta$ there exist $U, G$ such that*

1. *$U$ is a compact neighborhood of $0$ in $\mathbb{R}^m$ and $U \cap S$ is compact;*
2. *$G$ is a CCA set-valued map from $\overline{x} + U \cap S$ to the $\delta$-neighborhood $\Lambda^\delta$ of $\Lambda$ in $\mathbb{R}^{n \times m}$;*
3. *$G(x) \cdot (x - \overline{x}) \subseteq F(x) - \overline{y}$ for every $x - \overline{x} \in U \cap S$*

A multifunction $F$ may have many GDQs at $(\overline{x}, \overline{y})$. However we have the following:

**Theorem 2.6 (Minimality Theorem [9]).** *If the set of GDQs of a set-valued map $F$ at a point $(\overline{x}, \overline{y})$ in the direction $S$ is not empty, then there exists in this set at least one minimal GDQ at this point in the direction $S$ in the sense of inclusion of sets.*

**Corollary 2.7.** *Every element $\Lambda$ of $GDQ(F; \overline{x}, \overline{y}; S)$ contains a minimal element of $GDQ(F; \overline{x}, \overline{y}; S)$ in the sense of inclusion of sets.*

**Example 2.8.** *Let $F : \mathbb{R} \twoheadrightarrow \mathbb{R}$ be a set-valued map such that*

$$F(x) = \begin{cases} [-|x|, |x|] & \text{if } x \neq 0 \\ \{0\} & \text{if } x = 0 \end{cases}$$

Then any singleton $\{a\}$ for $a \in [-1, 1]$ is a minimal GDQ of $F$ at the point $(0, 0)$.

Let $X$ be a normed space. Recall that the *contingent cone* (the "Bouligand cone") to a set $C \subset X$ at $x$ is defined by

$$T_C(x) = \left\{ w \in X : \liminf_{t \downarrow 0} \frac{dist(x + tw, C)}{t} = 0 \right\}.$$

**Definition 2.9.** *Let $F : K \twoheadrightarrow Y$, $K \subset X$, and $F(x)$ be nonempty for all $x \in K$. The* contingent derivative *$DF(x_0, y_0)$ of $F$ at $x_0 \in K$ and $y_0 \in F(x_0)$ is a set-valued map from $X$ to $Y$ whose graph is the contingent cone $T_{Gr(F)}(x_0, y_0)$ to the graph of $F$ at $(x_0, y_0)$.*

In other words, $v_0 \in DF(x_0, y_0)(u_0) \iff (u_0, v_0) \in T_{Gr(F)}(x_0, y_0)$.

**Theorem 2.10 ([10]).** *Let $F : \mathbb{R} \twoheadrightarrow \mathbb{R}^n$, $Do(F) = \mathbb{R}$, and $\Lambda \in GDQ(F; x_0, y_0; \mathbb{R}_+)$. If $\Lambda$ is minimal, then $\Lambda \subseteq DF(x_0, y_0)(1)$.*

**Corollary 2.11.** *Consider $F : \mathbb{R} \twoheadrightarrow \mathbb{R}$. If $F$ is GDQ-differentiable at the point $(x, y)$ in the direction of $\mathbb{R}_+$ ($\mathbb{R}_-$) then there exists the contingent derivative $DF(x, y)(1)$ ($DF(x, y)(-1)$).*

**Definition 2.12.** *We say that* $F : G \twoheadrightarrow \mathbb{R}^n$, *where* $G \subset T \times \mathbb{R}^n$, *is a Scorza-Dragoni set-valued map if for every* $\varepsilon > 0$ *there exists a closed set* $T_\varepsilon \subset T$ *such that* $\lambda(T \backslash T_\varepsilon) \leq \varepsilon$ *and the multifunction* $(t, y) \longmapsto F(t, y)$ *is upper semi-continuous (u.s.c.) on* $(T_\varepsilon \times \mathbb{R}^n) \cap G$.

Let us define a set-valued map $SGDQ(K; t, x; \mathbb{R}_+)$ as the closure of the union of all minimal GDQs of $K$ at $(t, x) \in Gr(K)$ in the direction $\mathbb{R}_+$.

**Definition 2.13.** *Let* $K : T \twoheadrightarrow \mathbb{R}^n$, *where* $T = [a, b]$ *for* $a, b \in \mathbb{R}$. *We say that* $K$ *is GDQ-regular if*

1. $K$ *is GDQ-differentiable in the direction* $R_+$ *at every* $t \in [a, b)$ *and every* $y \in K(t)$
2. $K$ *is left u.s.c on* $T$
3. $K$ *has closed values*
4. $(t, x) \longmapsto SGDQ(K; t, x; \mathbb{R}_+)$ *is a Scorza-Dragoni set-valued map.*

The following theorem is often used in the proofs of viability theorems. It allows to reduce the original problem with weak assumptions on $F$ to another one, where a modified orientor field $F_0$ is Scorza-Dragoni.

**Theorem 2.14 (Jarnik-Kurzweil, [13]).** *Assume that* $G \subseteq \mathbb{R} \times Y$, $F : G \twoheadrightarrow \mathbb{R}^n$, *where* $Do(F) = G$, *and that for almost all* $t$ *the map* $F(t, \cdot)$ *is upper semicontinuous with compact convex values. Then there exists a Scorza-Dragoni multifunction* $F_0 : G \twoheadrightarrow \mathbb{R}^n$ *with compact convex values satisfying* $F_0(t, y) \subseteq F(t, y)$ *for* $(t, y) \in G$, *and such that if* $T \subset \mathbb{R}$ *is measurable* , $u : T \to Y$ *and* $v : T \to \mathbb{R}^n$ *are measurable maps such that* $v(t) \in F(t, u(t))$, *for almost all* $t \in T$, *then* $v(t) \in F_0(t, u(t))$ *for almost all* $t \in T$.

### 2.1. Viability result

Let $K : T \twoheadrightarrow \mathbb{R}^n$, where $Do(K) = [0, 1] = T \subset \mathbb{R}$, be a constraint multifunction and $F : Gr(K) \twoheadrightarrow \mathbb{R}^n$, where $Do(F) = Gr(K)$, be an orientor field (i.e., multivalued vector field). Consider the multivalued Cauchy problem as follows:

$$\begin{cases} \dot{x}(t) \in F\big(t, x(t)\big), & \text{a.e. on } T \\ x(t_0) = x_0 \, . \end{cases} \qquad (2.2)$$

We impose the following assumptions:

**H(K):** for any $t_0 \in [0, 1)$ and $x_0 \in K(t_0)$ there exists $K_{(t_0, x_0)} : [t_0, 1] \twoheadrightarrow \mathbb{R}^n$ such that $K_{(t_0, x_0)}$ is GDQ-regular and $K_{(t_0, x_0)}(t) \subseteq K(t)$ for every $t \in [t_0, 1]$.

**H(F):**   (i) $F : GrK \twoheadrightarrow \mathbb{R}^n$ has closed convex values;

(ii) for any measurable $\gamma(\cdot)$ the multifunction $t \longmapsto F(t, \gamma(t))$ is measurable;

(iii) for any $t_0 \in [0, 1]$ and $x_0 \in K(t_0)$ there exists $\alpha \in L^1(t_0, 1)$ such that for any $(t, x) \in GrK_{(t_0, x_0)}$, $\|F(t, x)\| \leq \alpha(t)(1 + \|x\|)$;

(iv) for any $(t_0, x_0) \in GrK$ and for any $t \in [t_0, 1]$ the multifunction $x \longmapsto F\big|_{GrK_{(t_0, x_0)}}(t, x)$ is u.s.c.

**H:** for any $(t_0, x_0) \in GrK$, for almost every $t \in [0, 1]$ and for any $x \in K_{(t_0, x_0)}(t)$,

$$F(t, x) \cap SGDQ\big(K_{(t_0, x_0)}; t, x; \mathbb{R}_+\big) \neq \emptyset \, .$$

Now we can state the main result of this paper.

**Theorem 2.15.** *Assume that $H(K)$, $H(F)$ and $H$ hold. Then for any $(t_0, y_0) \in GrK$ the multivalued Cauchy problem (2.2) has a solution $y : [t_0, 1] \to \mathbb{R}^n$, which is an absolutely continuous function satisfying $y(t) \in K(t)$ for every $t \in [t_0, 1]$.*

The idea of the proof of Theorem 2.15 is similar to that of [10] and [12], and consists of reducing the problem to the 'almost u.s.c case' studied in [5].

**Example 2.16.** Observe that for $K$ and $F$ from Example 2.1 the assumptions of Theorem 2.15 are satisfied. In particular, for $t_0 \in [-1, 0]$ and $x_0 \in K(t_0)$ one can take

$$K_{(t_0, x_0)}(t) = \begin{cases} \{|t \sin \frac{1}{t}|\} \cup \{0\} & \text{if } t \in [t_0, \tau_0] \\ 0 & \text{if } t \in (\tau_0, 0] \end{cases}$$

where $\sin \frac{1}{\tau_0} = 0$. Then $F \mid_{K_{(t_0, x_0)}}$ is bounded by an integrable function.

# References

[1] J. P. Aubin, *Viability Theory*, Birkhäuser, Boston Basel Berlin, 1991.

[2] J. P. Aubin, *Contingent Derivatives of Set-Valued Maps and Existence of Solutions to Nonlinear Inclusions and Differential Inclusions*, in: L. Nachbin (ed.), Adv. Math., Suppl. Stud., Academic Press, Orlando, (1981), 160–232.

[3] J. P. Aubin, A. Cellina, *Differential Inclusions*, Springer-Verlag, Berlin Heidelberg New York Tokyo 1984.

[4] J. P. Aubin, H. Frankowska, *Set-Valued Analysis*, Birkhäuser, 1990.

[5] D. Bothe, *Multivalued differential equations on graphs*, J. Nonlinear Analysis, TAMS **18** (1992), 245–252.

[6] K. Deimling, *Mutivalued Differential Equations*, Walter de Gruyter, Berlin New York, 1992.

[7] H. Frankowska, S. Plaskacz, T. Rzeżuchowski, *Measurable viability theorems and the Hamilton-Jacobi-Bellman Equation*, Journal of Differential Equations **116** (1995), 265–305.

[8] E. Girejko, *On generalized differential quotients of set-valued maps*, Rendiconti del Seminario Matematico dell'Universita' e del Politecnico di Torino **63** no. 4 (2005), 357–362.

[9] E. Girejko, B. Piccoli, *On some concepts of generalized differentials* to appear in Set-valued analysis, 2006.

[10] E. Girejko, Z. Bartosiewicz, *Viability and generalized differential quotients*, to appear in Control and Cybernetics, 2006.

[11] Sh. Hu, S. N. Papageorgiou, *Handbook of Multivalued Analysis. Vol. I: Theory*, Kluwer Academic Publishers, Dordrecht/Boston/London 1997.

[12] Sh. Hu, S. N. Papageorgiou, *Handbook of Multivalued Analysis. Vol. II: Applications*, Kluwer Academic Publishers, Dordrecht/Boston/London 1997.

[13] J. Jarnik, J. Kurzweil, *On conditions on right hand sides of differentials relations*, Casopis Pest. Mat. **102**, 1968, 334–349.

[14] T. Rzeżuchowski, *Scorza-Dragoni type theorem for upper-semicontinuous multivalued functions*, Bull. Acad. Polonaise des Science **28**, no. 1–2, 1980, 61–65.

[15] H. J. Sussmann, *New theories of set-valued differentials and new version of the maximum principle of optimal control theory*, in Nonlinear Control in the Year 2000, A. Isidori, F. Lamnabhi-Lagarrigue and W. Respondek, Eds., Springer-Verlag, London 2000, 487–526.

[16] H. J. Sussmann, *Warga derivate containers and other generalized differentials*, Proceedings of the 41stIEEE 2002 Conference on Decision and Control, Las Vegas, Nevada, December 10–13, 2002 , 1101–1106.

Ewa Girejko and Zbigniew Bartosiewicz
Department of Computer Science
Białystok Technical University
Zwierzyniecka 14
PL-15-333 Białystok
Poland
e-mail: `egirejko@pb.bialystok.pl`
       `bartos@pb.bialystok.pl`

Progress in Nonlinear Differential Equations
and Their Applications, Vol. 75, 231–240

# Nonlinear Prediction in Riverflow – the Paiva River Case

Rui Gonçalves, Alberto A. Pinto, and Francisco Calheiros

*Dedicated to Arrigo Cellina and James Yorke*

**Abstract.** We exploit ideas of nonlinear dynamics in a non-deterministic dynamical setting. Our object of study is the observed riverflow time series of the Portuguese Paiva river whose water is used for public supply. The Takens delay embedding of the daily riverflow time series revealed an intermittent dynamical behaviour due to precipitation occurrence. The laminar phase occurs in the absence of rainfall. The nearest neighbour method of prediction revealed good predictability in the laminar regime but we warn that this method is misleading in the presence of rain. The correlation integral curve analysis, Singular Value Decomposition and the Nearest Neighbour Method indicate that the laminar regime of flow is in a small neighbourhood of a one-dimensional affine subspace in the phase space. The Nearest Neighbour method attested also that in the laminar phase and for a data set of 53 years the information of the current runoff is by far the most relevant information to predict future runoff. However the information of the past two runoffs is important to correct non-linear effects of the riverflow as the MSE and MRE criteria results show. The results point out that the Nearest Neighbours method fails when used in the irregular phase because it does not predict precipitation occurrence.

**Mathematics Subject Classification (2000).** Primary 93C57; Secondary 93C10.

**Keywords.** Dynamical systems, phase-space reconstruction, hydrology.

## 1. Introduction

The direct link between deterministic dynamical systems theory and the real world is the analysis of real systems time series in terms of nonlinear dynamics. Great advances have been made to exploit ideas of dynamical systems theory in cases where the system is not necessarily deterministic but it displays a structure not captured by classical stochastic methods. The application of dynamical systems methods found a firm ground on the reconstruction theorem of Takens [16] and in the probabilistic justification due to Sauer, Yorke and Casdagli, [14]. The motivation for researchers such as, [2, 10, 11, 15] to apply methods of deterministic systems in riverflow time series lies in the natural tendency of river systems to present recurrent behaviour.

We start by doing a Takens delay coordinates reconstruction, [14, 16] of the daily flow series. When considering the entire data set, the false nearest neighbours show a low percentage of false neighbours for embedding dimensions above or equal to 6 being the interval on one day the best delay time. The correlation curve analysis and the non-linear prediction results revealed the existence of two dynamically different regimes. These findings led us to conclude that the Paiva river is an intermittent system. This intermittent dynamical behaviour is not of a deterministic type because rainfall is a stochastic and not a deterministic forcing. The laminar phase occurs in the absence of rainfall and the irregular phase occurs under the action of rain. The nearest neighbour method of prediction revealed good predictability in the laminar regime. However since 75% of data is laminar the use of nonlinear methods can be misleading when both dynamical regimes are considered. These features of data are already visible in the Histogram plot. The Non-parametric Nearest Neighbour Method of prediction indicates that the laminar regime of flow is in a small neighbourhood of a one-dimensional affine subspace in the phase space. The prediction results for the laminar phase revealed that it is essential to know the current runoff to predict future values. Also the small improvement in prediction when the former two runoffs are used is a consequence of nonlinear effects as shown by the MSE and MRE criteria. The Nearest Neighbour method show better performance when compared with other methods applied in former works [2]. The results of the Nearest Neighbours show also that it can still be improved by tuning the neighbourhood radius or the local number of neighbours.

## 1.1. Data and preliminary analysis

The most relevant data for this work consist of the time series of mean daily runoff of the Paiva river, measured at Fragas da Torre section, North of Portugal. They are available for download in the *Instituto Nacional da Água* webpage[1]. The sample period runs from 1st of October of 1946 to 30th of September of 1999 for a total of 19358 observations (see chronogram of Figure 1). The riverflow of Paiva is the closest to a natural flow one might expect. The Paiva river has a small basin of about $700Km^2$ and it is not an runoff intermittent river in the sense that at the referred location and in the 53 years of observation the surface stream never disappeared. In Figure 1 is shown a partial chronogram of the time series.

The daily river flow descriptive statistics, Table 1 shows the strong asymmetry of the data. The Paiva basin does not have regulators such as dams or glaciers and is also a mountain river with a rocky bed reacting very fast to rainfall.

---

[1]http://www.inag.pt

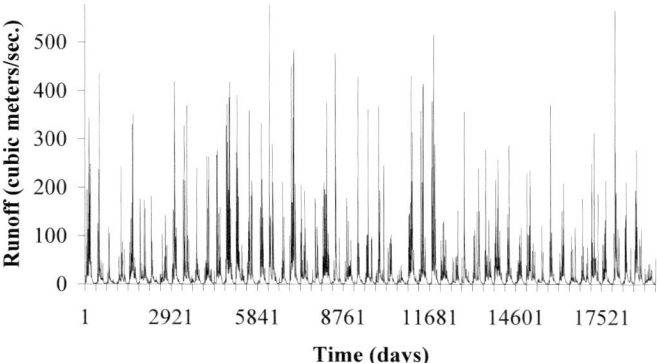

FIGURE 1. Chronogram of the daily mean riverflow of Paiva measured at Fragas da Torre 1946–99.

TABLE 1. Descriptive statistics for the daily mean riverflow series of Paiva (1946-99) measured at Fragas da Torre.

| Statistic | Value |
|-----------|-------|
| Mean | 20.73 $m^3/s$ |
| Median | 5.66$m^3/s$ |
| Skewness | 5.3 |
| Kurtosis | 45.98 |
| Maximum | 920.0 $m^3/s$ |
| Minimum | 0.06$m^3/s$ |

The average and the standard deviation for each day of the year (Figure 3)[2] explains the behaviour of the sample autocorrelation function (ACF). The sample ACF (Figure 2) is characterized by seasonality but is it locally very irregular due to the differences between the years.

## 2. Dimension estimation

We start by doing a reconstruction embedding [16], using the Paiva river 1946–99 daily mean data for several embedding dimensions. Our goal is to understand the dynamics of riverflow. The dynamic characterization includes invariant estimation and in this direction we do a Correlation-Integral (CI) Analysis for all the data and then we consider only the runoffs of the laminar phase (less than $20m^3/s$) which represents about 75% of the data corresponding mainly to the periods without rain, Figure 6. We realized that the CI slopes are close to 1 for laminar runoffs.

---

[2]The observations of the days 29th of February were deleted from the time series.

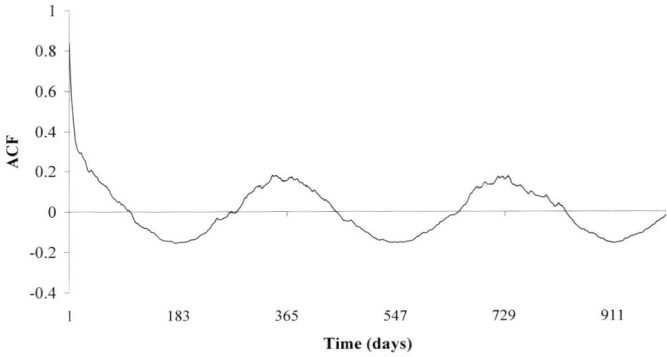

FIGURE 2. Sample Autocorrelation Function of the daily mean
runoff series (1946-99) of the Paiva river at Fragas da Torre.

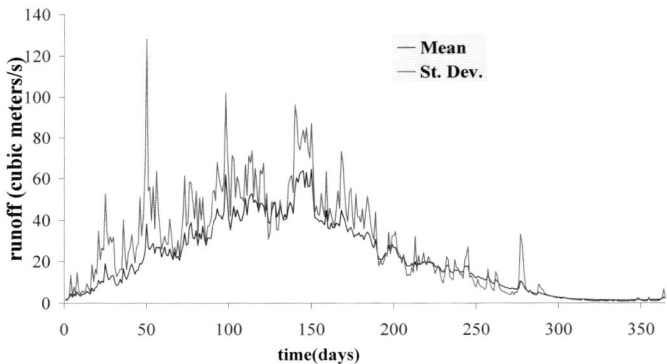

FIGURE 3. Mean and standard deviation for each day of the year.

This fact is confirmed by SVD analysis showing that the dynamics of runoffs of
the laminar phase are close to a segment line.

### 2.1. Correlation-integral analysis

The sample CI, $C_N^{(m)}(\varepsilon)$ of a reconstructed system is defined by,

$$C_N^{(m)}(\varepsilon) = \frac{2}{N(N-1)}\Theta\{(i,j) : 1 \leq i < j \leq N, \|X_i - X_j\| < \varepsilon\} \qquad (2.1)$$

where $(X_t, X_{t+1}, \ldots, X_{t+m-1})$ is a vector reconstructed with the values of the
time series, $\{X_t\}_{t=1}^{N}$, $N$ is the number of data points of the series, $\Theta$ the Heaviside
function, $\varepsilon$ the neighbourhood radius and $m$ the embedding dimension of the
reconstructed phase space. The sample CI as a statistic for the estimation of
the correlation dimension was proposed by [3]. The sample CI is the fraction of
reconstruction vectors at a distance smaller than $\varepsilon$ in the reconstructed phase

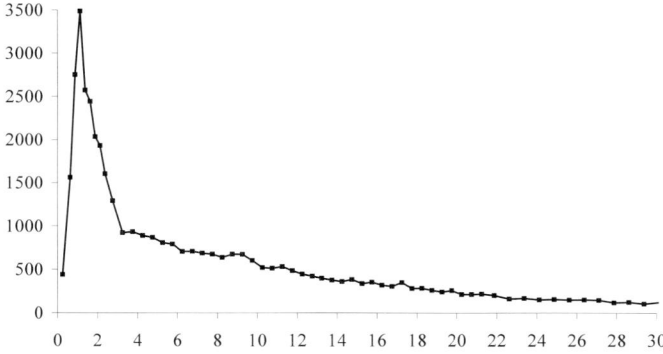

FIGURE 4. Histogram of the mean daily runoff series of Paiva river.

space. The sum, (2.1), is computed for a set of distances, $\varepsilon_1, \ldots, \varepsilon_n$ evenly spaced on a logarithmic scale. A *scaling range* is said to exists if for such a range of values the sample correlation integral behaves like a power law or the same to say like a line on a $\log - \log$ scale. In practice there is a cut-off on the radius size due to data size restrictions.

$$C_N^{(m)}(\varepsilon) \sim \alpha\varepsilon^{D_C}, \, \varepsilon \to 0, \, N \to \infty \tag{2.2}$$

For a constant $\alpha$, $d(N, \varepsilon)$ is the slope of the CI curve for a certain range and $D_C$ is then the estimate of the correlation dimension,

$$d(N, \varepsilon) = \frac{\partial \ln C_N^{(m)}(\varepsilon)}{\partial \ln \varepsilon} \quad \text{and} \quad D_C = \lim_{\varepsilon \to 0^+} \lim_{N \to \infty} d(N, \varepsilon). \tag{2.3}$$

The false nearest neighbours analysis, [8], using the parameter value suggested by the same authors, indicates as adequate an embedding dimension above 5, Figure 5.

On Figure 6 we present the correlation integral slopes. We can see three different behaviours in the correlation-integral curve for different ranges of the radius, $\varepsilon$. The runoff values larger than $30m^3/s$ no scaling range exists. For runoffs in the interval $[5 - 30m^3/s]$ there is a scaling range which indicates dimension 1 for the attractor. This dimension is not fractal and indicates that the behaviour of riverflow for that range is close to that of a curve. This indicates the existence in the reconstructed phase-space of a one-dimensional manifold to which all the laminar phase orbits are close. It may be said that the orbits near this one-dimensional manifold constitute a $\varepsilon$-neighbourhood.

## 2.2. Singular value decomposition analysis

This information given by this analysis is highly relevant for the understanding of the correlation integral curve. We take as vector variable the reconstruction vectors, $(X_t, X_{t+1}, \ldots, X_{t+m-1})$, where $X_t$ is the daily mean runoff at day $t$. Using

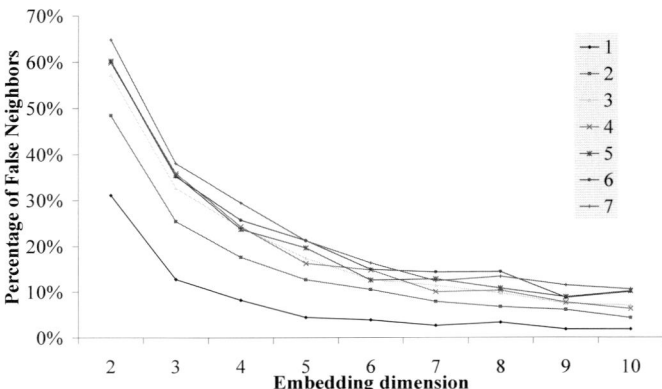

FIGURE 5. False nearest neighbours curves for the mean daily runoff series of Paiva river and for several delay times.

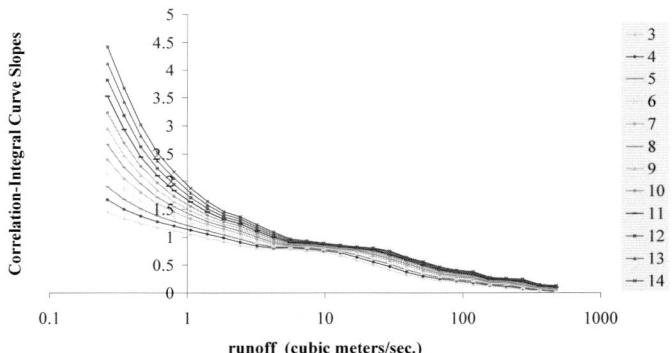

FIGURE 6. Slopes of the sample correlation integral curve of the Paiva river data and for several embedding dimensions.

the *SPAD* statistical package to perform the SVD one computes the principal directions of the data set and corresponding weights and also calculated the principal factors for the covariance matrix for the laminar phase runoff series for different embedding dimensions. The values are presented on Table 2.

The correlations between the principal components and the original variables are presented on the Table 3. In the laminar phase there exists a principal component explaining more than 90% of the variance. This is explained by a laminar dynamic close to a segment line, Figure 6. According to the usual criteria to quantify the number of significant eigenvalues, (see [13]) the reconstruction vectors (*individuals*) of the last two data sets are almost one-dimensional.

TABLE 2. Percentage (%) of the total variance explained by the three largest eigenvalues of the Covariance Matrix for the laminar phase.

| Dimension | % | % | % |
|---|---|---|---|
| 3 | 96.34 | 2.69 | 0.96 |
| 4 | 95.01 | 3.23 | 1.13 |
| 5 | 93.82 | 3.70 | 1.33 |
| 6 | 92.75 | 4.13 | 1.50 |
| 7 | 91.75 | 4.52 | 1.66 |

TABLE 3. Correlation between the 1st Principal Component and the original variables at the laminar phase and for several embedding dimensions.

| Variable | 3 | 4 | 5 | 6 | 7 |
|---|---|---|---|---|---|
| $X_t$ | 0.98 | 0.97 | 0.96 | 0.95 | 0.94 |
| $X_{t+2}$ | 0.99 | 0.98 | 0.98 | 0.97 | 0.96 |
| $X_{t+3}$ | 0.98 | 0.98 | 0.98 | 0.98 | 0.97 |
| $X_{t+4}$ | - | 0.97 | 0.98 | 0.98 | 0.97 |
| $X_{t+5}$ | - | - | 0.96 | 0.97 | 0.97 |
| $X_{t+6}$ | - | - | - | 0.95 | 0.96 |
| $X_{t+7}$ | - | - | - | - | 0.94 |

## 2.3. Nonlinear prediction

Several authors used nonlinear prediction methods for river flow data locally in the phase space, [6, 7, 10, 12] and [4] among others.

In this work we use a different version of the nearest neighbours method proposed by [9] to predict the next day runoff for the years 1997/98 using the information of the historic series from 1946 to 1999. The prediction set is the phase space average of the neighbour's images. Other authors, [10] reported that predictors based on the average give better results than other local linear functions. Taking into account the findings of the former section we started by considering small embeddings and a time delay of one day. On Table 4 we present a summary of the usual fitting evaluation criteria for the one-step ahead prediction in the laminar phase.

Instead of using all the neighbours within a fixed radius we have used the the ten closest neighbours. For this data it gives better results due to statistical reasons. In Figure 7 we can see that for a two-dimensional embedding in the laminar phase the 10 closest neighbours are close to the central values of the sample conditional distributions of $X_{t+1}$ given $X_t = x$. This is still true for dimension 3 but for higher dimensions the prediction accuracy decreases.

TABLE 4. Mean Relative Error (MRE) and MSE for the one-step ahead prediction for the years 1995-99 and also for the laminar regime. Several embedding dimensions were considered.

| m | MRE | MSE |
|---|-----|-----|
| 1 | 0.0486 | 2.915 |
| 2 | 0.0328 | 1.004 |
| 3 | 0.0301 | 0.888 |
| 4 | 0.0310 | 1.034 |
| 5 | 0.0341 | 1.155 |
| 6 | 0.0378 | 1.193 |
| 7 | 0.0427 | 1.375 |
| 8 | 0.0472 | 1.651 |
| 9 | 0.0523 | 1.605 |
| 10 | 0.0584 | 2.063 |

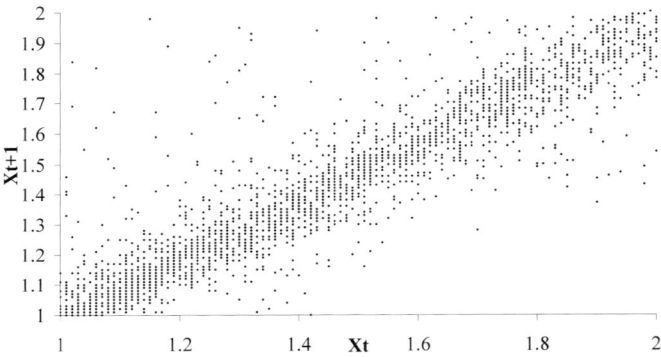

FIGURE 7. Phase space portrait for runoff values between 1 and 2 cubic meters/s.

The best Mean Square Error[3] (MSE) result was found for an embedding dimension 3 and for time delay of one day, (see Table 4) as predicted by the false nearest neighbours method. Here we should mention the paper on river flow prediction, [4] where the authors also obtained results of the same magnitude for the *MSE* for different embedding dimensions.

Predicting locally in the phase space with linear functions can be misleading if the system has an intermittent behaviour. In laminar phase the flow converges slowly to an equilibrium. This convergence is of stochastic nature but reveals strong

---

[3]The Mean Square Error of Prediction (MSE) is defined by MSE $= \frac{1}{n} \sum_{i=1}^{n} \left( X_t - \hat{X}_t \right)$. When several models are proposed for the same data the ultimate choice of one may depend on goodness of fit such as the MSE.

determinism. If a rain event occurs then the flow increases and the system starts a much more erratic behaviour outside the laminar phase.

## 3. Conclusions

A Dynamical analysis of the Paiva river data was performed using Takens method of dynamical reconstruction, [16]. Later we used the Nearest Neighbours method of prediction for one-step ahead prediction. The results indicate a gain on quality prediction when one considers only the laminar phase. These differences are due to the action of rain that seems to be unpredictable when we are dealing with daily mean runoff data. The prediction results also reveal close MSE for different embedding dimensions of the phase space and the dimension 3 has been proven to be the best. We noted also that the principal component analysis of the reconstruction vectors confirmed the correlation curve analysis. This means that the information given by recent past runoffs is not sufficient to improve prediction accuracy. For the Paiva systems the optimal embedding dimension for the laminar phase is dictated by statistical rather than dynamical reasons.

## References

[1] K. Alligood, T. Sauer, and J. Yorke *Chaos – An Introduction to Dynamical Systems*, Springer, New York, (1996)

[2] F. Calheiros and R. Gonçalves, *Previsão em Hidrologia*, Literacia e Estatística – Actas do X Congresso Anual da SPE, P. Brito, A. Figueiredo, F. Sousa, P. Teles, and F. Rosado, (2003), 229–241.

[3] P. Grassberger and I. Procaccia, *Measuring the strangeness of strange attractors*, Physica 9D, **9**, (1983), 189–208.

[4] S. Islam and B. Sivakumar *Characterization and prediction of runoff dynamics: a nonlinear dynamical view*, Journal of the American Water Resources Association, **25**, (2002), 179–190.

[5] A. Jayawardena, A. Gurung, *Noise reduction and prediction of hydrometereological time series dynamical systems approach vs stochastic approach*, Journal of Hydrology, **228**, (2000), 242–64.

[6] A. W. Jayawardena and F. Lai, *Analysis and prediction of chaos in rainfall and stream flow time series*, Journal of Hydrology, **153**, (1994), 23–52.

[7] A. Jayawardena, W. Li, and P. Xu, *Neighbourhood selection for local modelling and prediction of hydrological time series*, Journal of Hydrology, **258**, (2002), 40–57.

[8] M. Kennel, R. Brown, and H. Abarbanel, *Determining minimum embedding dimension for phase space reconstruction using a geometrical construction*, Phys. Rev. E **48(3)**, (1992), 1752–1763.

[9] H. Kantz and T. Schreiber, *Nonlinear Time Series Analysis*, Cambridge Univ. Press, (1997).

[10] Q. Liu, S. Islam, I. Rodriguez-Iturbe, and Y. Le *Phase-space analysis of daily stream-flow: characterization and prediction*, American Water Resources Association, **210**, (1998), 463–475.

[11] L. Porporato and L. Ridolfi *Nonlinear analysis of a river flow time sequences*, Water Resources Research, **33**, (1997), 1353–1367.

[12] A. Porporato and L. Ridolfi, *Multivariate nonlinear prediction of river flows*, Journal of Hydrology, **248**, (2001), 109–122.

[13] G. Saporta, *Probabilités, Analyse des Données Et Statistique*, Editions Technip, Paris, (1990).

[14] T. Sauer, J. Yorke, and M. Casdagli, *Embedology*, Journal of Statistical Physics, **65**, (1991), 579–616.

[15] B. Sivakumar, *Chaos in hydrology: important issues and interpretations*, J. of Hydrology, **227**, (2000), 1–20.

[16] F. Takens, *Detecting strange attractors in Turbulence*. In Lecture Notes in Mathematics, **898**, 366–81, Springer, D. A. Rand, L. Young, Editors (1980).

Rui Gonçalves
Faculdade de Engenharia
R. Dr. Roberto Frias,
P-4200-465 Porto
Portugal
e-mail: rjasg@fe.up.pt

Alberto A. Pinto
Faculdade de Ciências
R. do Campo Alegre,
P-4169-007 Porto
Portugal
e-mail: aapinto@fc.up.pt

Francisco Calheiros
Faculdade de Engenharia
R. Dr. Roberto Frias,
P-4200-465 Porto
Portugal
e-mail: xico@fe.up.pt

Progress in Nonlinear Differential Equations
and Their Applications, Vol. 75, 241–246

# Shadowing in Higher Dimensions

Judy Kennedy and James A. Yorke

**Abstract.** This paper presents methods using algebraic topology for showing that pseudo-trajectories are close to trajectories of a dynamical system. Our emphasis is the case where the trajectories are unstable in two or more dimensions. We develop the algebraic topology for guaranteeing the existence of such trajectories.

## 1. Introduction

When studying chaotic dynamical systems numerically, errors are introduced, so the result is called a pseudo-trajectory rather than an actual trajectory of the system being studied. Such errors will have an effect that grows exponentially fast with time. In some cases, the pseudo-trajectory that is observed is close to an actual trajectory of the system and sometimes it is far from all trajectories of the system being studied. Numerical investigations of chaotic systems over long times (like studies of climate) often aim at revealing statistical properties and the choice of initial condition is unimportant. In such situations, finding that a pseudo-trajectory is near some actual trajectory for a long time gives the investigator confidence in the pseudo-trajectory. Beginning with [2, 3, 5], a number of papers have worked to show that the observed pseudo-trajectories are close to a true trajectory over a long specified time interval. [5], for example, focused on solutions of ordinary differential equations. Such "shadowing" properties often fail. Indeed, Yuan and Yorke [7] showed that there is an open set of maps for which every point is not shadowable. The above papers have aimed at establishing such shadowing properties in cases where the dynamics are one-dimensionally unstable. Our goal here is to establish methods for showing that pseudo-trajectories are close to a trajectory in cases in which the dynamics are unstable in two directions. Our methods follow those we introduced in [4].

In writing this paper, we assume the reader has studied some cohomology theory, though not necessarily recently. We could have used homology theory but we prefer Čech–Alexander–Spanier cohomology theory (as presented by Spanier [6] and Eilenberg and Steenrod [1]) because of its stronger properties and have chosen to use it here.

We will say $(A, B)$ is a *pair* if $A$ and $B$ are compact and $B \subset A$. If $(C, D)$ is a pair we write $f : (A, B) \to (C, D)$ to mean $A$ is the domain of $f$ and $f(A) \subset C$ and $f(B) \subset D$. All maps are continuous in the paper.

It is perhaps easiest to think about the cohomology of a pair $(A, B)$ as the cohomology of the pair that results if the set $B$ is collapsed to a point. Hence, if $A = [0, 1]$ and $B$ is $\{0, 1\}$, identifying 0 with 1 results topologically in a circle or rather the pair $(S^1, \{b\})$ where $b \in S^1$.

If $A, B$ are compact and $B \supset A$, the corresponding *inclusion* map (for $A$ and $B$) is denoted $i : A \to B$, and is defined by $i(a) = a$ for all $a \in A$. Similarly, a *(pair) inclusion* $i : (A, B) \to (A', B')$ is defined if $A \subset A'$ and $B \subset B'$. We use cohomology groups with coefficients in the integers $\mathbf{Z}$. We also use the symbol $j$ to denote inclusion maps, as is customary, and in case several inclusion maps are being considered, we use subscripts (e.g., $i_1$ or $j_2$) to avoid confusion.

An *upper sequence* of groups is a sequence $(G^i, \phi^i)$ where for each $i$, $G^i$ is a group and $\phi^i : G^i \to G^{i-1}$ is a homomorphism. An upper sequence is *exact* if for each integer $i$, $\phi^i(G^i)$ is the kernel of $G^{i+1}$ with respect to $\phi^{i+1}$. The sequence is of *order* 2 if the composition of any two successive homomorphisms of the sequence yields the trivial homomorphism.

If $X$ is a space, define $(A, B) \times X := (A \times X, B \times X)$. Let $I$ denote the unit interval $[0, 1]$. Two maps $f, g : (A, B) \to (C, D)$ are said to be *homotopic* if there is a map $H : (A, B) \times I \to (C, D)$ such that $f(x) = H(x, 0)$ and $g(x) = H(x, 1)$ for each $x \in A$. For $t \in I$, $H_t$ denotes the map defined by $H_t(x) = H(x, t)$ for $x \in A$. A pair $(A, B)$ contained in a pair $(C, D)$ is called a *retract* of $(C, D)$ if there exists a map $r : (C, D) \to (A, B)$ such that $r(x) = x$ for each $x$ in $A$. The map $r$ is called a *retraction*. The pair $(A, B)$ is a *deformation retract* of $(C, D)$ if there is a retraction $r : (C, D) \to (A, B)$ and the composition $r \circ i$, where $i : (A, B) \to (C, D)$ is the identity, is homotopic to the identity map $(C, D) \to (A, B)$. The pair $(A, B)$ is a *strong deformation retract* of $(C, D)$ if the latter homotopy can be chosen to leave each point of $A$ fixed (i.e., $H(x, t) = x$ for $x \in A$). The pairs $(A, B)$ and $(C, D)$ are *homotopically equivalent* if there exist maps $f : (A, B) \to (C, D)$ and $g : (C, D) \to (A, B)$ such that $f \circ g$ is homotopic to the identity on $(C, D)$ and $g \circ f$ is homotopic to the identity on $(A, B)$.

For convenience, we list the axioms of cohomology and some other facts that we use ([1] and [6]): Suppose $(X, A)$, $(Y, B)$, and $(Z, C)$ are compact pairs. If $f : (X, A) \to (Y, B)$ is continuous, then for each integer $k$, $f$ induces a homomorphism $f_k^* : H^k(Y, B) \to H^k(X, A)$. As is customary, we depend on context to tell which of the homomorphisms induced by $f$ is intended, and write only $f^* : H^k(Y, B) \to H^k(X, A)$. For the pair $(X, A)$, and integer $q$, $H^q(X, A)$ is the *q-dimensional relative cohomology group* of $X$ mod $A$. Cohomology groups are abelian groups; our coefficient group is the group of integers $\mathbf{Z}$ (thus this is also suppressed in the notation).

**Axiom 1c.** If $f$ is the identity function on $(X, A)$, then $f^*$ is the identity iso-morphism.

**Axiom 2c.** If $f : (X, A) \to (Y, B)$ and $g : (Y, B) \to (Z, C)$, then $(g \circ f)^* = f^* \circ g^*$.

**Axiom 3c.** The coboundary operator, denoted by $\delta$, is a homomorphism from $H^{k-1}(A)$ to $H^k(X, A)$ with the property that $\delta \circ (f \mid A)^* = f^* \circ \delta$. (Again, the notation is ambiguous, and we rely on context to determine which groups and which homomorphism is intended.)

**Axiom 4c.** (*Partial exactness*) If $i : A \to X, j : X \to (X, A)$ are inclusion maps, then the upper sequence of groups and homomorphisms

$$\cdots \xrightarrow{i^*} H^{k-1}(A) \xrightarrow{\delta} H^k(X, A) \xrightarrow{j^*} H^k(X) \xrightarrow{i^*} H^k(A) \xrightarrow{\delta} \cdots$$

is of order 2. If $(X, A)$ is triangulable, the sequence is exact. This upper sequence is called the cohomology sequence of the pair $(X, A)$.

**Axiom 5c.** If the maps $f, g$ are homotopic maps from $(X, A)$ into $(Y, B)$, then $f^* = g^*$.

**Axiom 6c.** (*The excision axiom*) If $U$ is open in $X$, and $\overline{U}$ is contained in the interior of $A$, then the inclusion map $i : (X \backslash U, A \backslash U) \to (X, A)$ induces isomorphisms, i.e., $H^k(X, A) \cong H^k(X \backslash U, A \backslash U)$ for all $k$.

**Axiom 7c.** If $p$ is a point, then $H^k(\{p\}) = \{0\}$ for $k \neq 0$.

**Theorem [1].** Suppose $f : (X, A) \to (Y, B)$ and $g : (Y, B) \to (X, A)$. If $f$ and $g$ are homotopy equivalent, then $f$ and $g$ induce isomorphisms $f^* : H^k(Y, B) \to H^k(X, A)$ and $g^* : H^k(X, A) \to H^k(Y, B)$ with $(f^*)^{-1} = g^*$.

**Theorem [1].** If $(X', A')$ is a deformation retract of $(X, A)$, then the inclusion map $i : (X', A') \to (X, A)$ induces isomorphisms $i^* : H^k(X, A) \to H^k(X'A')$. Furthermore, if $r : (X, A) \to (X', A')$ is the associated retract, then $(i^*)^{-1} = r^*$.

In addition to the usual cohomology axioms and theorems above, Čech–Alexander–Spanier cohomology satisfies the following *strong excision property* and *weak continuity property*:

**Theorem [6].** (*Strong excision property*) Let $(X, A)$ and $(Y, B)$ be pairs, with $X$ and $Y$ paracompact Hausdorff and $A$ and $B$ closed. Let $f : (X, A) \to (Y, B)$ be a closed continuous map such that $f$ induces a one-to-one map of $X \backslash A$ onto $Y \backslash B$. Then, for all $k$, $f^* : H^k(Y, B) \to H^k(X, A)$ is an isomorphism.

**Theorem [6].** (*Weak continuity property*) Let $\{(X_\alpha, A_\alpha)\}_\alpha$ be a family of compact Hausdorff pairs in some space, directed downward by inclusion, and let $(X, A) = (\cap_{\alpha \in A} X_\alpha, \cap_{\alpha \in A} A_\alpha)$. The inclusion maps $i_\alpha : (X, A) \subset (X_\alpha, A_\alpha)$ induce an isomorphism

$$\{i_\alpha^*\} : \varinjlim H^\kappa(X_\alpha, A_\alpha) \to H^k(X, A).$$

**Definition 1.1.** *If $f : \mathbf{R}^m \to \mathbf{R}^m$ is continuous, then $\{\mathbf{y}_i\}_{i=j}^k$ is a $\delta$-pseudo-orbit or noisy orbit if $d(\mathbf{y}_{i+1}, f(\mathbf{y}_i)) \leq \delta$ for $j \leq i < k$. An exact orbit $\{\mathbf{x}_i\}_{i=j}^k$ for $f$ $\epsilon$-shadows the pseudo-orbit $\{\mathbf{y}_i\}_{i=j}^k$ if $d(\mathbf{x}_i, \mathbf{y}_i) \leq \epsilon$ for $j \leq i \leq k$.*

## 2. The results

Suppose that $f : \mathbf{R}^m \to \mathbf{R}^m$ is continuous, $k$ is a fixed integer such that $1 < k \leq m$, and $(X_i, B_i)_{i=0}^N, (Z_i, Y_i)_{i=0}^N$ are finite sequences of pairs of compact sets in $\mathbf{R}^m$ satisfying the following conditions:

(1). For $i = 0, \ldots, N$, $B_i := X_i \cap Y_i$, $Z_i := X_i \cup Y_i$ and $Z_i$ and $X_i$ are rectangles (products of intervals).

(2). For $i = 0, \ldots, N$, $B_i$ is homeomorphic to $S^{k-1} \times R_i$ (where $R_i$ is a rectangle) and is the union of some or all of the faces of $X_i$.

(3). For $i = 0, \ldots, N - 1$, $f(B_i) \subset Y_{i+1}$ and $f(X_i) \subset Z_{i+1}$.

We now assume a stronger version of (3):

(3*). For $i = 0, \ldots, N - 1$, $f$ maps $(X_i, B_i)$ into $(Z_{i+1}, Y_{i+1})$, and $f^* : H^k(Z_{i+1}, Y_{i+1}) \to H^k(X_i, B_i)$ is nontrivial.

**Definition 2.1.** *We say $E$ $k$-crosses the pair $(Z, Y)$ if $E$ is compact, $E \subset Z$, and the inclusion map $(E, E \cap Y) \to (Z, Y)$ induces a one-to-one homomorphism from $H^k(Z, Y)$ into $H^k(E, E \cap Y)$.*

**Lemma 2.2.** *Suppose $E$ $k$-crosses $(Z_i, Y_i)$ and suppose (1),(2), and (3*) hold for $f : \mathbf{R}^m \to \mathbf{R}^m$, $k$ a fixed integer such that $1 < k \leq m$, and finite sequences $(X_i, B_i)_{i=0}^N, (Z_i, Y_i)_{i=0}^N$ of pairs of compact sets in $\mathbf{R}^m$. Then there is a compact set $\widehat{E} \subset E$ such that $f|\widehat{E}$ induces a one-to-one homomorphism from $H^k(Z_{i+1}, Y_{i+1})$ into $H^k(\widehat{E}, \widehat{E} \cap Y_i)$.*

*Proof.* Let $\widehat{E} = E \cap X_i$. Then $\widehat{E} \cap Y_i = \widehat{E} \cap B_i$. The inclusion $i_1 : (\widehat{E}, (\widehat{E} \cap B_i)) \to (E, (E \cap Y_i))$ induces an isomorphism $i_1^* : H^k(E, (E \cap Y_i)) \to H^k(\widehat{E}, (\widehat{E} \cap B_i))$, by strong excision. Likewise, if $i_2$ denotes the inclusion from $(X_i, B_i)$ into $(Z_i, Y_i)$, $i_2^*$ is an isomorphism. By assumption, the inclusion $i_4 : (E, (E \cap Y_i)) \to (Z_i, Y_i)$, induces a one-to-one homomorphism $i_4^* : H^k(Z_i, Y_i) \to H^k(E, E \cap Y_i)$. Let $i_3$ denote the inclusion from $(\widehat{E}, (\widehat{E} \cap B_i))$ into $(X_i, B_i)$. Since $i_4 \circ i_1 = i_2 \circ i_3$, $i_1^* \circ i_4^* = (i_4 \circ i_1)^* = (i_2 \circ i_3)^* = i_3^* \circ i_2^*$. Since $i_2^*$ is an isomorphism, $i_1^* \circ i_4^* \circ (i_2^*)^{-1} = i_3^*$. Since $i_1^*$ and $(i_2^*)^{-1}$ are isomorphisms, and $i_4^*$ is a one-to-one homomorphism, $i_3^*$ is a one-to-one homomorphism from $H^k(X_i, B_i)$ into $H^k(\widehat{E}, (\widehat{E} \cap Y_i))$.

Note that $f|\widehat{E} = f \circ i_3$. Since the groups $H^k(X_i, B_i)$ and $H^k(Z_{i+1}, Y_{i+1})$ are isomorphic to the group of integers, and $f^*$ is not trivial, $f^*$ must be one-to-one. Since $i_3^*$ is a one-to-one homomorphism, $(f|\widehat{E})^*$ must be one-to-one. $\square$

**Proposition 2.3.** *Suppose (1),(2), and (3*) hold for $f : \mathbf{R}^m \to \mathbf{R}^m$, $k$ a fixed integer such that $1 < k \leq m$, and finite sequences $(X_i, B_i)_{i=0}^N, (Z_i, Y_i)_{i=0}^N$ of pairs of compact sets in $\mathbf{R}^m$. Further, suppose $E \subset Z_i$, and $f|E : (E, E \cap Y_i) \to (Z_{i+1}, Y_{i+1})$ induces a one-to-one homomorphism from $H^k(Z_{i+1}, Y_{i+1})$ into $H^k(E, E \cap Y_i)$. Then the inclusion $j : (f(E), f(E \cap Y_i)) \to (Z_{i+1}, Y_{i+1})$ induces a one-to-one homomorphism from $H^k(Z_{i+1}, Y_{i+1})$ into $H^k(f(E), f(E \cap Y_i))$.*

*Proof.* Note that $f|E$ can be regarded as a map from $(E, E \cap Y_i)$ into $(Z_{i+1}, Y_{i+1})$ and as a map from $(E, E \cap Y_i)$ onto $(f(E), f(E \cap Y_i))$. (The ranges are different.) To avoid ambiguity, denote the map $f|E : (E, E \cap Y_i) \to (f(E), f(E \cap Y_i))$ as $f_2$, while continuing to denote $f|E : (E, E \cap Y_i) \to (Z_{i+1}, Y_{i+1})$ as $f|E$. Then $f_2^* : H^k(f(E), f(E \cap Y_i)) \to H^k(E, E \cap Y_i)$, and $(f|E)^* : H^k(Z_{i+1}, Y_{i+1}) \to H^k(E, E \cap Y_i)$, with $(f|E)^*$ a one-to-one homomorphism. Since $f|E = j \circ f_2$, where $j : (f(E), f(E \cap Y_i)) \to (Z_{i+1}, Y_{i+1})$ denotes the inclusion, $j^*$ must be a one-to-one homomorphism. $\qquad\square$

The following is a simple version of a proposition that appeared in [4].

**Proposition 2.4 (The Covering Principle).** *Let $Q$ be a metric space. Assume $h : Q_0 \to Q$ is continuous, where $Q_0 \subset Q$ is compact. Let*

$$E_0, E_1 \subset P_1, E_2 \subset P_2, \ldots, E_n \subset P_n$$

*be $n+1$ nonempty compact subsets of $Q_0$ such that $h(E_i) \supset P_{i+1}$ for each $0 \leq i < n$. Then there exists a point $x_0 \in E_0$ such that $x_n := h^n(x_0) \in E_n$.*

*Proof.* Suppose $x_n \in E_n$. Since $E_n \subset P_n \subset h(E_{n-1})$, there is some $x_{n-1} \in E_{n-1}$ such that $h(x_{n-1}) = x_n$. Since $E_{n-1} \subset P_{n-1} \subset h(E_{n-2})$, there is some $x_{n-2} \in E_{n-2}$ such that $h(x_{n-2}) = x_{n-1}$. Then $h^2(x_{n-2}) = x_n$. We can continue this finite process until we reach $E_0$. Thus, there is some $x_0$ such that $h(x_0) = x_1$, $h^2(x_0) = x_2$, $\ldots$, and $h^n(x_0) = x_n$. $\qquad\square$

**Theorem 2.5 (Shadowing Containment Theorem).** *Suppose (1),(2), and (3*) hold for $f : \mathbf{R}^m \to \mathbf{R}^m$, $k$ a fixed integer such that $1 < k \leq m$, and finite sequences $(X_i, B_i)_{i=0}^N, (Z_i, Y_i)_{i=0}^N$ of pairs of compact sets in $\mathbf{R}^m$. Suppose $(\mathbf{y}_i)_{i=0}^N$ is a pseudo-trajectory with $\mathbf{y}_i \in Z_i$ over $i$, and let $\epsilon > 0$ be the maximum diameter of $Z_i$ over $i$. Then there is an $\epsilon$-shadow $(\mathbf{x}_i)_{i=0}^N$ of $(\mathbf{y}_i)_{i=0}^N$, i.e., $|\mathbf{x}_i - \mathbf{y}_i| \leq \epsilon$, $i = 0, \ldots, N$.*

*Proof.* Let $E_0 = X_0 \cup Y_0 = Z_0$. Then $E_0$ $k$-crosses $(Z_0, Y_0)$, and, by Lemma 1 and Proposition 1, there is a compact subset $E_1 = E_0 \cap X_0$ of $E_0$ such that the inclusion $j_1 : (f(E_1), f(E_1 \cap Y_0)) \to (Z_1, Y_1)$ induces a one-to-one homomorphism from $H^k(Z_1, Y_1)$ into $H^k(f(E_1), f(E_1 \cap Y_0))$. Thus, $f(E_1)$ $k$-crosses $(Z_1, Y_1)$. Proceeding inductively, if $f(E_l)$ $k$-crosses $(Z_l, Y_l)$ with $1 \leq l \leq N - 1$, there is a compact subset $E_{l+1} = f(E_l) \cap X_l$ of $f(E_l)$ such that the inclusion $j_l : (f(E_{l+1}), f(E_{l+1} \cap Y_l)) \to (Z_{l+1}, Y_{l+1})$ induces a one-to-one homomorphism from $H^k(Z_{l+1}, Y_{l+1})$ into $H^k(f(E_{l+1}), f(E_{l+1} \cap Y_l))$, and $f(E_{l+1})$ $k$-crosses $(Z_{l+1}, Y_{l+1})$. Let $E_{N+1} = f(E_N)$.

Note that the hypotheses of the preceding proposition are satisfied for $f$ and the sequence

$$E_1, E_2 \subset f(E_1), \ldots, E_{N+1} \subset f(E_N).$$

Then there is a nonempty compact subset $E'$ of $E_1 \subset Z_0$ such that if $\mathbf{x}_0 \in E'$, $f^l(\mathbf{x}_0) := \mathbf{x}_l \in E_{l+1} \subset X_l \subset Z_l$ for $0 \leq l \leq N$. Thus, $(\mathbf{x}_i)_{i=0}^N$ is an $\epsilon$-shadow of $(\mathbf{y}_i)_{i=0}^N$. $\qquad\square$

**Acknowledgement**

This research was supported under NSF Grant DMS 0616585

# References

[1] S. Eilenberg and N. E. Steenrod, *Foundations of Algebraic Topology*, Princeton University Press (1952).

[2] C. Grebogi, S. M. Hammel, J. A. Yorke, and T. Sauer, Shadowing of physical trajectories in chaotic dynamics: Containment and refinement, *Phys. Rev. Lett.* **65** (1990), 1527–1530.

[3] S. M. Hammel, J. A. Yorke, and C. Grebogi, Do numerical orbits of chaotic dynamical processes represent true orbits?, *J. of Complexity* **3** (1987) 136–145.

[4] J. A. Kennedy and J. A. Yorke, Generalized Hénon difference equations with delay, *Universitatis Jagellonicae Acta Mathematic* **41** (2003) 9–28.

[5] T. Sauer and J. A. Yorke, Rigorous verification of trajectories for the computer simulation of dynamical systems, *Nonlinearity* **4** (1991) 961–979.

[6] E. H. Spanier, *Algebraic Topology,* Springer (1994).

[7] G.-C. Yuan and J. A. Yorke, An open set of maps for which every point is absolutely nonshadowable, *Proc. Amer. Math. Soc.* **128** (1999) 909–918.

Judy Kennedy
Department of Mathematical Sciences
University of Delaware
Newark DE 19716
USA

James A. Yorke
Institute for Physical Science and Technology
University of Maryland
College Park MD 20742
USA

Progress in Nonlinear Differential Equations
and Their Applications, Vol. 75, 247–256
© 2007 Birkhäuser Verlag Basel/Switzerland

# Boundary Value Problems for Nonlinear Perturbations of Singular $\phi$-Laplacians

Jean Mawhin

*To Arrigo Cellina and Jim Yorke, for 130 years of excellence*

**Abstract.** We give existence and multiplicity theorems for solutions of various boundary value problems for $(\phi(u'))' = f(t, u, u')$, when $\phi : ] - a, a[ \to \mathbb{R}$ is an increasing homeomorphism.

**Mathematics Subject Classification (2000).** Primary 34B15; Secondary 47H10.

**Keywords.** $\phi$-Laplacian, boundary value problems, fixed points.

## 1. Introduction

Let $\phi : \mathbb{R} \to \mathbb{R}$ (*classical*), or $\phi : \mathbb{R} \to ] - a, a[$ (*bounded*), or $\phi : ] - a, a[ \to \mathbb{R}$ (*singular*) be an increasing homeomorphism such that $\phi(0) = 0$. Canonical examples are respectively

$$\phi(s) = |s|^{p-2}s \quad (p\text{-}Laplacian)\,,$$

$$\phi(s) = \frac{s}{\sqrt{1 + s^2}} \quad (curvature)\,,$$

$$\phi(s) = \frac{s}{\sqrt{1 - s^2}} \quad (special\ relativity)\,.$$

If $f : [0, T] \times \mathbb{R}^2 \to \mathbb{R}$ is continuous, we associate to $\phi$ and $f$ the quasilinear ordinary differential equation

$$\big(\phi(u')\big)' = f(t, u, u')\,. \tag{1.1}$$

By *solution* of (1.1) we mean a function $u$ of class $C^1$ whose range belongs to the domain of $\phi$ and such that $\phi \circ u'$ is of class $C^1$.

Like in the classical second order case, we can associate to (1.1) various *boundary conditions,* for example

$$u(0) = 0, \qquad u(T) = 0 \quad (\textit{Dirichlet}), \qquad\qquad (1.2)$$

$$u'(0) = 0, \qquad u'(T) = 0 \quad (\textit{Neumann}), \qquad\qquad (1.3)$$

$$u(0) = u(T), \qquad u'(0) = u'(T) \quad (\textit{periodic}). \qquad\qquad (1.4)$$

At first sight one could think that the increasing order of difficulty in treating a given boundary value problem with respect to the class of $\phi$ is classical, bounded, singular. But this is not the case and papers [3–5] show that, indeed, the increasing order of difficulty is singular, classical, bounded. In this paper, we concentrate on the singular case, and describe the results of [2] and [5]. The classical and bounded cases are treated in [2–4]. All results presented here are joint work with C. Bereanu.

## 2. Dirichlet problem

The first step is to reduce the problem to a fixed point problem in a suitable function space. Let $C = C([0, T])$, $C^1 = C^1([0, T])$, with the respective norms

$$\|u\|_\infty = \max_{[0,T]} |u|, \quad \|u\| = \|u\|_\infty + \|u'\|_\infty,$$

and let

$$C_0^1 := \left\{ u \in C^1 : u(0) = 0 = u(T) \right\}.$$

The construction of the fixed point operator requires the following result, valid for $\phi :\, ] - a, a[ \to \mathbb{R}$, with $0 < a \le +\infty$.

**Lemma 2.1.** *For each $h \in C$, there is a unique $\alpha := Q_\phi(h)$ such that*

$$\int_0^T \phi^{-1}\big(h(t) - \alpha\big)\, dt = 0.$$

*Furthermore $Q_\phi : C \to \mathbb{R}$ is continuous.*

For $\phi = I$, $Q_I(h)$ is the mean value of $h$ and hence $Q_\phi(h)$ is called the *$\phi$-mean value* of $h$. Let $H : C \to C^1$ be defined by

$$H u(t) = \int_0^t u(s)\, ds \quad (t \in [0, T]),$$

and let $F : C^1 \to C$ be a continuous mapping which takes bounded sets into bounded sets. For example, this is the case for the *Nemytski operator*

$$N_f : C^1 \to C, \qquad u \mapsto f\big(\cdot, u(\cdot), u'(\cdot)\big)$$

associated to a continuous function $f : [0, T] \times \mathbb{R}^2 \to \mathbb{R}$. The following lemma is easily proved. See [10] for early references and proof.

**Lemma 2.2.** *u is a solution of*

$$\left(\phi(u')\right)' = F(u), \quad u(0) = 0 = u(T) \tag{2.1}$$

*if and only if $u \in C_0^1$ is a fixed point of $M_0$, with*

$$M_0(u) := H \circ \phi^{-1} \circ (I - Q_\phi) \circ HF(u). \tag{2.2}$$

*Furthermore, $M_0$ is completely continuous on $C_0^1$ and $\|(M_0(u))'\|_\infty < a$ for all $u \in C_0^1$.*

We then obtain a *universal* existence theorem for Dirichlet problem when $\phi$ is singular.

**Theorem 2.3.** *If $\phi$ is singular, problem (2.1) has at least one solution for any F.*

*Proof.* For each $u \in C_0^1$, we have

$$\left\|(M_0(u))'\right\|_\infty < a,$$

and hence $M_0$ maps $C_0^1$ into the closed ball $\overline{B_{a(T+1)}}$ of $C_0^1$. The conclusion follows from Schauder's fixed point theorem. □

**Remark 2.4.** For the Dirichlet problem associated to a singular $\phi$ :

1. Existence of a solution is guaranteed for *any* right-hand member $F$, without any *growth condition* with respect to $u$, like in the classical or bounded case.
2. For any reasonable spectral theory for $\phi$ singular with Dirichlet boundary conditions (if any !), *the spectrum is empty.*

## 3. Neumann problem: sign condition

The problem

$$\left(\phi(u')\right)' = 1, \quad u'(0) = 0 = u'(T)$$

has *no* solution, and hence there is no universal existence theorem for *Neumann* boundary conditions.

To deal with the Neumann problem, let

$$C_\#^1 := \left\{u \in C^1 : u'(0) = 0 = u'(T)\right\},$$

and define the linear projectors $P, Q : C \to C^1$ respectively by

$$Pu := u(0), \quad Qu := \frac{1}{T} \int_0^T u(s).$$

If again $F : C^1 \to C$ is continuous and takes bounded sets into bounded sets, it is easy to prove the following lemma (see [10] for references and proof).

**Lemma 3.1.** *u is a solution of*

$$\left(\phi(u')\right)' = F(u), \quad u'(0) = 0 = u'(T) \tag{3.1}$$

*if and only if $u \in C^1_\#$ is a fixed point of $M_\#$, with*

$$M_\#(u) := Pu + QF(u) + H \circ \phi^{-1} \circ H(I - Q)F(u).$$

*Furthermore, $M_\#$ is completely continuous on $C^1_\#$ and $\|(M_\#(u))'\|_\infty < a$ for all $u \in C^1_\#$.*

For $u \in C$, let

$$u_L := \min_{[0,T]} u, \quad u_M := \max_{[0,T]} u.$$

We denote by $d_{LS}$ the Leray–Schauder degree et by $d_B$ the Brouwer degree (see, e.g., [6]). We have the following existence theorem.

**Theorem 3.2.** *If there exists $R > 0$ and $\epsilon \in \{-1, 1\}$ such that*

$$\epsilon (\operatorname{sign} u) QF(u) > 0 \tag{3.2}$$

*whenever $u_L \geq R$ or $u_M \leq -R$, and $\|u'\|_\infty < a$, then:*

1. *Problem (3.1) has at least one solution.*
2. *Each fixed point $u$ of $M_\#$ satisfies $\|u\| < R + a(T + 1)$.*
3. *$d_{LS}[I - M_\#, B_{R+a(T+1)}, 0] = -\epsilon$.*

*Proof.* (Sketched). We introduce the homotopy

$$\mathcal{M}(\lambda, u) := Pu + QF(u) + H \circ \phi^{-1} \circ \lambda H(I - Q)F(u) \quad \left(\lambda \in [0, 1]\right). \tag{3.3}$$

Each possible fixed point of $M(\lambda, \cdot)$ satisfies the *a priori* bound

$$\|u'\|_\infty < a \tag{3.4}$$

and the identity $QF(u) = 0$. Consequently, we have $u_M > -R$ and $u_L < R$, which, together with (3.4), implies the second conclusion. Using degree invariance under a homotopy and some reduction formula, we get

$$d_{LS}\left[I - \mathcal{M}(1, \cdot), B_{R+(a+1)T}, 0\right] = d_{LS}\left[I - \mathcal{M}(0, \cdot), B_{R+a(T+1)}, 0\right]$$
$$= d_B\left[-QF|_{\mathbb{R}}, \right] - R - a(T + 1), R + a(T + 1)[, 0]$$
$$= -\epsilon,$$

which is the third conclusion, and the first one follows from existence property of degree. □

**Example 3.3.** If $b \neq 0$ and $h \in C$, problem

$$\left(\frac{u'}{\sqrt{1 - u'^2}}\right)' = b \arctan u + h(t), \quad u'(0) = 0 = u'(T)$$

has at least one solution if and only if

$$-\frac{|b|\pi}{2} < Qh < \frac{|b|\pi}{2}.$$

**Remark 3.4.** A standard approximation argument shows that the existence conclusion in Theorem 3.2 survives if the sign condition is weakened to

$$\epsilon \, (sign \ u) QF(u) \geq 0 \tag{3.5}$$

whenever $u_L \geq R$ or $u_M \leq -R$, and $\|u'\|_\infty < a$.

**Corollary 3.5.** *If $f : [0, T] \times \mathbb{R}^2 \to \mathbb{R}$ is continuous and such that*

$$\lim_{u \to \pm\infty} \epsilon f(t, u, v) = \pm\infty$$

*for some $\epsilon \in \{-1, 1\}$, uniformly in $(t, v) \in [0, T] \times \, ] - a, a[$, problem*

$$\big(\phi(u')\big)' = f(t, u, u'), \quad u'(0) = 0 = u'(T)$$

*has at least one solution, and $d_{LS}[I - M_\#, B_\rho, 0] = -\epsilon$ for large $\rho > 0$.*

This is in particular the case for $f(t, u, v) = \mu u + h(t)$ with $\mu \neq 0$ and $h \in C$.

**Remark 3.6.** For Neumann problem (3.1):
1. Existence holds under the *sign condition* (3.5) only upon $F$, which reduces to the necessary and sufficient condition $Qh = 0$ when $F(u)(t) = h(t)$.
2. There is no need of any *growth condition* with respect to $u$ like in the classical or the bounded case.
3. For any reasonable spectral theory for singular $\phi$ with Neumann boundary conditions (if any), the *spectrum* is $\{0\}$.

**Example 3.7.** If $b \neq 0$ and $p > 0$, problem

$$\left(\frac{u'}{\sqrt{1 - u'^2}}\right)' = b|u|^{p-1}u + h(t), \quad u'(0) = 0 = u'(T)$$

has at least one solution for each $h \in C$.

**Remark 3.8.** Similar results hold for the case of *periodic boundary conditions*. See [5] for details.

## 4. Neumann problem: lower and upper solutions

We extend the classical *method of lower and upper solutions* to the Neumann problem with singular $\phi$.

$$\big(\phi(u')\big)' = f(t, u, u'), \quad u'(0) = 0 = u'(T), \tag{4.1}$$

where $f : [0, T] \times \mathbb{R}^2 \to \mathbb{R}$ is continuous.

**Definition 4.1.** $\alpha \in C^1$ with $\phi \circ \alpha' \in C^1$ is a lower solution *for* (4.1) *if* $\|\alpha'\|_\infty < a$,

$$\big(\phi(\alpha'(t))\big)' \geq f\big(t, \alpha(t), \alpha'(t)\big) \quad (t \in [0, T]), \tag{4.2}$$

*and* $\alpha'(0) \geq 0 \geq \alpha'(T)$. $\beta \in C^1$ with $\phi \circ \beta' \in C^1$ is an upper solution *for* (4.1) *if* $\|\beta'\|_\infty < a$,

$$\big(\phi(\beta'(t))\big)' \leq f\big(t, \beta(t), \beta'(t)\big) \quad (t \in [0, T]), \tag{4.3}$$

*and* $\beta'(0) \leq 0 \leq \beta'(T)$.

The fundamental existence theorem goes as follows.

**Theorem 4.2.** *If* $\phi$ *is singular and* (4.1) *has a lower solution* $\alpha$ *and an upper solution* $\beta$ *such that*

$$\alpha(t) \leq \beta(t) \quad (t \in [0, T]),$$

*then* (4.1) *has at least one solution verifying*

$$\alpha(t) \leq u(t) \leq \beta(t) \quad (t \in [0, T]). \tag{4.4}$$

*Proof.* (Sketched). It follows the classical scheme introduced in [11] for $\phi = I$. Define

$$\gamma(t, u) = \begin{cases} \beta(t) & \text{if} \quad u > \beta(t) \\ u & \text{if} \quad \alpha(t) \leq u \leq \beta(t) \\ \alpha(t) & \text{if} \quad u < \alpha(t), \end{cases}$$

and consider the modified problem

$$\big(\phi(u')\big)' = f\big(t, \gamma(t, u), u'\big) + u - \gamma(t, u), \quad u'(0) = 0 = u'(T). \tag{4.5}$$

The Equation (4.5) has a solution from Corollary 3.5. Using a maximum principe argument one then shows that each possible solution of (4.5) satisfies $\alpha(t) \leq u(t) \leq \beta(t)$ for all $t \in [0, T]$, and hence is a solution of (4.1). $\qquad\square$

**Remark 4.3.** For a singular $\phi$, there is no need of any *Nagumo condition*, like in the cases of $u'' = f(t, u, u')$ or of $(|u'|^{p-2}u')' = f(t, u, u')$.

Lower and upper solutions are called *strict* if the strict inequality holds in (4.2) and (4.3).

**Theorem 4.4.** *If the lower and upper solutions* $\alpha$ *and* $\beta$ *are strict in Theorem 4.2, then each solution* $u$ *of* (4.1) *verifying* (4.4) *is such that*

$$\alpha(t) < u(t) < \beta(t) \quad (t \in [0, T])$$

*and* $d_{LS}[I - M_{\#}, \Omega_{\alpha,\beta}, 0] = -1$, *where* $\Omega_{\alpha,\beta}$ *is the open bounded set*

$$\Big\{ u \in C_{\#}^1 : \alpha(t) < u(t) < \beta(t) \ (t \in [0, T]), \ \|u'\|_\infty < a \Big\}.$$

*Proof.* (Sketched). The strict inequality for the solution $u$ of the modified problem follows from the definition of strict lower and upper solutions and a maximum principle argument. The computation of the degree is based upon the excision property. $\qquad\square$

**Remark 4.5.** Theorem 4.2 also holds if $f : [0, T] \times \,]0, +\infty[ \,\times \mathbb{R} \to \mathbb{R}$ is continuous and $\alpha(t) > 0$ for all $t \in [0, T]$.

The existence of a solution survives when *unordered* lower and upper solutions exist. The argument of the proof is inspired from the one introduced by Amann–Ambrosetti–Mancini [1] for the semilinear elliptic problem $\Delta u + g(x, u) = 0$ with zero Dirichlet conditions and $g$ bounded.

**Theorem 4.6.** *If $\phi$ is singular and (4.1) has a lower solution $\alpha$ and an upper solution $\beta$, then it has at least one solution.*

*Proof.* (Sketched). Decompose any $u \in C^1_\#$ in the form $u = u_0 + u_1$ ($u_0 = u(0)$, $u_1(0) = 0$), and let $\widetilde{C^1_\#} = \{u \in C^1_\# : u(0) = 0\}$. The set $\mathcal{S}$ of the solutions $(u_0, u_1) \in \mathbb{R} \times \widetilde{C^1_\#}$ of problem

$$\left(\phi(u'_1)\right)' = f(t, u_0 + u_1, u'_1) - QN_f(u_0 + u_1), \quad u'_1(0) = 0 = u'_1(T) \qquad (4.6)$$

contains a continuum $\mathcal{C}$ whose projection on $\mathbb{R}$ is $\mathbb{R}$ and projection on $\widetilde{C^1_\#}$ is contained in the ball $B_{a(T+1)}$. This follows from the fact that, for each fixed $u_0 \in \mathbb{R}$, problem (4.6) is equivalent to the fixed point problem in $\widetilde{C^1_\#}$

$$u_1 = H \circ \phi^{-1} \circ H(I - Q)N_f(u_0 + u_1),$$

and from continuation arguments of Leray–Schauder's type (see [5] for details). If there is some $(u_0, u_1) \in \mathcal{C}$ such that $QN_f(u_0 + u_1) = 0$, then $u_0 + u_1$ solves (4.1). If $QN_f(u_0 + u_1) > 0$ for all $(u_0, u_1) \in \mathcal{C}$, then for $(\alpha_M + aT, u_1) \in \mathcal{C}$, $\alpha_M + aT + u_1(t) \geq \alpha(t)$ for all $t \in [0, T]$ is an upper solution for (4.1), and the existence of a solution to (4.1) follows from Theorem 4.2. Similarly in the opposite case. $\qquad \square$

**Remark 4.7.** Similar results hold for the *periodic boundary conditions*. See [5] for details.

## 5. Neumann problem: Ambrosetti–Prodi type result

One can use the results on lower and upper solutions to prove an *Ambrosetti–Prodi type result* for the Neumann problem

$$\left(\phi(u')\right)' = f(t, u, u') - s, \quad u'(0) = 0 = u'(T), \qquad (5.1)$$

with singular $\phi$, $f : [0, T] \times \mathbb{R}^2 \to \mathbb{R}$ continuous, *coercive* or *anticoercive* in $u$, and $s \in \mathbb{R}$.

**Theorem 5.1.** *If $\phi$ is singular and*

$$\lim_{|u| \to \infty} f(t, u, v) = +\infty \quad (resp. -\infty) \qquad (5.2)$$

*uniformly in $(t, v) \in [0, T] \times \,] - a, a[$, there exists $s_1$ such that problem (5.1) has no solution (resp. at least two solutions) if $s < s_1$, at least one solution if $s = s_1$ and at least two solutions (resp. no solution) if $s > s_1$.*

*Proof.* (Sketched). Follows the scheme used in [7] when $\phi = I$ and in [12] when $\phi(s) = |s|^{p-2}s$, based upon a combination of lower and upper solutions techniques, *a priori* estimates and Leray–Schauder degree. See [5] for details. $\qquad \square$

**Example 5.2.** Write $h \in C$ as $h(t) = \overline{h} + \widetilde{h}(t)$ where $\overline{h} = Qh$, and $\widetilde{C} = \{h \in C : \overline{h} = 0\}$. For each $b > 0$ (resp. $b < 0$), $p > 0$, and $\widetilde{h} \in \widetilde{C}$, there exists $s_1 \in \mathbb{R}$ such that problem

$$\left(\frac{u'}{\sqrt{1 - u'^2}}\right)' + b|u|^p = \overline{h} + \widetilde{h}(t), \quad u'(0) = 0 = u'(T)$$

has no solution (resp. at least two solutions) if $\overline{h} < s_1$, at least one solution if $\overline{h} = s_1$, and at least two solutions (resp. no solution) if $\overline{h} > s_1$.

**Remark 5.3.** Similar results hold for the *periodic boundary conditions*. See [5] for details.


## 6. Singular forces

We now discuss Neumann problems for equations with *singular restoring forces* and $\phi$ singular, which were considered, in the periodic case, by Lazer–Solimini [9] when $\phi = I$ and by Jebelean–Mawhin [8] and Rachunková–Tvrdý [13] for classical $\phi$. We first consider the case of *attractive forces*.

**Theorem 6.1.** *If $h \in C$, $g : [0, T] \times ]0, +\infty[ \times \mathbb{R} \to ]0, +\infty[$ is continuous and*

$$\lim_{u \to 0+} g(t, u, v) = +\infty, \qquad \lim_{u \to +\infty} g(t, u, v) = 0,$$

*uniformly in $(t, v) \in [0, T] \times ] - a, a[$, problem*

$$\left(\phi(u')\right)' + g(t, u, u') = h(t), \quad u'(0) = 0 = u'(T) \tag{6.1}$$

*has at least one positive solution if and only if $Qh > 0$.*

*Proof.* (Sketched). *Necessity.* If $u$ solves (6.1), then $Qh = QN_g(u) > 0$.
*Sufficiency.* For $\epsilon > 0$ such that $g(t, \epsilon, 0) - h(t) > 0$ for all $t \in [0, T]$, $\alpha \equiv \epsilon$ is a strict lower solution for (6.1). Take now $w$ such that

$$\left(\phi(w')\right)' = h(t) - Qh, \quad w'(0) = 0 = w'(T).$$

For $\delta > 0$ such that $\beta(t) := \delta + w(t) > \alpha(t)$ and $g(t, \beta(t), \beta'(t)) < Qh$ for all $t \in [0, T]$, $\beta$ is a strict upper solution of (6.1). The result follows from Theorem 4.2 and Remark 4.5. □

**Example 6.2.** If $b > 0$, $p < 0$ and $h \in C$, problem

$$\left(\frac{u'}{\sqrt{1 - u'^2}}\right)' + bu^p = h(t), \quad u'(0) = 0 = u'(T)$$

has at least one positive solution if and only if $Qh > 0$.

Consider now the case of *repulsive forces*, discussed, for periodic boundary conditions, by Lazer–Solimini [9] when $\phi = I$, and by Jebelean–Mawhin [8] and Rachunková–Tvrdý [13] for classical $\phi$.

**Theorem 6.3.** *If $h \in C$, $g : ]0, +\infty[ \to ]0, +\infty[$ is continuous and*

$$\lim_{u \to 0+} g(u) = +\infty, \quad \lim_{u \to +\infty} g(u) = 0, \quad \int_0^1 g(s)\,ds = +\infty,$$

*problem*

$$\left(\phi(u')\right)' - g(u) = h(t), \quad u'(0) = 0 = u'(T) \tag{6.2}$$

*has at least one positive solution if and only if $Qh < 0$.*

*Proof.* (Sketched). A basic ingredient in the proof is to show that there exists $R > \epsilon > 0$ such that for all $\lambda \in ]0,1]$, any positive solution $u$ of

$$\left(\phi(u')\right)' - \lambda g(u) = \lambda h(t), \quad u'(0) = 0 = u'(T) \tag{6.3}$$

satisfies $R > u(t) > \epsilon$ for all $t \in [0, T]$. The upper bound $R$ follows from the relations

$$\int_0^T g\big(u(s)\big)\,ds = -TQh, \quad \left\|\left(\phi(u')\right)'\right\|_1 \leq T|Qh| + \|h\|_1 \leq 2\|h\|_1,$$

which gives, for some $t_0 \in [0, T]$ and $R_0 > 0$,

$$u(t_0) < R_0, \quad \|\phi(u')\|_\infty \leq 2T\|h\|_1,$$

where $\|\cdot\|_1$ denotes the $L^1$-norm. For the lower bound $\epsilon$, the energy identity associated to (6.2) is used, explaining the more restricted class of restoring forces. The proof ends by showing that $|d_{LS}[I - M_\#, \Omega, 0]| = 1$, with

$$\Omega = \left\{ u \in C_\#^1 : \epsilon < u(t) < R \ (t \in [0, T]), \ \|u'\|_\infty < a \right\}. \qquad \square$$

**Example 6.4.** If $b < 0$, $p \leq -1$ and $h \in C$, problem

$$\left(\frac{u'}{\sqrt{1 - u'^2}}\right)' + bu^p = h(t), \quad u'(0) = 0 = u'(T)$$

has at least one positive solution if and only if $Qh < 0$.

**Remark 6.5.** Similar results hold for the periodic case. See [5] for details.

# References

[1] H. Amann, A. Ambrosetti, G. Mancini, *Elliptic equations with noninvertible Fredholm linear part and bounded nonlinearities,* Math. Z. **158** (1978), 179–194.

[2] C. Bereanu, J. Mawhin, *Nonlinear Neumann boundary value problems with $\phi$-Laplacian operators,* An. Ştiinţ. Univ. Ovidius Constanţa Ser. Mat. **12** (2004), 73–92.

[3] C. Bereanu, J. Mawhin, *Boundary-value problems with non-surjective $\phi$-Laplacian and one-sided bounded nonlinearity,* Adv. Differential Equations **11** (2006), 35–60.

[4] C. Bereanu, J. Mawhin, *Periodic solutions of nonlinear perturbations of $\phi$-Laplacians with possibly bounded $\phi$,* Nonlinear Analysis to appear.

[5] C. Bereanu, J. Mawhin, *Existence and multiplicity results for some nonlinear problems with singular $\phi$-Laplacian,* J. Differential Equations to appear.

[6] K. Deimling, *Nonlinear Functional Analysis,* Springer, Berlin, 1985.

[7] C. Fabry, J. Mawhin, M. Nkashama, *A multiplicity result for periodic solutions of forced nonlinear second order ordinary differential equations,* Bull. London Math. Soc. **18** (1986), 173–180.

[8] P. Jebelean, J. Mawhin, *Periodic solutions of singular nonlinear perturbations of the ordinary p-Laplacian,* Adv. Nonlinear Stud. **2** (2002), 299–312.

[9] A. C. Lazer, S. Solimini, *On periodic solutions of nonlinear differential equations with singularities,* Proc. Amer. Math. Soc. **99** (1987), 109–114.

[10] R. Manásevich, J. Mawhin, *Boundary value problems for nonlinear perturbations of vector p-Laplacian-like operators,* J. Korean Math. Soc. **37** (2000), 665–685.

[11] J. Mawhin, *Points fixes, points critiques et problèmes aux limites,* Sémin. Math. Sup. No. 92, Presses Univ. Montréal, Montréal, 1985.

[12] J. Mawhin, *The periodic Ambrosetti-Prodi problem for nonlinear perturbations of the p-Laplacian,* J. Eur. Math. Soc. **8** (2006), 375-388.

[13] I. Rachunková, M. Tvrdý, *Periodic singular problem with quasilinear differential equations,* Math. Bohem. **131** (2006), 321–336.

Jean Mawhin
Department of Mathematics
Université Catholique de Louvain
B-1348 Louvain-la-Neuve
Belgium
e-mail: mawhin@math.ucl.ac.be

Progress in Nonlinear Differential Equations
and Their Applications, Vol. 75, 257–267
© 2007 Birkhäuser Verlag Basel/Switzerland

# Existence, Nonexistence and Multiplicity Results for Some Beam Equations

Feliz Manuel Minhós

*Dedicated to Arrigo Cellina and James Yorke*

**Abstract.** This paper studies the fourth order nonlinear fully equation

$$u^{(4)}(x) + f\left(x, u(x), u'(x), u''(x), u'''(x)\right) = s\, p(x)$$

for $x \in [a, b]$, $f : [a, b] \times \mathbb{R}^4 \to \mathbb{R}$, $p : [a, b] \to \mathbb{R}^+$ continuous functions and $s \in \mathbb{R}$, with the boundary conditions

$$u(a) = A, \quad u'(a) = B,$$
$$k_1\, u''(a) - k_2\, u'''(a) = C, \quad k_3\, u''(b) + k_4\, u'''(b) = D$$

for $A, B, C, D, k_1, k_3 \in \mathbb{R}$, $k_2, k_4 \geq 0$ such that $k_1^2 + k_2 > 0$ and $k_3^2 + k_4 > 0$. This problem models several phenomena, such as, a cantilevered beam with a linear relation between the curvature and the shear force at both endpoints. For some values of the real constants, it will be presented an Ambrosetti–Prodi type discussion on $s$. The arguments used apply lower and upper solutions technique, *a priori* estimations and topological degree theory.

**Mathematics Subject Classification (2000).** Primary 34B15; Secondary 34B18; 34L30; 47H10; 47H11.

**Keywords.** Nagumo-type conditions, lower and upper solutions, Leray–Schauder degree, Ambrosetti–Prodi problems, beam equation.

## 1. Introduction

In this paper it is studied the fourth order nonlinear fully equations

$$u^{(iv)}(x) + f\left(x, u(x), u'(x), u''(x), u'''(x)\right) = sp(x), \tag{$\text{E}_s$}$$

for $f : [a, b] \times \mathbb{R}^4 \to \mathbb{R}$ and $p : [a, b] \to \mathbb{R}^+$ continuous functions and $s$ a real parameter, with several types of two-point boundary conditions.

This work was partially supported by CRUP, Acção E-99/06.

If

$$u(a) = A, \quad u'(a) = B,$$
$$k_1 \, u''(a) - k_2 \, u'''(a) = C, \quad k_3 \, u''(b) + k_4 \, u'''(b) = D \tag{1.1}$$

for $A, B, C, D, k_1, k_3 \in \mathbb{R}$, $k_2, k_4 \geq 0$ such that $k_1^2 + k_2 > 0$ and $k_3^2 + k_4 > 0$ it will be proved the existence of solutions to problem $(E_s)$–(1.1) for the values of $s$ such that there are lower and upper solutions.

In Section 3 it is considered, for clearness, a particular case of the above boundary conditions in $[0, 1]$:

$$u(0) = 0, \quad u'(0) = 0$$
$$k_1 \, u''(0) - k_2 \, u'''(0) = 0, \quad k_3 \, u''(1) + k_4 \, u'''(1) = 0 \tag{1.2}$$

with $k_1, k_2, k_3, k_4 \geq 0$ such that $k_1 + k_2 > 0$ and $k_3 + k_4 > 0$ and the existence of solution for the problem $(E_s)$–(1.2) will depend on $s$.

Taking, in (1.2), $k_2 = k_4 = 0$ and $k_1, k_3 > 0$, the two-point boundary conditions are

$$u(0) = u'(0) = u''(0) = u''(1) = 0 \tag{1.3}$$

and it is obtained in Section 4 an Ambrosetti–Prodi type result, that is, there are $s_0, s_1 \in \mathbb{R}$ such that $(E_s)$–(1.3) has no solution if $s < s_0$, it has at least one solution if $s = s_0$ and $(E_s)$–(1.3) has at least two solutions for $s \in ]s_0, s_1]$.

As far as we know these Ambrosetti–Prodi results were never applied to fourth order nonlinear fully equations. The arguments used were suggested by several papers namely [2], applied to second order periodic problems, [10], to third order three points boundary value problems, [1] for two-point boundary value problems, and make use of Nagumo-type growth condition, [9], upper and lower solutions technique for higher order boundary value problems, [3,4,7], and degree theory, [6].

We point out that the localization of solutions provided by lower and upper solutions method, combined with Ambrosetti–Prodi type results or by itself, can be useful to prove the existence of positive solutions (if the lower function is non-negative) or multiple solutions (if there are solutions in two disjoint branches). In fact, this property is a sharp tool in some applications where bounds on the solution or its derivatives are important, as it is illustrated in last section.

## 2. Existence and non-existence results

In the following, $C^k([a, b])$ denotes the space of real valued functions with continuous $i$-derivative in $[a, b]$, for $i = 1, ..., k$, equipped with the usual norms. The nonlinearity of $(E_s)$ must verify some growth conditions, given by next definition, providing also an *a priori* estimate for $u'''$, if some bounds on $u$, $u'$ and $u''$ are verified.

**Definition 2.1.** *A continuous function $g : [a, b] \times \mathbb{R}^4 \to \mathbb{R}$ is said to satisfy Nagumo-type conditions in*

$$E = \left\{ (x, y_0, y_1, y_2, y_3) \in [a, b] \times \mathbb{R}^4 : \gamma_i(x) \leq y_i \leq \Gamma_i(x), \ i = 0, 1, 2 \right\},$$

*with $\gamma_i(x)$ and $\Gamma_i(x)$ continuous functions such that $\gamma_i(x) \leq \Gamma_i(x)$, for each $i$ and every $x \in [a, b]$, if there exists a continuous function $h_E : \mathbb{R}_0^+ \to [k, +\infty]$, for some fixed $k > 0$, such that*

$$|g(x, y_0, y_1, y_2, y_3)| \leq h_E(|y_3|), \quad \forall (x, y_0, y_1, y_2, y_3) \in E, \tag{2.1}$$

*with*

$$\int_0^{+\infty} \frac{\xi}{h_E(\xi)} \, d\xi = +\infty. \tag{2.2}$$

**Lemma 2.2.** *Let $f : [a, b] \times \mathbb{R}^4 \to \mathbb{R}$ be a continuous function that satisfies Nagumo-type conditions (2.1) and (2.2) in*

$$E = \left\{ (x, y_0, y_1, y_2, y_3) \in [a, b] \times \mathbb{R}^4 : \gamma_i(x) \leq y_i \leq \Gamma_i(x), \ i = 0, 1, 2 \right\},$$

*where $\gamma_i(x)$ and $\Gamma_i(x)$ are continuous functions. Then there is $r > 0$ such that every solution $u(x)$ of $(E_s)$ verifying $\gamma_i(x) \leq u^{(i)}(x) \leq \Gamma_i(x)$, for $i = 0, 1, 2$ and every $x \in [a, b]$, satisfies $\|u'''\| < r$.*

The proof is contained in [4].

To apply upper and lower solutions method it will be considered the following type of functions:

**Definition 2.3.** *Consider $A, B, C, D, k_1, k_2, k_3, k_4 \in \mathbb{R}$ such that $k_2, k_4 \geq 0$, $k_1^2 + k_2 > 0$ and $k_3^2 + k_4 > 0$.*

*A function $\alpha(x) \in C^4(]a, b[) \cap C^3([a, b])$ is a lower solution of $(E_s)$–(1.1) if*

$$\alpha^{(4)}(x) + f\big(x, \alpha(x), \alpha'(x), \alpha''(x), \alpha'''(x)\big) \geq s \, p(x), \tag{2.3}$$

*for $x \in ]a, b[$, and*

$$\alpha(a) \leq A, \quad \alpha'(a) \leq B,$$
$$k_1 \, \alpha''(a) - k_2 \, \alpha'''(a) \leq C, \quad k_3 \, \alpha''(b) + k_4 \, \alpha'''(b) \leq D.$$

*A function $\beta(x) \in C^4(]a, b[) \cap C^3([a, b])$ is an upper solution if the inequalities are reversed.*

For $s$ such that there are upper and lower solutions of $(E_s)$–(1.1) with the second derivatives "well ordered", it is obtained not only an existence result but also some information concerning the location of the solution of $(E_s)$–(1.1) and its derivatives.

**Theorem 2.4.** *Let $f : [a, b] \times \mathbb{R}^4 \to \mathbb{R}$ be a continuous function. Suppose that there are lower and upper solutions of $(E_s)$–(1.1), $\alpha(x)$ and $\beta(x)$, respectively, such that, $\alpha''(x) \leq \beta''(x)$, for $x \in [a, b]$, and $f$ satisfies Nagumo-type conditions in*

$$E_* = \left\{ (x, y_0, y_1, y_2, y_3) \in [a, b] \times \mathbb{R}^4 : \alpha^{(i)}(x) \leq y_i \leq \beta^{(i)}(x), i = 0, 1, 2 \right\}.$$

*If f verifies*

$$f\big(x, \alpha(x), \alpha'(x), y_2, y_3\big) \leq f(x, y_0, y_1, y_2, y_3) \leq f\big(t, \beta(x), \beta'(x), y_2, y_3\big), \qquad (2.4)$$

*for fixed* $(x, y_2, y_3) \in [a, b] \times \mathbb{R}^2$ *and* $\alpha(x) \leq y_0 \leq \beta(x)$, $\alpha'(x) \leq y_1 \leq \beta'(x)$ *then* $(E_s)$–$(1.1)$ *has at least a solution* $u(x) \in C^4([a, b])$ *satisfying*

$$\alpha(x) \leq u(x) \leq \beta(x), \quad \alpha'(x) \leq u'(x) \leq \beta'(x), \quad \alpha''(x) \leq u''(x) \leq \beta''(x), \quad \forall x \in [a, b].$$

The proof is a particular case of the main result of [4].

A first discussion on $s$ about the existence and nonexistence of a solution will be done, for clearness, in [0,1] with $A = B = C = D = 0$ and $k_1, k_2, k_3, k_4 \geq 0$ with $k_1 + k_2 > 0$, $k_3 + k_4 > 0$, that is, for problem $(E_s)$–$(1.2)$. Lower and upper solutions definition for this problem are obtained considering these restrictions:

**Definition 2.5.** *For* $k_1, k_2, k_3, k_4$ *nonnegative real numbers such that* $k_1 + k_2 > 0$ *and* $k_3 + k_4 > 0$, *a function* $\alpha(x) \in C^4(]0, 1[) \cap C^3([0, 1])$ *is a lower solution of* $(E_s)$–$(1.2)$ *if it verifies* $(2.3)$ *in* $]0, 1[$ *and*

$$\alpha(0) \leq 0, \quad \alpha'(0) \leq 0,$$
$$k_1\, \alpha''(0) - k_2\, \alpha'''(0) \leq 0, \quad k_3\, \alpha''(1) + k_4\, \alpha'''(1) \leq 0.$$

*A function* $\beta(x)$ *is an upper solution if it satisfies the reversed inequalities.*

**Theorem 2.6.** *Let* $f : [0, 1] \times \mathbb{R}^4 \to \mathbb{R}$ *be a continuous function satisfying Nagumo-type condition and such that:*

**(H₁)** *$f$ is nondecreasing on the second and third variables;*
**(H₂)** *$f$ is nonincreasing on the fourth variable;*
**(H₃)** *there are $s_1 \in \mathbb{R}$ and $r > 0$ such that*

$$\frac{f(x, 0, 0, 0, 0)}{p(x)} < s_1 < \frac{f(x, y_0, y_1, -r, 0)}{p(x)}, \qquad (2.5)$$

*for every* $x \in [0, 1]$ *and every* $y_0 \leq -r$ *and* $y_1 \leq -r$. *Then there is* $s_0 < s_1$ *(with the possibility that* $s_0 = -\infty$*) such that: for* $s < s_0$, $(E_s)$–$(1.2)$ *has no solution; for* $s_0 < s \leq s_1$, $(E_s)$–$(1.2)$ *has at least one solution.*

*Proof.* Defining $s^* = \max\{f(x, 0, 0, 0, 0)/p(x), x \in [0, 1]\}$, by $(2.5)$, there exists $x^* \in [a, b]$ such that

$$\frac{f(x, 0, 0, 0, 0)}{p(x)} \leq s^* = \frac{f(x^*, 0, 0, 0, 0)}{p(x^*)} < s_1, \quad \forall x \in [0, 1],$$

and, by the first inequality, $\beta(x) \equiv 0$ is an upper solution of $(E_{s^*})$–$(1.2)$.

The function $\alpha(x) = -r\, x^2/2$ is a lower solution of $(E_{s^*})$–$(1.2)$ and, by Theorem 2.4, there is $s^* < s_1$ and a solution of $(E_{s^*})$–$(1.2)$ with $s^* < s_1$. Suppose that $(E_\sigma)$–$(1.2)$ has a solution $u_\sigma(x)$. For $s$ such that $\sigma \leq s \leq s_1$,

$$u_\sigma^{(4)}(x) \leq s\, p(x) - f\big(x, u_\sigma(x), u_\sigma'(x), u_\sigma''(x), u_\sigma'''(x)\big)$$

and so $u_\sigma(x)$ is an upper solution of $(E_s)$–$(1.2)$ for every $s$ such that $\sigma \leq s \leq s_1$.

For $r > 0$ given by (2.5) take $R \geq r$ large enough such that

$$u_\sigma''(0) \geq -R, \quad u_\sigma''(1) \geq -R \quad \text{and} \quad \min_{x \in [0,1]} u_\sigma'(x) \geq -R \qquad (2.6)$$

the function $\alpha(x) = -Rx^2/2$ is a lower solution of (E$_s$)–(1.2) for $s \leq s_1$. If there is $x \in [0,1]$ such that $u_\sigma''(x) < -R$, define

$$\min_{x \in [0,1]} u_\sigma''(x) := u_\sigma''(x_0) \ (< -R),$$

then, by (2.6), $x_0 \in \,]0,1[$, $u_\sigma'''(x_0) = 0$, $u_\sigma^{(4)}(x_0) \geq 0$. By $(H_1)$, $(H_2)$, (2.6) and (2.5), the following contradiction is obtained

$$0 \leq u_\sigma^{(4)}(x_0) \leq \sigma \, p(x_0) - f\big(x_0, u_\sigma(x_0), u_\sigma'(x_0), -R, 0\big)$$
$$\leq s_1 \, p(x_0) - f(x_0, -R, -R, -R, 0) < 0.$$

So $-R \leq u_\sigma''(x)$, for every $x \in [0,1]$, and, by Theorem 2.4, problem (E$_s$)–(1.2) has at least a solution $u(x)$ for every $s$ such that $\sigma \leq s \leq s_1$. Let $S = \{s \in \mathbb{R} : $ (E$_s$)–(1.2) has at least a solution$\}$. As $s^* \in S$ then $S \neq \emptyset$. Defining $s_0 = \inf S$, therefore, $s_0 \leq s^* < s_1$ and (E$_s$)–(1.2) has at least a solution for $s \in \,]s_0, s_1]$ and it has no solution for $s < s_0$. Observe that if $s_0 = -\infty$ then (E$_s$)–(1.2) has a solution for every $s \leq s_1$. $\qquad \square$

## 3. Multiplicity results

In the particular case of boundary conditions (1.1) where $k_2 = k_4 = A = B = C = D = 0$ and $k_1, k_3 > 0$ it will be proved the existence of a second solution for problem (E$_s$)–(1.3) as a consequence of a non null degree for the same operator in two disjoint sets.

The arguments are based on strict lower and upper solutions an some new assumptions on the nonlinearity.

**Definition 3.1.** *Consider* $\alpha, \beta : [0,1] \to \mathbb{R}$ *such that* $\alpha, \beta \in C^3(]0,1[) \cap C^2([0,1])$. *A function* $\alpha(x)$ *is a strict lower solution of (E$_s$)–(1.3 ) if*

$$\alpha^{(4)}(x) + f\big(x, \alpha(x), \alpha'(x), \alpha''(x), \alpha'''(x)\big) > s \, p(x), \quad \text{for } x \in ]0,1[, \qquad (3.1)$$
$$\alpha(0) \leq 0, \quad \alpha'(0) \leq 0, \quad \alpha''(0) < 0, \quad \alpha''(1) < 0.$$

*A function* $\beta(x)$ *is a strict upper solution of (E$_s$)–(1.3) if the reversed inequalities hold.*

Define the set $X = \{y \in C^2([0,1]) : y(0) = y'(0) = y''(0) = y''(1) = 0\}$ and the operators $L : dom \, L \to C([a,b])$, with $dom \, L = C^4([0,1]) \cap X$, given by $Lu = u^{(4)}$ and, for $s \in \mathbb{R}$, $N_s : C^3([0,1]) \cap X \to C([0,1])$ given by

$$N_s u = f\big(x, u(x), u'(x), u''(x), u'''(x)\big) - s \, p(x).$$

For an open and bounded set $\Omega \subset X$, the operator $L + N_s$ is $L$-compact in $\overline{\Omega}$, [6]. Remark that in $dom \, L$ the equation $Lu + N_s u = 0$ is equivalent to problem (E$_s$)–(1.3).

Next result will be an important tool to evaluate the Leray–Schauder topological degree.

**Lemma 3.2.** *Consider a continuous function* $f : [a, b] \times \mathbb{R}^3 \to \mathbb{R}$ *verifying a Nagumo-type condition and* $(H_1)$. *If there are strict lower and upper solutions of* $(E_s)$–(1.3), $\alpha(x)$ *and* $\beta(x)$, *respectively, such that*

$$\alpha''(x) < \beta''(x), \quad \forall x \in [0, 1], \tag{3.2}$$

*then there is* $\rho_3 > 0$ *such that* $d(L + N_s, \Omega) = \pm 1$ *for*

$$\Omega = \left\{ y \in domL : \alpha^{(i)}(x) < y^{(i)}(x) < \beta^{(i)}(x), i = 0, 1, 2, \|y'''\| < \rho_3 \right\}.$$

**Remark 3.3.** The set $\Omega$ can be taken the same, independent of $s$, as long as $\alpha$ and $\beta$ are strict lower and upper solutions for $(E_s)$–(1.3) and $s$ belongs to a bounded set.

*Proof.* For $i = 0, 1, 2$ define the truncations

$$\delta_i(x, y_i) = \max \left\{ \alpha^{(i)}(x), \min \left\{ y_i, \beta^{(i)}(x) \right\} \right\}, \quad \forall(x, y_i) \in [0, 1] \times \mathbb{R}$$

consider the modified problem

$$\begin{cases} u^{(4)}(x) + F\left(x, u(x), u'(x), u''(x), u'''(x)\right) = s\, p(x) \\ u(0) = u'(0) = u''(0) = u''(1) = 0 \end{cases} \tag{3.3}$$

with $F : [0, 1] \times \mathbb{R}^4 \to \mathbb{R}$ a continuous function given by

$$F(x, y_0, y_1, y_2, y_3) = f\left(x, \delta_0(x, y_0), \delta_1(x, y_1), \delta_2(x, y_2), y_3\right) - y_2 + \delta_2(x, y_2)$$

and define the operator $F_s : C^3([0, 1]) \cap X \to C([0, 1])$ by

$$F_s u = F\left(x, u(x), u'(x), u''(x), u'''(x)\right) - s\, p(x).$$

With these definitions problem (3.3) is equivalent to the equation $Lu + F_s u = 0$ in $dom\ L$. For $\lambda \in [0, 1]$ and $u \in dom\ L$ consider the homotopy

$$\mathcal{H}_\lambda u := Lu - (1 - \lambda)\ u'' + \lambda\ F_s u$$

and take $\rho_2 > 0$ large enough such that, for every $x \in [0, 1]$,

$$-\rho_2 \le \alpha''(x) < \beta'' \le \rho_2,$$

$$s\, p(x) - f(x, \alpha(x), \alpha'(x), \alpha''(x), 0) - \rho_2 - \alpha''(x) < 0$$

and

$$s\, p(x) - f\left(x, \beta(x), \beta'(x), \beta''(x), 0\right) + \rho_2 - \beta''(x) > 0.$$

Following similar arguments to the proof of Theorem 2.4, there is $\rho_3 > 0$ such that every solution $u(x)$ of $\mathcal{H}_\lambda u = 0$ satisfies $\|u''\| < \rho_2$ and $\|u'''\| < \rho_3$, independently of $\lambda \in [0, 1]$. Defining $\Omega_1 = \{y \in dom\ L : \|y''\| < \rho_2, \|y'''\| < \rho_3\}$ then, every solution $u$ of $\mathcal{H}_\lambda u = 0$ belongs to $\Omega_1$ for every $\lambda \in [0, 1]$, $u \notin \partial\Omega_1$ and the degree $d(\mathcal{H}_\lambda, \Omega_1)$ is well defined, for every $\lambda \in [0, 1]$.

For $\lambda = 0$ the equation $\mathcal{H}_0 u = 0$, that is, the linear problem

$$\begin{cases} u^{(4)}(x) - u''(x) = 0 \\ u(0) = u'(0) = u''(0) = u''(1) = 0 \end{cases}$$

has only the trivial solution and, by degree theory, $d(\mathcal{H}_0, \Omega_1) = \pm 1$. By the invariance under homotopy

$$\pm 1 = d(\mathcal{H}_0, \Omega_1) = d(\mathcal{H}_1, \Omega_1) = d(L + F_s, \Omega_1). \tag{3.4}$$

By (3.4), there is $u_1(x) \in \Omega_1$ solution of $Lu + F_s u = 0$. Assume, by contradiction, that there is $x \in [0, 1]$ such that $u_1''(x) \leq \alpha''(x)$ and define

$$\min_{x \in [0,1]} \left[ u_1''(x) - \alpha''(x) \right] := u_1''(x_1) - \alpha''(x_1) \ (\leq 0).$$

From (3.1) $x_1 \in ]0, 1[$, $u_1'''(x_1) - \alpha'''(x_1) = 0$ and $u_1^{(4)}(x_1) - \alpha^{(4)}(x_1) \geq 0$. By $(H_1)$, the following contradiction is achieved

$$
\begin{aligned}
u_1^{(4)}(x_1) &= s\, p(x_1) - F\big(x_1, u_1(x_1), u_1'(x_1), u_1''(x_1), u_1'''(x_1)\big) \\
&\leq s\, p(x_1) - f\big(x_1, \alpha(x_1), \alpha'(x_1), \alpha''(x_1), \alpha'''(x_1)\big) + u_1''(x_1) - \alpha''(x_1) \\
&\leq s\, p(x_1) - f\big(x_1, \alpha(x_1), \alpha'(x_1), \alpha''(x_1), \alpha'''(x_1)\big) < \alpha^{(4)}(x_1).
\end{aligned}
$$

Therefore $u_1''(x) > \alpha''(x)$, for $x \in [0, 1]$. By a similar way it can be proved that $u_1''(x) < \beta''(x)$, for every $x \in [0, 1]$. By integration and (1.3), $u_1 \in \Omega$.

As the equations $Lu + F_s u = 0$ and $Lu + N_s u = 0$ are equivalent on $\Omega$ then

$$d(L + F_s, \Omega_1) = d(L + F_s, \Omega) = d(L + N_s, \Omega) = \pm 1,$$

by (3.4) and the excision property of the degree.                          $\square$

The main result is attained assuming that $f$ is bounded from below and it satisfies some adequate condition of monotonicity-type which requires different "speeds" of growth.

**Theorem 3.4.** *Let $f : [0, 1] \times \mathbb{R}^4 \to \mathbb{R}$ be a continuous function such that the assumptions of Theorem 2.6 are fulfilled. Suppose that there is $M \in \mathbb{R}$ such that every solution $u$ of $(\mathrm{E}_s)$–(1.3), with $s \leq s_1$, satisfies*

$$u''(x) < M, \quad \forall x \in [0, 1], \tag{3.5}$$

*and there exists $m \in \mathbb{R}$ such that*

$$f(x, y_0, y_1, y_2, y_3) \geq m\, p(x), \tag{3.6}$$

*for every $(x, y_0, y_1, y_2, y_3) \in [0, 1] \times [-r, |M|]^2 \times [-r, M] \times \mathbb{R}$, with $r$ given by (2.5). Then $s_0$, provided by Theorem 2.6, is finite and: if $s < s_0$, $(\mathrm{E}_s)$–(1.3) has no solution; if $s = s_0$, $(\mathrm{E}_s)$–(1.3) has at least one solution.*

*Moreover, let $M_1 := \max\{r, |M|\}$ and assume that there is $\theta > 0$ such that, for every $(x, y_0, y_1, y_2, y_3) \in [0, 1] \times [-M_1, M_1]^3 \times \mathbb{R}$ and $0 \leq \eta_0, \eta_1 \leq 1$,*

$$f(x, y_0 + \eta_0\, \theta, y_1 + \eta_1\, \theta, y_2\, \theta, y_3) \leq f(x, y_0, y_1, y_2, y_3). \tag{3.7}$$

*Then for $s \in ]s_0, s_1]$, $(\mathrm{E}_s)$–(1.3) has at least two solutions.*

*Proof.* If, by contradiction, there are $s \in ]s_0, s_1]$, $u$ solution of $(E_s)$–(1.3) and $x_2 \in [0,1]$ such that

$$u''(x_2) := \min_{x \in [0,1]} u''(x) \leq -r.$$

By (1.3), $x_2 \in ]0, 1[$, $u'''(x_2) = 0$ and $u^{(4)}(x_2) \geq 0$. By $(H_2)$,

$$0 \leq u^{(4)}(x_2) \leq s_1 \, p(x_2) - f\big(x_2, u(x_2), u'(x_2), -r, 0\big).$$

If $u(x_2) < -r$ and $u'(x_2) < -r$, from (2.5) the following contradiction is obtained

$$0 \leq s_1 \, p(x_2) - f\big(x_2, u(x_2), u'(x_2), -r, 0\big) < 0.$$

If $u(x_2) \geq -r$ and $u'(x_2) \leq -r$ (the case $u(x_2) \leq -r$ and $u'(x_2) \geq -r$ is similar), from $(H_1)$ and (2.5), this contradiction is achieved

$$0 \leq s_1 \, p(x_2) - f\big(x_2, u(x_2), u'(x_2), -r, 0\big) \leq s_1 \, p(x_2) - f\big(x_2, u(x_2), -r, -r, 0\big) < 0.$$

If $u(x_2) \geq -r$ and $u'(x_2) \geq -r$ then

$$0 \leq s_1 \, p(x_2) - f\big(x_2, u(x_2), u'(x_2), -r, 0\big) \leq s_1 \, p(x_2) - f(x_2, -r, -r, -r, 0) < 0.$$

Therefore, every solution $u$ of $(E_s)$–(1.3), with $s_0 < s \leq s_1$, verifies $u''(x) > -r$, for $x \in [0,1]$, and, by (3.5), $-r < u''(x) < M$, for every $x \in [0,1]$. Integrating on $[0, x]$, it is obtained $-r \leq -rx < u'(x) < M \, x \leq |M|$, $\forall x \in [0,1]$.

Suppose that $s_0 = -\infty$, that is, by Theorem 2.6, for every $s \leq s_1$ problem $(E_s)$–(1.3) has at least a solution. Define

$$p_1 := \min_{x \in [0,1]} p(x) > 0$$

and take $s$ sufficiently small such that

$$m - s > 0 \quad \text{and} \quad \frac{(m - s) \, p_1}{16} > M.$$

If $u(x)$ is a solution of $(E_s)$–(1.3), then, by (3.6), $u^{(4)}(x) \leq (s - m) \, p(x)$ and, by (1.3), there is $x_3 \in ]0, 1[$ such that $u'''(x_3) = 0$. For $x < x_3$

$$u'''(x) = -\int_x^{x_3} u^{(4)}(\xi) \, d\xi \geq \int_x^{x_3} (m - s) \, p(\xi) \, d\xi \geq (m - s)(x_3 - t) \, p_1.$$

For $x \geq x_3$

$$u'''(x) = \int_{x_3}^x u^{(4)}(\xi) \, d\xi \leq (s - m)(x - x_3) \, p_1.$$

Choose $I = \left[0, \frac{1}{4}\right]$, or $I = \left[\frac{3}{4}, 1\right]$, such that $|x_3 - t| \geq \frac{1}{4}$, for every $x \in I$. If $I = \left[0, \frac{1}{4}\right]$, then $u'''(x) \geq (m - s)p_1/4$, for $x \in I$, and if $I = \left[\frac{3}{4}, 1\right]$, then $u'''(x) \leq (s - m) \, p_1/4$, for $x \in I$. In the first case,

$$0 = \int_0^{\frac{1}{4}} u'''(x) \, dx + \int_{\frac{1}{4}}^1 u'''(x) \, dx \geq \int_0^{\frac{1}{4}} \frac{(m - s) \, p_1}{4} \, dx - u''\left(\frac{1}{4}\right)$$

$$= \frac{1}{16}(m - s) \, p_1 - u''\left(\frac{1}{4}\right) > M - u''\left(\frac{1}{4}\right),$$

which is in contradiction with (3.5).

For $I = \left[\frac{3}{4}, 1\right]$ a similar contradiction is achieved. Therefore, $s_0$ is finite.

By Theorem 2.6, for $s_{-1} < s_0$, $(E_{s_{-1}})$–(1.3) has no solution. By Lemma 2.2, consider $\rho_3 > 0$ large enough such that the estimate $\|u'''\| < \rho_3$ holds for every $u$ solution of $(E_s)$–(1.3), with $s \in [s_{-1}, s_1]$.

Define $M_1 := \max\{r, |M|\}$ and the set

$$\Omega_2 = \{y \in \text{dom } L : \|y''\| < M_1, \|y'''\| < \rho_3\}.$$

Then $d(L + N_{s_{-1}}, \Omega_2) = 0$. If $u$ is a solution of $(E_s)$–(1.3), with $s \in [s_{-1}, s_1]$, then $u \notin \partial\Omega_2$. Defining the convex combination of $s_1$ and $s_{-1}$ as $H(\lambda) = (1-\lambda)s_{-1} + \lambda s_1$ and considering the corresponding homotopic problems $(E_{H(\lambda)})$–(1.3), the degree $d(L + N_{H(\lambda)}, \Omega_2)$ is well defined for every $\lambda \in [0, 1]$ and for every $s \in [s_{-1}, s_1]$. Therefore, by the invariance of the degree

$$0 = d(L + N_{s-1}, \Omega_2) = d(L + N_s, \Omega_2), \tag{3.8}$$

for $s \in [s_{-1}, s_1]$. Let $\sigma \in ]s_0, s_1] \subset [s_{-1}, s_1]$ and $u_\sigma(x)$ be a solution of $(E_\sigma)$–(1.3), which exists by Theorem 2.6. Take $\varepsilon > 0$ such that

$$|u''_\sigma(x) + \varepsilon| < M_1, \quad \forall x \in [0, 1]. \tag{3.9}$$

Then $\widetilde{u}(x) := u_\sigma(x) + \varepsilon\frac{x^2}{2}$ is a strict upper solution of $(E_s)$–(1.3), with $\sigma < s \leq s_1$. In fact, by (3.7) with $\theta = \varepsilon$, $\eta_0 = \frac{x^2}{2}$ and $\eta_1 = x$ for such $\sigma$,

$$\widetilde{u}^{(4)}(x) = u_\sigma^{(4)}(x) < s\, p(x) - f\left(x, u_\sigma(x), u'_\sigma(x), u''_\sigma(x), \widetilde{u}'''(x)\right)$$

$$\leq s\, p(x) - f\left(x, u_\sigma(x) + \varepsilon\frac{x^2}{2}, u'_\sigma(x) + \varepsilon x, \widetilde{u}''(x) + \varepsilon, \widetilde{u}'''(x)\right)$$

$$= s\, p(x) - f\left(x, \widetilde{u}(x), \widetilde{u}'(x), \widetilde{u}''(x), \widetilde{u}'''(x)\right);$$

$$\widetilde{u}(0) = 0, \quad \widetilde{u}'(0) = 0, \quad \widetilde{u}''(0) = \widetilde{u}''(1) = \varepsilon > 0.$$

Moreover $\alpha(x) := -r\frac{x^2}{2}$ is a strict lower solution of $(E_s)$–(1.3), for $s \leq s_1$. Indeed, by (2.5) and $(H_1)$,

$$\alpha^{(4)}(x) = 0 > s_1\, p(x) - f(x, -r, -r, -r, 0) \geq s\, p(x) - f\left(x, -r\frac{x^2}{2}, -rx, -r, 0\right);$$

$$\alpha(0) = \alpha'(0) = 0, \quad \alpha''(0) = \alpha''(1) = -r < 0.$$

As $-r < u''_\sigma(x)$ for every $x \in [0, 1]$ and therefore $-r < u''_\sigma(x) + \varepsilon$, $\forall x \in [0, 1]$, that is, $\alpha''(x) < \widetilde{u}''(x)$. Integrating on $[0, x]$, $\alpha'(x) \leq \alpha'(x) - \alpha'(0) < \widetilde{u}'(x) - \widetilde{u}'(0) = \widetilde{u}'(x)$, for every $x \in [0, 1]$. Then, by (3.9), Lemma 3.2 and Remark 3, there is $\overline{\rho}_3 > 0$, independent of $s$, such that for

$$\Omega_\varepsilon = \left\{y \in domL : \alpha^{(i)}(x) < y^{(i)}(x) < \widetilde{u}^{(i)}(x), i = 0, 1, 2, \|y'''\| < \overline{\rho}_3\right\}$$

the degree of $L + N_s$ in $\Omega_\varepsilon$ satisfies

$$d(L + N_s, \Omega_\varepsilon) = \pm 1, \quad \text{for } s \in ]\sigma, s_1]. \tag{3.10}$$

Taking $\rho_3$ in $\Omega_2$ large enough such that $\Omega_\varepsilon \subset \Omega_2$, by (3.8), (3.9) and the additivity of the degree, we obtain

$$d(L + N_s, \Omega_2 - \overline{\Omega_\varepsilon}) = \mp 1, \quad \text{for } s \in ]\sigma, s_1]. \tag{3.11}$$

So, problem $(E_s)$–(1.3) has at least two solutions $u_1, u_2$ such that $u_1 \in \Omega_\varepsilon$ and $u_2 \in \Omega_2 - \overline{\Omega_\varepsilon}$, for $s \in ]s_0, s_1]$, since $\sigma$ is arbitrary in $]s_0, s_1]$.

Consider a sequence $(s_m)$ with $s_m \in ]s_0, s_1]$ and $\lim s_m = s_0$. By Theorem 2.6, for each $s_m$, $(E_{s_m})$–(1.3) has a solution $u_m$. Using the estimates of Step 1, $\|u_m^{(i)}\| < M_1$, $i = 0, 1, 2$, independently of $m$. As there is $\widetilde{\rho}_3 > 0$ large enough such that $\|u_m'''\| < \widetilde{\rho}_3$, independently of $m$, t hen sequences $(u_m)$, $(u_m')$ and $(u_m'')$, $m \in \mathbb{N}$, are bounded in $C([0,1])$. By the Arzèla-Ascoli theorem, we can take a subsequence of $(u_m)$ that converges in $C^3([0,1])$ to a solution $u_0(x)$ of $(E_{s_0})$–(1.3). Hence, there is at least one solution for $s = s_0$. $\qquad\square$

# References

[1] C. de Coster and L. Sanchez, *Upper and lower solutions, Ambrosetti–Prodi problem and positive solutions for a fourth order O.D.E.*, Riv. Mat. Pura Appl., **14** (1994), 57–82.

[2] C. Fabry, J. Mawhin, and M. N. Nkashama, *A multiplicity result for periodic solutions of forced nonlinear second order ordinary differential equations*. Bull. London Math. Soc., **18** (1986), 173-180.

[3] D. Franco, D. O'Regan, and J. Perán, *Fourth-order problems with nonlinear boundary conditions*, J. Comput. Appl. Math. 174 (2005) 315–327.

[4] M. R. Grossinho and F. Minhós, *Upper and lower solutions for some higher order boundary value problems*, Nonlinear Studies, **12** (2005) 165–176.

[5] T. F. Ma and J. da Silva, *Iterative solutions for a beam equation with nonlinear bondary conditions of third order*, Appl. Math. Comp., 159 (2004) 11–18.

[6] J. Mawhin, *Topological degree methods in nonlinear boundary value problems*, Regional Conference Series in Mathematics, 40, American Mathematical Society, Providence, Rhode Island, 1979.

[7] F. Minhós, T. Gyulov, and A. I. Santos, *Existence and location result for a fourth order boundary value problem*, Proc. of the $5^{th}$ International Conference on Dynamical Systems and Differential Equations, Disc. Cont. Dyn. Syst. (2005) 662–671.

[8] F. Minhós, T. Gyulov, and A. I. Santos, *On an elastic beam fully equation with nonlinear boundary conditions*, Proc. of the Conference on Differential & Difference Equations and Applications (2005) (To appear).

[9] M. Nagumo, *Über die Differentialgleichung $y'' = f(t, y, y')$*, Proc. Phys.-Math. Soc. Japan 19, (1937), 861–866.

[10] M. Šenkyřík, *Existence of multiple solutions for a third order three-point regular boundary value problem*, Mathematica Bohemica, **119**, n°2 (1994), 113–121.

Feliz Manuel Minhós
Departmento de Matemática
Universidade de Évora
Centro de Investigação em Matemática e Aplicações da U.E
Colégio Luis António Verney
Rua Romão Ramalho, 59
P-7000-671 Évora
Portugal
e-mail: `fminhos@uevora.pt`

Progress in Nonlinear Differential Equations
and Their Applications, Vol. 75, 269–283

# Reducing a Differential Game to a Pair of Optimal Control Problems

Ştefan Mirică

*To Arrigo Cellina and James Yorke*

**Abstract.** Author's recent concepts and results are extended here to the more general case of *non-autonomous* (i.e., time-dependent) differential games.

The main idea is to introduce first a concept of *admissible pair of (multi-valued) feedback strategies* to which one may associate a *value function* and which reduce the differential game to a pair of symmetric, non-smooth optimal control problems for differential inclusions; secondly, one introduces the concepts of *bilaterally-optimal* pairs of feedback strategies and one proves an *abstract verification theorem* containing necessary and sufficient optimality conditions; next, this approach is made more realistic by the proof of several "practical verification theorems" containing corresponding differential inequalities and regularity hypotheses on the value function that imply the optimality.

As a method for constructing simultaneously, the value function, the pair of optimal feedback strategies and also the optimal trajectories, certain natural extensions of Cauchy's Method of Characteristics to non-smooth Hamilton–Jacobi equations are suggested.

**Mathematics Subject Classification (2000).** Primary 49N70; Secondary 49N35; 49N90; 91A23; 93B52.

**Keywords.** Differential game, feedback strategy, value function, verification theorem, Hamiltonian flow.

## 1. Introduction

The aim of this paper is to extend to the more general case of *non-autonomous* (i.e., time-dependent) differential games, author's recent concepts and results developed in [23, 24, 27, 29, etc.] for autonomous differential games; though the main ideas are practically the same as in the autonomous case, some formulations (including the verification theorems, the associated Hamilton–Jacobi equation, etc.) may take different forms in the non-autonomous case.

We start from the fact that a (possibly multi-valued) feedback strategy of one player defines an optimal control problem for the other player and these two optimal control problems may be studied in the framework of Dynamic Programming; in this context, we introduce the concept of *admissible pair of (multi-valued) feedback strategies* to which one may associate a *value function* and which should coincide with the *common value function* of the two symmetric optimal control problems; we thus arrive at the concept of **bilaterally-optimal pair of feedback strategies** in which each member is an optimal feedback for the corresponding optimal control problem.

Therefore, an admissible pair of feedback strategies *reduces the differential game to a pair of symmetric optimal control problems* which are *non-smooth* since the feedbacks are usually so and also are defined by corresponding *differential inclusions* since the feedbacks may be multi-valued.

This approach has the great advantage of allowing the direct use of the existing experience in Optimal Control Theory (e.g., [5, 15–22, 26, 28–30], etc.), to the study of differential games; in particular, one may use suitable extensions of Cauchy's Method of Characteristics to non-smooth Hamilton–Jacobi equations to find "candidates" for value functions, optimal feedback strategies and the corresponding "optimal trajectories" and then one may use suitable "verification theorems" to prove the optimality.

The efficiency of this approach is illustrated by the possibility to obtain complete rigorous solutions to some classes of differential games in the literature, such as, for instance, those considered in R. Isaacs' book [12] (see [27]).

Using recent developments in Non-smooth Analysis and author's experience in optimal control (e.g., [15–30]), the approach in this paper may be considered as an attempt to refine and to make "theoretically acceptable" the rather heuristical but **pragmatical** approach of R. Isaacs in [12], in a different direction than those pursued by Krassovskii and Subbotin (e.g., [13, 32, 33], etc.), Berkovitz (e.g., [3, 4], etc.) and other authors.

In our setting there are two essential mathematical problems:

1) *find efficient methods and procedures for describing (characterizing, etc.) admissible, possibly optimal, pairs of feedback strategies, accompanied by the corresponding value functions and trajectories;*
2) *provide "practical" verification theorems that contain "verifiable" sufficient conditions for optimality.*

For the first problem we are suggesting the use of certain recent extensions and generalizations of Cauchy's Method of Characteristics to non-smooth Hamilton–Jacoby equations (see [16, 22, 26, 29], etc.), to describe (more or less explicitly) plausible "candidates" for optimal pairs of feedback strategies, corresponding value functions and optimal trajectories; in contrast with Isaacs' approach in [12] (which rather improperly used the classical Method of Characteristics for non-differentiable Hamiltonians) this procedure is more precisely described by the introduction of certain "generalized Hamiltonian systems", "generalized Hamiltonian flows" and certain parameterized, finite-dimensional optimization problems that may define, simultaneously, the value function, the feedback strategies and the corresponding trajectories which are "suspects" of being optimal in a precisely defined sense.

As already stated, for the second essential problem we are using the experience in optimal control to obtain first *an abstract verification theorem* containing necessary and sufficient optimality conditions then we use some concepts and results from non-smooth analysis to obtain *easier verifiable sufficient optimality conditions* expressed in terms of certain differential inequalities accompanied by suitable "regularity properties" of the value functions.

One should also note that the two problems are closely related since some properties of the Hamiltonian flow and certain evaluations of the contingent derivatives of the involved "marginal functions" may facilitate the applications of suitable verification theorems.

Returning to the "pragmatical" point of view of R. Isaacs in [12], we consider that the only possibility of acting "optimally" in a "two-person zero-sum" differential game is that of following the trajectories generated by "previously calculated (described, etc.)" *feedback strategies* that are "mutually (relatively) optimal" in a certain sense that should be precisely defined; this approach is obviously different from the one based on the so-called "non-anticipative" strategies and the corresponding VREK (Varaiya–Roxin–Elliot–Kalton) value functions and viscosity solutions (e.g., [2, 6, 9–11, 14, 31, 34], etc.); in fact, besides being very abstract, the non-anticipative strategies are obviously "non-computable" and "non-playable" while the VREK value functions, even in the form of viscosity solutions, are of little use for any of the two players.

The paper is organized as follows: in Section 2 we introduce the concepts of (set-valued) admissible and, respectively, of relatively optimal feedback strategies and of the associated value function, in Section 3 we present the abstract verification theorem and in Section 4 the statements of several verification theorems of Dynamic Programming type are presented; in the last section, we present very shortly a generalization of Cauchy's Method of Characteristics that may be used to construct the relatively optimal feedback strategies, the associated value functions and also the optimal trajectories.

## 2. Admissible and optimal pairs of feedback strategies

In a rather "vague formulation", a non-autonomous differential game (DG) is stated first in the following *"standard form"*

**Problem 2.1 ($(DG)$-vague formulation). Find**

$$\inf_{u(.)\in\mathcal{U}_\alpha} \sup_{v(.)\in\mathcal{V}_\alpha} \mathcal{C}\big(s,y;u(.),v(.)\big) \quad \forall\, (s,y)\in E_0 \tag{2.1}$$

*subject to*

$$\mathcal{C}\big(s,y;u(.),v(.)\big) = g\big(\tau,x(\tau)\big) + \int_s^\tau f_0\big(t,x(t),u(t),v(t)\big)dt\,, \quad (s,y)\in E_0 \tag{2.2}$$

$$x'(t) = f\big(t,x(t),u(t),v(t)\big) \text{ a.e.}(s,\tau)\,, \quad x(s)=y \tag{2.3}$$

$$u(t)\in U\,, \quad v(t)\in V \text{ a.e.}(s,\tau)\,, \quad \big(u(.),v(.)\big)\in\mathcal{U}_\alpha\times\mathcal{V}_\alpha \tag{2.4}$$

$$\widehat{x}(.) = \big(x(.),x_0(.)\big)\in\Omega_\alpha\,, \quad x_0(t) := \int_s^t f_0\big(r,x(r),u(r),v(r)\big)dr \tag{2.5}$$

$$\big(t,x(t)\big)\in E_0 \quad \forall\, t\in[s,\tau)\,, \quad \big(\tau,x(\tau)\big)\in E_1\,, \quad E_0\cap E_1=\emptyset\,. \tag{2.6}$$

Obviously, *the data of the problem* are the following:

- the *sets of all control parameters* $U,V$, of the two players, assumed to be at least topological spaces and, usually, subsets of some Euclidean spaces;
- the *set of initial phases (time-state)* $E_0\subset R\times R^n$ and the *set of terminal phases*, $E_1\subset\partial E_0$;
- the *terminal payoff function* $g(.,.) : E_1\to R$ and the "running payoff function", $f_0(.,.,.,.) : D\times U\times V\to R$,
- the *parameterized vector field* $f(.,.,.,.) : D\times U\times V\to R^n$, that defines the "dynamics" in (2.3) of the differential game;
- the class $P_\alpha=\mathcal{U}_\alpha\times\mathcal{V}_\alpha\in\{P_1,P_\infty,P_r\}$ of *admissible control-pairs* $(u(.),v(.))$ that generate the corresponding class $\Omega_\alpha\in\{\Omega_1,\Omega_\infty,\Omega_r\}$ of *admissible trajectories*;
- the *payoff (cost) functionals* in (2.2).

**Remark 2.2.** As it is apparent from this succinct formulation, the "terminal time" $\tau=\tau(s,y,x(.))>s$ is **free** and defines in an obvious way the "terminating rule" in (2.6), i.e., $\tau>s$ is the first moment at which the graph of the trajectory $x(.)$ reaches the terminal set, $E_1$ ; also, the mappings $\widehat{f}(.,x(.),u(.),v(.))$ should be (Lebesgue) integrable if $\Omega_\alpha=\Omega_1=AC$ (while $P_\alpha=P_1$ denotes the set of mappings $(u(.),v(.))$ for which this property holds), should be "essentially bounded" if $\Omega_\alpha=\Omega_\infty$ and should be "regular" if $\Omega_a=\Omega_r$; the classes $P_\alpha\in\{P_1,P_\infty,P_r\}$ of *admissible control pairs* become "relevant" classes of *measurable mappings* in the case $U,V$ are topological spaces and the mapping $\widehat{f}(.)=(f(.),f_0(.))$ is continuous.

An important particular case is that of *fixed terminal-time problems* in which $\tau=T\in R$ *is fixed*, and in which the terminal set is of the form $E_1=T\times Y_1$, $Y_1\subseteq R^n$ and $E_0\subset(-\infty,T)\times Y_0$, $Y_0\subseteq R^n$, that is studied in a large number of books and papers; another important particular case is that of *autonomous problems* in which *the time-variable*, $t$, is absent from all the data and therefore the formulation of the problem may be simplified taking always $s=0$ as the "initial moment", a set $Y_0 \subset R^n$ as the set of initial states and $Y_1 \subset \partial Y_0$ as the set of terminal states (e.g., [23, 24, 29], etc.).

One may note also that in some concrete problems the sets of "usable control parameters" may be "multi-functions" of the type $U(t, x) \subseteq U$, $V(t, x) \subseteq V$ for which all the results to follow remain valid; in order to simplify the exposition we consider here only the case $U(t, x) \equiv U$, $V(t, x) \equiv V$.

The rather vague formulation of Problem 2.1, in which *the aim* in (2.1) is not associated to any "information pattern", should be replaced by the following, more precise one, in which the **feedback-type** information pattern has been chosen:

**Problem 2.3** (*DG*-**accurate formulation**). *Given the data of Problem* 2.1, **find the feedback strategies** $\widetilde{U}(t, x) \subset U$, $\widetilde{V}(t, x) \subset V$, $(t, x) \in \widetilde{E}_0 \subseteq E_0$ with the following properties

(A) The pair $(\widetilde{U}(.,.), \widetilde{V}(.,.))$ is **admissible** in the sense that for any $(s, y) \in \widetilde{E}_0$, the set $\widetilde{\Omega}_\alpha(s, y)$ of trajectories $x_{s,y}(.) \in \Omega_\alpha(s, y)$ of the differential inclusion

$$x' \in f(t, x, \widetilde{U}(t, x), \widetilde{V}(t, x)), \quad x(s) = y \qquad (2.7)$$

that satisfy the constraints

$$(t, x_{s,y}(t)) \in \widetilde{E}_0 \ \forall \ t \in [s, \tau), \quad (\tau, x_{s,y}(\tau)) \in \widetilde{E}_1 \subseteq E_1, \quad \tau = \tau(x_{s,y}(.)) > s \ (2.8)$$

is not empty; moreover, if $\widetilde{P}_\alpha(s, y)$, $(s, y) \in \widetilde{E}_0$ are the corresponding sets of *control mappings*, $(u_{s,y}(.), v_{s,y}(.)) \in P_\alpha$ that satisfy

$$\begin{aligned} x'_{s,y}(t) &= f(t, x_{s,y}(t), u_{s,y}(t), v_{s,y}(t)), \quad x(s) = y, \\ u_{s,y}(t) &\in \widetilde{U}(t, x_{s,y}(t)), \quad v_{s,y}(t) \in \widetilde{V}(t, x_{s,y}(t)) \ a.e.(s, \tau) \end{aligned} \qquad (2.9)$$

then *there exists the associated value function defined by*

$$\begin{aligned} \widetilde{W}_0(s, y) &= C(s, y; u_{s,y}(.), v_{s,y}(.)) \quad \forall \ (u_{s,y}(.), v_{s,y}(.)) \in \widetilde{P}_\alpha(s, y), \\ \widetilde{W}(s, y) &= \begin{cases} \widetilde{W}_0(s, y) \ if \ (s, y) \in \widetilde{E}_0 \\ g(s, y), \ if \ (s, y) \in \widetilde{E}_1 \end{cases} \end{aligned} \qquad (2.10)$$

where the set of effective end-points is defined by: $\widetilde{E}_1 = \{(\tau, x_{s,y}(\tilde{\tau})); x_{s,y}(.) \in \widetilde{\Omega}_\alpha(s, y), (s, y) \in \widetilde{E}_0\}$;

(B) The feedback strategies $\widetilde{U}(.,.), \widetilde{V}(.,.)$ are **bilaterally ("relatively") optimal** for the restriction $DG|\widetilde{E}_0$ in the following sense

$$\widetilde{W}_0(s, y) = \underline{W}_{\widetilde{V}}(s, y) := \inf_{u(.), \overline{v}(.)} C(s, y; u(.), \overline{v}(.)) \qquad (2.11)$$

**subject to**

$$x'(t) = f(t, x(t), u(t), \overline{v}(t)), \quad u(t) \in U, \quad \overline{v}(t) \in \widetilde{V}(t, x(t)) \qquad (2.12)$$

$$x(s) = y, \quad (t, x(t)) \in \widetilde{E}_0 \ \forall \ t \in [s, \tau), \quad (\tau, x(\tau)) \in \widetilde{E}_1 \qquad (2.13)$$

and also

$$\widetilde{W}_0(s,y) = \overline{W}_{\widetilde{U}}(s,y) = \sup_{\overline{u}(.),v(.)} \mathcal{C}\big(s,y;\overline{u}(.),v(.)\big) \quad \forall (s,y) \in \widetilde{E}_0 \tag{2.14}$$

**subject to (2.13) and to**

$$x'(t) = f\big(t, x(t), \overline{u}(t), v(t)\big), \quad \overline{u}(t) \in \widetilde{U}\big(t, x(t)\big), \quad v(t) \in V. \tag{2.15}$$

(C) Either $\widetilde{E}_0 = E_0$ or $\widetilde{E}_0 \subset E_0$ is *strongly invariant with respect to the control system in* (2.3) *and* (2.4) in the sense that $(t, x(t)) \in \widetilde{E}_0 \ \forall t \in [s, \tau)$ for any admissible trajectory.

**Remark 2.4.** If $\widetilde{E}_0$ it is not strongly invariant in this sense then the "solution" $\widetilde{U}(.,.), \widetilde{V}(.,.), \widetilde{W}(.,.)$ above may not be "satisfactory" since at least one of the players may choose controls producing trajectories that leave the set $\widetilde{E}_0$; in this case one should use additional "ad hoc" arguments to decide whether $DG|\widetilde{E}_0$ it is either a "proper" or an "apparent" restriction of the original problem, $DG$; in particular, one may try to solve the corresponding **game of kind** characterized by the existence of an **evading feedback strategy**, $\widetilde{V}_e(.,.) : \widetilde{E}_0^e \subseteq E_0 \to \mathcal{P}(V)$, of player **V**, such that none of the problems in (2.12), (2.13), for $(s,y) \in \widetilde{E}_0^e$, has an admissible trajectory; in this case, using the "conventions" $\sup \emptyset = -\infty$, $\inf \emptyset = +\infty$, one has

$$\underline{W}_{\widetilde{V}_e}(s,y) = \inf_{u(.),\overline{v}(.)} \mathcal{C}\big(s,y;u(.),\overline{v}(.)\big) = +\infty \quad \forall (s,y) \in \widetilde{E}_0^e$$

and therefore the problem does not have a value function on $\widetilde{E}_0^e$.

**Remark 2.5.** We note that the "feedback differential system" in (2.7) is allowed to have several admissible trajectories through some initial points $(s,y) \in \widetilde{E}_0$, which, however, "produce" the same value, $\widetilde{W}_0(s,y)$, of the payoff functional in (2.2).

On the other hand, the "bilateral" (or "relative" ) optimality conditions in (2.11)–(2.15) may be interpreted as follows in the case condition (C) is satisfied:

- **if player V** *chooses the feedback strategy* $\widetilde{V}(.,.)$ *then* $\widetilde{U}(.,.)$ *is the best choice for player* **U** and, symmetrically,
- *if player* **U** *chooses the feedback strategy* $\widetilde{U}(.,.)$ *then* $\widetilde{V}(.,.)$ *is the best choice for player* **V**; in the case the strict subset $\widetilde{E}_0 \subset E_0$ it is not strongly invariant with respect to the control system in (2.3) and (2.4) then the same conclusions remain valid, provided each player "voluntarily" restricts his own set of control parameters such that all the trajectories remain in the subset $\widetilde{E}_0$, which seems rather improbable.

**Remark 2.6.** As it is apparent from the formulations of Problems 2.1, 2.3, **the admissible trajectories** are at least AC (i.e., absolutely continuous) solutions (of class $\Omega_\alpha$) of one of the differential inclusions:

$$x' \in f\big(t, x, \widetilde{U}(t,x), \widetilde{V}(t,x)\big), \quad x' \in f\big(t, x, U, \widetilde{V}(t,x)\big), \quad x' \in f\big(t, x, \widetilde{U}(t,x), V\big),$$

that satisfy the constraints in (2.6), (2.8), (2.13) and for which there exist corresponding (measurable) selections, $u(.), v(.)$, (of class $\mathcal{P}_\alpha$) satisfying relations of the form in (2.9), (2.12), (2.15).

We recall that in his original approach, Isaacs [12], (taken-up also by Subbotin [33]) used *single-valued* feedback strategies, $\widetilde{U}(t,x) = \{\widetilde{u}(t,x)\}$, $\widetilde{V}(t,x) = \{\widetilde{v}(t,x)\}$, and a certain type of *approximating trajectories* of the feedback differential systems in (2.16), generated by the so called "K(Karlin)-strategies".

Later on, in order to make this approach more rigorous, Krassovskii and Subbotin [13], used *uniform limits* of Isaacs' approximate trajectories and conjectured that under suitable hypotheses they are AC trajectories of certain associated ("convexified") differential inclusions; a more detailed study of this topic may be found in [30] where the associated differential inclusions are more precisely described.

As it is easy to see, the concepts and results in this paper remain valid if the AC trajectories of the differential inclusions in (2.16) are replaced by corresponding *limiting trajectories*, either of Isaacs–Krassovskii–Subbotin type or of Euler type.

**Remark 2.7.** The approach in this paper (and, for that matter, that of Isaacs [12], Krassovskii–Subbotin [13], Subbotin [32, 33], etc.) is, obviously, essentially different from the approach using the so called **NA(non-anticipative)-strategies, VREK value functions and viscosity solutions** adopted in the last 15–20 years by a majority of the authors in the field of Differential Games.

## 3. The abstract verification theorem

Assuming that a pair of *admissible feedback strategies* $(\widetilde{U}(.,.), \widetilde{V}(.,.))$ with the *associated value function* $\widetilde{W}_0$ in (2.10) is given, *the problem is to find criteria* ("*verification theorems*") for their "bilateral optimality" in the sense of statement (B) of Problem 2.3.

To solve this problem one may use the experience in Optimal Control Theory since the associated value function in (2.10) coincides with each of the value function of two **symmetric Bolza optimal control problems**: the first one, $\mathcal{B}_1$, consists in the **minimization of the functionals**:

$$\widetilde{W}_0(s,y) = \inf_{u(.),\overline{v}(.)} \mathcal{C}\big(s,y; u(.), \overline{v}(.)\big) \quad \forall\, (s,y) \in \widetilde{E}_0 \tag{3.1}$$

subject to:

$$x' \in F_{\widetilde{V}}(t,x) := f\big(t,x,U,\widetilde{V}(t,x)\big), \quad x(s) = y \tag{3.2}$$

and to the phase and end-point constrains in (2.13), the other one, $\mathcal{B}_2$, consisting in the **maximization of the functionals**:

$$\widetilde{W}_0(s,y) = \sup_{\overline{u}(.),v(.)} \mathcal{C}\big(s,y; \overline{u}(.), v(.)\big) \quad \forall (s,y) \in \widetilde{E}_0 \tag{3.3}$$

**subject to (2.13) and to:**

$$x' \in F_{\widetilde{U}}(t,x) := f\big(t,x,\widetilde{U}(t,x),V\big), \quad x(s) = y \tag{3.4}$$

where the controls $u(.), \overline{v}(), \overline{u}(.), v(.)$ are (measurable) selections (of class $P_\alpha$) of the multi-functions in (2.12), (2.15), respectively.

Thus, the differential game $DG$ in Problems 2.1, 2.3 **is reduced** to the **two symmetric (non-smooth) optimal control problems** $\mathcal{B}_1$, $\mathcal{B}_2$ described in (3.1), (3.2) and, respectively, (3.3), (3.4).

Using the well known fact that the (completed) value function in (2.10) of an optimal control problem has certain monotonicity properties along admissible trajectories (e.g., Cesari [2], Lupulescu and Mirică [15], Mirică [25, 26], etc.), we obtain the following basic result providing a (rather abstract) **necessary and sufficient optimality condition** that may easily be proved directly:

**Theorem 3.1 (Abstract verification theorem).** *If $\widetilde{U}(.,.), \widetilde{V}(.,.)$ is an admissible pair of feedback strategies with the associated value functions $\widetilde{W}_0(.,.), \widetilde{W}(.,.)$ in (2.10), then it is a bilaterally optimal pair in the sense of statement $(B)$ of Problem 2.3* **iff** *it satisfies the following two monotonicity conditions:*

- **"monotonicity condition"** $(M_1)$: *the real function*

$$\omega_{u,\overline{v}}(t) := \widetilde{W}\big(t, x(t)\big) + \int_s^t f_0\big(r, x(r), u(r), \overline{v}(r)\big) dr \tag{3.5}$$

*is increasing on the interval $[s, \tau]$ for any $u(.), \overline{v}(.), x(.)$ satisfying (2.12) and:*
- **"monotonicity condition"** $(M_2)$: *the real function*

$$\omega_{\overline{u},v}(t) := \widetilde{W}\big(t, x(t)\big) + \int_s^t f_0\big(r, x(r), \overline{u}(r), v(r)\big) dr \tag{3.6}$$

*is decreasing on $[s, \tau]$ for any $\overline{u}(.), v(.), x(.)$ satisfying (2.15).*

For the necessity part of this statement one may see the proof of Th. 4.1 in Mirică [25] (see also Prop. 4.5.i in Cesari [5]) while the sufficiency part is rather obvious: e.g., $\widetilde{W}(s, y) = \omega_{u,\overline{v}}(0) \leq \omega_{u,\overline{v}}(t_1) = \mathcal{C}_1(s, y; u(.), \overline{v}(.)) \ \forall \ u(.) \in \mathcal{U}_\alpha$.

The necessary monotonicity conditions in Mirică [18] expressed in terms of the Dini derivatives of the real functions $\omega_{u,\overline{v}}(.)$, $\omega_{\overline{u},v}(.)$ in (3.5) and (3.6), respectively, and suitable "chain rules" for the composite functions $\widetilde{W}(., \overline{x}(.))$ above may lead to certain *necessary optimality conditions* of Dynamic Programming type, including, in particular, the fact that, under certain hypotheses, the value function $\widetilde{W}(.,.)$ in (2.10) is a *viscosity solution* of the associated "Isaacs' main equation" in (4.5) below; however, in this paper we are concerned only with the *sufficient optimality conditions* of the same type which, as noted in Lupulescu and Mirică [15], Mirică [19], in some cases may be weaker than the necessary ones.

## 4. Practical verification theorems

As in Optimal Control Theory (e.g., Cesari [5], Lupulescu and Mirică [15], Mirică [24, 26], etc.), starting from the "abstract verification Theorem" 3.1, more realistic ("practical") verification theorems may be obtained under different types of hypotheses on the value function $\widetilde{W}(.,.)$ in (2.10) using corresponding "monotonicity results" for real-valued functions in Mirică [18, 26], etc. and the suitable "chain rules" in Lemmas 4.1, 4.2 above for the composite functions $\widetilde{W}(.,x(.))$ in (3.5) and (3.6).

Due to the difficulties of "checking" the needed "generalized differential inequalities" that may imply the monotonicity properties $(M_1),(M_2)$ in Th. 3.1, a reasonable approach is that of choosing the weakest type of such inequalities for each type of regularity properties of the value function in (2.10); on the other hand, as simple examples show, the value function may have "weaker" regularity properties on the "effective terminal set", $\widetilde{E}_1$ so the needed differential inequalities and regularity assumptions may involve only **the proper value function**, $\widetilde{W}_0(.,.) := \widetilde{W}(.,.)|\widetilde{E}_0$.

From this point of view, the simplest case is that in which the "proper value function" $\widetilde{W}_0(.,.)$ in (2.10) is differentiable on its *open domain* $\overset{\circ}{\widetilde{E}}_0 = Int(\widetilde{E}_0) \subseteq E_0$, considered rather heuristically in Isaacs [12] (Th. 4.4.1) and also in Subbotin [33] (Section 11.5).

A first natural and significant extension of this case may be obtained in the case *the proper value function* $\widetilde{W}_0(.,.) := \widetilde{W}(.,.)|\widetilde{E}_0$ is *differentiably stratified* in the sense that its domain, $\widetilde{E}_0$, has a countable partition into differentiable manifolds (of different dimensions) such that the restriction of $\widetilde{W}_0(.,.)$ to each manifold ("stratum") is differentiable in the classical sense (see Mirică [22–24, 26, 29], etc.); in this case we shall use the following notations:

$$U_T(t,x;v) := \left\{u \in U;\ \big(1, f(t,x,u,v)\big) \in T_{(t,x)}\widetilde{E}_0\right\}, \quad v \in V, \qquad (4.1)$$

$$V_T(t,x;u) := \left\{v \in V;\ \big(1, f(t,x,u,v)\big) \in T_{(t,x)}\widetilde{E}_0\right\}, \quad u \in U, \quad (t,x) \in \widetilde{E}_0.$$

where $T_{(t,x)}\widetilde{E}_0$ denotes the "stratified tangent space" to the stratified set $\widetilde{E}_0$; using the simple properties of stratified sets and mappings we obtain:

**Theorem 4.1 (Verification theorem for stratified value functions).** *Let $\widetilde{U}(.,.)$, $\widetilde{V}(.,.)$ be a pair of admissible feedback strategies in the sense of Problem 2.3 and assume that the associated value function $\widetilde{W}(.,.)$ in (2.10) has the following properties:*

(i) *$\widetilde{W}(.,.)$ is continuous at the terminal points in $\widetilde{E}_1$ in the sense that:*

$$\exists \lim_{\widetilde{E}_0 \ni (s,y) \to (\tau,\xi)} \widetilde{W}(s,y) = \widetilde{W}(\tau,\xi) = g(\tau,\xi) \quad \forall\ (\tau,\xi) \in \widetilde{E}_1; \qquad (4.2)$$

(ii) *the restriction* $\widetilde{W}_0(.,.) := \widetilde{W}(.,.)|\widetilde{E}_0$ *is differentiably stratified and its "strat-ified derivatives" satisfy the following pair of* **differential inequalities**:

$$\inf_{u\in U_T(t,x;\overline{v}),\ \overline{v}\in\widetilde{V}(t,x)} \left[D\widetilde{W}_0(t,x).\big(1, f(x,u,\overline{v})\big) + f_0(t,x,u,\overline{v})\right] \geq 0\,;\qquad (4.3)$$

$$\sup_{v\in V_T(t,x;\overline{u}),\ \overline{u}\in\widetilde{U}(t,x)} \left[D\widetilde{W}_0(t,x).\big(1, f(t,x,\overline{u},v)\big) + f_0(t,x,\overline{u},v)\right] \leq 0\,;\qquad (4.4)$$

(iii) *either* $\widetilde{W}_0(.,.)$ *is continuous and the admissible controls are* **regulated** (*i.e.,* $P_\alpha = P_r$) *or* $\widetilde{W}_0(.,.)$ *is locally Lipschitz.*

   *Then* $\widetilde{U}(.,.),\widetilde{V}(.,.)$ *are bilaterally optimal in the sense of statement* (B) *in Problem* 2.3.

   As in Lupulescu and Mirică [15], Mirică [24, 26], etc., this theorem follows from the "abstract verification Theorem" 3.1 proving that the inequalities in (4.3), (4.4) imply corresponding differential inequalities for the real functions $\omega_{u,\overline{v}}(.),\ \omega_{\overline{u},v}(.)$ in (3.5), (3.6) which, in turn, imply the corresponding monotonicity properties using either a Corollary of the Zygmund's Lemma (in the case $\widetilde{W}_0(.,.)$ is continuous) or the Lebesgue monotonicity theorem in the case $\widetilde{W}_0(.,.)$ is locally Lipschitz since in this case $\omega_{u,\overline{v}}(.),\omega_{\overline{u},v}(.)$ are locally AC.

**Remark 4.2.** One may note that at the points in the open strata (at which $\widetilde{W}_0(.,.)$ is differentiable), the differential inequalities in (4.3), (4.4) are equivalent with *Isaacs' basic Equation* (4.2.1) in Isaacs [12]:

$$\min_{u\in U}\max_{v\in V}\left[D\widetilde{W}_0(t,x).\big(1, f(t,x,u,v)\big) + f_0(t,x,u,v)\right]$$

$$= \max_{v\in V}\min_{u\in U}\left[D\widetilde{W}_0(t,x).\big(1, f(t,x,u,v)\big) + f_0(t,x,u,v)\right]$$

$$= D\widetilde{W}_0(t,x).\big(1, f(t,x,\overline{u},\overline{v})\big) + f_0(t,x,\overline{u},\overline{v}) = 0$$

which is usually written in the "compact" form

$$\frac{\partial\widetilde{W}}{\partial t}(t,x) + H\left(t,x,\frac{\partial\widetilde{W}}{\partial x}\right) = 0\,,\quad (t,x)\in\widetilde{E}_0\,,\quad \widetilde{W}(\tau,\xi) = g(\tau,\xi)\,,\quad (\tau,\xi)\in\widetilde{E}_1$$

$$(4.5)$$

where **the Isaacs Hamiltonian** $H(.,.,.)$ is the function defined by:

$$H(t,x,p) := \min_{u\in U}\max_{v\in V}\big[ <p, f(t,x,u,v)> + f_0(t,x,u,v)\big]$$

$$= \max_{v\in V}\min_{u\in U}\big[ <p, f(t,x,u,v)> + f_0(t,x,u,v)\big]\,,\qquad (4.6)$$

$$(t,x,p)\in Z := dom\,H(.,.,.)\,.$$

Obviously, Theorem 4.1 may be considered as a significant refinement of Isaacs' ("elementary") verification Theorem 4.4.1 in Isaacs [12], (see also Section 11.5 in Subbotin [33]) in which the value function $\widetilde{W}(.,.)$ is assumed to be of class $C^1$ on its open domain $\widetilde{E} = \widetilde{E}_0 \cup \widetilde{E}_1$; we recall that, besides a rather "loose" proof,

Isaacs' Theorem 4.4.1 has been used (in a rather improper way) to study problems in which the value function does not satisfy its hypotheses; rigorous solutions of the examples in Isaacs [12], in which the domain, $\widetilde{E}_0$ in (2.10) of the "proper" value function $\widetilde{W}_0(.,.)$ is no more open and/or the restriction $\widetilde{W}_0(.,.) = \widetilde{W}(.,.)|\widetilde{E}_0$ is no more differentiable, may be obtained only using refinements and extensions of Isaacs' ("elementary") theorem, like Theorem 4.1 above; such extensions and refinements may be obtained using further concepts and results from non-smooth Analysis.

Another significant extension of Isaacs' theorem may be obtained in the case *the proper value function* $\widetilde{W}_0(.,.) := \widetilde{W}(.,.)|\widetilde{E}_0$ *is contingent differentiable*; in this case the "stratified derivatives" in Th. 4.1 are replaced by the "contingent derivatives" and the "stratified tangent spaces" are replaced by the "contingent cones" $K_{(t,x)}^{\pm}\widetilde{E}_0$ while the proof remains practically the same.

It is interesting to note that in the case of locally-Lipschitz value functions, the differential inequalities that should be satisfied by "contingent differentiable" value functions may be replaced by much simpler conditions; using the following notations for each $(t,x) \in \widetilde{E}_0$:

$$U_K^{\pm}(t,x;v) := \left\{ u \in U;\ \left(1, f(t,x,u,v)\right) \in K_{(t,x)}^{\pm}\widetilde{E}_0 \right\}, \quad v \in V\,,$$

$$V_K^{\pm}(t,x;u) := \left\{ v \in V;\ \left(1, f(t,x,u,v)\right) \in K_{(t,x)}^{\pm}\widetilde{E}_0 \right\}, \quad u \in U\,,$$

$$U_K(t,x;v) := U_K^{+}(t,x;v) \cap U_K^{-}(t,x;v)\,, \quad V_K(t,x;u) := V_K^{+}(t,x;u) \cap V_K^{-}(t,x;u)$$

we obtain the following:

**Theorem 4.3 (Verification theorem for locally-Lipschitz value functions).** *Let* $\widetilde{U}(.,.), \widetilde{V}(.,.)$ *be a pair of admissible feedback strategies in the sense of Problem* 2.3 *and assume that the associated value function* $\widetilde{W}(.,.)$ *in* (2.10) *has the following properties:*

(i) $\widetilde{W}(.,.)$ *is continuous at the points in* $\widetilde{E}_1$ *in the sense of* (4.2);
(ii) *the restriction* $\widetilde{W}_0(.,.) := \widetilde{W}(.,.)|\widetilde{E}_0$ *is locally-Lipschitz and its extreme contingent derivatives satisfy the following pair of differential inequalities:*

$$\inf_{u\in U_K(t,x;\overline{v}),\ \overline{v}\in\widetilde{V}(t,x)} \left[ \overline{D_K^{+}}\widetilde{W}_0\Big((t,x);\big(1, f(t,x,u,\overline{v})\big)\Big) + f_0(t,x,u,\overline{v}) \right] \geq 0\,; \quad (4.7)$$

$$\sup_{v\in V_K(t,x;\overline{u}),\ \overline{u}\in\widetilde{U}(t,x)} \left[ \underline{D_K^{-}}\widetilde{W}_0\Big((t,x);\big(1, f(t,x,\overline{u},v)\big)\Big) + f_0(t,x,\overline{u},v) \right] \leq 0\,. \quad (4.8)$$

*Then* $\widetilde{U}(.,.), \widetilde{V}(.,.)$ *are bilaterally optimal in the sense of statement* (B) *in Problem* 2.3.

As in Lupulescu and Mirică [15], Mirică [24,26], etc., this theorem follows from the "abstract verification Theorem" 3.1 proving that the inequalities in (4.8), (4.9) imply corresponding differential inequalities for the real functions $\omega_{u,\overline{v}}(.)$, $\omega_{\overline{u},v}(.)$ in

(3.5), (3.6) which, in turn, imply the corresponding monotonicity properties using the Lebesgue monotonicity theorem since in this case $\widetilde{W}_0(.,.)$ is locally Lipschitz.

In the case the associated value function $\widetilde{W}(.,.)$ in (2.10) it is only semi-continuous (in particular, continuous), the possible verification theorems have more complicated statements that depend essentially on the class $P_\alpha = \mathcal{U}_\alpha \times \mathcal{V}\alpha$ of *admissible control mappings* specified in the statement of Problem 2.1; in the general case of "bounded measurable" controls $P_\alpha = P_\infty$ and semi-continuous value functions, due to certain monotonicity conditions for semi-continuous real functions (e.g., Mirică [18, 26], etc.) and to the "weak forms" of the chain rules for the extreme contingent derivatives, we have to use the "u.s.c.-relaxed" differential inclusions in Mirică [17, 28], etc. that contain the sets of contingent directions to the admissible trajectories.

Moreover, using some more "sophisticated" monotonicity criteria in Mirică [18], one may obtain similar verification theorems for arbitrary, "strongly discontinuous" value functions, that are neither lower nor upper semi-continuous, as in the case of the optimal control problems in Lupulescu and Mirică [15]; in such cases, certain more restrictive differential inequalities should be added.

**Remark 4.4.** One may note also that the concepts and results above may be extended to the case in which the "usual (AC) trajectories" are replaced by the KS-trajectories in Remark 2.6; in this case, the corresponding differential inequalities should be expressed in terms of the "directions" in the associated "u.s.c.-relaxed multi-functions"

## 5. Construction of the value function, optimal strategies, and optimal trajectories

**The "playable solution"** of Problem 2.3 obviously consists in the triplet $(\widetilde{U}(.,.),$ $\widetilde{V}(.,.), \widetilde{W}(.,.))$ satisfying the admissibility properties (A) as well as the optimality properties (B), (C) in the statement of Problem 2.3, accompanied by the "optimal trajectories" in (2.9).

Using the experience in optimal control (e.g., Mirică [26]) as well as Isaacs' rather artisanal procedures in [12], we are suggesting the use of a suitable **generalization of Cauchy's Method of Characteristics** for the (usually non-smooth) **Isaacs' Hamiltonian** in (4.6) in the following way:

1) Integrate "backwardly" (for $t \le \tau$) a suitable **generalized Hamiltonian system**:

$$(x', p') \in d^\# H(t, x, p), \quad (x(\tau), p(\tau)) = (\xi, q) \in Z_1^*(\tau) \tag{5.1}$$

where $Z_1^*(\tau)$ is given by the usual "transversality conditions", to construct a **generalized Hamiltonian flow** $X^*(.,.) = (X(.,.), P(.,.))$, and a **generalized Characteristic flow** $C^*(.,.) = (X^*(.,.), C(.,.))$, where $X^*(.,a)$, $a = (\tau, \xi, q)$ is a solution of (5.1) and the function $C(.,.)$ is given by:

$$C(t,a) := g(\tau,\xi) + \int_{\tau}^{t} \Big[ \langle P(r,a), X'(r,a) \rangle - H\big(r, X^*(r,a)\big) \Big] dr$$

$$\text{if} \quad a = (\tau,\xi,q) \in Z_1^* \,.$$

For instance, in the case $H(.,.,.)$ is differentiably-stratified, one may take:

$$d^{\#} H(t,x,p) = d_s^{\#} H(t,x,p) := \Big\{ (x',p') \in T_{(t,x,p)} Z; \, x' \in f\big(t, x, \widehat{U}(t,x,p),$$

$$\widehat{V}(t,x,p)\big), \langle x', \overline{p} \rangle - \langle p', \overline{x} \rangle = DH(t,x,p).(0,\overline{x},\overline{p}) \; \forall \; (0,\overline{x},\overline{p}) \in T_{(t,x,p)} Z \Big\}$$

which on open strata (at differentiability points) coincides with the *classical Hamiltonian field*:

$$d_s^{\#} H(t,x,p) = \Big\{ \Big( \frac{\partial H(t,x,p)}{\partial p}, -\frac{\partial H(t,x,p)}{\partial x} \Big) \Big\};$$

in more general cases one may use either the extreme contingent derivatives or Clarke's generalized gradient to define (a possible larger) generalized Hamiltonian field (e.g., Mirică [26, 29], etc.).

Similarly, in the case the "terminal payoff", $g(.,.) : E_1 \to R$ is differentiably-stratified, the sets of "terminal transversality points" are defined by:

$$Z_1^*(\tau) := \Big\{ (\xi,q); \, (\tau,\xi,q) \in Z, \, (\tau,\xi) \in E_1, \, \langle q, \overline{\xi} \rangle - \overline{\tau} H(\tau,\xi,q) =$$

$$Dg(\tau,\xi)(\overline{\tau},\overline{\xi}) \; \forall \; (\overline{\tau},\overline{\xi}) \in T_{(\tau,\xi)} E_1 \Big\}$$

while in the general case one may use the extreme contingent derivatives of $g(.,.)$ to define similar sets.

The experience shows that the "optimal value function", $\widetilde{W}(.,.)$ may be **one** of the "minimal" or the "maximal" value functions:

$$W_m(s,y) := \min_{X(s;\tau,\xi,q)=y} C(s;\tau,\xi,q),$$

$$W_M(s,y) := \max_{X(s;\tau,\xi,q)=y} C(s;\tau,\xi,q). \tag{5.2}$$

While the pair of mutually-optimal feedback strategies, $\widetilde{U}(.,.),\widetilde{V}(.,.)$, is the corresponding pair of feedback strategies, $U_m(.,.),V_m(.,.)$, respectively, $U_M(.,.),V_M(.,.)$ defined in a certain way by the corresponding (finite-dimensional) optimization problem in (5.2); moreover, the first components, $X(.,a)$, of the Hamiltonian flow, that correspond to the "optimal parameters", $a = (\tau,\xi,q)$ in (5.2), define the optimal trajectories in (2.9), generated by the chosen feedback strategies $\widetilde{U}(.,.),\widetilde{V}(.,.)$; this procedure is presented in more detail in Mirică [29] in the case of autonomous differential games.

# References

[1] J. P. Aubin and A. Cellina, *Differential Inclusions*, Springer, Berlin (1984).

[2] M. Bardi and I. Capuzzo-Dolcetta, *Optimal Control and Viscosity Solutions for Hamilton–Jacobi–Bellman Equations*, Birkhäuser, Boston (1997).

[3] L. D. Berkovitz, *The existence of value and saddle points in games of fixed duration*, SIAM J. Control Optim., **23** (1985), 172–196; Errata and addenda in *SIAM J. Control Optim.*, **26** (1988), 740–742.

[4] L. D. Berkovitz, *Characterization of the value of differential games*, Appl. Math. Opt., **17**, 1 (1988), 77–183.

[5] L. Cesari, *Optimization – Theory and Applications*, Springer, New York (1983).

[6] M. G. Crandall and P. L. Lions, *Viscosity solutions of Hamilton–Jacobi equations*, Trans. AMS, **277** (1983), 1–42.

[7] F. H. Clarke, Yu. Ledyaev, R. J. Stern, and P. Wolensky, *Nonsmooth Analysis and Control Theory*, Springer, New York (1998).

[8] F. H. Clarke, Yu. Ledyaev, and A. I. Subbotin , *Universal feedback control via proximal aiming in problems of control under disturbances and differential games*, Trudy Mat. Inst. Steklova, **224** (1999), 165–180 (Russian; translation of Rapport CRM-2386, Univ. Montréal, 1994).

[9] R. J. Elliot, *Viscosity solutions and optimal control*, Pitman Research Notes in Mathematics, Longman, Harlow (1987).

[10] R. J. Elliot and N. J. Kalton, *The existence of value in differential games*, Mem. A.M.S. **126** (1972).

[11] A. Friedman, *Differential Games*, Wiley, New York (1971).

[12] R. Isaacs, *Differential Games*, Wiley, New York (1965).

[13] N. N. Krassovskii and A. I. Subbotin, *Positional Differential Games*, Nauka, Moskva (1974) (Russian; English translation: Springer, New York, 1988).

[14] P. L. Lions and P. E. Souganidis, *Differential games and directional derivatives of viscosity solutions of Bellman's and Isaacs' equations*, SIAM J. Control Optim., **23** (1985), 566–583.

[15] V. Lupulescu and Şt. Mirică, *Verification theorems for discontinuous value functions in optimal control*, Math. Reports, **2** (52) (2000), 299–326.

[16] Şt. Mirică, *Generalized solutions by Cauchy's Method of Characteristics*, Rend. Sem. Mat. Univ. Padova, **77** (1987), 317–349.

[17] Şt. Mirică, *Tangent and contingent directions to trajectories of differential inclusions*, Revue Roum. Math. Pures Appl. **37** (1992), 421–444.

[18] Şt. Mirică, *Monotonicity of real functions without regularity assumptions*, Anal. Univ. Bucureşti, Mat., **44** (1995), 49–60.

[19] Şt. Mirică, *Invariance and monotonicity with respect to autonomous differential inclusions*, Studii Cerc. Mat., **44** (1995), 179–204.

[20] Şt. Mirică, *A Dynamic Programming solution to Bernoulli's Brachistochrone problem*, Bull. Math. Soc. Sci. Math. Roumanie, **39** (87) (1996), 201–222.

[21] Şt. Mirică, *Optimal feedback control in closed form via Dynamic Programming*, Revue Roum. Math. Pures Appl., **42** (1997), 621–648.

[22] Şt. Mirică, *Hamilton–Jacobi equations on possibly non-symplectic differentiable manifolds*, Bull. Math. Soc. Sci. Math. Roumanie, **41** (87) (1998), 23–39.

[23] Şt. Mirică, *A constructive Dynamic Programming approach to differential games*, Anal. Univ. Bucureşti, Mat., **51** (2002), 51–70.

[24] Şt. Mirică, *Verification theorems for optimal feedback strategies in differential games*, Int. Game Theory Review, **5** (2003), 167–189.

[25] Şt. Mirică, *Monotonicity and differential properties of the value functions in optimal control*, Int.J. Math. Math. Systems, **2004–65**, 3513–3540.

[26] Şt. Mirică, *Constructive Dynamic Programming in Optimal Control. Autonomous problems*, Editura Academiei Romane, Bucuresti, (2004).

[27] Şt. Mirică, *On the solution of Isaacs' differential game of pursuit in the half-plane*, Math. Reports, **6** (56), (2004), 233–257.

[28] Şt. Mirică, *Tangent, quasitangent and peritangent directions to the trajectories of differential inclusions*, J. Dyn. Control Systems, **10** (2004), 227–245.

[29] Şt. Mirică, *User's guide to Dynamic Programming for Differential Games and Optimal Control*, Revue Roumaine Math. Pure Appl., **40** (2004), 501–528.

[30] Şt. Mirică, *Approximate, limiting and generalized trajectories to feedback differential systems*, Math. Nachr., **278** (2005), 451–459.

[31] E. Roxin, *Axiomatic approach in differential games*, J. Opt. Theory Appl., **3** (1969), 153–163.

[32] A.I. Subbotin, *Generalization of the main equation of differential games*, J. Opt. Theory Appl., **43** (1984), 103–134.

[33] A.I. Subbotin, *Generalized solutions of first order PDEs; The dynamical optimization perspective*, Birkhäuser, Boston (1995).

[34] P. P. Varaiya, *On the existence of solutions to a differential game*, SIAM J. Control, **5** (1967), 153–162.

Ştefan Mirică
University of Bucharest
Faculty of Mathematics and Informatics
Academiei 14
RO-010014 Bucharest
Romania
e-mail: mirica@fmi.unibuc.ro

Progress in Nonlinear Differential Equations
and Their Applications, Vol. 75, 285–303
© 2007 Birkhäuser Verlag Basel/Switzerland

# Optimal Control of Nonconvex Differential Inclusions

## B. S. Mordukhovich

*Dedicated to Arrigo Cellina and James Yorke*

**Abstract.** The paper deals with dynamic optimization problems of the Bolza and Mayer types for evolution systems governed by nonconvex Lipschitzian differential inclusions in Banach spaces under endpoint constraints described by finitely many equalities and inequalities with generally nonsmooth functions. We develop a variational analysis of such problems mainly based on their discrete approximations and the usage of advanced tools of generalized differentiation. In this way we establish extended results on stability of discrete approximations and derive necessary optimality conditions for nonconvex discrete-time and continuous-time systems in the refined Euler–Lagrange and Weierstrass–Pontryagin forms accompanied by the appropriate transversality inclusions. In contrast to the case of geometric endpoint constraints in infinite dimensions, the necessary optimality conditions obtained in this paper do not impose any nonempty interiority/finite codimension/normal compactness assumptions.

**Mathematics Subject Classification (2000).** 49J53, 49J52, 49J24, 49M25, 90C30.

**Keywords.** Variational analysis, dynamic optimization and optimal control, differential inclusions, infinite dimension, discrete approximations, generalized differentiation, necessary optimality conditions.

## 1. Introduction

The paper is devoted to the study of dynamic optimization problems governed by differential inclusions in infinite-dimensional spaces. We pay the main attention to variational analysis of the following *generalized Bolza problem* $(P)$ for differential

Research was partially supported by the USA National Science Foundation under grants DMS-0304989 and DMS-0603846 and by the Australian Research Council under grant DP-0451168.

inclusions in Banach spaces with endpoint constraints described by finitely many equalities and inequalities.

Let $X$ be a Banach *state space* with the *initial state* $x_0 \in X$, and let $T :=$ $[a, b] \subset \mathbb{R}$ be a fixed *time interval*. Given a set-valued mapping $F \colon X \times T \rightrightarrows X$ and real-valued functions $\varphi_i \colon X \to \mathbb{R}$ as $i = 0, \ldots, m+r$ and $\vartheta \colon X \times X \times T \to \mathbb{R}$, consider the problem:

$$\text{minimize } J[x] := \varphi_0\big(x(b)\big) + \int_a^b \vartheta\big(x(t), \dot{x}(t), t\big)\, dt \tag{1.1}$$

subject to *dynamic constraints* governed by the differential inclusion [1]

$$\dot{x}(t) \in F\big(x(t), t\big) \quad \text{a.e.} \quad t \in [a, b] \quad \text{with} \quad x(a) = x_0 \tag{1.2}$$

with *functional endpoint constraints* of the inequality and equality types given by

$$\varphi_i\big(x(b)\big) \leq 0, \qquad\qquad i = 1, \ldots, m, \tag{1.3}$$

$$\varphi_i\big(x(b)\big) = 0, \qquad\qquad i = m+1, \ldots, m+r. \tag{1.4}$$

Dynamic optimization problems for differential inclusions with the *finite-dimensional* state space $X = \mathbb{R}^n$ have been intensively studied over the years, especially during the last decade, mainly from the viewpoint of deriving necessary optimality conditions; see [3,8,11,13,15,17] for various results, methods, and more references. Dynamic optimization problems governed by infinite-dimensional *evolution equations* have also been much investigated, motivating mainly by applications to optimal control of partial differential equations; see, e.g., the books [7,9] and the references therein. To the best of our knowledge, deriving necessary optimality conditions in dynamic optimization problems for evolution systems governed by differential inclusions in *infinite-dimensional spaces* has not drawn attention in the literature till the very recent time.

In [13], the author developed the method of *discrete approximations* to study optimal control problems of minimizing the Bolza functional (1.1) over appropriate solutions to evolution systems governed by infinite-dimensional differential inclusions of type (1.2) with endpoint constrains given in the *geometric form*

$$x(b) \in \Omega \subset X \tag{1.5}$$

via closed subsets of Banach spaces satisfying certain requirements. The major assumption on $\Omega$ made in [13] is the so-called *sequential normal compactness* (SNC) property of $\Omega$ at the optimal endpoint $\bar{x}(b) \in \Omega$; see [12] for a comprehensive theory for this and related properties, which play a major role in infinite-dimensional variational analysis. Loosely speaking, the SNC property means that $\Omega$ should be "sufficiently fat" around the reference point; e.g., it never holds for singletons unless $X$ is finite-dimensional, where the SNC property is satisfied for every nonempty set. For *convex* sets in infinite-dimensional spaces, the SNC property always holds when $\text{int}\, \Omega \neq \emptyset$. Furthermore, it happens to be closely related [13] to the so-called "finite-codimension" property of convex sets, which is known to be essential for the fulfillment of an appropriate counterpart of the Pontryagin maximum principle for infinite-dimensional systems of optimal control; see, e.g., [7, 9].

In this paper we show that the dynamic optimization problem $(P)$ formulated above, with the *functional* endpoint constraints (1.3) and (1.4), admits necessary optimality conditions in the extended Euler–Lagrange form accompanied by the corresponding Weierstrass–Pontryagin/maximum and transversality relations with *no SNC* and similar assumptions imposed on the underlying endpoint constraint set in infinite dimensions. Moreover, the case of endpoint constraints (1.3) and (1.4) allows us to partly avoid some other rather restrictive assumptions (like "strong coderivative normality," which may not hold in infinite-dimensional spaces; see Sections 6, 7 for more details) imposed in [13] in the general case of geometric constraints (1.5). Our approach is based, in addition to [13], on refined properties of appropriate *subdifferentials* of locally Lipschitzian functions on infinite-dimensional spaces, as well as on *dual/coderivative characterizations* of Lipschitzian and metric regularity properties of set-valued mappings.

The rest of the paper is organized as follows. In Section 2 we formulate the standing assumptions on the initial data of $(P)$, and discuss the relaxation procedure used for some results and proofs in the paper. The main attention in this paper is paid to the so-called *intermediate local minimizers*, which occupy an intermediate position between the classical weak and strong minima.

In Section 3 we construct a sequence of the *well-posed discrete approximations* $(P_N)$ to the original Bolza problem $(P)$ involving *consistent perturbations* of the endpoint constraints in the discrete approximation procedure. Then we present a major result on the *strong stability* of discrete approximations that justifies the $W^{1,2}$-norm convergence of optimal solutions for $(P_N)$ to the fixed local minimizer for the original problem $(P)$.

Section 4 contains an overview of the basic tools of *generalized differentiation* needed to perform the subsequent variational analysis of the discrete-time and continuous-time evolution systems under consideration in infinite-dimensional spaces. Most of the material in this section is taken from the author's book [12], where the reader can find more results and commentaries in this direction and related topics.

Section 5 is devoted to deriving necessary optimality conditions for the constrained *discrete-time* problems arising from the discrete approximation procedure. These problems are reduced to constrained problems of mathematical programming in infinite dimensions, which happen to be *intrinsically nonsmooth* and involve finitely many functional and geometric constraints generated by those in (1.2)–(1.4) via the discrete approximation procedure. Variational analysis of such problems requires applications of the full power of the generalized differential calculus in infinite-dimensional spaces developed in [12].

In Section 6 we derive necessary optimality conditions of the extended *Euler–Lagrange* type for *relaxed* intermediate minimizers to the original Bolza problem $(P)$ by passing to the limit from those obtained for discrete-time problems. It worth emphasizing that the realization of the limiting procedure requires not only the strong convergence of optimal trajectories to discrete approximation problems but also justifying an appropriate convergence of *adjoint trajectories* in necessary optimality conditions for discrete-time systems. The latter becomes passible due to specific properties of the basic generalized differential constructions reviewed in Section 4, which include complete *dual characterizations* of Lipschitzian and metric regularity properties of set-valued mappings.

The concluding Section 7 concerns necessary optimality conditions for arbitrary (*non-relaxed*) intermediate minimizers to problem $(P)$ that are established in terms of the *extended Euler–Lagrange* inclusion accompanied by the *Weierstrass–Pontryagin/maximum* and transversality relations without imposing any SNC assumptions on the target/endpoint constraint set. The approach is based on an additional approximation procedure that allows us to reduce $(P)$ to an unconstrained Bolza problem of the type treated in Section 6 for which *any* intermediate local minimizer happens to be a relaxed one.

Our notation is basically standard; cf. [12,13]. Unless otherwise stated, all the spaces considered are Banach with the norm $\| \cdot \|$ and the canonical dual pairing $\langle \cdot, \cdot \rangle$ between the space in question, say $X$, and its topological dual $X^*$ whose weak* topology is denoted by $w^*$. We use the symbols $I\!B$ and $I\!B^*$ to signify the closed unit balls of the space under consideration and its dual, respectively. Given a set-valued mapping $F \colon X \rightrightarrows X^*$, its *sequential Painlevé-Kuratowski upper/outer limit* at $\bar{x}$ is defined by

$$\underset{x \to \bar{x}}{\mathrm{Lim}\,\mathrm{sup}}\, F(x) := \left\{ x^* \in X^* \middle|\; \exists \quad \text{sequences} \quad x_k \to \bar{x},\; x_k^* \overset{w^*}{\longrightarrow} x^* \text{ with} \right.$$
$$\left. x_k^* \in F(x_k) \text{ as } k \in I\!N := \{1, 2, \ldots\} \right\}. \tag{1.6}$$

## 2. Bolza problem for differential inclusions

Just for brevity and simplicity, we consider in this paper the Bolza problem $(P)$ with *autonomous* (time-independent) data. By a *solution* to inclusion (1.2) we understand (as in [1,6] a mapping $x \colon T \to X$, which is Fréchet differentiable for a.e. $t \in T$ satisfying (1.2) and the *Newton–Leibniz formula*

$$x(t) = x_0 + \int_a^t \dot{x}(s)\, ds \quad \text{for all} \quad t \in T,$$

where the integral in taken in the BOCHNER SENSE.

Recall that a Banach space $X$ is *Asplund* if any of its separable subspaces has a separable dual. This is a major subclass of Banach spaces that particularly includes every space with a *Fréchet differentiable renorm* off the origin (i.e., every *reflexive* space), every space with a separable dual, etc.; see [4] for more details, characterizations, and references. There is a deep relationship between spaces having the *Radon-Nikodým property* (RNP) and Asplund spaces, which is used in what follows: *given a Banach space $X$, the dual space $X^*$ has the RNP if and only if $X$ is Asplund.*

It has been well recognized that differential inclusions (1.2), which are certainly of their own interest, provide a useful generalization of *control systems* governed by differential/evolution *equations* with control parameters:

$$\dot{x} = f(x, u), \quad u \in U, \tag{2.1}$$

where the control sets $U(\cdot)$ may also depend on time and *state* variables via $F(x, t) = f(x, U(x, t), t)$. In some cases, especially when the sets $F(\cdot)$ are convex, the differential inclusions (1.2) admit parametric representations of type (2.1), but in general they cannot be reduced to parametric control systems and should be studied for their own sake; see [1]. Note also that the *ODE form* (2.1) in Banach spaces is strongly related to various control problems for evolution *partial differential equations* of parabolic and hyperbolic types, where solutions may be understood in some other appropriate senses [7,9].

In what follows, we pay the main attention to the study of *intermediate local minimizers* for problem $(P)$ introduced in [11]. Recall that a *feasible arc* to $(P)$ is a solution to the differential inclusion (1.2) for which $J[x] < \infty$ in (1.1) and the endpoint constraints (1.3) and (1.4) are satisfied.

**Definition 2.1 (Intermediate local minimizers).** *A feasible arc* $\bar{x}(\cdot)$ *is an* INTERMEDIATE LOCAL MINIMIZER *(i.l.m.) of rank* $p \in [1, \infty)$ *for* $(P)$ *if there are numbers* $\epsilon > 0$ *and* $\alpha \geq 0$ *such that* $J[\bar{x}] \leq J[x]$ *for any feasible arcs to* $(P)$ *satisfying the relationships*

$$\|x(t) - \bar{x}(t)\| < \epsilon \quad \text{for all} \quad t \in [a, b] \quad \text{and} \tag{2.2}$$

$$\alpha \int_a^b \|\dot{x}(t) - \dot{\bar{x}}(t)\|^p \, dt < \epsilon. \tag{2.3}$$

In fact, relationships (2.2) and (2.3) mean that we consider a neighborhood of $\bar{x}(\cdot)$ in the Sobolev space $W^{1,p}([a, b]; X)$ with the norm

$$\|x(\cdot)\|_{W^{1,p}} := \max_{t \in [a,b]} \|x(t)\| + \left( \int_a^b \|\dot{x}(t)\|^p \, dt \right)^{1/p},$$

where the norm on the right-hand side is taken in the space $X$. If there is only the requirement (2.2) in Definition 2.1, i.e., $\alpha = 0$ in (2.3), then we get the classical *strong* local minimum corresponding to a neighborhood of $\bar{x}(\cdot)$ in the norm topology of $C([a, b]; X)$. If instead of (2.3) one puts the more restrictive requirement

$$\|\dot{x}(t) - \dot{\bar{x}}(t)\| < \epsilon \quad \text{a.e.} \quad t \in [a, b],$$

then we have the classical *weak* local minimum in the framework of Definition 2.1. Thus the introduced notion of i.l.m. takes, for any $p \in [1, \infty)$, an *intermediate* position between the classical concepts of strong ($\alpha = 0$) and weak ($p = \infty$) local minima, being indeed different from both classical notions; see various examples in [13, 18]. Clearly all the necessary conditions for i.l.m. automatically hold for strong (and hence for global) minimizers.

Considering the autonomous Bolza problem $(P)$ in this paper, we impose the following *standing assumptions* on its initial data along a given intermediate local minimizer $\bar{x}(\cdot)$:

**(H1)** There are a open set $U \subset X$ and a number $\ell_F > 0$ such that $\bar{x}(t) \in U$ for all $t \in [a, b]$, the sets $F(x)$ are nonempty and compact for all $x \in U$ and satisfy the inclusion

$$F(x) \subset F(u) + \ell_F \|x - u\| I\!\!B \quad \text{whenever} \quad x, u \in U, \tag{2.4}$$

which implies the uniform boundedness of the sets $F(x)$ on $U$, i.e., the existence of some constant $\gamma > 0$ such that

$$F(x) \subset \gamma I\!\!B \quad \text{for all} \quad x \in U.$$

**(H2)** The integrand $\vartheta$ is Lipschitzian continuous on $U \times (\gamma I\!\!B)$.
**(H3)** The endpoint functions $\varphi_i$, $i = 0, \ldots, m + r$, are locally Lipschitzian around $\bar{x}(b)$ with the common Lipschitz constant $\ell > 0$.

Observe that (2.4) is equivalent to say that the set-valued mapping $F$ is *locally Lipschitzian* around $\bar{x}(\cdot)$ with respect to the classical Hausdorff metric on the space of nonempty and compact subsets of $X$. In what follows, along with the original problem $(P)$, we consider its "relaxed" counterpart significantly used in some results and proofs of the paper. Roughly speaking, the relaxed problem is obtained from $(P)$ by a *convexification* procedure with respect to the *velocity* variable. It follows the route of Bogolyubov and Young in the classical calculus of variations and of Gamkrelidze and Warga in optimal control; see the books [1, 13] and the references therein for more details and commentaries.

To construct an appropriate relaxation of the Bolza problem $(P)$ under consideration, we first consider the extended-real-valued function

$$\vartheta_F(x, v) := \vartheta(x, v) + \delta\big(v; F(x)\big),$$

where $\delta(\cdot; \Omega)$ is the *indicator function* of the set $\Omega$. Denote

$$\widehat{\vartheta}_F(x, v) := \big(\vartheta_F\big)_v^{**}(x, v), \quad (x, v) \in X \times X,$$

the *biconjugate/bypolar* function to $\vartheta_F(x, \cdot)$, i.e., the greatest proper, convex, and lower semicontinuous (l.s.c.) function with respect to $v$ that is majorized by $\vartheta_F$. Then the *relaxed problem* $(R)$ to $(P)$, or the *relaxation* of $(P)$, is defined as follows:

$$\text{minimize} \ \ \widehat{J}[x] := \varphi\big(x(b)\big) + \int_a^b \widehat{\vartheta}_F\big(x(t), \dot{x}(t)\big)\, dt \tag{2.5}$$

over a.e. differentiable arcs $x: [a, b] \to X$ that are Bochner integrable on $[a, b]$ together with $\vartheta_F\big(x(t), \dot{x}(t)\big)$, satisfy the Newton–Leibniz formula and the endpoint constraints (1.3), (1.4).

Note that the feasibility requirement $\widehat{J}[x] < \infty$ in (2.5) is fulfilled only if $x(\cdot)$ is a solution to the *convexified differential inclusion*

$$\dot{x}(t) \in \text{clco}\, F\big(x(t), \dot{x}(t)\big) \quad \text{a.e.} \ \ t \in [a, b] \quad \text{with} \quad x(a) = x_0, \tag{2.6}$$

where "clco" stands for the convex closure of a set in $X$. Thus the relaxed problem $(R)$ can be considered under explicit dynamic constraints given by the convexified differential inclusion (2.6). Any trajectory for (2.6) is called a *relaxed trajectory* for (1.2), in contrast to the *ordinary* (or *original*) trajectories for the latter inclusion.

There are deep relationships between relaxed and ordinary trajectories for differential inclusions, which reflect the fundamental *hidden convexity* inherent in continuous-time (nonatomic measure) dynamic systems defined by differential and integral operators. In particular, *any relaxed trajectory* of (1.2) under assumption (H1) can be *uniformly approximated* (in the $C([a,b];X)$-norm) by a sequence of ordinary trajectories; see, e.g., [1, 6, 16]. We need the following version [5] of this approximation/density property involving not only differential inclusions but also minimizing functionals.

**Lemma 2.2 (Approximation property for the relaxed Bolza problem).** *Let $x(\cdot)$ be a relaxed trajectory for the differential inclusion (1.2) with a separable state space $X$, where $F$ and $\vartheta$ satisfy assumptions (H1) and (H2), respectively. Then there is sequence of the ordinary trajectories $x_k(\cdot)$ for (1.2) such that $x_k(\cdot) \to x(\cdot)$ in $C([a,b];X)$ as $k \to \infty$ and*

$$\liminf_{k \to \infty} \int_a^b \vartheta\big(x_k(t), \dot{x}_k(t)\big)\, dt \leq \int_a^b \widehat{\vartheta}_F\big(x(t), \dot{x}(t)\big)\, dt\,.$$

Note that Theorem 2.2 does *not* assert that the approximating trajectories $x_k(\cdot)$ satisfy the endpoint constraints (1.3) and (1.4). Indeed, there are examples showing that the latter may not be possible and, moreover, the property of *relaxation stability*

$$\inf(P) = \inf(R) \tag{2.7}$$

is violated; in (2.7) the infima of the cost functionals (1.1) and (2.5) are taken over all the feasible arcs in $(P)$ and $(R)$, respectively.

An obvious sufficient condition for the relaxation stability is the *convexity* of the sets $F(x,t)$ and of the integrand $\vartheta$ in $v$. However, the relaxation stability goes *far beyond* the standard convexity due to the *hidden convexity* property of continuous-time differential systems. In particular, Theorem 2.2 ensures the relaxation stability of nonconvex problems $(P)$ with no constraints on the endpoint $x(b)$. There are various efficient conditions for the relaxation stability of nonconvex problems with endpoint and other constraint; see [13, Subsection 6.1.2] with the commentaries therein for more details, discussions, and references.

A *local* version of the relaxation stability property (2.7) regarding intermediate minimizers for the Bolza problem $(P)$ is postulated as follows.

**Definition 2.3 (Relaxed intermediate local minimizers).** *A feasible arc $\bar{x}(\cdot)$ to the Bolza problem $(P)$ is a* RELAXED INTERMEDIATE LOCAL MINIMIZER *(r.i.l.m.) of rank $p \in [1, \infty)$ for $(P)$ if it is an intermediate local minimizer of this rank for the relaxed problem $(R)$ providing the same value of the cost functionals: $J[\bar{x}] = \widehat{J}[\bar{x}]$.*

It is not hard to observe that, under the standing assumptions formulated above, the notions of intermediate local minima and relaxed intermediate local minima do not actually depend on rank $p$. In what follows we always take (unless otherwise stated in Section 7) $p = 2$ and $\alpha = 1$ in (2.3) for simplicity.

The principal method of our study in this paper involves *discrete approximations* of the original Bolza problem $(P)$ for constrained continuous-time evolution inclusions by a family of dynamic optimization problems of the Bolza type governed by discrete-time inclusions with endpoint constraints. We show that this method generally leads to necessary optimality conditions for *relaxed* intermediate local minimizers of $(P)$. Then an additional approximation procedure allows us to establish necessary optimality conditions for *arbitrary* (non-relaxed) intermediate local minimizers by reducing them to problems, which are *automatically* stable with respect to relaxation.

## 3. Discrete approximations

In this section we present basic constructions of the method of discrete approximations in the theory of necessary optimality conditions for differential inclusions following the scheme of [11, 13] developed for the case of geometric constraints, with certain modifications required for the functional endpoint constraints (1.3) and (1.4).

For simplicity we use the replacement of the derivative by the *uniform Euler scheme*:
$$\dot{x}(t) \approx \frac{x(t+h) - x(t)}{h}, \quad h \to 0.$$

To formalize this process, we take any natural number $N \in \mathbb{N}$ and consider the *discrete grid/mesh* on $T$ defined by
$$T_N := \{a, a + h_N, \dots, b - h_N, b\}, \quad h_N := (b-a)/N,$$

with the *stepsize of discretization* $h_N$ and the *mesh points* $t_j := a + j h_N$ as $j = 0, \dots, N$, where $t_0 = a$ and $t_N = b$. Then the differential inclusion (1.2) is replaced by a sequence of its *finite-difference/discrete approximations*
$$x_N(t_{j+1}) \in x_N(t_j) + h_N F\big(x_N(t_j)\big), \quad j = 0, \dots, N-1, \quad x(t_0) = x_0. \quad (3.1)$$

Given a discrete trajectory $x_N(t_j)$ satisfying (3.1), we consider its *piecewise linear extension* $x_N(t)$ to the continuous-time interval $T = [a, b]$, i.e., the *Euler broken lines*. We also define the *piecewise constant extension* to $T$ of the corresponding *discrete velocity* by
$$v_N(t) := \frac{x_N(t_{j+1}) - x_N(t_j)}{h_N}, \quad t \in [t_j, t_{j+1}), \quad j = 0, \dots, N-1.$$

It follows from the very definition of the Bochner integral that
$$x_N(t) = x_0 + \int_a^t v_N(s)\, ds \quad \text{for} \quad t \in T.$$

The next result establishes the *strong $W^{1,2}$-norm approximation* of *any* trajectory for the differential inclusion (1.2) by extended trajectories of the sequence of discrete inclusions (3.1). Note that the norm convergence in $W^{1,2}([a,b];X)$ implies the *uniform* convergence of the trajectories on $[a,b]$ and the *pointwise*, for a.e. $t \in [a,b]$, convergence of (some subsequence of) their *derivatives*. The latter is crucial for the study of *nonconvex*-valued differential inclusions. The proof of this result in given in [13, Theorem 6.4], which is an infinite-dimensional counterpart of the one in [11, Theorem 3.1].

**Lemma 3.1 (Strong $W^{1,2}$-approximation by discrete trajectories).** *Let $\bar{x}(\cdot)$ be an arbitrary solution to the differential inclusion (1.2) under the assumptions in (H1), where $X$ is a general Banach space. Then there is a sequence of solutions $\widehat{x}_N(t_j)$ to the discrete inclusions (3.1) such that their extensions $\widehat{x}_N(t)$, $a \le t \le b$, converge to $\bar{x}(t)$ strongly in the space $W^{1,2}([a,b];X)$ as $N \to \infty$.*

Now fix an *intermediate local minimizer* $\bar{x}(\cdot)$ for the Bolza problem $(P)$ and construct a sequence of discrete approximation problems $(P_N)$, $N \in \mathbb{N}$, admitting optimal solutions $\bar{x}_N(\cdot)$ whose extensions converge to $\bar{x}(\cdot)$ in the norm topology of $W^{1,2}([a,b];X)$ as $N \to \infty$.

To proceed, we take a sequence of the discrete trajectories $\widehat{x}_N(\cdot)$ approximating by Lemma 3.1 the given local minimizer $\bar{x}(\cdot)$ to $(P)$ and denote

$$\eta_N := \max_{j\in\{1,\ldots,N\}} \|\widehat{x}_N(t_j) - \bar{x}(t_j)\| \to 0 \quad \text{as} \quad N \to \infty. \tag{3.2}$$

Having $\epsilon > 0$ from relations (2.2) and (2.3) for the intermediate minimizer $\bar{x}(\cdot)$ with $p = 2$ and $\alpha = 1$, we always suppose that

$$\bar{x}(t) + \epsilon/2 \in U \quad \text{for all} \quad t \in [a,b],$$

where $U$ is a neighborhood of $\bar{x}(\cdot)$ from (H1). Let $\ell > 0$ be the common Lipschitz constant of $\varphi_i$, $i = 1, \ldots, m+r$, from (H3). Construct problems $(P_N)$, $N \in \mathbb{N}$, as follows: minimize

$$J_N[x_N] := \varphi_0\big(x_N(t_N)\big) + h_N \sum_{j=0}^{N-1} \vartheta\left(x_N(t_j), \frac{x_N(t_{j+1}) - x_N(t_j)}{h_N}\right)$$

$$+ \sum_{j=0}^{N-1} \int_{t_j}^{t_{j+1}} \left\|\frac{x_N(t_{j+1}) - x_N(t_j)}{h_N} - \dot{\bar{x}}(t)\right\|^2 dt \tag{3.3}$$

over discrete trajectories $x_N = x_N(\cdot) = \big(x_0, x_N(t_1), \ldots, x_N(t_N)\big)$ for the difference inclusions (3.1) subject to the constraints

$$\varphi_i\big(x_N(t_N)\big) \le \ell\eta_N \qquad \text{for } i = 1, \ldots, m, \tag{3.4}$$

$$-\ell\eta_N \le \varphi_i\big(x_N(t_N)\big) \le \ell\eta_N \qquad \text{for } i = m+1, \ldots, m+r, \tag{3.5}$$

$$\|x_N(t_j) - \bar{x}(t_j)\| \le \frac{\epsilon}{2} \qquad \text{for } j = 1, \ldots, N, \quad \text{and} \tag{3.6}$$

$$\sum_{j=0}^{N-1} \int_{t_j}^{t_{j+1}} \left\| \frac{x_N(t_{j+1}) - x_N(t_j)}{h_N} - \dot{\bar{x}}(t) \right\|^2 dt \leq \frac{\epsilon}{2}. \tag{3.7}$$

Considering in the sequel (without mentioning any more) piecewise linear extension of $x_N(\cdot)$ to the whole interval $[a, b]$, we observe the relationships:

$$\begin{cases} x_N(t) = x_0 + \int_a^t \dot{x}_N(s)\, ds \quad \text{for all } t \in [a, b] \quad \text{and} \\[2mm] \dot{x}_N(t) = \dot{x}_N(t_j) \in F\big(x_N(t_j)\big), \quad t \in [t_j, t_{j+1}), \quad j = 0, \ldots, N-1. \end{cases} \tag{3.8}$$

In the next theorem, we establish that the given *relaxed* intermediate local minimizer (r.i.l.m.) $\bar{x}(\cdot)$ to $(P)$ can be *strongly* in $W^{1,2}$ approximated by *optimal solutions* to $(P_N)$; the latter implies the a.e. *pointwise* convergence of the derivatives significant for the main results of the paper. To justify such an approximation, we need to impose the *Asplund* structure on *both* $X$ and its dual $X^*$, which is particularly the case when $X$ is *reflexive*.

**Theorem 3.2 (Strong convergence of discrete optimal solutions).** *Let $\bar{x}(\cdot)$ be an r.i.l.m. for the Bolza problem $(P)$ under the standing assumptions* (H1)–(H3) *in the Banach state space $X$, and let $(P_N)$, $N \in I\!N$, be a sequence of discrete approximation problems built above. The following hold:*

(i) *Each $(P_N)$ admits an optimal solution.*

(ii) *If in addition both $X$ and $X^*$ are Asplund, then any sequence $\{\bar{x}_N(\cdot)\}$ of optimal solutions to $(P_N)$ converges to $\bar{x}(\cdot)$ strongly in $W^{1,2}\big([a,b]; X\big)$.*

The proof of this theorem follows the arguments in [12, Theorem 6.13] and the estimates

$$\big|\varphi_i\big(\hat{x}_N(t_N)\big) - \varphi_i\big(\bar{x}(b)\big)\big| \leq \ell\,\|\hat{x}(t_N) - \bar{x}(t_N)\| \leq \ell\eta_N \quad \text{for all} \quad i = 1, \ldots, m+r$$

due to (3.2), which are needed for (3.4) and (3.5).

The strong convergence result of Theorem 3.2 *makes a bridge* between the original continuous-time dynamic optimization problem $(P)$ and its discrete-time counterparts $(P_N)$, which allows us to derive necessary optimality conditions for $(P)$ by passing to the limit from those for $(P_N)$. The latter ones are *intrinsically nonsmooth* and require appropriate tools of generalized differentiation for their variational analysis.

## 4. Generalized differentiation

In this section, we define the main constructions of generalized differentiation used in what follows. Since our major framework in this paper is the class *Asplund spaces*, we present simplified definitions and some properties held in this setting. All the material reviewed and employed below is taken from the author's book [12], where the reader can find more details and references.

We start with generalized normals to closed sets $\Omega \subset X$. Given $\bar{x} \in \Omega$, the (basic, limiting) *normal cone* to $\Omega$ at $\bar{x}$ is defined by

$$N(\bar{x}; \Omega) := \operatorname{Lim\,sup}_{x \to \bar{x}} \widehat{N}(x; \Omega), \tag{4.1}$$

where "Lim sup" stands for the *sequential* upper/outer limit (1.6) of the *Fréchet normal cone* (or the *prenormal cone*) to $\Omega$ at $x \in \Omega$ given by

$$\widehat{N}(x; \Omega) := \left\{ x^* \in X^* \,\middle|\, \limsup_{u \xrightarrow{\Omega} x} \frac{\langle x^*, u - x \rangle}{\|u - x\|} \leq 0 \right\}, \tag{4.2}$$

where $x \xrightarrow{\Omega} \bar{x}$ signifies that $x \to \bar{x}$ with $x \in \Omega$, and where $\widehat{N}(x; \Omega) := \emptyset$ for $x \notin \Omega$.

Given a set-valued mapping $F \colon X \rightrightarrows Y$ of closed graph

$$\operatorname{gph} F := \left\{ (x, y) \in X \times Y \,\middle|\, y \in F(x) \right\},$$

define its *normal coderivative* and *Fréchet coderivative* at $(\bar{x}, \bar{y}) \in \operatorname{gph} F$ by, respectively,

$$D^* F(\bar{x}, \bar{y})(y^*) := \left\{ x^* \in X^* \,\middle|\, (x^*, -y^*) \in N\big((\bar{x}, \bar{y}); \operatorname{gph} F\big) \right\}, \tag{4.3}$$

$$\widehat{D}^* F(\bar{x}, \bar{y})(y^*) := \left\{ x^* \in X^* \,\middle|\, (x^*, -y^*) \in \widehat{N}\big((\bar{x}, \bar{y}); \operatorname{gph} F\big) \right\}. \tag{4.4}$$

If $F = f \colon X \to Y$ is *strictly differentiable* at $\bar{x}$ (in particular, if $f \in C^1$), then

$$D^* f(\bar{x})(y^*) = \widehat{D}^* f(\bar{x})(y^*) = \left\{ \nabla f(\bar{x})^* y^* \right\}, \quad y^* \in Y^*,$$

i.e., both coderivatives (4.3) and (4.4) are positively homogeneous extensions of the classical *adjoint* derivative operator to nonsmooth and set-valued mappings.

Finally, consider a function $\varphi \colon X \to I\!R$ *locally Lipschitzian* around $\bar{x}$; in this paper we do not use more general functions. Then the (basic, limiting) *subdifferential* of $\varphi$ at $\bar{x}$ is

$$\partial \varphi(\bar{x}) := \operatorname{Lim\,sup}_{x \to \bar{x}} \widehat{\partial} \varphi(x), \tag{4.5}$$

where the sequential outer limit (1.6) of the *Fréchet subdifferential* mapping $\widehat{\partial} \varphi(\cdot)$ is

$$\widehat{\partial} \varphi(x) := \left\{ x^* \in X^* \,\middle|\, \frac{\varphi(u) - \varphi(x) - \langle x^*, u - x \rangle}{\|u - x\|} \geq 0 \right\}. \tag{4.6}$$

We are not going to review in this section appropriate properties of the generalized differential constructions (4.1)–(4.6) used in Sections 5–7: these properties will be invoked with the exact references to [12] in the corresponding places of the proofs in the subsequent sections. Just note here that our basic/limiting constructions (4.1), (4.3), and (4.5) enjoy *full calculus* in the framework of Asplund spaces, while the Fréchet-like ones (4.2), (4.4), and (4.6) satisfy certain rules of "fuzzy calculus."

## 5. Necessary conditions for discrete inclusions

In this section we derive necessary optimality conditions for the sequence of discrete approximation problems $(P_N)$ defined in (3.1) and (3.3)–(3.7). We only present results in the "fuzzy" form, which are more convenient to derive necessary conditions for the original problem $(P)$ by the limiting procedure in Section 6.

Observe first that each discrete optimization problem $(P_N)$ can be equivalently written in a special form of *constrained mathematical programming* (*MP*):

$$
\begin{cases}
\text{minimize } \psi_0(z) \text{ subject to} \\
\psi_j(z) \le 0, \quad j = 1, \ldots, s, \\
f(z) = 0, \\
z \in \Theta_j \subset Z, \quad j = 1, \ldots, l,
\end{cases}
$$

where $\psi_j$ are real-valued functions on some Banach space $Z$, where $f \colon Z \to E$ is a mapping between Banach spaces, and where $\Theta_j \subset Z$. To see this, we let

$$
z := (x_1, \ldots, x_N, v_0, \ldots, v_{N-1}) \in Z := X^{2N},
$$

$$
E := X^N, \; s := N + 2 + m + 2r, \; l := N - 1
$$

and rewrite $(P_N)$ as an $(MP)$ problem (5) with the following data:

$$
\psi_0(z) := \varphi_0(x_N) + h_N \sum_{j=0}^{N-1} \vartheta(x_j, v_j) + \sum_{j=0}^{N-1} \int_{t_j}^{t_{j+1}} \|v_j - \dot{\bar{x}}(t)\|^2 \, dt,
$$

$$
\psi_j(z) :=
\begin{cases}
\|x_{j-1} - \bar{x}(t_{j-1})\| - \epsilon/2, \; j = 1, \ldots, N+1, \\[2mm]
\displaystyle\sum_{i=0}^{N-1} \int_{t_i}^{t_{i+1}} \|v_i - \dot{\bar{x}}(t)\|^2 \, dt - \epsilon/2, \; j = N+2, \\[2mm]
\varphi_i(x_N) - \ell \eta_N, \; j = N + 2 + i, \; i = 1, \ldots, m+r, \\[1mm]
-\varphi_i(x_N) - \ell \eta_N, \; j = N + 2 + m + r + i, \; i = m+1, \ldots, m+r;
\end{cases}
$$

$$
\begin{cases}
f(z) = \big(f_0(z), \ldots, f_{N-1}(z)\big) \quad \text{with} \\
f_j(z) := x_{j+1} - x_j - h_N v_j, \quad j = 0, \ldots, N-1,
\end{cases}
$$

$$
\Theta_j := \Big\{ z \in X^{2N} \Big| \, v_j \in F(x_j) \Big\} \quad \text{for } j = 0, \ldots, N-1.
$$

The next theorem establishes necessary optimality conditions for each problem $(P_N)$ in the *approximate/fuzzy* form of refined *Euler–Lagrange* and *transversality inclusions* expressed in terms of Fréchet-like normals and subgradients. The proof is based on applying the corresponding *fuzzy calculus* rules and *neighborhood criteria* for *metric regularity and Lipschitzian behavior* of set-valued mappings; cf. [13, Theorem 6.19].

**Theorem 5.1 (Fuzzy Euler–Lagrange conditions for discrete approximations).** *Let $\bar{x}_N(\cdot) = \{\bar{x}_N(t_j) | \, j = 0, \ldots, N\}$ be local optimal solutions to problems $(P_N)$ as*

$N \to \infty$ under the assumptions (H1)–(H3) with the Asplund state space $X$. Consider the quantities

$$\theta_{Nj} := 2 \int_{t_j}^{t_{j+1}} \left\| \frac{\bar{x}_N(t_{j+1}) - \bar{x}_N(t_j)}{h_N} - \dot{\bar{x}}(t) \right\| dt, \quad j = 0, \dots, N-1.$$

Then there exists a sequence $\varepsilon_N \downarrow 0$ along some $N \to \infty$, and there are sequences of Lagrange multipliers $\lambda_{iN}$, $i = 0, \dots, m+r$, and adjoint trajectories $p_N(\cdot) = \{p_N(t_j) \in X^* \mid j = 0, \dots, N\}$ satisfying the following relationships:-

– the sign and nontriviality conditions

$$\lambda_{iN} \geq 0 \quad for \ all \quad i = 0, \dots, m+r, \quad \sum_{i=0}^{m+r} \lambda_{iN} = 1;$$

– the complementary slackness conditions

$$\lambda_{iN} \left[ \varphi_i(\bar{x}_N(t_N)) - \ell\eta_N \right] = 0 \quad for \quad i = 1, \dots, m;$$

– the extended Euler–Lagrange inclusion in the approximate form

$$\left( \frac{p_N(t_{j+1}) - p_N(t_j)}{h_N}, p_N(t_{j+1}) - \lambda_{0N} \frac{\theta_{Nj}}{h_N} b^*_{Nj} \right)$$

$$\in \lambda_{0N} \widehat{\partial}\vartheta \left( \bar{x}_N(t_j), \frac{\bar{x}_N(t_{j+1}) - \bar{x}_N(t_j)}{h_N} \right)$$

$$+ \widehat{N} \left( \left( \bar{x}_N(t_j), \frac{\bar{x}_N(t_{j+1}) - \bar{x}_N(t_j)}{h_N} \right); \mathrm{gph}\, F \right) + \varepsilon \mathbb{B}^* \quad with$$

$$b^*_{Nj} \in \mathbb{B}^*, \quad j = 0, \dots, N-1;$$

– the approximate transversality inclusion

$$- p_N(t_N) \in \sum_{i=0}^{m} \lambda_{iN} \widehat{\partial}\varphi_i(\bar{x}_N(t_N))$$

$$+ \sum_{i=m+1}^{m+r} \lambda_{iN} \left[ \widehat{\partial}\varphi_i(\bar{x}_N(t_N)) \bigcup \widehat{\partial}(-\varphi_i)(\bar{x}_N(t_N)) \right] + \varepsilon \mathbb{B}^* .$$

## 6. Euler–Lagrange conditions for relaxed minimizers

This section contains necessary optimality conditions in the refined forms of the extended Euler–Lagrange and transversality inclusions for *relaxed* intermediate local minimizers of the original problem $(P)$. The proof is based on the passing to the limit from the necessary optimality conditions for discrete approximation problems obtained in Section 5 and on the usage of the *strong stability* of discrete approximations established in Section 3. A crucial part of the proof involves the justification of an appropriate convergence of *adjoint arcs*; the latter becomes possible due to the *coderivative characterization* of Lipschitzian set-valued mappings; cf. [13, Theorem 6.21]

**Theorem 6.1 (extended Euler–Lagrange and transversality inclusions for relaxed intermediate minimizers).** *Let $\bar{x}(\cdot)$ be a relaxed intermediate local minimizer for the Bolza problem $(P)$ given in $(1.1)$–$(1.4)$ under the standing assumptions of Section 2, where the spaces $X$ and $X^*$ are Asplund. Then there are nontrivial Lagrange multipliers $0 \neq (\lambda_0, \ldots, \lambda_{m+r}) \in I\!\!R^{m+r+1}$ and an absolutely continuous mapping $p \colon [a, b] \to X^*$ such that the following necessary conditions hold:*

– *the sign conditions*

$$\lambda_i \geq 0 \quad \text{for all} \quad i = 0, \ldots, m+r \,, \tag{6.1}$$

– *the complementary slackness conditions*

$$\lambda_i \varphi_i\big(\bar{x}(b)\big) = 0 \quad \text{for} \quad i = 1, \ldots, m \,, \tag{6.2}$$

– *the extended Euler–Lagrange inclusion, for a.e., $t \in [a, b]$,*

$$\begin{aligned}
\dot{p}(t) \in \mathrm{clco}\Big\{ u \in X^* \Big| \; &(u, p(t)) \in \lambda_0 \partial \vartheta\big(\bar{x}(t), \dot{\bar{x}}(t)\big) \\
&+ N\Big(\big(\bar{x}(t), \dot{\bar{x}}(t)\big); \mathrm{gph}\, F\Big)\Big\} \,,
\end{aligned} \tag{6.3}$$

– *and the transversality inclusion*

$$-p(b)) \in \sum_{i=0}^{m} \lambda_i \partial \varphi_i\big(\bar{x}(b)\big) + \sum_{i=m+1}^{m+r} \lambda_i \Big[ \partial \varphi_i\big(\bar{x}(b)\big) \bigcup \partial\big(-\varphi_i\big)\big(\bar{x}(b)\big) \Big] \,. \tag{6.4}$$

Note that the results obtained in Theorem 6.1 are different from those derived in [13, Subsection 6.1.5] not only by the *absence* of any SNC-like assumptions on the target/constraint set but also by *not* imposing the "coderivative normality" property on $F$ needed in [13] in similar settings. Observe also that the arguments developed above allow us to provide the correspondent improvements in the case of Lipschitzian endpoint constraints of the Euler–Lagrange type necessary optimality conditions derived in [14] for evolution models governed by *semilinear inclusions*

$$\dot{x}(t) \in Ax(t) + F\big(x(t), t\big) \,, \tag{6.5}$$

where $A$ is an *unbounded* infinitesimal generator of a *compact $C_0$-semigroup* on $X$, and where continuous solutions to (6.5) are understood in the *mild* sense.

## 7. Necessary conditions without relaxation

In this section we establish necessary optimality conditions for intermediate local minimizers $\bar{x}(\cdot)$ of evolution inclusions *without any relaxation*. It can be done under certain more restrictive assumptions on the initial data in comparison with those in Theorem 6.1. For simplicity, consider here the *Mayer version* $(P_M)$ of problem

$(P)$ with $\vartheta = 0$ in (1.1). In this case, the *Euler–Lagrange inclusion* (6.3) admits the *coderivative form*

$$\dot{p}(t) \in \operatorname{clco} D^* F\big(\bar{x}(t), \dot{\bar{x}}(t)\big)\big(-p(t)\big) \quad \text{a.e.} \quad t \in [a, b], \tag{7.1}$$

which easily implies, due to the extremal property for coderivatives of convex-valued mappings given in [12, Theorem 1.34], the *Weierstrass–Pontryagin maximum condition*

$$\big\langle p(t), \dot{\bar{x}}(t) \big\rangle = \max_{v \in F(\bar{x}(t))} \big\langle p(t), v \big\rangle \quad \text{a.e.} \quad t \in [a, b] \tag{7.2}$$

provided that the sets $F(x)$ are *convex* near $\bar{x}(t)$ for a.e. $t \in [a, b]$. Our goal is to justify the above Euler–Lagrange and Weierstrass–Pontryagin conditions, together with the other necessary optimality conditions of Theorem 6.1, for intermediate minimizers of the Mayer problem $(P_M)$ subject to the Lipschitzian endpoint constraints (1.3) and (1.4), *without any convexity or relaxation* assumptions and with *no SNC-like* requirements imposed on the endpoint constraint set. To accomplish this goal, we employ a certain approximation technique involving *Ekeland's variational principle* combined with other advanced results of variational analysis and generalized differentiation, which allow us to reduce the constrained problem under consideration to an unconstrained (and thus *stable with respect to relaxation*) Bolza problem studied in Section 6. However, this requires additional assumptions on the initial data of $(P_M)$ imposed in what follows.

Recall that a set-valued mapping $F \colon X \rightrightarrows Y$ is *strongly coderivatively normal* at $(\bar{x}, \bar{y}) \in \operatorname{gph} F$ if its normal coderivative (4.3) admits the representation

$$D^* F(\bar{x}, \bar{y})(y^*) = \Big\{ x^* \in X^* \,\Big|\, \exists \text{ sequences } (x_k, y_k) \to (\bar{x}, \bar{y}), \ x_k^* \xrightarrow{w^*} x^*, \text{ and}$$

$$y_k^* \to y^* \quad \text{with} \quad y_k \in F(x_k) \quad \text{and} \quad x_k^* \in \widehat{D}^* F(x_k, y_k)(y_k^*) \text{ as}$$

$$k \to \infty \Big\} =: D_M^* F(\bar{x}, \bar{y})(y^*), \tag{7.3}$$

where $D_M^* F(\bar{x}, \bar{y})$ is called the *mixed coderivative* of $F$ at $(\bar{x}, \bar{y})$. Observe that the only difference between the normal and mixed coderivatives of $F$ at $(\bar{x}, \bar{y})$ is that the *mixed* weak* convergence of $x_k^* \xrightarrow{w^*} x^*$ and the norm convergence of $y_k^* \to y^*$ is used for $D_M^* F(\bar{x}, \bar{y})$ in (7.3), in contrast to the weak* convergence of *both* components $(x_k^*, y_k^*) \xrightarrow{w^*} (x^*, y^*)$ for $D^* F(\bar{x}, \bar{y})$ in (4.3) via (4.1). Besides the obvious case of $\dim Y < \infty$, the strong coderivative normality holds in many important infinite-dimensional settings, and the property is preserved under various compositions; see [12, Proposition 4.9] describing major classes of mappings satisfying this property.

A mapping $F \colon X \rightrightarrows Y$ is called *sequentially normally compact* (SNC) at $(\bar{x}, \bar{y}) \in \operatorname{gph} F$ if for any sequences $(x_k, y_k) \xrightarrow{\operatorname{gph} F} (\bar{x}, \bar{y})$ and $(x_k^*, y^*) \in \widehat{N}\big((x_k, y_k); \operatorname{gph} F\big)$ one has

$$(x_k^*, y_k^*) \xrightarrow{w^*} 0 \Longrightarrow \|(x_k^*, y_k^*)\| \to 0 \quad \text{as} \quad k \to \infty.$$

As discussed in Section 1, this property is a far-going extension of the "finite-codimension" and other related properties of sets and mappings. It always holds in finite dimensions, while in reflexive spaces agrees with the "compactly epi-Lipschitzian" property by Borwein and Strójwas; see [12] for more details, discussions, and calculus.

Finally, recall that the given norm on a Banach space $X$ is *Kadec* if the strong and weak convergences agree on the boundary of the unit sphere of $X$. It is well known that every reflexive space admits an equivalent Kadec norm.

**Theorem 7.1 (Euler–Lagrange and Weierstrass–Pontryagin conditions for intermediate local minimizers with no relaxation).** *Let $\bar{x}(\cdot)$ be an intermediate local minimizer for the Mayer problem $(P_M)$ in (1.1)–(1.4) under the standing hypotheses (H1) and (H3) on $F$ and $\varphi_i$. Assume in addition that:*

**(a)** *the state space $X$ is separable and reflexive with the Kadec norm on it;*
**(b)** *the velocity mapping $F$ is SNC at $\big(\bar{x}(t), \dot{\bar{x}}(t)\big)$ and strongly coderivatively normal with weakly closed graph around this point for a.e. $t \in [a, b]$.*

*Then there are nontrivial Lagrange multipliers $0 \neq (\lambda_0, \ldots, \lambda_{m+r}) \in I\!\!R^{m+r+1}$ and an absolutely continuous mapping $p \colon [a, b] \to X^*$ satisfying the following relationships:*

- *the sign and complementarity slackness conditions in (6.1) and (6.2);*
- *the Euler–Lagrange inclusion (7.1), where the closure operation is redundant;*
- *the Weierstrass–Pontryagin maximum condition (7.2); and*
- *the transversality inclusion (6.4).*

*Proof.* Denote

$$\varphi_0^+(x, \nu) := \max\big\{\varphi_0(x) - \nu, 0\big\}, \quad \varphi_i^+(x) := \max\big\{\varphi_i(x), 0\big\}, \ i = 1, \ldots, m, \quad (7.4)$$

and, by the *method of metric approximations* [10], consider the parametric cost functional

$$\theta_\nu[x] := \left[(\varphi_0^+)^2\big(x(b), \nu\big) + \sum_{i=1}^{m} (\varphi_i^+)^2\big(x(b)\big) + \sum_{i=m+1}^{m+r} \varphi_i^2\big(x(b)\big)\right]^{1/2}, \ \nu \in I\!\!R, \quad (7.5)$$

over trajectories for (1.1) with *no endpoint constraints*. Since $\bar{x}(\cdot)$ is an *intermediate local minimizer* for $(P_M)$ and due to the constructions in (7.4) and (7.5), we have

$$\theta_\nu[x] > 0 \quad \text{for any} \quad \nu < \bar{\nu} := \varphi_0\big(\bar{x}(b)\big)$$

provided that $x(\cdot)$ is a trajectory for (1.2) belonging to the prescribed $W^{1,1}$-neighborhood of the given intermediate local minimizer and such that $x(t) \in U$ for all $t \in [a, b]$, where the open set $U \subset X$ is taken from the requirements in (H1) imposed on $\bar{x}(\cdot)$.

Then following the proof of [13, Theorem 6.27], we find an absolute continuous arc $x_\varepsilon(\cdot)$ satisfying the estimate

$$\int_a^b \|\dot{\bar{x}}(t) - \dot{x}_\varepsilon(t)\| \, dt \leq \varepsilon$$

and such that $x_\varepsilon$ provides an intermediate minimum to the *unconstrained* Bolza problem with *Lipschitzian* data:

$$\text{minimize} \quad \varphi_\varepsilon\big(x(b)\big) + \int_a^b \vartheta_\varepsilon\big(x(t), \dot{x}(t), t\big)\, dt \tag{7.6}$$

over absolutely continuous arcs $x(\cdot)$ satisfying $x(a) = x_0$ and lying in a $W^{1,1}$-neighborhood of $\bar{x}(\cdot)$, where the functions $\varphi_\varepsilon \colon X \to I\!R$ and $\theta_\varepsilon \colon X \times X \times [a,b] \to I\!R$ are given by

$$\varphi_\varepsilon(x) := \left[ (\varphi_0^+)^2(x, \nu_\varepsilon) + \sum_{i=1}^{m} \left(\varphi_i^+\right)^2(x) + \sum_{i=m+1}^{m+r} \varphi_i^2(x) \right]^{1/2}, \tag{7.7}$$

$$\vartheta_\varepsilon(x, v, t) := \eta\sqrt{1 + \ell_F^2}\, \text{dist}\big((x, v); \text{gph}\, F\big) + \sqrt{\varepsilon}\|v - \dot{x}_\varepsilon(t)\|. \tag{7.8}$$

Applying the optimality conditions of Theorem 6.1 to problem (7.6) with the initial data (7.7) and (7.8), for all small $\varepsilon > 0$ we find an absolutely continuous adjoint arc $p_\varepsilon \colon [a,b] \to X^*$ satisfying

$$\dot{p}_\varepsilon(t) \in \text{co}\left\{ u \in X^* \,\middle|\, (u, p_\varepsilon(t)) \in \mu\, \partial\text{dist}\big((x_\varepsilon(t), \dot{x}_\varepsilon(t)); \text{gph}\, F\big) + \sqrt{\varepsilon}(0, I\!B^*) \right\} \tag{7.9}$$

for a.e. $t \in [a,b]$ with $\mu := \eta\sqrt{1 + \ell_F^2}$ and

$$-p_\varepsilon(b) \in \partial\left[ (\varphi_0^+)^2(\cdot, \nu_\varepsilon) + \sum_{i=1}^{m} \left(\varphi_i^+\right)^2(\cdot) + \sum_{i=m+1}^{m+r} \varphi_i^2(\cdot) \right]^{1/2} \big(x_\varepsilon(b)\big). \tag{7.10}$$

Passing to the limit in (7.9), (7.10) and using the calculus rules of generalized differentiation as in the proof of [13, Theorem 6.27], we arrive at the Euler–Lagrange and transversality conditions of the theorem *without any relaxation*.

Observe that in the general nonconvex setting the Euler–Lagrange inclusion (7.1) does not automatically imply the maximum condition (7.2). To establish the latter condition supplementing the other necessary conditions of the theorem, we follow the proof of [17, Theorem 7.4.1] given for a Mayer problem of the type $(P_M)$ involving nonconvex differential inclusions in finite-dimensional spaces; it holds with minor changes in infinite-dimensions under the assumptions imposed. The proof of the latter theorem is based on reducing the constrained Mayer problem for nonconvex differential inclusions to an unconstrained Bolza (finite Lagrangian) problem, which in turn is reduced to a problem of optimal control with *smooth dynamics* and *nonsmooth endpoint constraints* first treated in [10] via the nonconvex normal cone (4.1) and the corresponding subdifferential (4.5) introduced therein to describe the appropriate transversality conditions in the maximum principle. $\qquad\square$

## Acknowledgment
Many thanks to our T$_E$X-pert for developing this class file.

# References

[1] J.-P. Aubin and A. Cellina, *Differential Inclusions*, Grundlehren Series (Fundamental Principles of Mathematical Sciences) **264**, Springer, Berlin, 1984.

[2] J. M. Borwein and Q. J. Zhu, *Techniques of Variational Analysis*, CMS Books in Mathematics, Springer, New York, 2005.

[3] F. H. Clarke, *Necessary conditions in dynamic optimization*, Memoirs of Amer. Math. Soc. **173** Number 816 (2005).

[4] J. Diestel and J. J. Uhl, Jr., *Vector Measures*, American Mathematical Society, Providence, RI, 1977.

[5] F. S. De Blasi, G. Pianigiani, and A. A. Tolstonogov, *A Bogolyubov type theorem with a nonconvex constraint in Banach spaces*, SIAM J. Control Optim. **43**, 466–476 (2004).

[6] K. Deimling, *Multivalued Differential Equations*, De Gruyter, Berlin, 1992.

[7] H. O. Fattorini, *Infinite Dimensional Optimization and Control Theory*, Cambridge University Press, Cambridge, UK, 1999.

[8] A. D. Ioffe, *Euler–Lagrange and Hamiltonian formalisms in dynamic optimization*, Trans. Amer. Math. Soc. **349** (1997), 2871–2900.

[9] X. Li and J. Yong, *Optimal Control Theory for Infinite-Dimensional Systems*, Birkhäuser, Boston, 1995.

[10] B. S. Mordukhovich, *Maximum principle in problems of time optimal control with nonsmooth constraints*, J. Appl. Math. Mech. **40** (1976), 960–969.

[11] B. S. Mordukhovich, *Discrete approximations and refined Euler–Lagrange conditions for nonconvex differential inclusions*, SIAM J. Control Optim. **33** (1995), 882–915.

[12] B. S. Mordukhovich, *Variational Analysis and Generalized Differentiation, I: Basic Theory*, Grundlehren Series (Fundamental Principles of Mathematical Sciences) **330**, Springer, Berlin, 2006.

[13] B. S. Mordukhovich, *Variational Analysis and Generalized Differentiation, II: Applications*, Grundlehren Series (Fundamental Principles of Mathematical Sciences) **331**, Springer, Berlin, 2006.

[14] B. S. Mordukhovich and D. Wang, *Optimal control of semilinear evolution inclusions via discrete approximations*, Control Cybernet. **34** (2005), 849–870.

[15] G. V. Smirnov, *Introduction to the Theory of Differential Inclusions*, American Mathematical Society, Providence, RI, 2001.

[16] A. A. Tolstonogov, *Differential Inclusions in a Banach Space*, Kluwer, Dordrecht, The Netherlands, 2000.

[17] R. B. Vinter, *Optimal Control*, Birkhäuser, Boston, 2000.

[18] R. B. Vinter and P. D. Woodford, *On the occurrence of intermediate local minimizers that are not strong local minimizers*, Systems Cont. Lett. **31** (1997), 235–342.

B. S. Mordukhovich
Department of Mathematics
Wayne State University
Detroit, Michigan 48202
USA
e-mail: boris@math.wayne.edu

Progress in Nonlinear Differential Equations
and Their Applications, Vol. 75, 305–315
© 2007 Birkhäuser Verlag Basel/Switzerland

# On Chaos of a Cubic $p$-adic Dynamical System

Farrukh Mukhamedov and José F. F. Mendes

*Dedicated to Arrigo Cellina and James Yorke*

**Abstract.** In the paper we describe basin of attraction of the $p$-adic dynamical system $f(x) = x^3 + ax^2$. Moreover, we also describe the Siegel discs of the system, since the structure of the orbits of the system is related to the geometry of the $p$-adic Siegel discs.

**Mathematics Subject Classification (2000).** Primary 37E99, 37B25; Secondary 54H20, 12J120.

**Keywords.** Attractor, Siegel disc, $p$-adic dynamics.

## 1. Introduction

It is known that the $p$-adic numbers were first introduced by the German mathematician K. Hensel. During a century after their discovery they were considered mainly objects of pure mathematics. Starting from 1980s various models described in the language of $p$-adic analysis have been actively studied. Applications of $p$-adic numbers in $p$-adic mathematical physics [6,12,20,30], quantum mechanics [16] and many others [17] stimulated increasing interest in the study of $p$-adic dynamical systems. In [17, 29] $p$-adic field have arisen in physics in the theory of superstrings, promoting questions about their dynamics. Also some applications of $p$-adic dynamical systems to some biological, physical systems were proposed in [3,4,18,31]. Other studies of non-Archimedean dynamics in the neighborhood of a periodic and of the counting of periodic points over global fields using local fields appeared in [9, 13, 19, 25]. Certain rational $p$-adic dynamical systems were investigated in [14, 21] which appear from problems of $p$-adic Gibbs measures [22–24]. In [7, 8] the Fatou set of a rational function defined over some finite extension of $\mathbb{Q}_p$ has been studied. Besides, an analogue of Sullivan's no wandering domains theorem for $p$-adic rational functions, which have no wild recurrent Julia critical points, was proved.

The most studied discrete $p$-adic dynamical systems (iterations of maps) are the so called monomial systems, i.e., $f(x) = x^n$. In [5] an asymptotic behavior of such a dynamical system over $p$-adic and $\mathbb{C}_p$ was investigated. In [18] perturbated monomial dynamical systems defined by $f_q(x) = x^n + q(x)$, where the perturbation term $q(x)$ was a polynomial whose coefficients had small $p$-adic absolute value, was studied. There it was shown a connection between monomial and perturbated monomial systems. Formulas for the number of cycles of a specific length to a given system and the total number of cycles of such dynamical systems were provided. These investigations show that the study of perturbated dynamical systems is important. Even for a quadratic function $f(x) = x^2 + c$, $c \in \mathbb{Q}_p$ its chaotic behavior is complicated (see [28, 29]). In [28] the Fatou and Julia sets of such a $p$-adic dynamical system were found. Unique ergodicity and ergodicity of monomial and perturbated dynamical systems have been considered in [1, 11].

The aim of this paper is to investigate the asymptotic behavior of a cubic $p$-adic dynamical system $f(x) = x^3 + ax^2$ at $|a|_p \neq 1$. Note that globally attracting sets play an important role in dynamics, restricting the asymptotic behavior to certain regions of the phase space. However, descriptions of the global attractor can be difficult as it may contain complicated chaotic dynamics. Therefore, in the paper we will investigate the basin of attraction of such a dynamical system. Moreover, we also describe the Siegel discs of the system, since the structure of the orbits of the system is related to the geometry of the $p$-adic Siegel discs (see [2]).

## 2. Preliminaries

### 2.1. $p$-adic numbers

Let $\mathbb{Q}$ be the field of rational numbers. Throughout the paper $p$ will be a fixed prime number. Every rational number $x \neq 0$ can be represented in the form $x = p^r \frac{n}{m}$, where $r, n \in \mathbb{Z}$, $m$ is a positive integer and $p, n, m$ are relatively prime. The $p$-adic norm of $x$ is given by $|x|_p = p^{-r}$ and $|0|_p = 0$. This norm satisfies so called the strong triangle inequality

$$|x + y|_p \leq \max\{|x|_p, |y|_p\}.$$

From this inequality one can infer that

$$\text{if} \quad |x|_p \neq |y|_p, \quad \text{then} \quad |x - y|_p = \max\{|x|_p, |y|_p\} \tag{2.1}$$

$$\text{if} \quad |x|_p = |y|_p, \quad \text{then} \quad |x - y|_p \leq |2x|_p. \tag{2.2}$$

This is a ultrametricity of the norm. The completion of $\mathbb{Q}$ with respect to $p$-adic norm defines the $p$-adic field which is denoted by $\mathbb{Q}_p$. Note that any $p$-adic number $x \neq 0$ can be uniquely represented in the canonical series:

$$x = p^{\gamma(x)}(x_0 + x_1 p + x_2 p^2 + \cdots), \tag{2.3}$$

where $\gamma = \gamma(x) \in \mathbb{Z}$ and $x_j$ are integers, $0 \leq x_j \leq p - 1$, $x_0 > 0$, $j = 0, 1, 2, \ldots$ (see more detail [10, 15]). Observe that in this case $|x|_p = p^{-\gamma(x)}$.

We recall that an integer $a \in \mathbb{Z}$ is called *a quadratic residue modulo p* if the equation $x^2 \equiv a(\mathrm{mod}\ p)$ has a solution $x \in \mathbb{Z}$.

**Lemma 2.1 ([30]).** *In order that the equation*

$$x^2 = a, \quad 0 \neq a = p^{\gamma(a)}(a_0 + a_1 p + \cdots), \quad 0 \leq a_j \leq p - 1, \quad a_0 > 0$$

*has a solution $x \in \mathbb{Q}_p$, it is necessary and sufficient that the following conditions are satisfied:*

(i) $\gamma(a)$ *is even;*
(ii) $a_0$ *is a quadratic residue modulo p if $p \neq 2$, if $p = 2$ besides $a_1 = a_2 = 0$.*

For any $a \in \mathbb{Q}_p$ and $r > 0$ denote

$$\bar{B}_r(a) = \{x \in \mathbb{Q}_p : |x - a|_p \leq r\}, \quad B_r(a) = \{x \in \mathbb{Q}_p : |x - a|_p < r\},$$
$$S_r(a) = \{x \in \mathbb{Q}_p : |x - a|_p = r\}.$$

A function $f : B_r(a) \to \mathbb{Q}_p$ is said to be *analytic* if it can be represented by

$$f(x) = \sum_{n=0}^{\infty} f_n(x - a)^n, \quad f_n \in \mathbb{Q}_p,$$

which converges uniformly on the ball $B_r(a)$.

Note the basics of *p*-adic analysis, *p*-adic mathematical physics are explained in $[10, 15, 27, 30]$

## 2.2. Dynamical systems in $\mathbb{Q}_p$

In this section we recall some known facts about dynamical systems $(f, B)$ in $\mathbb{Q}_p$, where $f : x \in B \to f(x) \in B$ is an analytic function and $B = B_r(a)$ or $\mathbb{Q}_p$.

Recall some standard terminology of the theory of dynamical systems (see for example [26]). Let $f : B \to B$ be an analytic function. Denote $x^{(n)} = f^n(x^{(0)})$, where $x^0 \in B$ and $f^n(x) = \underbrace{f \circ \cdots \circ f(x)}_{n}$. If $f(x^{(0)}) = x^{(0)}$ then $x^{(0)}$ is called a *fixed point*. A fixed point $x^{(0)}$ is called an *attractor* if there exists a neighborhood $U(x^{(0)})$ of $x^{(0)}$ such that for all points $y \in U(x^{(0)})$ it holds $\lim_{n \to \infty} y^{(n)} = x^{(0)}$, where $y^{(n)} = f^n(y)$. If $x^{(0)}$ is an attractor then its *basin of attraction* is

$$A\left(x^{(0)}\right) = \left\{y \in \mathbb{Q}_p : y^{(n)} \to x^{(0)}, \ n \to \infty\right\}.$$

A fixed point $x^{(0)}$ is called *repeller* if there exists a neighborhood $U(x^{(0)})$ of $x^{(0)}$ such that $|f(x) - x^{(0)}|_p > |x - x^{(0)}|_p$ for $x \in U(x^{(0)})$, $x \neq x^{(0)}$. For a fixed point $x^{(0)}$ of a function $f(x)$ a ball $B_r(x^{(0)})$ (contained in $B$) is said to be a *Siegel disc* if each sphere $S_\rho(x^{(0)})$, $\rho < r$ is an invariant sphere of $f(x)$, i.e., if $x \in S_\rho(x^{(0)})$ then all iterated points $x^{(n)} \in S_\rho(x^{(0)})$ for all $n = 1, 2 \ldots$ . The union of all Siegel discs with the center at $x^{(0)}$ is said to *a maximum Siegel disc* and is denoted by $SI(x^{(0)})$.

**Remark 2.2.** In non-Archimedean geometry, a center of a disc is nothing but a point which belongs to the disc, therefore, in principle, different fixed points may have the same Siegel disc.

Let $x^{(0)}$ be a fixed point of an analytic function $f(x)$. Set

$$\lambda = \frac{d}{dx} f\left(x^{(0)}\right).$$

The point $x^{(0)}$ is called *attractive* if $0 \le |\lambda|_p < 1$, *indifferent* if $|\lambda|_p = 1$, and *repelling* if $|\lambda|_p > 1$.

## 3. The map $f : x \to x^3 + ax^2$ and its fixed points

In this section we consider dynamical system associated with the function $f : \mathbb{Q}_p \to \mathbb{Q}_p$ defined by

$$f(x) = x^3 + ax^2, \quad a \in \mathbb{Q}_p. \tag{3.1}$$

Throughout the paper we will consider only the case $|a|_p \ne 1$, since the case $|a|_p = 1$ requires more investigations and it will be considered elsewhere.

Direct checking shows that the fixed points of the function (3.1) are the following ones

$$x_1 = 0 \quad \text{and} \quad x_{2,3} = \frac{-a \pm \sqrt{a^2 + 4}}{2}. \tag{3.2}$$

Here it should be noted that $x_{2,3}$ are the solutions of the following equation

$$x^2 + ax - 1 = 0. \tag{3.3}$$

Note that these fixed points are formal, because, basically in $\mathbb{Q}_p$ the square root does not always exist, but a full investigation of behavior of the dynamics of the function needs the existence of the fixed points $x_{2,3}$. To do end this it is enough to verify when $\sqrt{a^2 + 4}$ does exist. So, by verifying the conditions of Lemma 2.1 one can prove the following

**Proposition 3.1.** *The following assertions hold*

(i) *Let* $|a|_p < 1$, *then the expression* $\sqrt{a^2 + 4}$ *exists in* $\mathbb{Q}_p$ *if and only if either* $p \ge 3$ *or* $p = 2$ *and* $|a|_p \le 1/p^3$.

(ii) *Let* $|a|_p > 1$, *then the expression* $\sqrt{a^2 + 4}$ *exists in* $\mathbb{Q}_p$.

## 4. Attractors and Siegel discs

In this section we will establish behavior of the fixed points of the dynamical system. After, we are going to describe size of the attractors and Siegel discs of the system.

Before going to details let us formulate certain useful results.

Let us assume that $x^{(0)}$ be a fixed point of $f$. Then $f$ can be represented as follows

$$f(x) = f\left(x^{(0)}\right) + f'\left(x^{(0)}\right)\left(x - x^{(0)}\right) + \frac{f''\left(x^{(0)}\right)}{2}\left(x - x^{(0)}\right)^2$$
$$+ \frac{f'''\left(x^{(0)}\right)}{6}\left(x - x^{(0)}\right)^3. \tag{4.1}$$

From the above equality putting $\gamma = x - x_0$ we obtain

$$\left|f(x) - f\left(x^{(0)}\right)\right|_p = |\gamma|_p \left|f'\left(x^{(0)}\right) + \frac{f''\left(x^{(0)}\right)}{2}\gamma + \frac{f'''\left(x^{(0)}\right)}{6}\gamma^2\right|_p. \tag{4.2}$$

**Lemma 4.1.** *Let $x^{(0)}$ be a fixed point of the function $f$ given by (3.1). If*

$$\max\left\{\left|3x^{(0)} + a\right|_p \left|x - x^{(0)}\right|_p, \left|x - x^{(0)}\right|_p^2\right\} < \left|f'\left(x^{(0)}\right)\right|_p$$

*then*

$$\left|f(x) - f\left(x^{(0)}\right)\right|_p = \left|f'\left(x^{(0)}\right)\right|_p \left|x - x^{(0)}\right|_p. \tag{4.3}$$

The proof immediately follows from (4.2) and

$$f'(x) = 3x^2 + 2ax$$
$$f''(x) = 6x + 2a, \qquad f'''(x) = 6. \tag{4.4}$$

From Lemma 4.1 we get

**Corollary 4.2.** [5] *Let $x^{(0)}$ be a fixed point of the function $f$ given by (3.1). The following assertions hold:*

(i) *if $x^{(0)}$ is an attractive point of $f$, then it is an attractor of the dynamical system. If $r > 0$ satisfies the inequality*

$$\max\left\{\left|3x^{(0)} + a\right|_p r, r^2\right\} < 1 \tag{4.5}$$

*then $B_r(x^{(0)}) \subset A(x^{(0)})$;*

(ii) *if $x^{(0)}$ is an indifferent point of $f$ then it is the center of a Siegel disc. If $r$ satisfies the inequality (4.5) then $B_r(x^{(0)}) \subset SI(x^{(0)})$;*

(iii) *if $x^{(0)}$ is a repelling point of $f$ then $x^{(0)}$ is a repeller of the dynamical system.*

Now we are going to calculate norms of the fixed points and their behavior. Consider several distinct cases with respect to the parameter $a$.

From (4.4) one can easily conclude that the fixed point $x_1$ is attractive. Therefore, furthermore we will deal with $x_{2,3}$.

Taking into account that the fixed points $x_2$ and $x_3$ are the solutions of (3.3) from (4.4) we find

$$f'(x_\sigma) = 3 - ax_\sigma, \qquad \sigma = 2, 3 \tag{4.6}$$

and

$$x_2 + x_3 = -a, \qquad x_2 x_3 = -1. \tag{4.7}$$

**Case** $|a|_p > 1$. In this case from (4.7) one gets that $|x_2 + x_3|_p = |a|_p$, $|x_2|_p |x_3|_p = 1$. These imply that either $|x_2|_p > 1$ or $|x_3|_p > 1$. Without loss of generality we may assume that $|x_2|_p > 1$, which means that $|x_3|_p < 1$. From the condition $|a|_p > 1$ one finds that $|x_2|_p = |a|_p$ and $|x_3|_p = 1/|a|_p$.

From (4.6) we infer that

$$|f'(x_2)|_p = |a|_p |x_2|_p = |a|_p^2 > 1 \tag{4.8}$$

which implies that $x_2$ is repelling. Now (4.4) with $|x_3|_p = 1/|a|_p$ implies that $|f'(x_3)|_p = 1$, hence $x_3$ is a indifferent point.

Now let us find the basin of attraction of $x_1$, i.e., $A(x_1)$.

Denote

$$r_k = \frac{1}{|a|_p^k}, \quad k \geq 0.$$

Consider several steps along the description of $A(x_1)$.

(I) From Corollary 4.2 and (4.5) we find that $B_{r_1}(0) \subset A(x_1)$. Now take $x \in S_{r_1}(0)$, i.e., $|x| = r_1$. Then one gets

$$|f(x)|_p = |x|_p^2 |x + a|_p = |x|_p^2 |a|_p = r_1,$$

whence we infer that $|f^{(n)}(x)|_p = r_1$ for all $n \in \mathbb{N}$. This means that $x \notin A(x_1)$, hence $A(x_1) \cap S_{r_1}(0) = \emptyset$. As a consequence we have $f(S_{r_1}(0)) \subset S_{r_1}(0)$.

In the sequel we will assume that $\sqrt{|a|_p} \notin \{p^k, \ k \in \mathbb{N}\}$. Denote

$$A(\infty) = \left\{ x \in \mathbb{Q}_p : |f^{(n)}(x)|_p \to \infty \quad \text{as} \quad n \to \infty \right\}.$$

It is evident that $A(x_1) \cap A(\infty) = \emptyset$.

(II) Let us take $x \in S_r(0)$ with $r > |a|_p$. Then we have

$$|f(x)|_p = |x|_p^2 |x + a|_p = |x|_p^2 |x|_p = |x|_p^3,$$

which means that $x \in A(\infty)$, i.e., $S_r(0) \subset A(\infty)$ for all $r > |a|_p$.

(III) Now assume that $x \in S_r(0)$ with $r \in (r_1, r_0) \cup (r_0, |a|_p)$. Then we have $f(S_r(0)) \subset S_{r^2|a|_p}(0)$. If $r \in (r_0, |a|_p)$ then $r^2 |a|_p > |a|_p$, hence we have $S_r(0) \subset A(\infty)$. If $r \in (r_1, r_0)$ then according to our assumption we have $r^{2^n} |a|_p^{1+2+\cdots+2^{n-1}} \neq 1$ for every $n \in \mathbb{N}$, hence there is $n_0 \in \mathbb{N}$ such that $f^{(n_0)}(S_r(0)) \subset A(\infty)$, from this we infer that $S_r(0) \subset A(\infty)$. Consequently, we have $S_r(0) \subset A(\infty)$ for all $r \in (r_1, r_0) \cup (r_0, |a|_p)$.

(IV) If $x \in S_{r_0}(0)$, then one gets $f(S_{r_0}(0)) \subset S_{|a|_p}(0)$.

(V) Therefore, we have to consider $x \in S_{|a|_p}(0)$. From (3.1) we can write

$$|f(x)|_p = |a|_p^2 |x + a|_p. \tag{4.9}$$

From the last equality and the following decomposition

$$S_{|a|_p}(0) = \bigcup_{r=0}^{|a|_p} S_r(-a). \tag{4.10}$$

one concludes that we have to investigate behavior of $f$ on spheres $S_r(-a)$ $(r \in [0, |a|_p])$.

(VI) Let $x \in B_{r_3}(-a)$, i.e., $|x + a|_p < r_3$, then the equality (4.9) implies that $|f(x)|_p < r_1$ which means $x \in A(x_1)$. Hence $B_{r_3}(-a) \subset A(x_1)$. Moreover, taking into account (I) one gets $f(S_{r_3}(-a)) \subset S_{r_1}(0)$.

(VII) If $x \in S_{r_1}(-a)$ then from (4.9) we find that $f(x) \in S_{|a|_p}(0)$.

(VIII) If $x \in S_{r_2}(-a)$ then again using (4.9) one gets that $f(x) \in S_{r_0}(0)$. This with (IV) implies that $f^{(2)}(S_{r_1}(-a)) \subset S_{|a|_p}(0)$.

(IX) If $x \in S_r(-a)$ with $r \in (r_3, r_2) \cup (r_2, r_1) \cup (r_1, |a|_p]$. Then from (4.9) we obtain that $f(x) \in S_\rho(0)$, $\rho \in (r_1, r_0) \cup (r_0, |a|_p) \cup (|a|_p, |a|_p^3]$. Hence thanks to (II) and (III) we infer that $f(x) \in A(\infty)$.

Let us introduce some more notations. Given sets $A, B \subset \mathbb{Q}_p$ put

$$T_{f,A,B}(x) = \min \left\{ k \in \mathbb{N} : \ f^{(k)}(x) \in B \right\}, \quad x \in A, \tag{4.11}$$

$$D[A, B] = \left\{ x \in A : \ T_{f,A,B}(x) < \infty \right\}. \tag{4.12}$$

Taking into account (II)–(IX) and (4.11), (4.12) we can define $D[S_{r_0}(0) \cup S_{|a|_p}(0), B_{r_3}(-a)]$, which is non empty, since from (VI) one sees that $B_{r_3}(-a) \subset D[S_{r_0}(0) \cup S_{|a|_p}(0), B_{r_3}(-a)]$. Thus from (I)–(IX) we have $A(x_1) = B_{r_1}(0) \cup D[S_{r_0}(0) \cup S_{|a|_p}(0), B_{r_3}(-a)]$.

Now consider the other $x_\sigma$ ($\sigma = 2, 3$) fixed points. We already have known from above calculations that $|x_2|_p = |a|_p$ and $|x_3|_p = r_1$, this means $x_3$ is indifferent, so Corollary 4.2 with (4.5) yields that $B_{r_1}(x_3) \subset SI(x_3)$. It is clear that $0 \notin SI(x_3)$, therefore $SI(x_3) = B_{r_1}(x_3)$. From (I) we infer that $SI(x_3) \subset S_{r_1}(0)$.

Thus we have proved the following

**Theorem 4.3.** *Let $|a|_p > 1$ and $\sqrt{|a|_p} \notin \{p^k, \ k \in \mathbb{N}\}$. Then $x_1$ is attractor and $A(x_1) = B_{r_1}(0) \cup D[S_{r_0}(0) \cup S_{|a|_p}(0), B_{r_3}(-a)]$. For the other fixed points we have $|x_2|_p = |a|_p$ and $|x_3|_p = r_1$, hence $x_3$ is indifferent and $SI(x_3) = B_{r_1}(x_3)$.*

**Case $|a|_p < 1$.** Let $p \geq 3$ then from (3.2) and (4.7) one finds that $|x_\sigma|_p = 1$.

Let $p = 2$ then Proposition 3.1 implies that $a = p^k \varepsilon$ for some $k \geq 3$ and $|\varepsilon|_p = 1$. From this and taking into account (3.2) we have

$$|x_\sigma|_p = |p^{k-1}\varepsilon \pm \sqrt{p^{2(k-1)}\varepsilon^2 + 1}|_p = 1,$$

since $|p^{2(k-1)}\varepsilon^2 + 1|_p = 1$ and $k \geq 3$.

Now let us compute $|f'(x_\sigma)|_p$. If $p \neq 3$, then using (4.6) one gets

$$|f'(x_\sigma)|_p = |3 - ax_\sigma|_p = 1, \quad \sigma = 2, 3.$$

This means that the fixed points $x_\sigma, (\sigma = 2, 3)$ are indifferent.

If $p = 3$ then from (4.6) we easily obtain that $|f'(x_\sigma)|_p < 1$, $\sigma = 2, 3$, which implies that the fixed points are attractive.

Let us first consider $x_1$. According to Corollary 4.2 we immediately find that $B_1(0) \subset A(x_1)$. Take $x \in S_1(0)$ then $|f(x)|_p = |x|_p^2|x + a|_p = |x|^3 = 1$, hence $|f^{(n)}(x)|_p = 1$ for all $n \in \mathbb{N}$. This yields that $x \notin A(x_1)$, hence $A(x_1) = B_1(0)$.

In the sequel according to the above done calculations we will consider two possible situations when $p \neq 3$ and $p = 3$.

Assume $p \neq 3$. In this case $x_2$ and $x_3$ are indifferent, so Corollary 4.2 implies that $B_1(x_\sigma) \subset SI(x_\sigma)$, $\sigma = 2, 3$.

Let us take $x \in S_r(x_\sigma)$, $r \geq 1$, then put $\gamma = x - x_\sigma$. It is clear that $|\gamma|_p = r$. By means of (4.2) and (3.3) we find

$$
\begin{aligned}
|f(x) - f(x_\sigma)|_p &= |\gamma|_p |3x_\sigma^2 + 2ax_\sigma + (3x_\sigma + a)\gamma + \gamma^2|_p \\
&= r|\gamma^2 + 3x_\sigma\gamma + 3 + a(\gamma - x_\sigma)|_p \, .
\end{aligned}
\tag{4.13}
$$

If $r > 1$ then from we easily obtain that

$$
|f(x) - f(x_\sigma)|_p = |\gamma|_p^3
$$

since $|\gamma^2 + 3x_\sigma\gamma + 3|_p = |\gamma|_p^2$, $|a(\gamma - x_\sigma)|_p = |a|_p|\gamma|_p$. This implies that $SI(x_\sigma) \subset \bar{B}_1(x_\sigma)$.

**Lemma 4.4.** *Let $|a|_p < 1$ and $p \neq 3$. The equality $SI(x_\sigma) = \bar{B}_1(x_\sigma)$ holds if and only if for every $\gamma \in S_1(0)$ the equality*

$$
|\gamma^2 + 3x_\sigma\gamma + 3| = 1
\tag{4.14}
$$

*is valid.*

*Proof.* If (4.14) is satisfied for all $\gamma \in S_1(0)$ then from (4.13) we infer that $f(S_1(x_\sigma)) \subset S_1(x_\sigma)$, since $|a(\gamma - x_\sigma)|_p < 1$. This proves the assertion. Now suppose that $SI(x_\sigma) = \bar{B}_1(x_\sigma)$ holds. Assume that (4.14) is not valid, i.e., there is $\gamma_0 \in S_1(0)$ such that

$$
|\gamma_0^2 + 3x_\sigma\gamma_0 + 3| < 1 \, .
\tag{4.15}
$$

The last one with (4.13) implies that $|f(x_0) - x_\sigma|_p < 1$ for an element $x_0 = x_\sigma + \gamma_0$. But this contradicts to $SI(x_\sigma) = \bar{B}_1(x_\sigma)$. This completes the proof. $\square$

From the proof of Lemma 4.4 we immediately obtain that if there is $\gamma_0 \in S_1(0)$ such that (4.15) is satisfied then $SI(x_\sigma) = B_1(x_\sigma)$.

**Lemma 4.5.** *Let $|a|_p < 1$ and $p \neq 3$. The following conditions are equivalent:*
(i) $SI(x_\sigma) = B_1(x_\sigma)$;
(ii) *there is $\gamma_0 \in S_1(0)$ such that (4.15) is satisfied;*
(iii) $\sqrt{-3}$ *exists in $\mathbb{Q}_p$.*

*Proof.* The implication (i)$\Leftrightarrow$(ii) immediately follows from the proof of Lemma 4.14. Consider the implication (ii)$\Rightarrow$(iii). The condition (4.15) according to the Hensel Lemma yields that existence of the solution $z \in \mathbb{Q}_p$ of the following equation

$$
z^2 + 3x_\sigma z + 3 = 0
\tag{4.16}
$$

such that $|z - \gamma_0| < 1$ which implies that $|z|_p = 1$. Now assume that there is a solution $z_1 \in \mathbb{Q}_p$ of (4.16). Then from Vieta's formula we infer the existence of the other solution $z_2 \in \mathbb{Q}_p$ such that

$$
z_1 + z_2 = -3x_\sigma \, , \quad z_1 z_2 = 3 \, .
$$

From these equalities we obtain that $|z_1 + z_2|_p = 1$, $|z_1 z_2|_p = 1$ which imply $z_1, z_2 \in S_1(0)$. So putting $\gamma_0 = z_1$ one gets (4.15).

Let us now analyze when (4.16) has a solution belonging to $\mathbb{Q}_p$. We know that a general solution of (4.16) is given by

$$z_{1,2} = \frac{-3x_\sigma \pm \sqrt{-3 - 9ax_\sigma}}{2}, \tag{4.17}$$

here we have used (3.3). But it belongs to $\mathbb{Q}_p$ if $\sqrt{-3 - 9ax_\sigma}$ exists in $\mathbb{Q}_p$. Since $|-9ax_\sigma|_p = |a|_p < 1$ implies that $-9ax_\sigma = p^k \varepsilon$ for some $k \geq 1$ and $|\varepsilon|_p = 1$. Hence, $-3 - 9ax_\sigma = -3 + p^k \varepsilon$. Therefore according to Lemma 2.1 we conclude that $\sqrt{-3 - 9ax_\sigma}$ exists if and only if $\sqrt{-3}$ exists in $\mathbb{Q}_p$. The implication (iii)$\Rightarrow$(ii) can be proven along the reverse direction in the previous implication. □

From Lemmas 4.4 and 4.5 we conclude that $SI(x_\sigma)$ is either $B_1(x_\sigma)$ or $\bar{B}_1(x_\sigma)$. The equality (3.2) yields that

$$|x_2 - x_3|_p = |\sqrt{a^2 + 4}|_p = |2|_p, \tag{4.18}$$

which implies that $SI(x_2) \cap SI(x_3) = \emptyset$ when $p \geq 5$ and $SI(x_2) = SI(x_3)$ when $p = 2$, since any point of a ball is its center in non-archimedean setting.

Now consider the case $p = 3$. Then, we know that both fixed points $x_2$ and $x_3$ are attractive. Taking into account $|x_\sigma|_p < 1$ and $|a|_p < 1$ from Corollary 4.2 one finds that $B_1(x_\sigma) \subset A(x_\sigma)$, $\sigma = 2, 3$. From the equality (4.18) we have $|x_2 - x_3|_p = 1$ which implies that $S_1(x_\sigma)$ is not a subset of $A(x_\sigma)$.

Let us take $x \in S_r(x_\sigma)$ with $r \geq 1$, then putting $\gamma = x - x_\sigma$ from (4.13) with $|3x_\sigma \gamma + 3 + a(\gamma - x_\sigma)|_p < |\gamma|_p$ we get

$$|f(x) - x_\sigma|_p = |\gamma||\gamma^2 + 3x_\sigma \gamma + 3 + a(\gamma - x_\sigma)|_p = r^3,$$

which implies that $f(S_r(x_\sigma)) \subset S_{r^3}(x_\sigma)$ for every $r \geq 1$. Hence, in particular, we obtain $f(S_1(x_\sigma)) \subset S_1(x_\sigma)$.

Consequently we have the following

**Theorem 4.6.** *Let $|a|_p < 1$. The following assertions hold:*
  (i) *The fixed point $x_1$ is attractor and $A(x_1) = B_1(0)$;*
 (ii) *If $p \neq 3$ the fixed points $x_\sigma, \sigma = 2, 3$ are indifferent and $SI(x_\sigma) = B_1(x_\sigma)$ is valid if and only if $\sqrt{-3}$ exists in $\mathbb{Q}_p$. Otherwise $SI(x_\sigma) = \bar{B}_1(x_\sigma)$ holds.*
(iii) *If $p \geq 5$ then $SI(x_2) \cap SI(x_3) = \emptyset$, if $p = 2$ then $SI(x_2) = SI(x_3)$.*
 (iv) *If $p = 3$ then the fixed points $x_\sigma$, $\sigma = 2, 3$ are attractors and $A(x_\sigma) = B_1(x_\sigma)$.*

Note that if we consider our dynamical system over $p$-adic complex field $\mathbb{C}_p$ we will obtain different result from the formulated theorem, since $\sqrt{-3}$ always exists in $\mathbb{C}_p$.

**Acknowledgments**

F. Mukhamedov thanks the FCT (Portugal) grant SFRH/BPD/17419/2004. J. F. F. Mendes acknowledges projects POCTI/FAT/46241/2002, POCTI/MAT/46176/2002 and European research NEST project DYSONET/ 012911.

# References

[1] V. Anashin, Ergodic Transformations in the Space of p-adic Integers, in: *p-adic Mathematical Physics (2nd International Conference*, Belgrade, 15–21 September 2005), AIP Conference Proceedings, Vol. 826, Melville, New York, 2006, pp. 3–24.

[2] D. K. Arrowsmith and F. Vivaldi, *Geometry of p-adic Siegel discs.* Physica D **71** (1994), 222–236.

[3] V. A. Avetisov, A. H. Bikulov, S. V. Kozyrev, V. A. Osipov, *p-adic modls of ultra-metric diffusion constrained by hierarchical energy landscapes*, J. Phys. A: Math. Gen. **35**(2002), 177–189.

[4] S. Albeverio, A. Khrennikov, P. E. Kloeden, *Memory retrieval as a p-adic dynamical system*, BioSys. **49**(1999), 105–115.

[5] S. Albeverio, A. Khrennikov, B. Tirozzi, S. De Smedt, *p-adic dynamical systems*, Theor. Math. Phys. **114**(1998), 276–287.

[6] I. Ya. Aref'eva, B. Dragovich, P. H. Frampton, I. V. Volovich, *Wave function of the universe and p-adic gravity*, Int. J. Mod. Phys. A. **6**(1991) 4341–4358.

[7] R. Benedetto, *Hyperbolic maps in p-adic dynamics*, Ergod. Th.& Dynam. Sys. **21**(2001), 1–11.

[8] R. Benedetto, *p-adic dynamics and Sullivan's no wandering domains theorem*, Composito Math. **122**(2000), 281–298.

[9] G. Call and J. Silverman, *Canonical height on varieties with morphisms*, Composito Math. **89**(1993), 163–205.

[10] F. Q. Gouvea, *p-adic Numbers*, Springer, Berlin, 1991.

[11] V. M. Gundlach, A. Khrennikov, K. O. Lindahl, *On ergodic behavior of p-adic dynamical systems*, Infin. Dimen. Anal. Quantum Probab. Relat. Top. **4**(2001), 569–577.

[12] P. G. O. Freund and E. Witten, *Adelic string ampletudes*, Phys. Lett. B **199**(1987) 191–194.

[13] M. Herman and J.-C. Yoccoz, Generalizations of some theorems of small divisors tp non-Archimedean fields, in: *Geometric Dynamics* (Rio de Janeiro, 1981), Lec. Notes in Math. 1007, Springer, Berlin, 1983, pp. 408–447.

[14] M. Khamraev and F. M. Mukhamedov, *On a class of rational p-adic dynamical systems*, Jour. Math. Anal. Appl. **315**(2006), 76–89.

[15] N. Koblitz, *p-adic Numbers, p-adic Analysis and Zeta-Function*, Springer, Berlin, 1977.

[16] A. Yu. Khrennikov, *p-adic Valued Distributions in Mathematical Physics*, Kluwer, Dordreht, 1994.

[17] A. Yu. Khrennikov, *Non-Archimedean Analysis: Quantum Paradoxes, Dynamical Systems and Biological Models*, Kluwer, Dordreht, 1997.

[18] A. Yu. Khrennikov and M. Nilsson, *p-adic Deterministic and Random Dynamical Systems*, Kluwer, Dordreht, 2004.

[19] J. Lubin, *Nonarchimedean dynamical systems*, Composito Math. **94**(1994), 321–346.

[20] E. Marinary and G. Parisi, *On the p-adic five point function*, Phys. Lett. B **203**(1988) 52–56.

[21] F. M. Mukhamedov, *On a recursive equation over p-adic field*, Appl. Math. Lett. **20** (2007), 88–92 .

[22] F. M. Mukhamedov and U. A. Rozikov, *On Gibbs measures of p-adic Potts model on the Cayley tree*, Indag. Math. N. S. **15**(2004), 85–100.

[23] F. M. Mukhamedov and U. A. Rozikov, *On inhomogeneous p-adic Potts model on a Cayley tree*. Infin. Dimens. Anal. Quantum Probab. Relat. Top. **8**(2005), 277–290.

[24] F. M. Mukhamedov, U. A. Rozikov, J. F. F. Mendes, On phase transitions for *p*-adic Potts model with competing interactions on a Cayley Tree, in: *p-adic Mathematical Physics* (2nd International Conference, Belgrade, 15–21 September 2005), AIP Conference Proceedings, Vol. 826, Melville, New York, 2006, pp. 140–150.

[25] T. Pezda, *Polynomial cycles in certain local domains*, Acta Arith. **66** (1994), 11–22.

[26] H.-O. Peitgen, H. Jungers, D. Saupe, *Chaos Fractals*, Springer, Heidelberg, New York, 1992.

[27] A. M. Robert, *A Course of p-adic Analysis*, Springer, New York, 2000.

[28] B. Shabat, *p-adic entropies of logistic maps*, Proc. Steklov Math. Inst. **245**(2004), 257–263.

[29] E. Thiran, D. Verstegen, J. Weters, *p-adic dynamics*, J. Stat. Phys. **54**(3/4)(1989), 893–913.

[30] V. S. Vladimirov, I. V. Volovich, E. I. Zelenov, *p-adic Analysis and Mathematical Physics*, World Scientific, Singapour, 1994.

[31] C. F. Woodcock and N. P. Smart, *p-adic chaos and random number generation*, Exp. Math. **7:4**(1998), 333–342.

Farrukh Mukhamedov and José F. F. Mendes
Departamento de Fisica
Universidade de Aveiro
Campus Universitario de Santiago
3810-193 Aveiro, Portugal
e-mail: `farruh@fis.ua.pt`
    `jfmendes@fis.ua.pt`

Progress in Nonlinear Differential Equations
and Their Applications, Vol. 75, 317–344
© 2007 Birkhäuser Verlag Basel/Switzerland

# Some New Concepts of Dimension

Józef Myjak

*To Arrigo Cellina and James Yorke*

**Abstract.** This paper contains a review of recent results concerning some new concepts of analytical dimensions of sets and measures. We make in evidence the relationship between these new concepts and the classical once. In particular we give some results concerning estimates, variational principles and generic properties. Finally, we give some applications in the theory of iterated function systems and the theory of differential equations.

**Mathematics Subject Classification (2000).** Primary 28A80; Secondary 26A21; 26E25; 28A33; 28A78; 54E52; 60J05.

**Keywords.** Hausdorff, fractal, concentration, packing, information, thin and topology dimension, iterated function systems, differential equations, measure, generic properties, variational principle.

## 1. Introduction and basic notation

The idea of dimension of sets and measures has a long story. In 1911, H. Lebesgue proposed a first definition of a dimension of set, using some properties of covers. In 1912, Poincaré formulated a framework of the theory known actually as the theory of topological dimension. A precise definition of the topological dimension was formulated in 1922–1923 by P. Uryshon and K. Menger. In the next years this notion was intesively studied and in 1941 appeared the monography *Dimension Theory* by W. Hurewicz and P. Wallman, which contains rather complete theory of (inductive) topological dimension for separable metric spaces. In the next 40–50 years this theory was generalized to arbitrary metric spaces and the results was presented in several books (J. Nagata (1965), K. Nagami (1970), A. R. Pears (1975), R. Engelking (1977) and others).

The quite different approach to the dimension theory was proposed by F. Hausdorff in 1919. He used some properties of measures which are satisfyied for arbitrary metric space. The Hausdorff dimension has the important property, by mean of similarities with scale $s$, the measure of image of $d$-dimension set changes $s^d$ times. Unfortunately, this dimension is very difficult to be calculated even for simple sets. Moreover, the Hausdorff dimension is not invariant with respect to the homeomorphismes. For that reason for long time the Hausdorff approach was rather undervalued.

The situation radically changed after 1975, when B. Mandelbrot introduced a new class of sets, so called fractals, defined as the sets for which the Hausdorff dimension is strictle greater then the topological dimension. On the other hand in course of time, the dimension became a very important characteristic of attractors in the theory of dynamical systems. Since the Hausdorff dimension was not sufficient to cover all situation, the various notions of dimensions have been proposed: box dimension, packing dimension, correlation dimension, informatic dimension, entropy, capacity. All these concepts were very widely investigated and used, but unfortunately all of them are rather hard to calculate.

Recently two other concepts of dimensions have been proposed by A. Lasota, the concentration dimension and the thin dimension. The concentration dimension is defined by using the Lévy concentration function. This dimension is relatively easy to be calculated. Moreover, it is strongly related to the mass distribution principle and often simplified the calculation of the Hausdorff dimension of sets. The thin dimension is based on the notion of the thin function which is a kind of anti-concetration function to Lévy function. The thin dimension is related to the fractal dimension.

Through this paper $X$ denotes a metric space with metric $\rho$. By $B(x, r)$ we denote the open ball with centre $x \in X$ and radius $r > 0$. For $A \subset X$ and $x \in X$, $\overline{A}$ stands for the closure of $A$, $\mathrm{diam} A$ for the diameter of $A$, $\partial A$ for the boundary of $A$ and $\rho(x, A)$ for the distance from $x$ to $A$. As usual, by $\mathbb{R}$ we denote the set of reals, by $\mathbb{N}$ the set of all positive integers, by $\mathbb{Q}$ the set of all rational numbers and $\mathbb{R}_+ = [0, \infty)$.

By $\mathcal{B}(X)$ we denote the $\sigma$-algebra of Borel subsets of $X$ and by $\mathcal{M}(X)$ the family of all finite Borel measures on $X$. By $\mathcal{M}_1(X)$ we denote the space of all $\mu \in \mathcal{M}(X)$ such that $\mu(X) = 1$. The elements of $\mathcal{M}_1(X)$ are called *probability measures*. Finally, by $B(X)$ we denote the space of all bounded Borel measurable functions $f : X \to \mathbb{R}$ and by $C(X)$ the subspace of all continuous functions.

## 2. Some classical notions of dimensions

**a) Hausdorff dimension.** For $A \subset X$ and $s, \delta > 0$ we define

$$\mathcal{H}_\delta^s(A) = \inf \sum_{i=1}^{\infty} \left( \mathrm{diam}\, U_i \right)^s,$$

where the infimum is taken over all countable covers $\{U_i\}$ of the set $A$ such that $\mathrm{diam}\, U_i \leq \delta$. Then

$$\mathcal{H}^s(A) = \lim_{\delta \to 0} \mathcal{H}_\delta^s(A)$$

is a *Hausdorff s-dimensional measure*. Note that all Borel sets are $\mathcal{H}^s$-measurable. For $s$ sufficienly large $\mathcal{H}^s(A) = 0$ and for $s$ sufficiently small $\mathcal{H}^s(A) = \infty$. The *Hausdorff dimension* of the set $A$ is defined by

$$\dim_H A = \inf \left\{ s > 0 : \mathcal{H}^s(A) < \infty \right\}.$$

Observe that Hausdorff dimension is well defined for every subset of $X$.

The Hausdorff dimension of the measure $\mu \in \mathcal{M}_1(X)$ is defined by the formula

$$\dim_H \mu = \inf \left\{ \dim_H A : A \in \mathcal{B}(X), \, \mu(A) = 1 \right\}.$$

It is easy to see that the Hausdorf dimension satisfies the following properties:

(i)   $\dim_H A \leq \dim_H B$ if $A \subset B$;
(ii)  $\dim_H \bigcup_{n=1}^{\infty} A_n = \sup_{n \in \mathbb{N}} \dim_H A_n$;
(iii) $\dim_H A = 0$ if $A$ is a countable set;
(iv)  If $f : X \to X$ is lipschitzian function, then $\dim_H f(A) \leq \dim_H A$.

**b) Fractal dimension.** The *lower* and *upper fractal* (or *box*) *dimension* of the set $A$ is defined by

$$\underline{\dim}_F A = \liminf_{r \to 0} \frac{\log N(r)}{-\log r}, \quad \overline{\dim}_F A = \limsup_{r \to 0} \frac{\log N(r)}{-\log r}$$

where $N(r)$ denotes the smallest number of closed balls of radius $r$ needed to cover $A$. Note that in the above definitions $N(r)$ can be replaced by $M(r)$ – the smallest number of sets of diameter less then or equal to $r$ needed to cover $A$.

If $\underline{\dim}_F A = \overline{\dim}_F A$ then this common value is called the *fractal dimension* of $A$. It is easy to verify the following properties of fractal dimensions:

(i)   $\dim_H A \leq \underline{\dim}_F A$ for arbitrary $A \subset X$;
(ii)  $\dim_F A = 0$ if $A$ is a finite set;
(iii) $\dim_F \bigcup_{n=1}^{N} A_n = \max \left\{ \dim_F A_n : n = 1, \ldots, N \right\}$.

**Example 2.1.** Let $A = \{1/n : n \in \mathbb{N}\}$, $B = \{1/(\log n) : n \in \mathbb{N}, n \geq 2\}$, $C = \mathbb{Q} \cap [0, 1]$. Then $\dim_H A = \dim_H B = \dim_H C = 0$, $\dim_F A = 1/2$, $\dim_F B = 1$ and $\dim_F C = 1$.

**c) Correlation dimension.** Given a Borel probability measure $\mu$ the *lower* and *upper correlation* dimension of $\mu$ are given by

$$\underline{\dim}_C \mu = \liminf_{r \to 0} \frac{\log C_\mu(r)}{\log r}, \quad \overline{\dim}_C \mu = \limsup_{r \to 0} \frac{\log C_\mu(r)}{\log r},$$

where

$$C_\mu(r) = \int_X \mu\big(B(x, r)\big) \mu(dx).$$

**d) Information dimension.** The *lower* and *upper information dimension* of the measure $\mu \in \mathcal{M}_1(X)$ is given by

$$\underline{I}_\mu(r) = \liminf_{r \to 0} \int_X \frac{\log \mu\big(B(x,r)\big)}{\log r} \mu(dx), \quad \overline{I}_\mu(r) = \limsup_{r \to 0} \int_X \frac{\log \mu\big(B(x,r)\big)}{\log r} \mu(dx).$$

If $\underline{I}_\mu = \overline{I}_\mu$ then this common value is called *information dimension* of $\mu$ and it is denoted by $I_\mu$. These dimensions are special cases of *generalized Rényi dimensions* introduced by Hentchel and Procacia (see [14]).

**e) Packing dimension.** Let $A \subset X$. A family $\big\{B(x_i, r_i) : i \in \mathbb{N}\big\}$ is called a *centered $\delta$-packing* of $A$ if it is pairwise disjoint, $x_i \in A$ and $0 < r_i \le \delta$. Define

$$\widetilde{\mathcal{P}}_\delta^s(A) = \sup \sum_{i=1}^\infty (2r_i)^s,$$

where the supremum is taken over all centered $\delta$-packing of $A$. Set

$$\widetilde{\mathcal{P}}^s(A) = \inf_{\delta > 0} \widetilde{\mathcal{P}}_\delta^s(A).$$

The last function is not necessarily countably subadditive, but it allows us to define the countably subadditive function $\mathcal{P}^s$ by formula

$$\mathcal{P}^s(A) = \inf_{A \subset \bigcup_{i=1}^\infty A_i} \sum \widetilde{\mathcal{P}}^s(A_i)$$

where the supremum is taken over all countable coverings of $A$. The function $\mathcal{P}^s(A)$ is called the *s-dimensional packing measure*. The *packing dimension* of $A$ is defined by the formula

$$\dim_P A = \inf \big\{ s > 0 : \mathcal{P}^s(A) < \infty \big\}.$$

The packing dimension was introduced by Taylor and Tricot (see [48]).

**Remark 2.2.** If $X$ is a separable metric space, then for each $s > 0$ we have $\mathcal{H}^s(A) \le \mathcal{P}^s(A)$ for every subset $A$ of $X$. It follows immediately that

$$\dim_H A \le \dim_P A.$$

**f) Local and average dimension.** The *local upper dimension* of the measure $\mu$ at the point $x$ is defined by

$$\overline{d}_\mu(x) = \limsup_{r \to 0} \frac{\log \mu\big(B(x,r)\big)}{\log r}.$$

The *upper average dimension* of a measure $\mu \in \mathcal{M}_1(X)$ is defined by the formula

$$\overline{d}_\mu = \int_X \overline{d}_\mu(x) \mu(dx).$$

The *lower local dimension* and the *lower average dimension* are defined in the same way replacing lim sup by lim inf.

It is known that

$$\underline{d}_\mu(x) \leq \dim_H \operatorname{supp} \mu \quad \text{and} \quad \overline{d}_\mu(x) \leq \dim_P \operatorname{supp} \mu \,.$$

Moreover, Cutler proved (see [4]) that if the measure $\mu$ has compact support then

$$\underline{I}_\mu \leq \underline{d}_\mu \leq \overline{d}_\mu \leq \overline{I}_\mu \,.$$

**Theorem 2.3.** *Let $\mu$ be a finite Borel measure on $\mathbb{R}^n$. Assume that there exist positive constants $M$ and $\gamma$ such that*

$$\int_{\mathbb{R}^n} log^{1+\gamma}\left(1 + \|x\|\right)\mu(dx) \leq M \,.$$

*Then the upper information dimension $\overline{I}_\mu$ is finite and we have $\overline{I}_\mu \leq \overline{d}_\mu$. Moreover, if average information $d_\mu$ exists then the information dimension $I_\mu$ exits and we have $I_\mu = d_\mu$.*

The proof of the last theorem can be found in [33].

**Theorem 2.4.** *Let $\mu$ and $\nu$ be finite Borel measures on $\mathbb{R}^n$ and $\alpha$ and $\beta$ positive numbers. Then*

$$\underline{d}_{\alpha\mu+\beta\nu} = \alpha\underline{d}_\mu + \beta\underline{d}_\nu \,.$$

*Proof.* First, one can verify that

$$\underline{d}_{\mu+\nu} = \min\left\{\underline{d}_\mu, \underline{d}_\nu\right\} \,.$$

Now, set

$$A_1 = \left\{x \in \mathbb{R}^n \,:\, \underline{d}_\mu(x) < \underline{d}_\nu(x)\right\} \,,$$
$$A_2 = \left\{x \in \mathbb{R}^n \,:\, \underline{d}_\mu(x) > \underline{d}_\nu(x)\right\} \,,$$
$$A_0 = \mathbb{R}^n \setminus (A_1 \cup A_2) \,.$$

Simple calculation shows that

$$\underline{d}_{\mu+\nu} = \int_{\mathbb{R}^n} \underline{d}_{\mu+\nu}(x)(\mu + \nu)(dx)$$
$$= \int_{A_0} \underline{d}_\mu(x)\mu(dx) + \int_{A_0} \underline{d}_\nu(x)\nu(dx) + \int_{A_1} \underline{d}_\mu(x)\mu(dx) + \int_{A_2} \underline{d}_\nu(x)\nu(dx)$$
$$= \underline{d}_\mu + \underline{d}_\nu \,.$$

Since equality $\underline{d}_{\alpha\mu} = \alpha\underline{d}_\mu$, $\alpha > 0$ is obvious, the proof is complete.          □

We end this section with two important classical results which will be frequently used in the sequel: Frostman lemma and mass distribution principle (see [10, 43]).

**Theorem 2.5. (Frostman Lemma).** *Let $A$ be a subset of a metric space $X$. Assume that there are $d > 0$, $c > 0$ and a Borel measure $\mu$ supported on $A$ such that for almost all $x \in A$ and $r > 0$ we have*

$$\mu\left(B(x,r)\right) \leq cr^d \,.$$

*Then*
$$\dim_H A \geq d .$$

**Theorem 2.6. (Mass Distribution Principle).** *Let $E \subset \mathbb{R}^n$ and let $\mu$ be a finite Borel measure such that $\mu(E) > 0$. Suppose that there exist $s > 0$, $c > 0$ and $r_0 > 0$ such that*
$$\mu(U) \leq c(\operatorname{diam} U)^s$$
*for all sets $U$ with $\operatorname{diam} U \leq r_0$. Then*
$$\dim_H E \geq s .$$

## 3. Concentration dimension

Given a Borel measure $\mu \in \mathcal{M}_1(X)$ the *lower* and *upper concentration dimension* of $\mu$ are given by formulae
$$\underline{\dim}_L \mu = \liminf_{r \to 0} \frac{\log Q_\mu(r)}{\log r} , \qquad \overline{\dim}_L \mu = \limsup_{r \to 0} \frac{\log Q_\mu(r)}{\log r} ,$$
where
$$Q_\mu(r) = \sup_{x \in X} \mu(B(x,r)) . \tag{3.1}$$

If $\underline{\dim}_L \mu = \overline{\dim}_L \mu$ then this common value is called *concentration* or *Lasota* dimension of $\mu$ and it is denoted by $\dim_L \mu$.

The function $Q_\mu$ is called the Lévy concentration function and it is frequently used in the theory of stochastic processes (see $[15, 23]$).

It is easy to verify that for every $\mu \in \mathcal{M}_1(X)$ we have
$$\underline{\dim}_L \mu \leq \underline{\dim}_C \mu \leq 2\underline{\dim}_L \mu .$$

Analogous inequalities hold for the corresponding upper dimensions.

**Theorem 3.1.** *Let $\mu \in \mathcal{M}_1(X)$ and let $A \in \mathcal{B}(X)$ be such that $\mu(A) > 0$. Then*
$$\dim_H A \geq \underline{\dim}_L \mu .$$

.

*Proof.* Set $d = \underline{\dim}_L \mu$. Suppose $d > 0$. (If $d = 0$ the assertion is obvious). Choose $s \in (0, d)$. Then there exists $r_0 > 0$ such that
$$\mu(B(x,r)) \leq r^s \quad \text{for} \quad r \in (0, r_0) \quad \text{and} \quad x \in X .$$
According to Frostman Lemma we have $\dim_H A \geq s$. Since $s < d$ was arbitrary, the statement of theorem follows. $\qquad\square$

The concentration dimension of a closed set $A$ is defined by the formula
$$\dim_L A = \sup \left\{ \underline{\dim}_L \mu : \mu \in \mathcal{M}_1(X), \ \operatorname{supp} \mu \subset A \right\} .$$

From Theorem 3.1 follows immediately that
$$\dim_H X \geq \dim_L X .$$

## 4. Thin dimension

For $\mu \in \mathcal{M}_1(X)$ we define the *lower* and *upper thin* dimension of $\mu$ by

$$\underline{\dim}_T \mu = \liminf_{r \to 0} \frac{\log T_\mu(r)}{\log r}, \qquad \overline{\dim}_T \mu = \limsup_{r \to 0} \frac{\log T_\mu(r)}{\log r},$$

where

$$T_\mu(r) = \inf \left\{ \mu\big(B(x,r)\big) \ : \ x \in \operatorname{supp} \mu \right\}.$$

If $\underline{\dim}_T \mu = \overline{\dim}_T \mu$ then this common value is called *thin* dimension of the measure $\mu$ and it is denoted by $\dim_T \mu$ (see [19]).

The function $T_\mu : (0, \infty) \to [0, 1]$ is called the *thin function* corresponding to the measure $\mu$. Obviously, if $\operatorname{supp} \mu$ is a compact set, the values of $T_\mu$ are positive. In general $T_\mu$ is only nonnegative. For convenience we make $\log 0 = -\infty$.

**Theorem 4.1.** *If $A \subset \operatorname{supp} \mu$, $A \in \mathcal{B}(X)$ then*

$$\underline{\dim}_F A \leq \underline{\dim}_T \mu \quad \text{and} \quad \overline{\dim}_F A \leq \overline{\dim}_T \mu.$$

*Proof.* Let $d = \underline{\dim}_T \mu$, $A \subset \operatorname{supp} \mu$, $A \neq \emptyset$, $d < \infty$. Choose $s \in (d, \infty)$. Obviously there exists a sequence $\{r_n\}$ of positive numbers such that $r_n \to 0$ and

$$T_\mu(r_n) \geq r_n^s, \qquad n \in \mathbb{N}. \tag{4.1}$$

For fixed $n \in \mathbb{N}$ let $I_n = N_A(r_n/2)$ be the smallest number of closed balls of radius $r_n/2$ needed to cover $A$. Let $\{B(x_i, r_n/2) : i \in I_n\}$ be the corresponding covering. Obviously we can find $y_i \in A$ such that the family $\{B(y_i, r_n) : i \in I_n\}$ covers $A$. Let $J_n \subset I_n$ be such that $\{B(x_i, r_n) : i \in J_n\}$ are pairwise disjoint and the family $\{B(x_i, 4r_n) : i \in J_n\}$ covers $A$. Consequently $N_A(4r_n) \leq \operatorname{card} J_n$. On the other hand

$$\sum_{i \in J_n} \mu\big(B(y_i, r_n)\big) = \mu \left( \bigcup_{i \in J_n} B(y_i, r_n) \right) \leq 1.$$

Since $y_i \in A$ we have $\mu\big(B(y_i, r_n)\big) \geq T_\mu(r_n)$ which implies that $T_\mu(r_n) \cdot \operatorname{card} J_n \leq 1$. Consequently $N_A(4r_n) \leq r_n^{-s}$. From the last inequality it follows that $\underline{\dim}_F A \leq s$. Since $s \in (d, \infty)$ was arbitrary, the first inequality of Theorem 4.1 follows. The proof of the second one is similar. $\qquad \square$

## 5. Variational principles

It is well known that the dimension of a measure allows us to estimate the dimension of its support. Moreover, the estimation of a set $A$ can be obtained either as the greatest lower bound or as the least upper bound of the dimensions of measures supported on $A$. Such results are called variationals principles. They are closely related with Frostman Lemma and Mass Distribution Principle.

In particular variational pinciple for the Hausdorff and packing dimension of sets and the point dimension of measures were proved by C. Tricot [48] and C. D. Cutler [5]. The variational principle for the Hausdorff dimension and packing dimension of sets and Rényi dimension of measures were found by C. D. Cutler and L. Olsen [6]. Here we recall the variational principles for Hausdorf and fractal dimensions of sets and concentration and thin dimension of measures (see [19,23])

**Theorem 5.1.** *Let $K \subset X$ be a nonempty compact set. Then*
$$\dim_H K = \sup \underline{\dim}_L \mu \,,$$
*where the supremum is taken over all $\mu \in \mathcal{M}_1(X)$ such that $\mathrm{supp}\, \mu \subset K$.*

*Proof.* From Theorem 3.1 it follows that
$$\dim_H K \geq \sup \underline{\dim}_L \mu \,,$$
where the supremum is taken over all $\mu \in \mathcal{M}_1(X)$ such that $\mathrm{supp}\, \mu \subset K$.

We need to prove the opposite inequality. Let $d = \dim_H K$. Suppose $d > 0$ (for $d = 0$ the assertion is obvious). Choose $s \in (0, d)$. Obviously $\mathcal{H}^s(K) > 0$. It can prove that there is a measure $\mu \in \mathcal{M}_1(X)$ and constants $c > 0$ and $r_0 > 0$ such that
$$\mu\big(B(x,r)\big) \leq cr^s \quad \text{for} \quad 0 < r < r_0 \,.$$
Consequently $\underline{\dim}_L \mu \geq s$. Since $s \in (0, d)$ was arbitrary, $\dim_H K \geq d$. $\qquad\square$

**Theorem 5.2.** *Let $K$ be a nonempty compact subset of $X$. Then there exists a Borel probability measure $\mu$ such that $\mathrm{supp}\, \mu = K$ and*
$$\underline{\dim}_F K = \underline{\dim}_T \mu \,, \qquad \overline{\dim}_F K = \overline{\dim}_T \mu \,.$$

The proof can be found in [19].

## 6. Iterated Function Systems

An Iterated Function System (shortly IFS) is given by a family of continuous functions
$$w_i : X \to X \,, \quad i \in I \,,$$
where I={1, …,N}.

For $A \subset X$ set
$$F(A) = \bigcup_{i=1}^{N} w_i(A) \,. \tag{6.1}$$
Obviously $F$ maps compact sets in compact sets. If all $w_i$ are strictly contractive, then there exists a unique compact set such that
$$K = \bigcup_{i=1}^{N} w_i(K) \,. \tag{6.2}$$
Moreover, for every compact set $A \subset X$, $F^n(A) \to K$ in Hausdorff distance. The set $K$ is called *attractor* or *fractal* corresponding to IFS $\{w_i : i \in I\}$.

We say that given IFS $\{w_i : i \in I\}$ satisfies the *Moran condition* if the sets $w_1(K), \ldots, w_n(K)$ are pairwise disjoint, where $K$ is the attractor corresponding to this system (see [27]).

Let IFS $\{w_i : i \in I\}$ satisfies the Moran condition and

$$\rho\big(w_i(x), w_i(y)\big) \geq l_i \rho(x, y) \quad \text{for} \quad x, y \in X, \quad i \in I, \tag{6.3}$$

then the Hausdorff dimension of the attractor $K$ is greater then or equal to the unique positive number $d$ such that

$$l_1^d + \cdots + l_n^d = 1. \tag{6.4}$$

Moreover, if $w_i$ are lipschitzian function with Lipschitz constants $l_i$, then $\dim_H K \leq d$, where $d$ is the unique solution of Moran equation (6.4). In particular, if $w_i$ are similarities with the scaling factors $l_i$, $i = 1, \ldots, N$, then $\dim_H K = d$.

An operator $P : \mathcal{M}_1 \to \mathcal{M}_1$ is called *Markov*, if:

(i) $P(\lambda_1\mu_1 + \lambda_2\mu_2) = \lambda_1 P(\mu_1) + \lambda_2 P(\mu_2)$ for $\lambda_1, \lambda_2 \in \mathbb{R}_+$, $\mu_1, \mu_2 \in \mathcal{M}_1$;

(ii) $P\mu(X) = \mu(X)$.

A measure $\mu_* \in \mathcal{M}_1(X)$ is called *invariant* with respect to operator $P$ if $P\mu_* = \mu_*$. If in addition

$$\int_X f(x) P^n \mu(dx) \to \int_X f(x) \mu_*(dx) \quad \text{for every} \quad f \in C(X), \tag{6.5}$$

then operator $P$ is called *asymptotically stable*.

The family $\{(w_i, p_i) : i \in I\}$, where $w_i : X \to X$, $p_i : X \to (0,1)$, $i \in I$, are continuous functions and $\sum_{i \in I} p_i(x) = 1$ for all $x \in X$, is called IFS *with probabilities*.

Given an IFS $\{(w_i, p_i) : i \in I\}$ we can define a Markov operator $P : \mathcal{M}_1 \to \mathcal{M}_1$ by

$$P\mu(A) = \sum_{i \in I} \int_{w^{-1}(A)} p_i(x)\mu(dx), \quad A \in \mathcal{B}(X).$$

We say that IFS $\{(w_i, p_i) : i \in I\}$ is asymptotically stable if the corresponding Markov operator is asymptotically stable.

**Remark 6.1.** Assume that all functions $w_i$ are lipschitzian with corresponding Lipschitz constants $L_i$. If

$$\sum_{i \in I} p_i L_i < 1,$$

then IFS $\{(w_i, p_i) : i \in I\}$ is asymptotically stable.

**Remark 6.2.** Let IFS $\{(w_i, p_i) : i \in I\}$ be such that all functions $w_i$ are strictly contractive. Then

$$A_* = \operatorname{supp} \mu^*,$$

where $A_*$ is the attractor of IFS $\{w_i : i \in I\}$ and $\mu_*$ is the invariant measure with respect to IFS $\{(w_i, p_i) : i \in I\}$.

Now we will give a generalizaton of above notion of attractor. For this purpose we recall the notion of convergence in Kuratowski's sence of sequence of sets.

Let $\{A_n\}$ be a sequence of subsets of a metric space $X$. The *lower bound* $\operatorname{Li}A_n$ and the *upper bound* $\operatorname{Ls}A_n$ are defined by the following conditions. A point $x$

belongs to $\mathrm{Li}A_n$ if there exists a sequence $\{x_n\}$, $x_n \in X_n$, such that $x_n \to x$. A point $x$ belongs to $\mathrm{Ls}A_n$ if there exists a sequence $\{x_{n_k}\}$, $x_{n_k} \in X_{n_k}$, such that $x_{n_k} \to x$. Obviously $\mathrm{Li}A_n \subset \mathrm{Ls}A_n$. If $\mathrm{Li}A_n = \mathrm{Ls}A_n$, we say that the sequence $\{A_n\}$ is topologically convergent and we denote this common limits by $\mathrm{Lt}A_n$.

In the case when $X$ is a compact set, $\mathrm{Lt}A_n = A$ if and only if $\{A_n\}$ converges to $A$ in the Hausdorff distance.

Given an IFS $\{w_i : i \in I\}$ we define

$$H(A) = \overline{\bigcup_{i \in I} w_i(A)} \,.$$

A set $A_0$ such that $H(A_0) = A_0$ is called *invariant* with respect to IFS $\{w_i : i \in I\}$. If in addition, for every nonempty bounded subset $A$ of $X$, $\mathrm{Lt}H^n(A) = A$, the IFS is called *asymptotically stable* (on sets) and the set $A_0$ is called attractor of IFS $\{w_i : i \in I\}$.

Note that, if we consider $H$ on the class of compact sets, this definition of attractor coincides with that used before.

We said that IFS $\{w_i : i \in I\}$ is *regular* if there is a nonempty subset $I_0$ of $I$ such that IFS $\{w_i : i \in I_0\}$ is asymptotically stable. The attractor corresponding to IFS $\{w_i : i \in I_0\}$ will be called *nucleous*.

**Theorem 6.3.** *Let $\{w_i : i \in I\}$ be a regular IFS and $A_0$ be a nucleous of this system. Let*

$$A_* = \bigcup_{n=1}^{\infty} H^n(A_0) \,.$$

*Then,*

(i) $A_*$ *does not depend on the choice of the nucleous $A_0$;*
(ii) $A_*$ *is the smallest nonempty set such that $H(A_*) = A_*$;*
(iii) $\mathrm{Lt}H^n(A) = A_*$ *for every nonempty set $A \subset A_*$.*

The set $A_*$ is called *semiattractor* or *semifractal* corresponding to regular IFS $\{w_i : i \in I\}$.

**Theorem 6.4.** *Let $X$ be a Polish space. Assume that IFS $\{(w_i, p_i) : i \in I\}$ is asymptotically stable and $\{w_i : i \in I\}$ is regular. Then*

$$A_* = \operatorname{supp} \mu^* \,,$$

*where $A_*$ is the semiattractor of IFS $\{w_i : i \in I\}$ and $\mu_*$ is the invariant measure with respect to IFS $\{(w_i, p_i) : i \in I\}$.*

For proofs and further results see [21, 22].

# 7. Upper and lower bound for concentration dimension

We start with the following technical lemma.

**Lemma 7.1.** *Let $\alpha_i$, $\beta_i$, $L_i \in (0,1)$ for $i \in J$. Let $\Phi : \mathbb{R}_+ \to \mathbb{R}_+$ be a bounded increasing function. Suppose that*

$$\Phi(r) \geq \sup_J \alpha_i \Phi(r/L_i) \quad \text{for} \quad r \in (0,a), \quad a > 0. \tag{7.1}$$

*Then, there exists $c > 0$ such that*

$$\Phi(r) \geq cr^s \quad \text{for} \quad r \in (0,a),$$

*where*

$$s = \min_J \frac{\log \alpha_i}{\log L_i}.$$

*Proof.* For $i \in J$ define

$$c_i = a^{-s_i} \Phi(aL_i),$$

where

$$s_i = \log \alpha_i / (\log L_i).$$

We claim that for arbitrary $n \in \mathbb{N}$ we have

$$\Phi(r) \geq c_i r^{s_i} \quad \text{for} \quad r \in [L_i^n a, a). \tag{7.2}$$

Indeed, for $r \in [L_i a, a)$ by the definition of $c_i$ we have

$$\Phi(r) \geq \Phi(L_i a) = c_i a^{s_i} \geq c_i r^{s_i}.$$

Suppose that (7.2) holds for some $n \geq 1$. Since $r/L_i \in [L_i^n a, a)$ for $r \in [L_i^{n+1} a, L_i^n a)$, from (7.1), (7.2) and the definition of $s_i$ we have

$$\Phi(r) \geq \alpha_i \Phi(r/L_i) \geq \alpha_i c_i (r/L_i)^{s_i} = c_i r^{s_i} \quad \text{for} \quad r \in [L_i^{n+1} a, L_i^n a).$$

By virtue of the induction principle the condition (7.2) holds for every $n \in \mathbb{N}$. Since $L_i < 1$ this implies that

$$\Phi(r) \geq c_i r^{s_i} \quad \text{for} \quad r \in (0,a).$$

Since $i \in J$ is arbitrary, the statement of Lemma 7.1 follows. $\square$

**Theorem 7.2.** *Suppose that IFS $\{(w_i, p_i) : i \in I\}$ has an invariant measure $\mu$. Assume that all functions $w_i$ are lipschitzian with Lipschitz constants $L_i$ and the set $J = \{i \in I : L_i < 1\}$ is nonempty. Then*

$$\overline{\dim}_L \mu \leq \inf_J \frac{\log \alpha_i}{\log L_i},$$

*where*

$$\alpha_i = \inf_X p_i(x).$$

*Proof.* Since the measure $\mu$ is invariant, for arbitrary $i \in J$ we have

$$\mu(B(x,r)) \geq \alpha_i \mu(w_i^{-1}(B(x,r))) \quad \text{for} \quad x \in X, \quad r > 0.$$

Substituting $x = w_i(y)$ we obtain

$$\mu(B(w_i(y),r)) \geq \alpha_i \mu(B(y,r/L_i)) \quad \text{for} \quad x \in X, \quad r > 0.$$

This implies that

$$Q_\mu(r) \geq \alpha_i Q_\mu(r/L_i) \quad \text{for} \quad r > 0, \quad i \in J,$$

where $Q_\mu$ is given by (3.1). Consequently the function $Q_\mu$ satisfies the inequality

$$Q_\mu(r) \geq \sup_J \alpha_i Q_\mu(r/L_i) \quad \text{for} \quad r > 0.$$

From the last inequality and Lemma 7.1 it follows that

$$Q_\mu(r) \geq cr^s,$$

for some $c > 0$. Consequently

$$\overline{\dim}_L \mu \leq s,$$

which completes the proof of Theorem 7.1.                                      $\square$

To obtain a lower estimate of concentration dimension of measure we need more restrictive assumptions on transformations $w_i$. Let $I_1, \ldots, I_m$ be a partition of $I$ and let $K \subset X$ be a nonempty set. Define

$$K_j = \bigcup_{i \in I_j} w_i(K) \quad \text{for} \quad j = 1, \ldots, m. \tag{7.3}$$

We say that the family $\{w_i : i \in I\}$ satsfies the *mixed Moran condition* with respect to the set $K$ and the partition $I_1, \ldots, I_m$, if $K_j \subset K$ for $j = 1, \ldots, m$ and

$$\text{dist}\big(K_{j_1}, K_{j_2}\big) = \inf\big\{\rho(x, y) : x \in K_{j_1}, y \in K_{j_2}\big\} > 0$$

for arbitrary $j_1, j_2 \in \{1, \ldots, m\}$, $j_1 \neq j_2$.

Similarly as Lemma 7.1 one can prove the following lemma.

**Lemma 7.3.** *Let $m_j \in (0, 1)$ and $\beta_j > 0$ for $j = 1, \ldots, m$, be given. Suppose that $\Phi : \mathbb{R}_+ \to \mathbb{R}_+$ be a bounded increasing function such that*

$$\Phi(r) \leq \max_{1 \leq j \leq m} \beta_j \Phi(r/m_j) \quad \text{for} \quad r \in (0, a).$$

*Then, there is $c > 0$ such that*

$$\Phi(r) \leq cr^s \quad \text{for} \quad r \in (0, a),$$

*where*

$$s = \min_{1 \leq j \leq m} \frac{\log \beta_j}{\log m_j}.$$

**Theorem 7.4.** *Suppose that IFS $\{(w_i, p_i) : i \in I\}$ has an invariant measure $\mu$. Assume that the family $\{w_i : i \in I\}$ satisfies the mixed Moran condition with respect to the set $K = \text{supp}\,\mu$ and a partition $I_1, \ldots, I_m$. Moreover, assume that the functions $w_i$ satisfy the condition*

$$\rho\big(w_i(x), w_i(y)\big) \geq l_i \rho(x, y) \quad \text{for} \quad x, y \in X, \quad i \in I, \tag{7.4}$$

*where $l_i$ are constants such that*

$$0 < \inf_{I_j} < 1 \quad \text{for} \quad j = 1, \ldots m.$$

*Then,*

$$\underline{\dim}_L \mu \geq \min_{1 \leq j \leq m} \frac{\log \beta_j}{\log m_j},$$

*where*

$$\beta_j = \sum_{i \in I_j} \sup_X p_i(x) \quad \text{and} \quad m_j = \inf_{i \in I_j} l_i.$$

*Proof.* Let

$$a = \min \left\{ \mathrm{dist}\left(K_{j_1}, K_{j_2}\right); j_1, j_2 \in \{1, \ldots, m\}, \ j_1 \neq j_2 \right\},$$

where $K_j$ are given by (7.3).

Obviously

$$w_i^{-1}(x) \cap K = \emptyset \quad \text{for} \quad i \notin I_j \quad \text{and} \quad x \in X \quad \text{such that} \quad \rho(x, K_j) < a. \quad (7.5)$$

Since $\mu$ is invariant, from (6.6) it follows that

$$\mu(A) \leq \sum_{j=1}^{m} \sum_{i \in I_j} \gamma_i \mu\left(w_i^{-1}(A)\right) \quad \text{for} \quad A \in \mathcal{B}(X), \quad (7.6)$$

where $\gamma_i = \sup_{x \in X} p_i(x)$. Set

$$A_0 = X \setminus \bigcup_{j=1}^{m} \overline{K}_j.$$

From (7.5) it follows that $\mu\left(w_i^{-1}(A_0)\right) = 0$ for $i \in I$. Thus $A_0 \cap K = \emptyset$ and so $K \subset \bigcup_{j=1}^{m} \overline{K}_j$.

Let $A \in \mathcal{B}(X)$ be such that $\mathrm{diam}\, A \leq r < a$ and $\mu(A) > 0$. Then $A \cap \overline{K}_j = \emptyset$ for some $j \in \{1, \ldots, m\}$. Since $\mathrm{diam}\, A < a$, by virtue of (7.5) we have

$$w_i^{-1}(A) \cap K = \emptyset \quad \text{for} \quad i \notin I_j.$$

Consequently, the inequality (7.6) reduces to

$$\mu(A) \leq \sum_{i \in I_j} \gamma_i \mu\left(w_i^{-1}(A)\right) \quad \text{for} \quad A \in \mathcal{B}(X),$$

From (7.4) it follows that $\mathrm{diam}\, w_i^{-1}(A) \leq l_i \,\mathrm{diam}\, A \leq r/m_j$ for $i \in I_j$. Thus for $A \in \mathcal{B}(X)$ with $\mathrm{diam}\, w_i^{-1}(A) \leq r < a$ there is $j \in \{1, \ldots, m\}$ such that

$$\mu(A) \leq \sum_{i \in I_j} \gamma_i Q_\mu(r/m_j) = \beta_j Q_\mu(r/m_j).$$

Consequently

$$Q_\mu(r) \leq \max_{1 \leq j \leq m} \beta_j Q_\mu(r/m_j) \quad \text{for} \quad r \in (0, a)$$

and by Lemma 7.3

$$Q_\mu(r) \leq c r^s,$$

where

$$s = \min_{1 \leq j \leq m} \frac{\log \beta_j}{\log m_j}.$$

From the last inequality the statement of Theorem 7.4 follows immediately. $\qquad\square$

## 8. Small overlaps

The lower estimate of dimensions can be also obtained for some "small overlaps". We say that IFS $\{w_i : i \in I\}$ satisfies the *open set condition* if there exists an open set $G$ such that the sets $w_1(G), \ldots, w_N(G)$ are pairwise disjoint and $w_i(G) \subset G$ for $i = 1, \ldots, N$. Note that open set condition does not imply the Moran condition.

**Lemma 8.1.** *Let $l_i$, $p_i \in (0,1)$, for $i = 1, \ldots, N$. Assume that $p_1 + \cdots + p_N = 1$. Let $\Lambda$ be a family of sequences $(i_1, \ldots, i_m)$, $i_k \in \{1, \ldots, N\}$, $i_1 < \cdots < i_m$. Let $\Phi : \mathbb{R}_+ \to \mathbb{R}_+$ be a bounded function. Suppose that there exists $a > 0$ such that*

$$\Phi(r) \leq \max_{\Lambda} \sum_{k=1}^{m} p_{i_k} \Phi(r/l_{i_k}) \quad \text{for} \quad r \in (0, a).$$

*Then there is $c > 0$ such that*

$$\Phi(r) \leq cr^s, \tag{8.1}$$

*for $r \in (0, a)$, where $s$ is a positive solution of equation*

$$\max_{\Lambda} \sum_{k=1}^{m} \frac{p_{i_k}}{l_{i_k}^s} = 1. \tag{8.2}$$

*Proof.* Let $s$ satisfies (8.2). Since $\Phi$ is bounded, there exists $c > 0$ such that (8.1) holds for every $r \geq a$. Let $l = \max\{l_1, \ldots, l_N\}$. We claim that (8.1) holds for every $r \geq l^n a$. Indeed, for $n = 0$ the condition (8.1) holds by the choice of $c$. Suppose that (8.1) holds for $r \geq L^n a$. Let $r \geq l^{n+1} a$ since $r/l_i \geq l^n a$ for $i = 1, \ldots, N$, we have

$$\Phi(r) \leq \max_{\Lambda} \sum_{k=1}^{m} p_{i_k} \Phi(r/l_{i_k}) \leq \max_{\Lambda} \sum_{k=1}^{m} p_{i_k} \frac{cr^s}{l_{i_k}^s} \leq cr^s \max_{\Lambda} \sum_{k=1}^{m} \frac{p_{i_k}}{l_{i_k}^s} = cr^s.$$

Since $l < 1$, the statement of Lemma 8.1 follows. $\qquad\square$

**Theorem 8.2.** *Let $w_i$, $i = 1, \ldots, N$, be strictly contractive functions and let $K$ be the corresponding attractor. Assume that*

$$\bigcap_{i=1}^{N} w_i(K) = \emptyset.$$

*Let $\Lambda$ be the family of all sequences $(i_1, \ldots, i_m)$, $i_1 < i_2 < \cdots < i_m$, $i_k \in \{1, \ldots, N\}$ such that*

$$\bigcap_{i=1}^{m} w_{i_k}(K) \neq \emptyset.$$

*Then,*

$$\dim_H K \geq s\,,$$

*where s is the solution of equation*

$$\max_{\Lambda} \sum_{k=i}^{m} l_{i_k}^{d-s} = 1\,. \qquad (8.3)$$

*Proof.* For $x \in K$ define

$$\varphi(x) = \max_{y \in K} \min_{i \in I_y} \rho\big(x, w_i(K)\big)\,,$$

where $I_y = \{i \in \{1, \ldots, N\} : y \notin w_i(K)\}$. It is easy to verify that $\varphi$ is continuous and positive. Thus

$$a = \inf_{x \in K} \varphi(x) > 0\,.$$

Let $\mu$ be an invariant measure of IFS $\{(w_i, p_i) : i = 1, \ldots, N\}$, where $p_i = l_i^d$ with $d$ given by (6.4). By virtue of Theorem 6.4 we have $K = \mathrm{supp}\mu$. Simple calculation shows that

$$\mu\big(B(x, r)\big) \leq \sum_{i=1} l_i^d \mu\big(B(w_i^{-1}(x), r/l_i)\big) \quad \text{for} \quad r \in (0, a]\,.$$

Consequently

$$Q_\mu(r) \leq \max_{\Lambda} \sum_{k=1}^{m} p_{i_k} Q_\mu(r/l_{i_k})\,,$$

where $Q_\mu$ is the Lévy function given by Formula (3.1). By Lemma 8.1

$$Q_\mu(r) \leq c r^s \quad \text{for} \quad r > 0\,,$$

where $s$ is given by (8.2). From the last inequality and mass distribution principle the statement of Theorem 8.2 follows immediately. $\qquad\square$

**Corollary 8.3.** *Let $w_1, \ldots, w_n$ be strictly contractive mappings satisfying the Moran condition. Let $K$ be the attractor of this system. Moreover, suppose that condition (6.3) holds. Then,*

$$\dim_H K \geq d\,,$$

*where d is given by Moran equation (6.4).*

*Proof.* It is enough to note that $\Lambda = \{\{1\}, \ldots, \{N\}\}\,.$ $\qquad\square$

For proofs and related results see [36, 39, 47].

## 9. Condensation systems

The sequence of lipschitzian functions $w_i : X \to X$, $i = 1, \ldots, N$ together with the probability vector $(p_0, p_1, \ldots, p_N)$ and a probability measure $\nu$ is called a condensation system (on measures) and it is denoted by $\{(w_i, p_i), \nu\}$.

Given a condensation system we define a Markov operator $P : \mathcal{M}_1 \to \mathcal{M}_1$ by fomula

$$P\mu(A) = p_0\nu(A) + \sum_{i-1}^{N} p_i\mu\big(w_i^{-1}(A)\big) \quad \text{for} \quad A \in \mathcal{B}(X).$$

Let $L_i$ be a Lipschitz constant of $w_i$. If $p_i$ and $L_i$ satisfy the condition

$$\sum_{i=1}^{N} p_i L_i < 1,$$

then $P$ has a unique invariant pobability measure $\mu_*$ and $\{P^n \mu_*\}$ converges to $\mu_*$ for every $\mu \in \mathcal{M}_1(X)$.

Moreover, if $C = \operatorname{supp} \nu$ is a compact set and all $w_i$ are strictly contractive, then $K = \operatorname{supp} \mu_*$ is also a compact set and

$$K = C \cup \bigcup_{i=1}^{N} w_i(K). \tag{9.1}$$

**Theorem 9.1.** *Let a condensation system on measures $\{(w_i, p_i), \nu\}$ be given. Assume that all functions $w_i$, $i = 1, \ldots, N$, are strictly contractive. Let $\mu_*$ be the corresponding invariant measure. Then*

$$\overline{\dim}_T \mu_* \leq \max\left\{\overline{\dim}_T \nu, \ \max_{1 \leq i \leq N} \frac{\log p_i}{\log L_i}\right\}.$$

*Sketch of the proof.* First we can show that for arbitrary $s > \overline{\dim}_T \nu$ there is $r_0$ such that

$$T_\nu(r) \geq r^s \quad \text{for} \quad r \in (0, r_0].$$

Now, let $\varphi, \psi : \mathbb{R}_+ \to \mathbb{R}_+$ be such that

$$\varphi(r) \leq \min\left\{p_0 r^s, \ \min_{1 \leq i \leq N} p_i\varphi(r/L_i) \quad \text{for} \quad 0 < r \leq r_0\right\} \tag{9.2}$$

and

$$\psi(r) \geq \min\left\{p_0 r^s, \ \min_{1 \leq i \leq N} p_i\psi(r/L_i) \quad \text{for} \quad 0 < r \leq r_0\right\}. \tag{9.3}$$

Put $L = \inf_I L_i$. Using similar argument as in the proof of Lemma 7.1 one can show that if inequality $\varphi(r) \leq \psi(r)$ holds for $r \in [Lr_0, r_0)$, then it holds for every $r \in (0, r_0]$.

Observe that functions $\varphi(r) = cr^s$ ($c \leq p_0$) and $\psi(r) = T_\mu(r)$ satisfies the condition (9.2) and (9.3). Consequently, for $c$ sufficiety small $T_\mu(r) \geq cr^s$ for $r \in (0, r_0]$, whence the statement of Theorem 9.1 follows.

A system of functions $\{w_1, \ldots, w_N\}$ together with a nonempty compact set $C$ is called an *iterated function system with condensation* (on sets) and will be denoted $\{w_i, C\}$.

Given an condensation system $\{w_i, C\}$ we define the operator $F$ acting on the space of sets by

$$F(A) = C \cup \bigcup_{i=1}^{N} w_i(A).$$

It is easy to verify that, if all functions $w_i$ are strictly contractive, then there exists a unique nonempty set $K$ which satisfy condition (9.1). Moreover, for every nonempty compact subset $A$ of $X$ the sequence $\{F^n(A)\}$ converges to $K$ in the Hausdorff metric.

**Theorem 9.2.** *Let a condensation system $\{w_i, C\}$ be given. Assume that all functions $\{w_i\}$ are strictly contractive. Let $K$ be the attractor of this system. Then*

$$\overline{\dim}_F K \leq \max\{d_0, d\},$$

*where $d_0 = \overline{\dim}_F C$ and $d$ is the unique solution of (6.4).*

*Proof.* Let $\max_{i \in I} L_i < \theta < 1$ and let $d(\theta)$ be the solution of equation

$$L_1^d + \cdots + L_N^d = \theta.$$

Define

$$p_0(\theta) = 1 - \theta, \quad p_i(\theta) = L_i^{d(\theta)} \quad \text{for} \quad i = 1, \ldots, N.$$

According to Theorem 5.2 there exists a probability measure $\nu$ such that $C = \operatorname{supp} \nu$ and $\overline{\dim}_T \nu = \overline{\dim}_F C$.

Consider the condensation system $\{(w_i, p_i(\theta), C\}$. Let $\mu_\theta$ be the corresponding invariant measure. By Theorem 9.1

$$\overline{\dim}_T \mu_\theta \leq \max\left\{ \overline{\dim}_T \nu, \max_{1 \leq i \leq N} \frac{\log p_i(\theta)}{\log L_i} \right\} = \max\{d_0, d(\theta)\}.$$

By Theorem 4.1

$$\overline{\dim}_F K \leq \overline{\dim}_T \mu_\theta \leq \max\{d_0, d(\theta)\}.$$

Since $\theta \in (c, 1)$ is arbitrary and $\lim_{\theta \to 1} d(\theta) = d$, the statement of Theorem 9.2 follows. $\qquad\square$

## 10. Systems with the squeezing property

Let $A$ be a nonempty bounded subset of $X$. Consider two families of functions

$$w_i : A \to A \quad \text{and} \quad P_i : A \to \mathbb{R}^{k_i}, \quad i \in I.$$

We assume that

$$\|P_i(x) - P_i(y)\|_i \leq \rho(x, y) \quad \text{for} \quad x, y \in X, \quad i \in I, \tag{10.1}$$

where $\|\cdot\|_i$ denotes the norm in $\mathbb{R}^{k_i}$.

Moreover, assume that the function $w_i$ satisfies the following *squeezing property*:

$$\rho\big(w_i(x), w_i(y)\big) \leq \max\big\{L_i\rho(x,y),\ c_i\|P_i(x) - P_i(y)\|_i\big\}, \quad x, y \in X, \qquad (10.2)$$

for some $L_i \in (0,1)$ and $c_i \geq 0$.

The following covering condition is essential for further results:

**Condition A.** Let $L \in (0,1)$, $c \geq 0$ and $(\mathbb{R}^k, \|\cdot\|)$ be given. Then there exists an integer $m \geq 1$ such that for arbitrary set $B \subset \mathbb{R}^k$ with $\mathrm{diam}\, B < c$ there exist sets $\Delta_1, \ldots, \Delta_m$ such that

$$B \subset \bigcup_{j=1}^{m} \Delta_j \quad \text{and} \quad \mathrm{diam}\, \Delta_j < L, \quad \text{for} \quad j = 1, \ldots, m.$$

**Theorem 10.1.** *Let transformation $w_i$, $P_i$, $i \in I$ satisfying conditions (10.1) and (10.2) be given. Moreover, for each $L_i \in (0,1)$, $c_i > 0$ and $\mathbb{R}^{k_i}$ let integer $m_i$ be chosen according to Condition A. Let $A$ be such that*

$$A = \bigcup_{i=1}^{N} w_i(A).$$

*Then,*

$$\dim_F A \leq d,$$

*where $d$ is the unique solution of equation*

$$\sum_{i=1}^{N} m_i L_i^d = 1. \qquad (10.3)$$

*Sketch of the proof.* Let $r > 0$. One can show that for every $i \in I$ and $B \subset A$ with $\mathrm{diam}\, B \leq r$ there exist sets $D_j \subset A$, $j = 1, \ldots, m_i$, such that

$$w_i(B) \subset \bigcup_{j=1}^{N} D_j \quad \text{and} \quad \mathrm{diam}\, D_j \leq rL_i \quad \text{for} \quad j = 1, \ldots, m_i.$$

Let $s = \mathrm{diam}\, A$ and $L = \min_I L_i$. One can construct a covering of the set $A$ by sets of the diameter less then or equal to $rs$, such that

$$A \subset \bigcup_{T(r)} \bigcup_{j_1=1}^{m_1} \cdots \bigcup_{j_k=1}^{m_k} D_{j_1, \ldots, j_k}^{i_1, \ldots, i_k},$$

where

$$T(r) = \big\{(i_1, \ldots, i_k) : L_{i_1} \cdots L_{i_k} < r \leq L_{i_1} \cdots L_{i_{k-1}}\big\}.$$

Observe that number of sets of this covering is given by the formula

$$\mathcal{N}(r) = \sum_{T(r)} m_{i_1} \cdots m_{i_k}.$$

It can be proved that

$$\mathcal{N}(r) = \sum_{i=1}^{N} m_i \mathcal{N}(r/L_i) \quad \text{for} \quad r \in (0, L^2).$$

Simple calculation shows that the last equation is satisfied by

$$\mathcal{N}(r) = cr^{-d} \quad \text{for} \quad r \in (0, L^2),$$

where $d$ is given by (10.3). Consequently

$$\dim_F A \le \limsup_{r \to 0} \frac{\log \mathcal{N}(r)}{-\log (rs)} = d.$$

**Corollary 10.2.** *Let $X$ be a Hilbert space and let $A$ be the attractor of an iterated function system $\{w_1, \ldots, w_N; C\}$, where all functions $w_i$ are strictly contractive. Then,*

$$\dim_F A \le \max\{d, k\},$$

*where $d$ is the unique solution of the Moran equation (6.4) and $k$ is the dimension of linear subspace $X_k \subset X$ spanned by $C$.*

**Theorem 10.3.** *Suppose that hypotheses of Theorem 10.1 are satisfied. Moreover assume that are given numbers $p_i > 0$, $i = 1, \ldots, N$ such that $\sum_{i=1}^{N} p_i = 1$ and that IFS $\{(w_i, p_i) : i = 1, \ldots, N\}$ admits an invariant measure $\mu_*$, i.e.,*

$$\mu_*(A) = \sum_{i=1}^{N} p_i \mu_* (w_i^{-1}(A)) \quad \text{for} \quad A \in \mathcal{B}(X).$$

*Then,*

$$\overline{\dim}_L \mu_* \le \min_{i \in I} \frac{\log(p_i/m_i)}{\log(L_i)}.$$

**Remark 10.4.** *If in addition of hypotheses of Theorem 10.3 we assume that $p_i = m_i L_i^d$, then*

$$\overline{\dim}_L \mu_* \le d,$$

*where $d$ is the solution of (10.3).*

In Theorem 10.1 we assumed that all coefficients $L_i$ are positive. Now we extend this theorem to the case when some of the constants $L_i$ can be equal to zero. Let $I = \{1, \ldots, N + M\}$, $N \ge 1$, $M \ge 1$.

**Theorem 10.5.** *Suppose that $P_i$ satisfy condition (10.1) for $i = 1, \ldots, N + M$ and condition (10.2) for $i = 1, \ldots, N$. In addition assume that*

$$\rho(w_i(x), w_i(y)) \le c_i \|P_i(x) - P_i(y)\|_i, \quad \text{for} \quad i = N, \ldots, N + M. \tag{10.4}$$

*Moreover assume that the integers $m_i$, $i = 1, \ldots, N$, are chosen according to the Condition A. Let $A$ be such that*

$$A = \bigcup_{i=1}^{N+M} w_i(A).$$

*Then,*

$$\dim_F A \leq \max \left\{ d, k_{N=1}, \ldots, k_{N+M} \right\},$$

*where $d$ is the unique positive solution of* (10.3) *and $k_i$ denote the dimension of the space $R^{ki}$, $i = N + 1, \ldots, N + M$.*

**Theorem 10.6.** *Suppose that hypotheses of Theorem 10.5 are satisfied. Moreover assume that are given numbers $p_i > 0$, $i = 1, \ldots, N + M$ such that $\sum_{i=1}^{N+M} p_i = 1$ and that IFS $\{(w_i, p_i) : i = 1, \ldots, N + M\}$ admits an invariant measure $\mu_*$. Then*

$$\overline{\dim}_L \mu_* \leq \min \left\{ \frac{\log(p_1/m_1)}{\log L_1}, \ldots, \frac{\log(p_N/m_N)}{\log L_N}, k_{N+1}, \ldots, k_{N+M} \right\}.$$

For details and further generalisation see [11, 25].


# 11. Some generic properties of dimensions

Recall that a subset of a metric space $X$ is called *residual*, if its complement is of the first Baire category. If the set of all elements of $X$ satisfying some property $P$ is residual in $X$, then the property $P$ is called *generic* or *typical*. We say also that typical elements from $X$ has property $P$.

For $f \in B(X)$ and $\mu \in \mathcal{M}(X)$ we write

$$\langle f, \mu \rangle = \int_X f(x)\mu(dx).$$

We assume that the space $\mathcal{M}_1(X)$ is endowed with the Fortet–Mourier distance $d_{FM}$ defined by the formula

$$d_{FM}(\mu, \nu) = \sup \left\{ |\langle f, \mu \rangle - \langle f, \nu \rangle| : f \in \mathcal{L} \right\},$$

where $\mathcal{L}$ is the set of all $f \in C(X)$ such that $|f(x)| \leq 1$ and $|f(x) - f(y)| \leq \rho(x, y)$, for $x, y \in X$.

We say that a sequence $\{\mu_n\} \subset \mathcal{M}_1(X)$ converges weakly to a measure $\mu \in \mathcal{M}_1(X)$, if

$$\lim_{n \to \infty} \langle f, \mu_n \rangle = \langle f, \mu \rangle \quad \text{for every} \quad f \in C(X).$$

It is well known that the convergence in the Fortet–Mourier metric is equivalent to the weak convergence. Moreover $(\mathcal{M}_1(X), d_{FM})$ is a complete metric space.

Finally, assume that the space of compact subsets of $X$ is equipped with the Hausdorff distance (see [7]). It is well known that it is a complete metric space.

The following results is well known (see [8, 13, 29])

**Theorem 11.1.** *A typical compact subset of a complete metric space $X$ has Hausdorff and fractal dimension zero.*

*The smallest local lower* and *upper* fractal dimension of a set $A \subset X$ are defined by the formulae

$$\mathrm{sl} - \underline{\dim}_F A = \inf \left\{ \underline{\dim}_F \left( B(x, r) \cap A \right) : x \in A, \ r > 0 \right\},$$

$$\mathrm{sl} - \overline{\dim}_F A = \inf \left\{ \overline{\dim}_F \left( B(x, r) \cap A \right) : x \in A, \ r > 0 \right\}.$$

**Theorem 11.2.** *For a typical compact subset $A$ of a complete metric space $X$ we have*

$$\mathrm{sl} - \overline{\dim}_F A \ \geq \ \mathrm{sl} - \overline{\dim}_F X$$

The *lower* and the *upper fractal dimension* of a measure $\mu \in \mathcal{M}_1(X)$ is defined by the formulae

$$\underline{\dim}_F \mu = \lim_{\epsilon \to 0} \inf \left\{ \underline{\dim}_F A \ : \ A \in \mathcal{B}(X), \ \mu(A) \geq 1 - \epsilon \right\}$$

$$\overline{\dim}_F \mu = \lim_{\epsilon \to 0} \inf \left\{ \overline{\dim}_F A \ : \ A \in \mathcal{B}(X), \ \mu(A) \geq 1 - \epsilon \right\}$$

**Theorem 11.3.** *Let $X$ be a complete metric space. For a typical measure $\mu \in \mathcal{M}_1(X)$ we have:*

(i) $\underline{\dim}_F \mu \ = \ 0$;

(ii) $\inf \left\{ \overline{\dim}_F A \ : \ A \in \mathcal{B}(X), \ \mu(A) > 0 \right\} \ \geq \ \mathrm{sl} - \overline{\dim}_F X$;

(iii) $\overline{\dim}_F \mu \ \geq \ \mathrm{sl} - \overline{\dim}_F X$.

**Theorem 11.4.** *Let $X$ be a complete separable metric space. Then for a typical measure $\mu \in \mathcal{M}_1(X)$ we have:*

(i) $\mathrm{supp}\, \mu = X$;

(ii) $\underline{\dim}_C \mu \ = \ 0$;

(iii) $\underline{\dim}_F A \ = \ \underline{\dim}_F X$ and $\overline{\dim}_F A \ = \ \overline{\dim}_F X$, *for every $A \in \mathcal{B}(X)$ such that $\mu(A) = 1$*;

(iv) $\mathrm{sl} - \underline{\dim}_F X \ \leq \ \overline{\dim}_C \mu \ \leq \ \mathrm{sl} - \overline{\dim}_F X$

**Theorem 11.5.** *Let $K$ be a compact fractal set. Then for a typical measure $\mu \in \mathcal{M}_1(K)$ we have:*

(i) $\underline{\dim}_L \mu \ = \ 0$;

(ii) $\overline{\dim}_L \mu \ = \ \dim_H K$.

For details and connected results see [12, 28–33, 40, 46].

## 12. Relationship with topological dimension

Recall that for a separable metric space the three principal topological dimensions (small inductive dimension, large inductive dimension and covering dimension) coincide. We denote this common value by $\dim_t X$ and we call it *topological dimension* of $X$. The value $\dim_t X$ is an integer greater then or equal to $-1$, or $\infty$. It can be defined by the following recurent scheme:

(i) $\dim_t X = -1$, if and only if, $X = \emptyset$;

(ii) $\dim_t X \leq n$, $n = 0, 1, \ldots$, if for every $x \in X$ and every neighbourhood $U$ of $x$ there is a neighbourhood $V$ of $x$ such that $V \subset U$ and $\dim_t \partial V \leq n - 1$;

(iii) $\dim_t X = n$, if and only if, $\dim_t X \leq n$ and it is not true that $\dim_t X \leq n - 1$;

(iv) $\dim_t X = \infty$, if and only if, $\dim_t X \geq n$ for every $n \in \mathbb{N}$.

The relationship between the Hausdorff and the topological dimension in the case of $\mathbb{R}^n$ was made in evidence by V. G. Nöbeling and in a more general setting by Szpilrajn in 1937 (see [49]). Namely, he proved that if $X$ is a separable metric space then $\dim_H X \geq \dim_t X$ and

$$\dim_t X = \inf \left\{ \dim_H X' \ : \ X' \text{ is homeomorphic to } X \right\} . \tag{12.1}$$

The similar results for topological and packing dimension was proved by H. Joyce. Since $\dim_P X \geq \dim_H X$ then $\dim_P X \geq \dim_t X$. Joyce [17] proved that Formula (12.1) holds with $\dim_P X'$ in the place of $\dim_H X'$.

Here we will give a result of this type for topological and concentration dimension.

The following lemma (see [37]) is essential for further results.

**Lemma 12.1.** *Suppose that* $\dim_t X \geq d$, *where* $d \in \mathbb{N} \cup \{0\}$. *Then there exists a Borel measure* $\nu$ *such that*

$$\nu\big(B(x,r)\big) \leq cr^d \quad \text{for} \quad x \in X \quad \text{and} \quad r > 0 , \tag{12.2}$$

*where* $c > 0$ *is a positive constant independent on* $x$ *and* $r$.

**Theorem 12.2.** *Let* $X$ *be a Polish space with* $\dim_t X < \infty$. *Then there exists a measure* $\nu \in \mathcal{M}_1(X)$ *such that*

$$\underline{\dim}_L \nu \geq \dim_t X .$$

*Proof.* Set $d = \dim_t X$, $X \neq \emptyset$. Let $\mu$ be a measure given by Lemma 12.1. Define $\nu = \mu/(\mu(X))$. Clearly $\nu \in \mathcal{M}_1(X)$ and

$$\nu\big(B(x,r)\big) \leq cr^d/\mu(X) \quad \text{for} \quad r > 0 .$$

Hence,

$$Q_\nu(r) \leq cr^d/\mu(X) \quad \text{for} \quad r > 0 ,$$

where $Q_\nu$ is Lévy function given by (3.1). Consequently

$$\underline{\dim}_L \nu = \lim_{r \to 0} \inf \frac{\log Q_\nu(r)}{\log r} \geq \lim_{r \to 0} \inf \frac{\log cr^d - \log \mu(X)}{\log r} = d . \qquad \square$$

**Corollary 12.3.** *Let $X$ be a Polish space. Then*
$$\dim_L X \geq \dim_t X \,.$$

Indeed, if $\dim_t X < \infty$, the assertion is true by virtue of Theorem 12.2. If $\dim_t X = \infty$, by Lemma 12.1 for every $k \in \mathbb{N}$ there exists $\nu_k \in \mathcal{M}_1(X)$ such that
$$\nu_k\big(B(x,r)\big) \leq c_k r^k \quad \text{for} \quad x \in X \quad \text{and} \quad r > 0 \,.$$
Hence, for $\mu_k = \nu_k/\nu_k(X)$ we have $\underline{\dim}_L \mu_k \geq k$, whence it follows immediately that $\dim_L X = \infty$.

**Corollary 12.4.** *If $\dim_t X = \infty$, then there exists a measure $\mu \in \mathcal{M}_1(X)$ such that* $\dim_H \mu = \infty$.

Indeed, by above observation for $k \in \mathbb{N}$ there is $\mu_k \in \mathcal{M}_1(X)$ such that $\underline{\dim}_L \mu_k \geq k$. Define
$$\mu = \sum_{k=1}^{\infty} 2^{-k} \mu_k \,.$$
Since $\dim_H \mu \geq \dim_L \mu_k \geq \underline{\dim}_L \mu_k$, $n \in \mathbb{N}$, the assertion of Corollary 12.4 follows.

**Theorem 12.5.** *Let $X$ be a Polish space. Then*
$$\dim_t X = \inf\{\dim_L X' \,:\, X' \text{ is homeomorhic to } X\} \,.$$

*Proof.* Let $\nu$ be a Borel measure given by Lemma 12.1 and let $A$ be a compact set such that $\nu(A) > 0$ (such set there exists by virtue of Ulam Theorem). Obviously The condition (12.2) holds for every $x \in A$ and $r > 0$. Hence the assertion follows from Frostman lemma.                                                                    □

# 13. Applications to differential equations

Consider the Ważewska-Czyżewska equation
$$\frac{\partial u}{\partial t} + x \frac{\partial u}{\partial x} = \lambda \,, \qquad t \in \mathbb{R}_+ \,, \quad x \in [0,1] \,, \tag{13.1}$$
with initial condition
$$u(0,x) = v(x) \,, \tag{13.2}$$
where $v$ is a given continuously differentiable function such that $v(0) = 0$.

This equation describes the process of the reproduction of the red blood cells (see [20]). It is well known that solutions of this equation may exhibit quite irregular behaviour.

Let
$$V = \Big\{v \,:\, v \in C\big([0,1]\big), \, v(0) = 0\Big\} \,.$$
Assume that $V$ is endowed with the norm
$$\|v\| = \max\left\{\sup_{[0,1]} |v(x)|, \, \sup_{[0,1]} |v'(x)|\right\} \,.$$

Using the method of characteristics it is easy to see that the solution of Problem (13.1), (13.2) is given by

$$u(t,x) = S^t\big((v(x))\big) = e^{\lambda t}v(xe^{-t})\,.$$

Obviously $\{S^t\}_{t\geq 0}$ is the semigroup defined on the space $V$. If $\lambda < 1$, then $\|S^t v\| \leq e^{\lambda - 1}$ and so the system is asymptotically stable. It is well known that for $\lambda > 2$ the semigroup $\{S^t\}_{t\geq 0}$ has a nontrivial invariant ergodic measure. We are going to show that for $\lambda > 1$ the semigroup $\{S^t\}_{t\geq 0}$ admits an invariant measure with large concentration dimension. Note that according to G. Prodi [45] and C. Foias [9] the chaotic dynamical system can be described by the dimension of invariant measure.

Recall that a measure $\mu$ is called *invariant with respect to semigroup* $\{S^t\}_{t\geq 0}$ if

$$\mu\big(S^{-t}(A)\big) = \mu(A) \quad \text{for} \quad A \in \mathcal{B}(V)\,, \quad t \geq 0\,.$$

**Theorem 13.1.** *Let $\lambda > 1$. Then for every $m \in \mathbb{N}$ there exists a probability measure $\widetilde{\mu}$ invariant with respect to semigroup $\{S^t\}_{t\geq 0}$ such that*

$$\dim_L \widetilde{\mu} \geq m\,.$$

*Sketch of the proof.* Fix $N \in \mathbb{N}$ and for $i \in \{1,\ldots,N\}$ consider the maps $T_i : V \to V$ given by

$$T_i v(x) = \begin{cases} 2^\lambda v(2x)\,, & x \in \big[0, \tfrac{1}{2}\big] \\[2mm] 2^\lambda\Big[v(1) + v'(1)(2x - 1) + i(x - \tfrac{1}{2})^2\Big]\,, & x \in \big(\tfrac{1}{2}, 1\big]\,. \end{cases}$$

It easy to see that all transformations $T_i$ are similarities with the same scaling factor $2^{1-\lambda} < 1$ and satisfies the Moran condition with respect to the set $\{v \in V : \|v\| \leq N/(2^{\lambda-1} - 1)\}$.

Consider the IFS $\{(T_i, p_i) : i = 1,\ldots,N\}$, where $p_i = 1/N$. Let $\mu_*$ be the corresponding invariant measure. By Theorem 7.2 and 7.4 (with $L_i = l_i = 2^{1-\lambda}$, $p_i = 1/N$) we obtain

$$\dim_L \mu_* = \frac{1}{1-\lambda}\log_2 N\,.$$

Now, define

$$\widetilde{\mu}(A) = \frac{1}{\ln 2}\int_0^{\ln 2} \mu_*\big(S^{-t}(A)\big)\,dt\,.$$

It is easy to verify that the measure $\widetilde{\mu}$ is invariant with respect to the semigroup $\{S^t\}_{t\geq 0}$ and that for $t > 0$ the transformation $S^{-t}T_i$ are similarities with scaling factor $e^{(\lambda-1)(t-t_0)}$. Hence

$$\mathrm{diam}\big((S^t T_i)^{-1}(A)\big) = e^{(\lambda-1)(t-t_0)}\mathrm{diam}\, A \leq c\,\mathrm{diam}\, A\,.$$

where $c = e^{(\lambda-1)t_0}$. From the last inequality it follows that

$$Q_{\widetilde{\mu}}(r) \leq Q_{\widetilde{\mu}}(cr)\,.$$

Consequently

$$\dim_L \widetilde{\mu} \geq \dim_L \mu_*\,.$$

Since $N$ was arbitrary, the statement of Theorem 13.1 follows.

Consider now the more general version of Ważewska-Czyżewska equation

$$\frac{\partial u}{\partial t} + c(x)\frac{\partial u}{\partial x} = f(x, u), \quad \text{for} \quad (t, x) \in [0, \infty) \times [0, 1], \tag{13.3}$$

with initial condition

$$u(0, x) = v(x), \quad \text{for} \quad x \in [0, 1]. \tag{13.4}$$

We assume that $c : [0, 1] \to \mathbb{R}$ and $f : [0, 1] \times [0, \infty) \to \mathbb{R}$ are given continuously differentiable functions satisfying $c(0) = 0$, $c(x) > 0$ for $x \in (0, 1]$, $f(x, y) \leq k_1 y + k_2$, for $(x, y) \in [0, 1] \times \mathbb{R}_+$ and $f(x, 0) = 0$, $f_y(x, 0) > 0$ for $x \in [0, 1]$.

Using the method of characteristics one can prove that the Cauchy Problem (13.3), (13.4) generates a semigroup $\{S^t\}_{t \geq 0}$ of operators acting on the space $V_+ = \{v \in V : v \text{ is nonnegative}\}$. A. Lasota [18] proved that under suitable condition the semigroup $\{S^t\}_{t \geq 0}$ is chaotic in the sense of Auslander–Yorke [1]. This means that every point $v \in V$ is unstable and that for some $v \in V$ the trajectory $\{S^i v : t \geq 0\}$ is dense in $V$. Moreover, it was proved that the semigroup $\{S^t\}_{t \geq 0}$ has invariant measures with interesting ergodic properties. Recently, using new concepts on dimensions, A. Lasota and T. Szarek [24] obtained another characterization of the asymptotic behaviour of the semigroup $\{S^t\}_{t \geq 0}$. Namely they obtained the following results.

**Theorem 13.2.** *Let $\{S^t\}_{t \geq 0}$ be semigroup generated by Problem (13.3), (13.4). Then:*

(i) *For every $m \in \mathbb{N}$ there exists a probability measure $\widetilde{\mu}$ invariant with respect to semigroup $\{S^t\}_{t \geq 0}$ such that $\underline{\dim}_L \widetilde{\mu} \geq m$ and $\overline{\dim}_L \widetilde{\mu}$ is finite.*

(ii) *There exists a probability measure $\widetilde{\mu}$ invariant with respect to semigroup $\{S^t\}_{t \geq 0}$ such that $\dim_H \widetilde{\mu} = \infty$.*

(iii) *There exists a probability measure $\widetilde{\mu}$ invariant with respect to semigroup $\{S^t\}_{t \geq 0}$ such that $\dim_L \widetilde{\mu} = \infty$.*

## References

[1] J. Auslander, J. Yorke, *Intervals maps, factor of maps and chaos*, Tôhoku Math. J., **32** (1980), 177–1888.

[2] M. F. Barnsley, *Fractals Everywhere*, Academic Press, Boston, 1993.

[3] W. Chin, B. Hunt, J. A. Yorke, *Correlation dimension for iterated function systems*, Trans. Amer. Math. Soc., **349** (1989), 1783–1796.

[4] C. D. Cutler, *Some results on the behaviour of the fractal dimensions of distribution on attractors*, J. Statist. Phys., **62** (1991), 651–708.

[5] C. D. Cutler, *Strong and weak duality principles for fractal dimension in Euclidean space*, Math. Proc. Camb. Soc., **118** (1995), 393–410.

[6] C. D. Cutler, L. Olsen, *A variational principle for the Hausdorff dimension of fractal sets*, Math. Scand., **74** (1994), 64–72.

[7] R. M. Dudley, *Probabilities and Metrics*, Lecture Notes, Ser., **45**, Aarhus University, 1978.

[8] D. Feng, J. Wu, *Category and dimension of compact subset of* $\mathbb{R}^n$, Chinese Sci. Bull., **42**(1997), 1680–1683.

[9] C. Foias, *Statistical study of Navier-Stokes equations*, Rend. Sem. Mat. Univ. Padova, **49** (1973), 9–123.

[10] O. Frostman, *Potential d'équilibre et capacité des ensemble avec quelques application á la théorie des fonctions*, Maddel. Lunds Univ. Mat. Sem., **3** (1935), 1–118.

[11] H. Gacki, A. Lasota, J. Myjak, *Upper estimate of concentration and thin dimensions of measures*, (to appear).

[12] J. Genyuk, *A typical measure typically has no local dimension*, Real Anal. Exchange, **23** (1997/1998), 525–537.

[13] P. M. Gruber, *Upper estimate of concentration and thin dimensions of measures*, Mh. Math., **108** (1989), 149–164.

[14] H. Hentschel, I. Procaccia, *The infinite number of generalized dimensions of fractals and strange attractors*, Phys. D, **8** (1983), 435–444.

[15] W. Hengartner, R. Theodorescu, *Concentration Functions*, Academic Press, New York, London, 1973.

[16] W. Hurewicz, H. Wallman, *Dimension Theory*, Princeton, 1941.

[17] H. Joyce, *A relationship between packing and topological dimensions*, Mathematika, **45** (1998), 43–53.

[18] A. Lasota, *Stable and chaotic solutions of a first order partial differential equations*, Nonlinear Anal. TMA, **5** (1981), 1181–1193.

[19] A. Lasota, *A variational principle for fractal dimensions*, Nonlinear Anal., **1** (2005), 1–11.

[20] A. Lasota, M. Mackey, M. Ważewska-Czyżewska, *Minimizing therepeutically induced anemia*, J. Math. Biol., **13** (1981), 1181–1193.

[21] A. Lasota, J. Myjak, *Semifractals on Polish spaces*, Bull. Pol. Ac.: Math., **46** (1998), 179–196.

[22] A. Lasota, J. Myjak, *Attractors of multifunctions*, Bull. Pol. Ac.: Math., **50** (2000), 221–235.

[23] A. Lasota, J. Myjak, *On a dimension of measures*, Bull. Pol. Ac.: Math., **48** (2002), 319–334.

[24] A. Lasota, T. Szarek, *Dimensions of measures invariant with respect to the Ważewska partial differential equation*, J. Differential Equations, **196** (2004), 448–465.

[25] A. Lasota, J. Traple, *Dimension of invariant set for mappings with the squeezing property*, Chaos, Solitons and Fractals, **28** (2006), 1271–1280.

[26] A. Lasota, J. A. Yorke, *Lower bounded technique for Markov operators and iterated function systems*, Random Comput. Dynamics, **2** (1994), 41–77.

[27] P. A. P. Moran, *Additive functions of intervals and Hausdorff measure*, Proc. Cambrigge. Philos. Soc., **42** (1946), 15–23.

[28] J. Myjak, *Some typical properies of dimensions of sets and measures*, Abstract and Applied Analysis, **3** (2005), 329–333.

[29] J. Myjak, R. Rudnicki, *Box and packing dimension of typical compact sets*, Monatsh. Math., **131** (2000), 223–226.

[30] J. Myjak, R. Rudnicki, *On the typical structure of compact sets*, Arch. Math., **76** (2001), 119–126.

[31] J. Myjak, R. Rudnicki, *On the box dimension of typical measures*, Monatsh. Math., **136** (2002), 143–150.

[32] J. Myjak, R. Rudnicki, *Typical properties of correlation dimension*, Real Anal. Exchange, **28** (2003), 269–278.

[33] J. Myjak, R. Rudnicki, *Dimension and of typical compact sets, continua and curves*, Bolletino U.M.I., (8), 10-B (2007), 357–364.

[34] J. Myjak, T. Szarek, *A lower estimation of the Hausdorff dimension for attractors with overlaps*, J. Statist. Phys., **105** (2001), 649–657.

[35] J. Myjak, T. Szarek, *Estimates of capacity and self-similar measures*, Ann. Polon. Math., **78** (2002), 141–157.

[36] J. Myjak, T. Szarek, *On Hausdorff dimension of invariant measures arising from non-contractive iterated function systems*, **181** (2002), 223–237.

[37] J. Myjak, T. Szarek, *Szpilrajn type theorem for concentration dimension*, Fund. Math., **172** (2002), 19–25.

[38] J. Myjak, T. Szarek, *Capacity of invariant measures related to Poisson-driven stochastic differential equations*, Nonlinearity, **16** (2003), 441–455.

[39] J. Myjak, T. Szarek, *On the Hausdorff dimension of Cantor like sets with overlaps*, Chaos, Solitons and Fractals, **18** (2003), 329–333.

[40] J. Myjak, T. Szarek, *Some generic properties of concentration dimension of measure*, Boll. Unione Mat. Ital. Sez. B, **6** (2003), 211–219.

[41] J. Myjak, T. Szarek, M. Ślęczka, *Szpilrajn–Marczewski type theorem for concentration dimension on Polish spaces*, Canad. Bull. ,**49** (2006), 247–255.

[42] L. Olsen, *A multifractal formalism*, Adv. in Math., **116** (1995), 82–196.

[43] Y. B. Pesin, *Dimension Theory in Dynamical Systems. Contemporary views and applications*, University of Chicago Press, Chicago (1997).

[44] L. Procaccia, P. Grassberger, H. G. E. Hentschel, *On the characterisation of chaotic motions*, Lecture Notes in Phys., **179**(1983), 212–222.

[45] G. Prodi, *Teoremi ergodici per le equazioni della idrodinamica*, C.I.M.E., Roma, 1960.

[46] T. D. Sauer, J. A. Yorke, *Are the dimensions of a set and its image equal under typical smooth functions*, Ergodic Theory Dyna. Systems, **17** (1997), 941–956.

[47] K. Simon, B. Solomyak, *Correlation dimension for self-similar Cantor sets wirth overlaps*, Fund. Math., **155** (1998), 293–300.

[48] C. Tricot, *Two definitions of fractal dimensions*, Math. Proc. Camb. Philos. Soc., **91** (1982), 57–74.

[49] E. Szpilrajn, *La dimension e la measure*, Fund. Math., **27** (1937), 81–89.

[50] L. S. Young, *Dimension, entropy and Lyapunov exponents*, Ergodic Theory Dynam. Systems, **2** (1982), 109–124.

Józef Myjak
WMS AGH
Al. Mickiewicza 30
PL-30059 Kraków
Poland
e-mail: myjak@univaq.it

Progress in Nonlinear Differential Equations
and Their Applications, Vol. 75, 345–356
© 2007 Birkhäuser Verlag Basel/Switzerland

# Degree Theory and Almost Periodic Problems

Rafael Ortega

*Dedicated to Arrigo Cellina and James Yorke*

**Abstract.** Periodic boundary value problems can be reduced to fixed point equations of integral type. After this reduction the Leray–Schauder degree becomes an useful tool for proving the existence of periodic solutions. The purpose of this paper is to discuss the analogous approach for almost periodic problems and to show by means of examples that degree theory is not applicable in this setting.

**Mathematics Subject Classification (2000).** Primary 34C27; Secondary 47H11.

**Keywords.** Fixed points, trigonometric solutions, module of frequencies.

## 1. Introduction

This paper deals with differential equations which are almost periodic in time. Examples of these equations are

- the pendulum equation with quasi-periodic forcing

$$\ddot{u} + c\dot{u} + \sin u = \sin t + \sin \sqrt{2}t$$

- the limit-periodic Riccati equation

$$\dot{u} = u^2 - 1 + \sum_{n=1}^{\infty} 3^{-n} \sin(2^{-n}t).$$

Since the beginning of the last century, the development of a theory for these equations has been inspired by the known results in the simpler class of periodic equations. This is shown in the following quotation, taken from the introduction of the paper by Bohl [1],

Supported by D. G. I. MTM2005-03483, Ministerio de Educación y Ciencia, Spain.

> Les méthodes de détermination des solutions périodiques ont été
> notablement perfectionées, ces derniers temps, grâce surtout aux
> travaux bien connus de M. Poincaré. On ne peut pas en dire autant
> à ce qu'il semble, des solutions trigonométriques plus générales.

By the time the paper [1] appeared, the notion of almost periodic function had
not been introduced and Bohl just referred to *trigonometric solutions*. After the
definition of almost periodicity by Bohr, many results for linear and nonlinear
equations were obtained. Information on many of these results can be found in [5,
8, 11].

Going back to the periodic problem, we notice that nowadays there are several
methods for proving the existence of periodic solutions. One of the most popular
consists in a combination of Functional Analysis and Degree Theory. After reducing
the periodic problem to a fixed point equation in a space of periodic functions, it
applies the theory of Leray and Schauder. We refer to the work by Krasnoselskii
and his school [9] and by Mawhin [10] for more information. The purpose of this
paper is to discuss what should be the analogous approach in the almost periodic
case and to present some examples which seem to indicate that the degree theory
is not applicable in this setting. In this context it is interesting to mention the
related discussion by Fink in [5] Chapter 8, Section 3.

The paper is organized in two parts. First we will consider a second order
equation of Newtonian type and transform the almost periodic problem in a fixed
point equation of the type

$$u = \mathcal{K}u, \quad u \in AP,$$

where $AP$ is the Banach space of almost periodic functions. In general the operator $\mathcal{K}$ is not compact on bounded sets and so the Leray–Schauder theory is not
applicable. We will present an example where $\mathcal{K}$ has no fixed points and maps the
unit ball into its interior; that is

$$\mathcal{K}(\overline{B}) \subset B, \quad B = \{u \in AP : \ ||u||_\infty < 1\}.$$

This shows that $\mathcal{K}$ cannot belong to any class of maps for which it is possible to
define a degree with the standard properties. As a corollary it will be proved that
the well known homotopy method for periodic problems does not extend to the
almost periodic case.

The second part of the paper deals with a first order almost periodic equation
having a prescribed module of frequencies. Given an additive subgroup of $\mathbb{R}$, which
will be denoted by $\Omega$, the space $AP(\Omega)$ is composed by those almost periodic
functions having all their frequencies in $\Omega$. The search of solutions in $AP(\Omega)$ leads
to a fixed point equation in this space. It will be shown that, unless $\Omega$ is cyclic, the
Schauder principle does not hold and so the degree is not applicable. The cases
of cyclic groups correspond to periodic functions of a fixed period and so this
result shows that periodic problems are rather special. At this point I would like
to express my gratitude to Professor Corduneanu, for these results on prescribed
frequencies were motivated by a question that he posed to me.

All the proofs in this paper are based on previous results in [13] and [14]. These papers, which were joint work with M. Tarallo, were inspired by some of the constructions by Opial [12], Fink and Frederickson [6], Zhikov and Levitan [15] and Johnson [7].

Before ending this introduction it is worth mentioning that there are several theories of weak almost periodic functions, leading to reinterpretations of the notion of *trigonometric solution*. It is conceivable that the change of $AP$ and $AP(\Omega)$ by larger functional spaces could allow the use of degree theory.

## 2. Equations of the second order

For a fixed number $c > 0$ we consider the Newtonian equation with friction

$$\ddot{u} + c\dot{u} = f(t, u)\,. \tag{2.1}$$

We shall restrict to the scalar case and assume that $f : \mathbb{R} \times \mathbb{R} \to \mathbb{R}$ is continuous. When $f$ has the appropriate time dependence, the periodic and almost periodic problems are well defined and can be transformed in fixed point equations.

### 2.1. The periodic problem

Fix the period $T > 0$ and assume that $f$ is periodic in time, that is

$$f(t + T, u) = f(t, u) \quad \text{for each} \quad (t, u)\,.$$

We work with the Banach space

$$C_T = \{u : \mathbb{R} \to \mathbb{R}/u \quad \text{is continuous and } T\text{-periodic}\}\,,$$

endowed with the $L^\infty$-norm

$$||u||_\infty = \sup_{t \in \mathbb{R}} |u(t)|\,.$$

Given $k > 0$, the linear equation

$$\ddot{u} + c\dot{u} = ku + p(t)\,, \quad p \in C_T$$

has a unique solution in $C_T$. This solution can be expressed in terms of the Green function as

$$u(t) = -\int_0^T G(t, s)p(s)ds\,.$$

We do not need the explicit form of $G$ and just recall that $G = G(t, s, k)$ is continuous and positive. After fixing $k$ (2.1) can be rewritten as

$$\ddot{u} + c\dot{u} = ku + g(t, u) \tag{2.2}$$

with $g(t, u) = f(t, u) - ku$. The search of $T$-periodic solutions of (2.1) or (2.2) becomes equivalent to

$$u = Ku\,, \quad u \in C_T$$

where

$$K : C_T \to C_T\,, \quad Ku(t) = -\int_0^T G(t, s)g(s, u(s))ds\,.$$

It is well known that the operator $K$ is compact on bounded sets and the Leray–Schauder degree is applicable to $id - K$. Since $G$ and $g$ depend upon $k$, we have a family of operators $K = K(u, k)$. This is irrelevant from the point of view of degree theory since the degree is independent of $k$. Indeed, all operators $K(\cdot, k)$ have the same fixed points and the formula $K = K(u, k)$ defines a homotopy on any bounded and open set $\mathcal{U} \subset C_T$ without fixed points on the boundary. This implies that

$$deg\big(id - K(\cdot, k_1), \mathcal{U}\big) = deg\big(id - K(\cdot, k_2), \mathcal{U}\big)$$

for all $k_1, k_2 > 0$.

### 2.2. The almost periodic problem

We start with the Banach spaces

$$BC = \{u : \mathbb{R} \to \mathbb{R}/u \quad \text{is continuous and bounded}\}$$
$$AP = \{u : \mathbb{R} \to \mathbb{R}/u \quad \text{is almost periodic}\}$$

endowed with the $L^\infty$-norm. For each $T > 0$, $C_T$ is contained in $AP$ and this space can be characterized as the smallest closed linear subspace of $BC$ which contains $\bigcup_{T>0} C_T$. Alternatively $AP$ can be characterized as the closure in $BC$ of the space of trigonometric polynomials

$$u(t) = a_0 + \sum_{n=1}^{N} \{a_n \cos \omega_n t + \sin \omega_n t\}$$

where the frequencies $\omega_n$ are arbitrary. We refer to [4] for more information on the definition of almost periodic function.

The dependence of $f$ with respect to $t$ will be almost periodic. This means that

$(i)$ $\qquad\qquad\qquad f(\cdot, u) \in AP \quad$ for each $\quad u \in \mathbb{R}$.

In order to apply the methodology of Functional Analysis we need the composition property

$$u \in AP \quad \Rightarrow \quad f\big(\cdot, u(\cdot)\big) \in AP.$$

However it is well known that the almost periodicity in $t$ is not sufficient to guarantee this property (see [5], page 16). We say that $f$ is in the class $UAP$ if it satisfies $(i)$ and the additional condition

$(ii)$ For each $r > 0$ the family of functions $\{f(t, \cdot)\}_{t \in \mathbb{R}}$ is equicontinuous on $[-r, r]$.

The composition property holds if $f \in UAP$ (see [5], Chapter 2).

Our task will be to adapt the discussion on the periodic problem to this new setting. Again we fix $k > 0$ and observe that the linear equation

$$\ddot{u} + c\dot{u} = ku + p(t), \quad p \in AP$$

has a unique solution in $AP$. It can be expressed in terms of a Green function, but this time the integral is extended over the whole real line. Namely,

$$u(t) = - \int_{-\infty}^{\infty} \tilde{G}(t, s) p(s) ds$$

with

$$\tilde{G}(t,s) = \begin{cases} \frac{1}{r_+ - r_-} e^{r_-(t-s)} & \text{if } t \geq s \\ \frac{1}{r_+ - r_-} e^{r_+(t-s)} & \text{if } t \leq s, \end{cases}$$

and $r_\pm = \frac{-c \pm \sqrt{c^2 + 4k}}{2}$.

We observe that $\tilde{G}$ is continuous and positive. It is interesting to observe that the periodic Green function $G$ can be obtained from $\tilde{G}$. Indeed,

$$G(t,x) = \sum_{n=-\infty}^{\infty} \tilde{G}(t, s + nT).$$

More information about linear almost periodic equations can be found in [3].

With the help of the Green function the search of almost periodic solutions of (2.1) or (2.2) is equivalent to

$$u = \mathcal{K}u, \quad u \in AP$$

with

$$\mathcal{K} : AP \to AP, \quad \mathcal{K}u(t) = -\int_{-\infty}^{\infty} \tilde{G}(t,s)g(s, u(s))ds.$$

### 2.3. Non-applicability of Schauder's Principle

Typically the operator $\mathcal{K}$ is not compact on bounded sets. We show this in the particular case $g(t,u) = u$. Now $\mathcal{K}$ is linear and an easy computation leads to

$$\mathcal{K}\left(e^{i\omega t}\right) = \frac{1}{-(\omega^2 + k) + ic\omega} e^{i\omega t}$$

for each $\omega \in \mathbb{R}$. In this way we have obtained an uncountable set of eigenvalues and so $\mathcal{K}$ cannot be compact. This observation explains why the theory of Leray and Schauder is not applicable to $id - \mathcal{K}$. Next we shall prove that a tentative version of Schauder Fixed Point Theorem cannot hold for $\mathcal{K}$. This excludes the possibility of defining a degree of $id - \mathcal{K}$.

**Theorem 2.1.** *For each $c > 0$ there exists $f \in UAP$ and $k > 0$ such that the associated operator $\mathcal{K}$ has no fixed points and*

$$\mathcal{K}(\overline{B}) \subset B$$

*where*

$$B = \{u \in AP/\ ||u||_\infty < 1\}.$$

The proof will be inspired by well known results for the periodic problem. In the periodic case, the region of $C_T$ lying between a lower and an upper solution is invariant under $K$ and contains a fixed point. Here we are assuming that $k$ is large enough. We refer to [2] for more details. However it was proved in [13] that the method of upper and lower solutions fails in the almost periodic case. This will be the starting point for the proof.

*Proof of Theorem 2.1.* Given $u, v \in BC$ we introduce the notations

$$u < v \quad \text{if} \quad v(t) - u(t) > 0 \quad \text{for each} \quad t \in \mathbb{R},$$
$$u \ll v \quad \text{if} \quad \inf_{t \in \mathbb{R}} \left( v(t) - u(t) \right) > 0.$$

In the spaces $C_T$ both notions are equivalent but not in $AP$. Notice that the strong inequality does not hold for the functions $u(t) = \sin t + \sin \sqrt{2}t$ and $v(t) = 2$. We observe that the unit ball in $AP$ can be expressed as

$$B = \{u \in AP / \ -1 \ll u \ll 1\}.$$

According to Theorem 10 in [13] it is possible to find $c > 0$, $f \in UAP$ and $\alpha < \beta$ such that (2.1) has no almost periodic solutions and

$$f(\cdot, \alpha) \ll 0 \ll f(\cdot, \beta). \tag{2.3}$$

This inequality means that the numbers $\alpha$ and $\beta$ are strict lower and upper solutions.

We must prove the Theorem for arbitrary $c$ positive but it is enough to do it for a concrete value $c$. This is sufficient because we can change the friction coefficient by a re-scaling of time $t \mapsto \lambda t$ with $\lambda > 0$. Also, after translation and dilation of $u$ we can assume that $\alpha = -1$ and $\beta = 1$. We do this but for convenience we keep the notation $\alpha$ and $\beta$, which now represent the numbers $-1$ and $1$.

According to [13] the function $f$ is smooth and the derivative $\frac{\partial f}{\partial u}$ is also in $UAP$. This implies that $|\frac{\partial f}{\partial u}(t, u)|$ is uniformly bounded in regions of the type $t \in \mathbb{R}, |u| \leq M$. We select $k > 0$ large enough so that the function $u \in [\alpha, \beta] \mapsto g(t, u) = f(t, u) - ku$ is decreasing for each $t \in \mathbb{R}$. The definition of $\mathcal{K}$ and the positivity of $\tilde{G}$ imply that $\mathcal{K}$ is monotone on $\overline{B}$; that is

$$\alpha \leq u \leq v \leq \beta \Rightarrow \mathcal{K}u \leq \mathcal{K}v.$$

Now we are going to use the condition (2.3). Given $\delta > 0$ with $f(t, \alpha) \leq -\delta$ everywhere,

$$\mathcal{K}\alpha(t) = -\int_{-\infty}^{\infty} \tilde{G}(t, s)\left[f(s, \alpha) - k\alpha\right]ds \geq (\delta + k\alpha)\int_{-\infty}^{\infty} \tilde{G}(t, s)ds = \frac{\delta}{k} + \alpha.$$

This implies that $\mathcal{K}\alpha \gg \alpha$. In the same way one obtains $\mathcal{K}\beta \ll \beta$. We are ready to prove that $\mathcal{K}(\overline{B}) \subset B$. Indeed, given $u \in \overline{B}$,

$$-1 = \alpha \ll \mathcal{K}\alpha \leq \mathcal{K}u \leq \mathcal{K}\beta \ll \mathcal{K}\beta = 1.$$

We know that (2.1) has no almost periodic solutions and so $\mathcal{K}$ has no fixed points. This remark completes the proof. □

## 2.4. The continuation method

A crucial property of the degree is the invariance under homotopies. The application of this property to the periodic problem leads to general continuation principles. As an example we consider the unit ball in $C_T$,

$$B_T = \{u \in C_T / \ ||u||_\infty < 1\},$$

and the family of equations

$$\ddot{u} + c\dot{u} = \lambda f(t, u) + (1 - \lambda)u, \quad \lambda \in [0, 1]. \tag{2.4}$$

Our initial equation (2.1) appears for $\lambda = 1$ while the equation is linear and has the unique $T$-periodic solution $u = 0$ for $\lambda = 0$. Assume that $f$ is continuous and $T$-periodic in $t$ and there are no $T$-periodic solutions of (2.4) in $\partial B_T$. The invariance under homotopies implis that (2.1) has a $T$-periodic solution lying in $B_T$.

As the reader probably expects, the analogous principle does not hold for almost periodic problems. Going back to the proof of Theorem 2.1 we take the same function $f \in UAP$ with $\alpha = -1$, $\beta = 1$ and observe that the functions

$$f_\lambda(t, u) = \lambda f(t, u) + (1 - \lambda)u$$

are also in $UAP$ and satisfy

$$f_\lambda(\cdot, \alpha) \ll 0 \ll f_\lambda(\cdot, \beta).$$

Moreover these strong inequalities are uniform in $\lambda$. This allows us to repeat the strategy of the proof of Theorem 2.1 for each $\lambda$ and conclude that

$$\mathcal{K}_\lambda(\overline{B}) \subset B \quad \text{for each} \quad \lambda \in [0, 1].$$

The operator $\mathcal{K}_\lambda$, defined on $AP$, refers to (2.4) for fixed $\lambda$. The choice of the constant $k$ is made independently of $\lambda$, this means that $u \in [\alpha, \beta] \mapsto f_\lambda(t, u) - ku$ is decreasing for all $t$ and $\lambda$. We sum up the above discussions.

**Corollary 2.2.** *There exist $f \in UAP$ and $k > 0$ such that the operator $\mathcal{K}_\lambda$ associated to (2.4) satisfies*

- *$\mathcal{K}_1$ has no fixed points*
- *$\mathcal{K}_0$ is linear and has a unique fixed point at $u = 0$*
- *$\mathcal{K}_\lambda(\overline{B}) \subset B$ for each $\lambda \in [0, 1]$.*

In the theory of almost periodic equations it is customary to impose the assumptions not only on the original equation but on all the equations lying in the hull. We will prove that this makes no difference, but first we recall the notion of hull. Given $f \in UAP$ we say that $f^\star : \mathbb{R} \times \mathbb{R} \to \mathbb{R}$ is in the hull of $f$, denoted by $f^\star \in \mathcal{H}(f)$, if there exists a sequence of real numbers $\{h_n\}$ such that

$$f(t + h_n, u) \to f^\star(t, u) \quad \text{as} \quad n \to \infty,$$

and the convergence is uniform in $(t, u) \in \mathbb{R} \times [-M, M]$ for each $M > 0$. It is easy to prove that $\mathcal{H}(f)$ is contained in $UAP$ but the hull of a smooth function $f$ can contain non-smooth functions.[1] By a passage to the limit we observe that the following two properties are inherited by each $f^\star \in \mathcal{H}(f)$,

$$f^\star(\cdot, \alpha) \ll 0 \ll f^\star(\cdot, \beta)$$

$$u \in [\alpha, \beta] \mapsto f^\star(t, u) - ku \quad \text{is monotone non-increasing.}$$

---

[1] The function $f(t) = \sqrt{2 - \sin t - \sin \sqrt{2}t}$ is smooth but $f^\star(t) = \sqrt{2 - \cos t - \cos \sqrt{2}t}$ is not.

This allows to improve the conclusion of Corollary 2.2 with

$$\mathcal{K}_{\lambda,f^\star}\left(\overline{B}\right) \subset B$$

for each $\lambda \in [0,1]$ and $f^\star \in \mathcal{H}(f)$. At this point one must observe that $\mathcal{K}_{\lambda,f^\star}$ has fixed points in and only if $\mathcal{K}_{\lambda,f}$ does.

## 3. First order equations: prescribing the module of frequencies

### 3.1. The space $AP(\Omega)$

Given $\Omega$, an additive subgroup of $\mathbb{R}$, the space $AP(\Omega)$ is defined as the closure in $AP$ of the class of trigonometric polynomials having frequencies in $\Omega$. These are polynomials of the type

$$a_0 + \sum_{n=1}^{N}\{a_n \cos\omega_n t + b_n \sin\omega_n t\}$$

with $\omega_1,\ldots,\omega_N \in \Omega$.

The space $AP$ becomes a commutative Banach algebra with the standard product of functions and each space $AP(\Omega)$ is a subalgebra. This can be deduced from the additive formulas for trigonometric functions.

Given an almost periodic function $u(t)$, the module $mod(u)$ is the smallest additive subgroup of $\mathbb{R}$ containing all the non vanishing Fourier coefficients of $u$. With the help of Fourier analysis one can prove that

$$AP(\Omega) = \left\{u \in AP : mod(u) \subset \Omega\right\}.$$

Next we identify the space $AP(\Omega)$ for some groups.

- The trivial group $\Omega_0 = \{0\}$, $AP(\Omega_0) = \{\text{constant functions}\}$
- Cyclic groups $\Omega_1 = T\mathbb{Z}$, $T > 0$, $AP(\Omega_1) = C_T$
- A free group with two generators $\Omega_2 = \{n + m\sqrt{2}/n, m \in \mathbb{Z}\}$, $AP(\Omega_2) = $ quasi-periodic functions with frequencies 1 and $\sqrt{2}$
- The rational numbers $\Omega_3 = \mathbb{Q}$, $AP(\Omega_3) = $ uniform limits of periodic functions having periods commensurable with $2\pi$.

### 3.2. The $\Omega$-almost periodic problem

We will study the equation

$$\dot{u} = f(t, u) \tag{3.1}$$

where $f$ is in $UAP$ and satisfies the additional condition

$$f(\cdot, u) \in AP(\Omega) \quad \text{for each} \quad u \in \mathbb{R}. \tag{3.2}$$

In this setting we discuss the existence of solutions in $AP(\Omega)$. We shall transform this problem into a fixed point equation in $AP(\Omega)$ but first we go back to the examples. For $\Omega_0$ the condition (3.2) says that $f$ is independent of $t$ and solutions in $AP(\Omega_0)$ are equilibria. For $\Omega_1$ we go back to the $T$-periodic problem. For $\Omega_2$ and $\Omega_3$ we have genuine almost periodic problems.

### 3.3. The fixed point equation

We start with the composition property, which is important to make the problem treatable with the methods of Functional Analysis.

**Lemma 3.1.** *Assume that* $f \in UAP$ *and the condition* (3.2) *holds. Then,*

$$u \in AP(\Omega) \Rightarrow f(\cdot, u(\cdot)) \in AP(\Omega).$$

This property depends essentially on the algebraic structure of $\Omega$. In principle one could define $AP(\Omega)$ for any non-empty subset of the real numbers and it would be a Banach space. The difference is that when $\Omega$ is not a group the space $AP(\Omega)$ is not an algebra and the composition property does not hold.

There is a proof of this Lemma using Fourier Analysis and we refer to [5]. We present a more direct approach.

*Proof of Lemma* 3.1. The space $AP(\Omega)$ is an algebra and so the composition property holds if $f$ is a polynomial in $u$ which is independent of $t$. Assume next that $f = f(u)$ is any continuous function, again independent of $t$. Given $u \in AP(\Omega)$ we approximate $f$ by polynomials $f_n$ converging uniformly to $f$ in the range of $u$. Then $f_n \circ u$ converges in $BC$ to $f \circ u$ and so $f \circ u \in AP(\Omega)$. It remains to prove that the composition property also holds when $f$ depends on $t$.

Given $\epsilon > 0$ we shall find $p_\epsilon \in AP(\Omega)$ such that

$$||f(\cdot, u(\cdot)) - p_\epsilon||_\infty < \epsilon.$$

Fix $R > ||u||_\infty$. As $\{f(t, \cdot)\}_{t \in \mathbb{R}}$ is equicontinuous in $[-R, R]$, we find $\delta > 0$ such that

$$|f(t, u) - f(t, v)| < \epsilon \quad \text{if} \quad t \in \mathbb{R}, |u - v| < \delta, |u|, |v| \leq R.$$

Next we find a partition $u_0 = -R < u_1 < \cdots < u_n = R$ such that $u_{i+1} - u_i < \delta$ for each $i$. We can construct a partition of unity on $[u_0, u_n]$ as follows. The functions $\chi_0, \chi_1, \dots, \chi_n$ are continuous on $[-R, R]$ and satisfy

$$\sum_{i=1}^n \chi_i(u) = 1, \quad \chi(u) \geq 0 \quad \text{for each} \quad u \in [-R, R],$$

$$\text{supp } \chi_i \subset [u_{i-1}, u_{i+1}], \quad \text{with the convention} \quad u_{-1} = u_0, u_{n+1} = u_n.$$

Finally we define

$$p_\epsilon(t) = \sum_{i=0}^n f(t, u_i) \chi_i(u(t)).$$

From the above discussions we know that $\chi_i \circ u$ belongs to $AP(\Omega)$. The structure of algebra and the condition (3.2) imply that $p_\epsilon \in AP(\Omega)$. The proof is complete, for it is clear that $||f(\cdot, u(\cdot)) - p_\epsilon||_\infty$ is less than $\epsilon$. $\qquad \square$

As in the case of second order equations we use an auxiliary linear equation. Given $k > 0$ we consider

$$\dot{u} = -ku + p(t), \quad p \in AP(\Omega).$$

This equation has a unique bounded solution given by

$$u(t) = \int_{-\infty}^{\infty} \mathcal{G}(t,s)p(s)ds$$

where

$$\mathcal{G}(t,s) = \begin{cases} 0 & \text{if } t \leq s \\ e^{k(s-t)} & \text{if } s \leq t. \end{cases}$$

Now it is easy to prove that $u \in AP(\Omega)$. First we observe that if $p_1, p_2$ are two functions in $AP(\Omega)$ then

$$||u_1 - u_2||_\infty \leq \frac{1}{k}||p_1 - p_2||_\infty.$$

By direct computation we observe that if $p(t)$ is a trigonometric polynomial with frequencies in $\Omega$ then $u(t)$ is in the same class.

The equation (3.1) is rewritten as

$$\dot{u} = -ku + g(t,u), \quad g(t,u) := f(t,u) + ku,$$

and the $\Omega$-problem becomes equivalent to

$$u = \mathcal{F}u,$$

with

$$\mathcal{F} : AP(\Omega) \rightarrow AP(\Omega), \quad \mathcal{F}u(t) = \int_{-\infty}^{\infty} \mathcal{G}(t,s)g(s,u(s))ds.$$

We know that degree theory can be applied to this operator when $\Omega$ is a cyclic group. Next we show that this is the only possible case.

**Theorem 3.2.** *Assume that $\Omega$ is not cyclic. Then there exist $f \in UAP$ satisfying (3.2) and $k > 0$ such that the associated operator $\mathcal{F}$ has no fixed points and*

$$\mathcal{F}\left(\overline{B}\right) \subset B$$

*where*

$$B = \{u \in AP(\Omega)/||u||_\infty < 1\}.$$

We will find a function $f$ and numbers $\alpha < \beta$ such that

$$f(\cdot, \alpha) \gg 0 \gg f(\cdot, \beta)$$

and (3.1) has no solutions in $AP$. From there the rest of the proof follows along the lines of Theorem 2.1. Yet we need to do some work to construct such an equation.

### 3.4. Primitives of functions in $AP(\Omega)$

Given a function $a \in C_T$ the primitive can be expressed as $A(t) = \bar{a}t + \tilde{A}(t)$, where $\bar{a}$ is the average and $\tilde{A}$ is $T$-periodic. The next result implies that such a result cannot be extended to any group $\Omega$ which is not cyclic.

**Lemma 3.3.** *Assume that $\Omega$ is not cyclic. Then there exists $a \in AP(\Omega)$ such that its primitives satisfy*

$$A(t) \rightarrow -\infty \quad as \quad |t| \rightarrow \infty.$$

*Proof.* It follows along the lines of Bohr's example (see [15], page 157). As $\Omega$ is not cyclic it must be dense in $\mathbb{R}$. In particular, for each integer $n \geq 1$ it is possible to find $\omega_n \in \Omega$ with

$$n^{-2/3} \leq \omega_n \leq 2n^{-2/3}.$$

Define

$$a(t) = \sum_{n=1}^{\infty} \omega_n^2 \sin \omega_n t.$$

The function is in $AP(\Omega)$ and the primitive with $A(0) = 0$ is given by

$$A(t) = \sum_{n=1}^{\infty} \omega_n (1 - \cos \omega_n t) = 2 \sum_{n=1}^{\infty} \omega_n \sin^2 \left( \frac{\omega_n t}{2} \right).$$

The inequality $|\sin x| \geq \frac{1}{2}|x|$ if $|x| \leq 1$ imply that

$$A(t) \geq \frac{t^2}{8} \sum_{n \in I(t)} \omega_n^3,$$

where $I(t) = \{n \in \mathbb{N}/n \geq 1,\ \omega_n |t| \leq 2\}$. The numbers satisfying $n \geq |t|^{3/2}$ are in the set $I(t)$ and so

$$A(t) \geq \frac{t^2}{8} \sum_{n \geq |t|^{3/2}} \frac{1}{n^2} \geq \frac{t^2}{8} \int_{|t|^{3/2}+1}^{\infty} \frac{ds}{s^2} \to \infty. \qquad \square$$

### 3.5. Linear equations, homoclinic solutions, and proof of Theorem 3.2

The function $a$ constructed in the previous lemma is such that all the solutions of

$$\dot{u} = a(t)u$$

are homoclinic to zero. This allows us to apply Theorem 2 in [14] and obtain the following result.

**Proposition 3.4.** *Assume that $\Omega$ is not cyclic. Then there exist functions $a, b \in AP(\Omega)$ such that for the linear equation*

$$\dot{u} = a(t)u + b(t)$$

*all the solutions are bounded but none of them is almost periodic.*

We can now apply the ideas of Section 5 in [13] and construct an equation of the type

$$\dot{u} = a(t)u + b(t) + D(u)$$

without almost periodic solutions. In this construction the function $D$ is $C^{\infty}$ and such that $-D(u)$ dominates the linear part at infinity. This implies that $h(t, u) := a(t)u + b(t) + D(u)$ satisfies

$$h(\cdot, -R) \gg 0 \gg h(\cdot, R)$$

for large $R$. Since $h$ also satisfies (3.2) the proof of Theorem 3.2 follows.

# References

[1] P. Bohl, *Sur certaines équations différentielles d'un type général utilizables en Mécanique.* Bull. Soc. Math. France **38** (1910), 5–138.

[2] C. De Coster, P. Habets, *Two-point boundary value problems: lower and upper solutions.* Elsevier, 2006.

[3] W. A. Coppel, *Dichotomies in stability theory, Lecture Notes in Mathematics* 629. Springer, 1978.

[4] C. Corduneanu, *Almost periodic functions.* John Wiley, 1968.

[5] A. M. Fink, *Almost periodic differential equations, Lecture Notes in Mathematics, Vol.* 377. Springer, 1974.

[6] A. M. Fink, P. Frederickson, *Ultimate boundedness does not imply almost periodicity.* J. Differential Equations **9** (1971), 280–284.

[7] R. Johnson, *A linear almost periodic equation with an almost automorphic solution.* Proc. Am. Math. Soc. **82** (1981), 199–205.

[8] M. A. Krasnosels'kii, V. S. Burd, Y. S. Kolesov, *Nonlinear almost periodic oscillations.* John Wiley, 1973.

[9] M. A. Krasnosels'kii, P. P. Zabreiko, *Geometrical methods of Nonlinear Analysis.* Springer-Verlag, 1984.

[10] J. Mawhin, *Topological degree methods in nonlinear boundary value problems* (CBMS Series in Mathematics no. 40). American Math. Soc., 1979.

[11] J. Moser, *On the theory of quasiperiodic motions.* SIAM Review **8** (1968), 145–172.

[12] Z. Opial, *Sur une équation différentielle presque-périodique sans solution presque-périodique.* Bull. Acad. Polon. Sci. Ser. Math. Ast. Phys. **IX** (1961), 673–676.

[13] R. Ortega, M. Tarallo, *Almost periodic upper and lower solutions.* J. Differential Equations **193** (2003), 343–358.

[14] R. Ortega, M. Tarallo, *Almost periodic linear differential equations with non-separated solutions.* J. Functional Analysis **237** (2006), 402–426.

[15] V. V. Zhikov, B. M. Levitan, *Favard theory.* Russian Math. Surveys **32** (1977), 129–180.

Rafael Ortega
Departamento de Matemática Aplicada
Facultad de Ciencias
Universidad de Granada
18071 Granada, Spain
e-mail: rortega@ugr.es

Progress in Nonlinear Differential Equations
and Their Applications, Vol. 75, 357–376
© 2007 Birkhäuser Verlag Basel/Switzerland

# $L^\infty$-Energy Method, Basic Tools and Usage

Mitsuharu Ôtani

*Dedicated to Professors Arrigo Cellina and James A. Yorke
on the occasion of their $65^{th}$ birthday*

**Abstract.** A new method called "$L^\infty$-Energy Method" is introduced. Several basic tools for this method are prepared and a couple of typical ways of usage of this method are exemplified for quasilinear parabolic equations.

**Mathematics Subject Classification (2000).** Primary 35K40; Secondary 35K65.

**Keywords.** $L^\infty$-space, energy method, quasilinear parabolic equations.

## 1. Introduction

It would be generally recognized that in the study of the nonlinear partial differential equations, the choice of the function spaces where one works is crucial for its high achievement. Normally such suitable function spaces should be carefully chosen according to the various nature of the equations to be considered and clearly there is no almighty function space good for all equations.

As for the energy method for the nonlinear parabolic equations, $L^2$-space or $L^p$-space $(1 \le p < \infty)$ is frequently used. On the other hand, $L^\infty$-space is not so often direcly used and it would be widely believed that it possesses no advantage over $L^p$-spaces because of its non-reflexivity and non-separability. So $L^\infty$-estimates are usually derived from $W^{m,p}$-estimates via Sobolev's embedding theorem.

However, the main purpose of this paper is to point out that as far as the uniqueness and local existence of solutions (in time) are concerned, $L^\infty$-space could be the most suitable space for a rather large class of nonlinear parabolic equations and systems, if we are equipped with "$L^\infty$-Energy Method".

Partially supported by Waseda University Grant for Special Research Projects #2003B-27 and the Grant-in-Aid for Scientific Research, #16340043 and #18654031, the Ministry of Education, Culture, Sports, Science and Technology, Japan.

This "$L^\infty$-Energy Method" is a device which makes it possible to derive energy estimates directly in $L^\infty$-spaces even when any energy methods could not work in $L^r$-spaces with $1 \le r < +\infty$.

The main objective of this paper is twofold. One is to fix some basic tools needed in carrying out $L^\infty$-Energy Method concerning the fundamental property of the $L^\infty$-norm and some implements to enable us to derive a priori bound or a lower bound for the blowing-up rate of the $L^\infty$-norm for solutions. The other one is to expose its basic idea and to exemplify its typical way of usage by using simple examples, i.e., the perturbation problems for the porous medium equations and the parabolic equations generated by the so-called p-Lapalacian.

Our $L^\infty$-Energy Method is quite useful not only for the case where the nonlinearity obstructing the energy estimates in $L^r$-spaces depends only on the solution $u$ itself but also for the case where the nonlinearity is composed of up to the first derivatives with respect to space variables such as parabolic systems with the nonlinearity similar to that of Navier-Stokes equations.

This method also provides us a very powerful tool in analyzing very complex systems of nonlinear parabolic equations such as quasilinear system of Chemotaxis and some doubly nonlinear parabolic equations.

These significant features will be discussed in our forthcoming paper [14]. (See also [13] and [9].)

In this paper we do not touch upon the technical aspects related with the derivation of a priori bounds in $L^\infty$-spaces for the space derivatives of solutions. However, such examples can be found in [12, 15] and [16].

## 2. Basic tools for $L^\infty$-Energy Method

In this section, we collect some basic tools needed in the procedure of the performing $L^\infty$-Energy Method.

In establishing the a priori bounds in $L^\infty$-spaces, our method provides the scheme which enables us to obtain a priori bounds directly in $L^\infty$-spaces without relying on any $W^{m,p}$-estimates nor Sobolev's embedding theorems. To realize this program, we utilize the following fact, which is well known for the case where the total measure $m(\Omega)$ of the domain $\Omega$ is finite and the function $u$ is known to be in $L^\infty(\Omega)$, (see Theorem 1, §3, Capt.1 of [17]).

**Lemma 2.1.** *Let $\Omega$ be any domain in $\mathbb{R}^N$ and assume that there exist a number $r_0 \ge 1$ and a constant $C$ independent of $r \in [r_0, \infty)$ such that*

$$|u|_{L^r(\Omega)} \le C \qquad \forall r \in [r_0, \infty),  \tag{2.1}$$

*then $u$ belongs to $L^\infty(\Omega)$ and the following property holds.*

$$\lim_{r \to \infty} |u|_{L^r(\Omega)} = |u|_{L^\infty(\Omega)}.  \tag{2.2}$$

*Conversely, assume that $u \in L^{r_0}(\Omega) \cap L^\infty(\Omega)$ for some $r_0 \in [1, \infty)$, then $u$ satisfies (2.2).*

*Proof.* We are going to show below that (2.1) assures (2.2).

**Case 1:** $m(\Omega) < +\infty, \quad u \in L^\infty(\Omega).$

For the completeness of the presentation, we here repeat the proof due to [17]. By the definition of $|u|_{L^\infty}$, for any $\varepsilon > 0$, there exists a null set $e \in \mathcal{N}$ such that $|u(x)| \leq |u|_{L^\infty} + \varepsilon \; \forall x \in \Omega \backslash e$. Then we get

$$\left( \int_\Omega |u(x)|^r dx \right)^{1/r} \leq \left( \int_{\Omega \backslash e} (|u(x)|_{L^\infty} + \varepsilon)^r dx \right)^{1/r}$$

$$\leq (|u|_{L^\infty} + \varepsilon) \, m(\Omega)^{1/r}.$$

Hence we obtain

$$\limsup_{r \to \infty} |u|_{L^r} \leq |u|_{L^\infty}. \tag{2.3}$$

Put $\Omega_\varepsilon = \{x \in \Omega; |u(x)| > |u|_{L^\infty} - \varepsilon\}$, then by definition $m(\Omega_\varepsilon) > 0$ for any $\varepsilon \in (0, |u|_{L^\infty})$. Therefore

$$\left( \int_\Omega |u(x)|^r dx \right)^{1/r} \geq \left( \int_{\Omega_\varepsilon} |u(x)|^r dx \right)^{1/r}$$

$$\geq (|u|_{L^\infty} - \varepsilon) m(\Omega_\varepsilon)^{1/r}. \tag{2.4}$$

Hence

$$\liminf_{r \to \infty} |u|_{L^r} \geq |u|_{L^\infty}. \tag{2.5}$$

Thus (2.3) and (2.5) yield (2.2).

**Case 2:** $m(\Omega) < +\infty.$

For the case that there is no knowing if $u \in L^\infty(\Omega)$, we fix $n$ so that $C < n$ and put

$$[u]_n(x) = g_n(u) = \begin{cases} u(x) & \text{if} \quad |u(x)| \leq n, \\ n & \text{if} \quad u(x) \geq n, \\ -n & \text{if} \quad u(x) \leq -n. \end{cases} \tag{2.6}$$

Since $[u]_n \in L^\infty(\Omega)$ and $|[u]_n|_{L^r} \leq |u|_{L^r} \leq C$, the argument above assures that

$$\big|[u]_n\big|_{L^\infty} = \lim_{r \to \infty} |[u]_n|_{L^r} \leq C < n,$$

which implies that $[u]_n(x) = u(x)$ for a.e. $x \in \Omega$, whence follows $u \in L^\infty(\Omega)$.

**Case 3:** $m(\Omega) = +\infty.$

Let $\Omega_n = \Omega \cap \{x \in \mathbb{R}^N; ||x|| < n\}$. Noting that $|u|_{L^r(\Omega_n)} \leq |u|_{L^r(\Omega)} \leq C$ and applying the above result, we find that $u \in L^\infty(\Omega_n)$ and $|u|_{L^\infty(\Omega_n)} \leq C$. Hence, for any $\varepsilon > 0$ and $n \in \mathbb{N}$, there exists $e_n \in \mathcal{N}$ such that

$$|u(x)| \leq C + \varepsilon \qquad \forall x \in \Omega_n \backslash e_n.$$

Consequently, we get

$$|u(x)| \leq C + \varepsilon \qquad \forall x \in \Omega \backslash e = \bigcup_{n=1}^\infty (\Omega_n \backslash e_n), \quad e = \bigcup_{n=1}^\infty e_n \in \mathcal{N},$$

which implies $u \in L^\infty(\Omega)$. Hence

$$|u|_{L^r(\Omega)} = \left( \int_{\Omega \setminus e} |u(x)|^{r-r_0} |u(x)|^{r_0} dx \right)^{1/r} \leq (|u|_{L^\infty} + \varepsilon)^{(r-r_0)/r} |u|_{L^{r_0}}^{r_0/r}.$$

Now letting $r \to \infty$, we again obtain (2.3).

Here, for any sufficiently small $\varepsilon > 0$, we put $\Omega_\varepsilon = \{x \in \Omega; |u(x)| > |u|_{L^\infty} - \varepsilon > 0\}$. Then the fact that $u \in L^{r_0}(\Omega)$ together with the definition of the $L^\infty$-norm assures that $0 < m(\Omega_\varepsilon) < +\infty$. Hence, by the same argument as in (2.4), we again obtain (2.5). Thus combining (2.3) and (2.5), we conclude (2.2).

Conversely, assume that $u \in L^{r_0}(\Omega) \cap L^\infty(\Omega)$ for some $r_0 \in [1, \infty)$, then we easily get $|u|_{L^r} \leq |u|_{L^{r_0}}^{r_0/r} |u|_{L^\infty}^{(r-r_0)/r} \leq C = \max(|u|_{L^{r_0}}, |u|_{L^\infty})$ $\forall r \in [r_0, \infty]$, which implies (2.1). Then the above arguments assure that (2.2) holds true.    □

As will be exemplified in the next section, with the help of the above result, it is often possible to see that $y(t) = |u(t)|_{L^\infty}$ satisfies a certain type of integral inequalities, which control the growth of $y(t)$. The following result is the implement which enables us to derive a priori bound for $y(t)$ from such a type of integral inequalities.

**Lemma 2.2.** *Let $y(t)$ be a bounded measurable non-negative function on $[0, T]$ and suppose that there exist $y_0 \geq 0$ and a monotone non-decreasing function $m(\cdot)$ : $[0, +\infty) \to [0, +\infty)$ such that*

$$y(t) \leq y_0 + \int_0^t m\big(y(s)\big)ds \qquad a.e. \ t \in (0, T). \tag{2.7}$$

*Then there exists a number $T_0 = T_0(y_0, m(\cdot)) \in (0, T]$ such that*

$$y(t) \leq y_0 + 1 \qquad a.e. \ t \in [0, T_0]. \tag{2.8}$$

*Proof.* Put $z(t) = y_0 + \int_0^t m(y(s))ds$, then $z(t) \in C([0, T]; \mathbb{R}^1)$ and $y(t) \leq z(t)$. So $z(t)$ satisfies

$$z(t) \leq y_0 + \int_0^t m\big(z(s)\big)ds \qquad \text{for all } t \in [0, T]. \tag{2.9}$$

We here claim that

$$z(t) \leq y_0 + 1 \qquad \text{for all } t \in [0, T_0], \quad T_0 = \min\left( \frac{1}{2m(y_0 + 1)}, T \right). \tag{2.10}$$

In fact, suppose that (2.10) does not hold, i.e., there exists $t_0 \in (0, T_0]$ such that $z(t_0) > y_0 + 1$, then since $z(t)$ is continuous on $[0, T]$ and $z(0) < y_0 + 1$, there exists $t_1 \in (0, t_0)$ such that $z(t_1) = y_0 + 1$ and $z(t) < y_0 + 1$ $\forall t \in [0, t_1)$. Then, by (2.9), we get

$$y_0 + 1 = z(t_1) \leq y_0 + \int_0^{t_1} m\big(z(s)\big)ds$$

$$\leq y_0 + m(y_0 + 1)T_0 \leq y_0 + \frac{1}{2},$$

which leads to a contradiction. Thus (2.10) is verified and hence (2.8) is derived from the fact that $y(t) \leq z(t)$ for all $t \in [0, T]$. □

If the function $m(y)$ given in the above lemma turns out to have at most linear growth order as $y \to +\infty$, we easily obtain time global a priori bound for $y(t) = |u(t)|_{L^\infty}$, which would lead us to the global existence of solutions. On the other hand, if it is not the case, the blowing-up of solution may occur at some finite time $T_m$. Even for such a case, it is also posssible to obtain some information on the blow up rate of solutions in terms of the $L^\infty$-norm within the framework of our method by the following results.

**Lemma 2.3.** *Let $y(t)$ be a bounded measurable positive function on $[0, T]$ for any $T \in (0, T_m)$ and let $\lim_{t \to T_m} y(t) = +\infty$. Suppose that there exists a monotone non-decreasing locally Lipschitz function $g : [0, +\infty) \to [0, +\infty)$ such that*

$$\int_0^{+\infty} \frac{1}{g(\tau)} d\tau = +\infty, \tag{2.11}$$

$$\int_a^{+\infty} \frac{1}{g(\tau)} d\tau < +\infty \quad \forall a > 0. \tag{2.12}$$

*Furthermore we assume*

$$y(s) \leq y(t) + \int_t^s g\big(y(\tau)\big) d\tau \qquad a.e.\ t, s \in [0, T_m)\ with\ t < s. \tag{2.13}$$

*Then the following estimate holds.*

$$y(t) \geq G^{-1}(t - T_m) \qquad for\ a.e.\ t \in [0, T_m), \tag{2.14}$$

*where $G^{-1}(\cdot)$ is the inverse function of $G(w) = -\int_w^{+\infty} \frac{1}{g(\tau)} d\tau$.*

*Proof.* Put $z(t, s) = y(t) + \int_t^s g(y(\tau)) d\tau$.
Then by (2.13), we get $y(s) \leq z(t, s)\ \forall s \in [t, T_m)$ and

$$z(t, s) \leq y(t) + \int_t^s g\big(z(t, \tau)\big) d\tau.$$

For any $t \in [0, T_m)$, let $t_0 = t$ and let $t_n$ be the first time in $(t_{n-1}, T_m)$ such that $y(t_n) = 2y(t_{n-1})$ for $n \in \mathbb{N}$. Then it is clear that $t_n \uparrow T_m$ as $n \to \infty$. Here we define another function $w_n(s)$ by the unique solution of the following problem.

$$dw_n(s)/ds = g\big(w_n(s)\big) \quad s \in (t_n, T_m), \quad w_n(t_n) = y(t_n). \tag{2.15}$$

Since $w_n(s)$ satisfies $w_n(s) = y(t_n) + \int_{t_n}^s g(w_n(\tau)) d\tau$, by comparison theorem, it is easy to see that $y(s) \leq z(t_n, s) \leq w_n(s)\ \forall s \in [t_n, T_m)$. Denote by $s_n$ the first time in $(t_n, T_m)$ such that $w_n(s_n) = y(t_{n+1}) = 2y(t_n)$, then we find that $t_n < s_n \leq t_{n+1}$. Here we put

$$G(w) = -\int_w^{+\infty} \frac{1}{g(\tau)} d\tau. \tag{2.16}$$

Then $G$ is a strictly monotone increasing continuous function from $[0, +\infty]$ onto $[-\infty, 0]$ such that $G(0) = -\infty, G(+\infty) = 0$ and $G'(w) = 1/g(w)$.

Therefore, we can integrate (2.15) in terms of $G$ on $[t_n, s_n]$ and get

$$t_{n+1} - t_n \geq s_n - t_n = G\big(w_n(s_n)\big) - G\big(w_n(t_n)\big) = G\big(y(t_{n+1})\big) - G\big(y(t_n)\big). \quad (2.17)$$

Summing up (2.17) from $n = 0$ up to $n = \infty$, we obtain $T_m - t \geq -G(y(t))$, since $G(y(t_{n+1})) \to G(\infty) = 0$ as $n \to +\infty$. Hence we get $G(y(t)) \geq (t - T_m)$, which implies (2.14). □

The dual version of Lemma 2.3 is stated as follows.

**Lemma 2.4.** *Let $y(t)$ be a bounded measurable positive function on $[0, T]$ for any $T \in (0, T_m)$ and let $\lim_{t \to T_m} y(t) = +\infty$. Suppose that there exists a monotone non-decreasing locally Lipschitz function $g : [0, +\infty) \to [0, +\infty)$ satisfying (2.11) and (2.12). Furthermore we assume*

$$y(s) \geq y(t) + \int_t^s g\big(y(\tau)\big) d\tau \qquad a.e.\ t, s \in [0, T_m)\ with\ t < s. \quad (2.18)$$

*Then the following estimate holds.*

$$y(t) \leq G^{-1}(t - T_m) \qquad for\ a.e.\ t \in [0, T_m), \quad (2.19)$$

*where $G^{-1}(\cdot)$ is the inverse function of $G(w) = -\int_w^{+\infty} \frac{1}{g(\tau)} d\tau$.*

*Proof.* Define $z(t, s)$ and $G(w)$ as in the proof of Lemma 2.3, and define $w(t, s)$ by the unique solution of the problem :

$$dw(t, s)/ds = g\big(w(t, s)\big) \quad s \in (t, T_m), \quad w(t, t) = y(t). \quad (2.20)$$

Then (2.18) assures $w(t, s) \leq z(t, s) \leq y(s) \quad \forall s \in [t, T_m)$. Hence, in stead of (2.17), by integrating (2.20) on $(t, s)$, we now have

$$s - t = G\big(w(t, s)\big) - G\big(w(t, t)\big) \leq G\big(y(s)\big) - G\big(y(t)\big) \quad \forall s \in (t, T_m). \quad (2.21)$$

Then, by letting $s \to T_m$ in (2.21), we obtain (2.19). □

If we put $g(r) = \lambda|r|^{q-2}r$ with $q \in (2, +\infty)$, then clearly $g$ is a monotone increasing locally Lipschitz continuous function and it is easy to see that $g$ satisfies the required conditions (2.11) and (2.12) in Lemma 2.3. Furthermore, by a simple calculation, we easily get $G(w) = \frac{1}{\lambda(2-q)} w^{2-q}$ and $G^{-1}(\tau) = (\lambda(2-q)\tau)^{\frac{-1}{q-2}}$. Thus we have the following two corollaries of Lemma 2.3 and Lemma 2.4.

**Corollary 2.5.** *Let $y(t)$ be a bounded measurable positive function on $[0, T]$ for any $T \in (0, T_m)$ and let $\lim_{t \to T_m} y(t) = +\infty$. Suppose that $y(\cdot)$ satisfies*

$$y(t) \leq y(s) + \lambda \int_s^t y(\tau)^{q-1} d\tau \qquad a.e.\ s, t \in [0, T_m)\ with\ s < t. \quad (2.22)$$

*Then the following estimate holds.*

$$y(t) \geq \big(\lambda(q - 2)\big)^{\frac{-1}{q-2}} (T_m - t)^{\frac{-1}{q-2}} \qquad for\ a.e.\ t \in [0, T_m). \quad (2.23)$$

**Corollary 2.6.** *Let $y(t)$ be a bounded measurable positive function on $[0, T]$ for any $T \in (0, T_m)$ and let $\lim_{t \to T_m} y(t) = +\infty$. Suppose that $y(\cdot)$ satisfies*

$$y(t) \geq y(s) + \lambda \int_s^t y(\tau)^{q-1} d\tau \qquad a.e. \ s, t \in [0, T_m) \ with \ s < t. \qquad (2.24)$$

*Then the following estimate holds.*

$$y(t) \leq \left(\lambda(q-2)\right)^{\frac{-1}{q-2}} (T_m - t)^{\frac{-1}{q-2}} \qquad for \ a.e. \ t \in [0, T_m). \qquad (2.25)$$

**Remark 2.7.** Above results give informations not only on the asymptotic behavior of $y(t)$ near $t = T_m$ but also on the a priori bounds for the blowing up time $T_m$. In fact, by putting $t = 0$ in (2.14), (2.19), (2.23) and (2.25), we can derive the estimates $T_m \geq -G(y(0))$, $T_m \leq -G(y(0))$, $T_m \geq 1/\left((\lambda(q-2)y(0)^{q-2})\right)$ and $T_m \leq 1/\left((\lambda(q-2)y(0)^{q-2})\right)$ respectively.

# 3. What is $L^\infty$-Energy Method

In this section, in order to describe the basic idea of $L^\infty$-Energy Method, we consider the following simple but typical quasilinear parabolic equations.

$$(\text{P}) \quad \begin{cases} u_t = \text{div}\left(|\nabla u|^{p-2}\nabla u\right) + |u|^{q-2}u & (x,t) \in \Omega \times (0, \infty), \\ u|_{\partial\Omega} = 0 \quad t \in (0, \infty), \qquad u(x,0) = u_0(x) \quad x \in \Omega, \end{cases}$$

$$(\text{E}) \quad \begin{cases} u_t = \text{div}\left(|u|^{\ell-2}\nabla u\right) + |u|^{q-2}u & (x,t) \in \Omega \times (0, \infty), \\ u|_{\partial\Omega} = 0 \quad t \in (0, \infty), \qquad u(x,0) = u_0(x) \quad x \in \Omega, \end{cases}$$

where $1 < p < \infty$, $2 < q < \infty$, $1 < \ell < \infty$ and $\Omega$ is a general domain in $\mathbb{R}^N$.

Then our method can provide the following results.

**Theorem 3.1.** *Suppose that $u_0 \in L^2(\Omega) \cap L^\infty(\Omega)$. Then there exists a positive number $T_0$ depending only on $|u_0|_{L^\infty}$ such that (P) admits a unique solution $u$ satisfying*

$$u \in C\left([0, T_0]; L^2(\Omega)\right) \cap L^\infty\left(0, T_0; L^\infty(\Omega)\right)$$
$$\cap L^p\left(0, T_0; W_0^{1,p}(\Omega)\right) \cap W_{loc}^{1,2}\left((0, T_0]; L^2(\Omega)\right). \qquad (3.1)$$

*Furthermore, the maximal existence time $T_m = \sup\{T_0 \in (0, +\infty); (\text{P}) \ admits \ a \ solution \ u \ on \ [0, T_0] \ satisfying \ (3.1)\}$ is finite if and only if $\lim_{t \to T_m} |u(t)|_{L^\infty(\Omega)} = +\infty$, more precisely,*

$$|u(t)|_{L^\infty(\Omega)} \geq (q-2)^{\frac{-1}{q-2}} (T_m - t)^{\frac{-1}{q-2}} \qquad for \ a.e. \ t \in [0, T_m). \qquad (3.2)$$

**Theorem 3.2.** *Suppose that $u_0 \in L^2(\Omega) \cap L^\ell(\Omega) \cap L^\infty(\Omega)$. Then there exists a positive number $T_0$ depending only on $|u_0|_{L^\infty}$ such that* (E) *admits a solution* $u$ *satisfying*

$$u \in L^\infty\left(0, T_0; L^2(\Omega) \cap L^\ell(\Omega) \cap L^\infty(\Omega)\right) \cap W^{1,2}\left([0, T_0]; H^{-1}(\Omega)\right) \tag{3.3}$$
$$\text{and} \quad |u|^{(\ell-2)/2}|\nabla u| \in L^2\left(0, T_0; L^2(\Omega)\right),$$

*where $H^{-1}(\Omega)$ denotes the dual space of $H_0^1(\Omega)$.*
*Here if $u_0 \in L^1(\Omega) \cap L^\infty(\Omega)$, then $u$ satisfies*

$$u \in L^\infty\left(0, T_0; L^1(\Omega) \cap L^\infty(\Omega)\right) \cap W^{1,2}\left([0, T_0] : H^{-1}(\Omega)\right) \tag{3.4}$$
$$\text{and} \quad |u|^{(\ell-2)/2}|\nabla u| \in L^2\left(0, T_0; L^2(\Omega)\right).$$

*Furthermore, the maximal existence time $T_m = \sup \{T \in (0, +\infty)$ ;* (E) *admits a solution $u$ on $[0, T_0]$ satisfying* (3.3) $\}$ *is finite if and only if $\lim_{t \to T_m} |u(t)|_{L^\infty} = +\infty$, more precisely,*

$$|u(t)|_{L^\infty} \geq (q-2)^{\frac{-1}{q-2}}(T_m - t)^{\frac{-1}{q-2}} \quad \text{for a.e. } t \in [0, T_m). \tag{3.5}$$

As for the uniqueness of solutions of (E), the following result holds.

**Theorem 3.3.** *Let $u_0 \in L^1(\Omega)$, then the solution of* (E) *satisfying* (3.4) *is unique.*

### 3.1. Proof of Theorem 3.1

In carrying out $L^\infty$-Energy Method for degenerate or singular quasilinear parabolic systems, we always need the approximate procedure, for which the following "cut-off technic" is quite often found to be very useful.

By using the same notation $[u]_n = g_n(u)$ as in (2.6), we introduce the following approximate problem $(P)_n$ for (P).

$$(P)_n \begin{cases} u_t = \text{div}\left(|\nabla u|^{p-2}\nabla u\right) + |[u]_n|^{q-2}u & (x, t) \in \Omega \times (0, \infty), \\ u|_{\partial\Omega} = 0 \quad t \in (0, \infty), \quad u(x, 0) = u_0(x) \quad x \in \Omega. \end{cases}$$

Let

$$X_p := \left\{u \in L^2(\Omega); \nabla u \in \left(L^p(\Omega)\right)^N\right\}$$

with

$$|u|_{X_p} := \left\{|u|_{L^2(\Omega)}^p + |\nabla u|_{L^p(\Omega)}^p\right\}^{1/p}$$

for all $u \in X_p$. Moreover let $V_p := \overline{C_0^\infty(\Omega)}^{X_p}$ with $|\cdot|_{V_p} := |\cdot|_{X_p}$. Then we find that $V_p$ is a uniformly convex Banach space, since $V_p$ is a closed subspace of $X_p$ which is a uniformly convex Banach space (see [1, 1.21, 22]). Moreover from the definition of $V_p$, it is easily seen that $V_p$ is embedded in $L^2(\Omega)$ with continuous injection. Furthermore we can verify that $V_p$ is dense in $L^2(\Omega)$.

We now define a function $\phi_p$ on $L^2(\Omega)$ by setting

$$\phi_p(u) = \begin{cases} \frac{1}{p}\int_\Omega |\nabla u(x)|^p dx + \frac{1}{2}\int_\Omega |u(x)|^2 dx & \text{if} \quad u \in V_p \,, \\ +\infty & \text{if} \quad u \in L^2(\Omega)\backslash V_p \,. \end{cases}$$

Then $\phi_p$ is obviously convex and the reflexivity of $V_p$ assures that $\phi_p$ is lower semicontinuous from $L^2(\Omega)$ into $[0, +\infty]$.

Furthermore, it follows that $\overline{D(\phi_p)}^{L^2} = \overline{V_p}^{L^2} = L^2(\Omega)$ and $\partial\phi_p(u) = -\Delta_p u + u = -\text{div}(|\nabla u|^{p-2}\nabla u) + u$ (see [2, p.53, Example 1]).

Since the map $B(u) : u \mapsto |[u]_n|^{q-2}u$ becomes Lipschitz continuous from $L^2(\Omega)$ into $L^2(\Omega)$, $(\text{P})_n$ can be regarded as the evolution equation in $L^2(\Omega)$ governed by the subdifferential operator $\partial\phi_p(u)$ with the Lipschitz perturbation $B(u) - u$. Then, by the standard argument from the maximal monotone operator theory, it is easy to see that for any $u_0 \in L^2(\Omega)$, $(\text{P})_n$ admits the unique global solution $u$ satisfying (3.1) with $T_0$ replaced by any $T \in (0, +\infty)$ (see [2] and [11]).

We here put $v^n(t) = e^{-n^{q-2}t}u(t)$, then $v^n$ satisfies

$$v_t^n = e^{(p-2)n^{q-2}t}\Delta_p v^n + \left(|g_n(u)|^{q-2} - n^{q-2}\right)v^n, \quad v^n(0) = u_0 \,.$$

Multiplying this equation by $[v^n(t) - n]^+ = \max(v^n(t) - n, 0)$, we easily see that $|[v^n(t) - n]^+|_{L^2} \le |[u_0 - n]^+|_{L^2}$ for a.e. $t \in [0, \infty)$, since $|g_n(u)|^{q-2} \le n^{q-2}$. Hence if $u_0 \in L^\infty(\Omega)$, fixing $n$ so that $|u_0|_{L^\infty} < n - 1$, we find that $|v^n(t)| \le n$, i.e., $u(t) \le ne^{n^{q-2}t}$. In particular, we observe that $u(t) \in L^\infty$ for a.e. $t \in [0, \infty)$.

Noticing that $|u|^{r-2}u \in L^2(\Omega)$ and $||[u]_n|^{q-2}u| \le |u|^{q-1}$, we multiply $(\text{P})_n$ by $|u|^{r-2}u$ to obtain

$$\frac{1}{r}\frac{d}{dt}|u(t)|_{L^r}^r + (r-1)\int_\Omega |\nabla u|^p |u|^{r-2} dx \le \int_\Omega |u|^{r+q-2} dx \,, \tag{3.6}$$

By factoring out $|u|_{L^\infty}^{q-2}$ from the right-hand side of (3.6), we get

$$\frac{1}{r}\frac{d}{dt}|u(t)|_{L^r}^r \le |u(t)|_{L^\infty}^{q-2}|u(t)|_{L^r}^r \,.$$

Divide both sides by $|u(t)|_{L^r}^{r-1}$ and integrate with respect to $t$ on $[0, T]$, then we get

$$|u(t)|_{L^r} \le |u_0|_{L^r} + \int_0^t |u(s)|_{L^\infty}^{q-2}|u(s)|_{L^r} ds \,.$$

Letting $r$ tend to $\infty$, we deduce by Lemma 2.1,

$$|u(t)|_{L^\infty} \le |u_0|_{L^\infty} + \int_0^t |u(s)|_{L^\infty}^{q-1} ds \,. \tag{3.7}$$

Here it should be noted that it is impossible to derive the estimate such as (3.7) in $L^p$-spaces$(1 \le p < \infty)$ from (3.6).

Thus applying Lemma 2.2, we find that there exist a positive number $T_0$ depending only on $|u_0|_{L^\infty}$ such that

$$|u(t)|_{L^\infty} \le |u_0|_{L^\infty} + 1 \le n \quad \text{for a.e. } t \in [0, T_0] \,.$$

Hence, by the definition of $[u]_n$, $u$ gives a solution of (P) on $[0, T_0]$.

As for the uniqueness of solutions, we multiply the difference of two equations for two solutions $u_1$ and $u_2$ by $w = u_1 - u_2$, then by the monotonicity of $u \mapsto -\Delta_p u$ and a priori bound for $|u|_{L^\infty}$, we easily deduce that $d|w(t)|_{L^2}^2/dt \le C|w(t)|_{L^2}^2$, whence follows the uniqueness at once.

Suppose here that $\lim_{t \to T_m} |u(t)|_{L^\infty} = +\infty$ does not hold, then there exist $C_0 > 0$ and $\{t_n\}$ such that $t_n \uparrow T_m$ and $|u(t_n)|_{L^\infty} \le C_0$. Since the dependence of $T_0$ given in Theorem 3.2 on the initial value $y_0 = |u_0|_{L^\infty}$ is monotone decreasing (see (2.10)), there exists a $T_0 > 0$ depending only on $C_0$ and $q$ but not on $n$ such that taking $u(t_n)$ as initial data, we can solve (E) in the interval $[t_n, t_n + T_0]$. Note that $t_n \uparrow T_m$ implies that there exists a sufficient large $n$ such that $t_n + T_0 > T_m$, i.e., we can extend the solution to the right of $T_m$, which leads to the contradiction to the definition of $T_m$.

In order to derive estimate (3.2), it suffices to apply Corollary 2.5 with $y(t) = |u(t)|_{L^\infty}$ to (3.7). $\qquad\square$

### 3.2. Proof of Theorem 3.2

As for the application of $L^\infty$-Energy Method to (E), it requires a more complicated approximation procedure than for (P). In fact, we prepare the approximate sequences $\Omega_n$ and $u_0^n$ such that $\Omega_n$ is bounded and approximating $\Omega$, $\partial\Omega_n$ and $u_0^n$ are sufficiently smooth, $u_0^n \to u_0$ in $L^2(\Omega)$ and $\limsup_{n \to \infty} |u_0^n|_{L^\infty(\Omega)} \le |u_0|_{L^\infty(\Omega)}$. We consider the following approximate equations.

$$(E)_n \quad \begin{cases} u_t = \Delta\left(\left(\dfrac{1}{n} + \varphi_n(u)\right)u\right) + |[u]_n|^{q-2}u \quad (x,t) \in \Omega_n \times (0, \infty), \\ u|_{\partial\Omega_n} = 0 \quad t \in (0, \infty), \qquad u(x,0) = u_0^n(x) \quad x \in \Omega_n, \end{cases}$$

where $\varphi_n(u) = \frac{1}{\ell-1}\left(u^2 + \frac{1}{n}\right)^{\frac{\ell-2}{2}}$.

Then the standard theory for quasilinear parabolic equations assures the existence of unique classical solutions $u_n$ of $(E)_n$ denoted by $u$ (see, e.g., [7, Theorem 12.14] or [6, Chapt. VI, §4]). In parallel with (3.6), multiplication of $(E)_n$ by $|u|^{r-2}u$ gives

$$\frac{1}{r}\frac{d}{dt}|u(t)|_{L^r(\Omega_n)}^r + (r-1)\int_{\Omega_n}\left(\frac{1}{n} + \varphi_n(u)\right)|u|^{r-2}|\nabla u|^2 dx$$

$$+ (r-1)\int_{\Omega_n}\frac{\ell-2}{\ell-1}(u^2 + \frac{1}{n})^{\frac{\ell-4}{2}}|u|^r|\nabla u|^2 dx \qquad (3.8)$$

$$\le \int_{\Omega_n}|u|^{r+q-2}dx.$$

Then as in the previous subsection, taking $n$ so that $|u_0|_{L^\infty} < n - 1$, we can conclude that there exist a positive number $T_0$ depending only on $|u_0|_{L^\infty}(\Omega)$ such that

$$|u(t)|_{L^\infty(\Omega_n)} \le |u_0|_{L^\infty(\Omega)} + 1 \le n \quad \text{for a.e. } t \in [0, T_0]. \qquad (3.9)$$

Hence, $u$ gives a solution of $(E)_n$ with $|[u]_n|^{q-2}[u]_n$ replaced by $|u|^{q-2}u$ and (3.8) with $r = 2$ implies

$$\sup_{0 \le t \le T_0} |u(t)|_{L^2(\Omega_n)} + \int_0^{T_0}\int_{\Omega_n} \left(\frac{1}{n} + \varphi_n(u)\right)|\nabla u|^2 dx dt \le C_0\,, \tag{3.10}$$

and (3.8) with $r = \ell \in (1, 2)$ yields

$$\sup_{0 \le t \le T_0} |u(t)|_{L^\ell(\Omega_n)} + \int_0^{T_0}\int_{\Omega_n} \left(u^2 + \frac{1}{n}\right)^{\ell-2}|\nabla u|^2 dx dt \le C_0\,, \tag{3.11}$$

where $C_0$ is a constant depending only on $|u_0|_{L^\infty(\Omega)}, |u_0|_{L^\ell(\Omega)}$ and $|u_0|_{L^2(\Omega)}$ but not on $n$.

Thus, in view of (3.9), (3.10) and (3.11), we find that

$$|u|_{L^\infty(0,T_0;L^r(\Omega_n))} \le C_0 \quad \forall r \in [\ell_*, \infty]\,, \quad \ell_* = \min(\ell, 2)\,, \tag{3.12}$$

$$|\nabla(\varphi_n(u)u)|_{L^2(0,T_0;L^2(\Omega_n))} \le C_0\,, \tag{3.13}$$

$$|\Delta(\varphi_n(u)u)|_{L^2(0,T_0;H^{-1}(\Omega_n))} \le C_0\,, \tag{3.14}$$

$$|\frac{1}{n}\Delta u|_{L^2(0,T_0;H^{-1}(\Omega_n))} \le C_0\frac{1}{\sqrt{n}}\,, \tag{3.15}$$

$$|u_t|_{L^2(0,T_0;H^{-1}(\Omega_n))} \le C_0\,. \tag{3.16}$$

In order to discuss the convergence of approximate solutions of $(E)_n$ denoted by $u_n$, we here need the following lemma due to Dubinskii [4] (for a proof, see [8, Theorem 12.1, Chapt. 1]).

**Lemma 3.4.** *Let $B$ and $B_1$ be Banach spaces such that $B$ is continuously embedded in $B_1$. Suppose that there exist a subset $S$ of $B$ and a function $M(\cdot)$ from $S$ into $[0, +\infty)$ such that $M(\lambda v) = |\lambda|M(v) \ \forall \lambda \in \mathbb{R} \ \forall v \in S$ and the set*

$$\{v; v \in S, M(v) \le 1\} \tag{3.17}$$

*is relatively compact in $B$.*

*Put*

$$\mathcal{F}_C = \left\{v\ ;\ v \in L^1_{loc}((0,T); B_1),\ \int_0^T M(v(t))^{p_0}dt \le C,\ |v_t|_{L^{p_1}(0,T;B_1)} \le C\right\}, \tag{3.18}$$

*where $p_0, p_1 \in (1, \infty)$ and $C \in (0, \infty)$. Then $\mathcal{F}_C \subset L^{p_0}(0,T; B)$ and $\mathcal{F}_C$ is relatively compact in $L^{p_0}(0,T; B)$ and in $C([0,T]; B_1)$.*

We apply this lemma with $B = L^p(\Omega_n)(p \ge \ell^* = \max(2, \ell))$, $B_1 = H^{-1}(\Omega_n)$, $S = \{v; |v|^{\frac{\ell-2}{2}}v \in H_0^1(\Omega_n), v \in L^\infty(\Omega_n)\}$, $p_0 = p, p_1 = 2$ and $T = T_0$.

Furthermore we put

$$M(v) = \left(\int_{\Omega_n}|\nabla(|v|^{\frac{p-2}{2}}v)|^2\right)^{1/p} + |v|_{L^\infty(\Omega_n)}\,. \tag{3.19}$$

Then, to apply Lemma 3.4, it suffices to check (3.17).

In fact, let $M(v_n) \leq 1$, then $|v_n|^{\frac{p-2}{2}} v_n$ forms a bounded set in $H_0^1(\Omega_n)$, so there exists a subsequence of $v_n$ denoted again by $v_n$ such that $v_n \to v$ weakly in $L^2(\Omega_n)$ and $|v_n|^{\frac{p-2}{2}} v_n \to |v|^{\frac{p-2}{2}} v$ strongly in $L^2(\Omega_n)$. Hence there exists a subsequence of $v_n$ denoted again by $v_n$ such that $v_n(x) \to v(x)$ for a.e. $x \in \Omega_n$. Then, since $|v_n|^p \leq 1$ for a.e. $x \in \Omega_n$, the dominant convergence theorem of Lebesgue assueres that $v_n \to v$ strongly in $L^p(\Omega_n)$, whence follows (3.17).

Thus there exists a subsequence $\{u_{n_k}\}$ of $\{u_n\}$ such that $u_{n_k}(x)$ converges for a.e. $x \in \Omega_n$. In view of (3.12) and (3.13), by the diagonal argument, we can extract a subsequence of $u_{n_k}$ denoted by $u_k$ such that

$$\forall n, \ \exists k_0(n) \quad \text{s.t.} \quad \hat{u}_k = u_k \quad \text{in} \quad \Omega_n \times [0, T_0] \qquad \forall k \geq k_0(n), \tag{3.20}$$

$$\hat{u}_k \to u \quad \text{weakly star in} \quad L^\infty(0, T_0; L^r(\Omega)) \qquad \forall r \in [\ell_*, \infty], \tag{3.21}$$

$$\hat{u}_k \to u \text{ strongly in} \quad L^p(0, T_0; L^p(\Omega_n)) \quad \forall p \in [\ell_*, \infty) \quad \forall n \in \mathbb{N}, \tag{3.22}$$

$$\hat{u}_k \to u \quad \text{a.e. in} \quad \Omega \times [0, T_0], \tag{3.23}$$

$$|\hat{u}_k|^{r-2} \hat{u}_k \to |u|^{r-2} u \quad \text{weakly in} \quad L^2\big(0, T_0; L^{\frac{2}{r-1}}(\Omega)\big) \quad \forall r \in (1, \infty), \tag{3.24}$$

$$\varphi_{n_k}(u_k) u_k = \left(\hat{u}_k^2 + \frac{1}{n_k}\right)^{\frac{\ell-2}{2}} \hat{u}_k \to |u|^{\ell-2} u \quad \text{weakly in } L^2\big(0, T_0; L^2(\Omega)\big), \tag{3.25}$$

$$\nabla\left(\left(\hat{u}_k^2 + \frac{1}{n}\right)^{\frac{\ell-2}{2}} \hat{u}_k\right) \to \nabla(|u|^{\ell-2} u) \quad \text{weakly in} \quad L^2\big(0, T_0; L^2(\Omega)\big), \tag{3.26}$$

where $\hat{u}_k = [u_k]^\wedge$ denotes the zero extension of $u_k$ into $\Omega \times [0, T_0]$.

Moreover, in view of (3.15), (3.20), (3.21) and (3.24), it is easy to see that $[(u_k)_t]^\wedge \to u_t, [\frac{1}{n}\Delta(u_k)]^\wedge \to 0$ and $[\Delta(\varphi_{n_k}(u_k) u_k]^\wedge \to \Delta(|u|^{\ell-2}u)$ in $\mathcal{D}'(\Omega \times [0, T_0])$.

Hence, by (3.24), we find that $u$ satisfies (E) in the sense of distribution. Here, since (3.26) assures that $|u|^{\ell-2}\nabla u \in L^2(0, T_0; L^2(\Omega))$, we find that $\nabla(|u|^{\ell-2}\nabla u) \in L^2(0, T_0; H^{-1}(\Omega))$, whence follows that $u_t \in L^2(0, T_0; H^{-1}(\Omega))$, in particular $u \in W^{1,2}(0, T_0; H^{-1}(\Omega)) \subset C([0, T_0]; H^{-1}(\Omega))$.

As for the case where $u_0 \in L^1(\Omega)$, by virtue of (3.8), we easily see that there exists a constant $C_0$ depending only on $|u_0|_{L^\infty(\Omega)}$ and $|u_0|_{L^1(\Omega)}$ such that

$$\sup_{0 \leq t \leq T_0} |u(t)|_{L^r(\Omega_n)} \leq C_0 \qquad \forall r \in (1, \infty]. \tag{3.27}$$

Here noting the relation

$$|u(t)|_{L^1(\Omega_n)} \leq |u(t)|_{L^r(\Omega_n)} |\Omega_n|^{\frac{r-1}{r}} \leq C_0 |\Omega_n|^{\frac{r-1}{r}} \qquad \text{for a.e. } t \in [0, T_0],$$

and letting $r \to 1$, we can derive the estimate (3.12) for all $r \in [1, \infty]$.  $\square$

### 3.3.  Proof of Theorem 3.3

There are so many works already done for the uniqueness of weak solutions for equations related with (E). However most of them treated the case where the perturbation term $|u|^{q-2}u$ is absent. As a matter of course, if the solutions are assumed to be in $W^{1,\infty}(\Omega)$, the standard arguments in $L^1(\Omega)$-space such as in [10] and [16] assures the uniqueness. Here we do not need such a strong regularity for solutions $u$ but assume the weaker regularity $|u|^{\frac{\ell-2}{2}}\nabla u \in L^2(\Omega)$. The crucial part of our argument consists of Theorems 3.5 and 3.6 stated below, which would have independent interest concerning the following equation.

$$(E)_f \begin{cases} u_t = \operatorname{div}\left(|u|^{\ell-2}\nabla u\right) + f(x,t) & (x,t) \in Q_T = \Omega \times (0,T), \\ u|_{\partial\Omega} = 0 \quad t \in (0,\infty), \qquad u(x,0) = u_0(x) \quad x \in \Omega, \end{cases} \tag{3.28}$$

where $1 < \ell < \infty$, $T \in (0,\infty)$ and $\Omega$ is a general domain in $\mathbb{R}^N$.

**Theorem 3.5.** *Let $f \in L^1_{loc}(Q_T)$ and let $u$ be a solution of $(E)_f$ satisfying*

$$u \in L^2(Q_T) \cap L^\ell(Q_T) \cap W^{1,2}\left([0,T] : H^{-1}(\Omega)\right)$$

*and*

$$|u|^{(\ell-2)/2}|\nabla u| \in L^2(Q_T). \tag{3.29}$$

*Then the solution $u$ is uniquely determined by its initial data $u_0 \in L^1(\Omega)$.*

**Theorem 3.6.** *Let $f_i \in L^1(0,T; L^1(\Omega) \cap L^{\ell^*}(\Omega))$, $u_{0,i} \in L^1(\Omega) \cap L^{\ell^*}(\Omega)$ $(i=1,2)$ with $\ell^* = \max(2,\ell)$ and $u_i$ be solutions of $(E)_{f_i}$ with $u_i(0) = u_{0,i}$ satisfying $u_i \in L^1(Q_T) \cap L^{\ell^*}(Q_T) \cap W^{1,2}([0,T_0]; H^{-1}(\Omega))$ and $|u_i|^{(\ell-2)/2}|\nabla u_i| \in L^2(Q_T)$, then $u_i$ satisfy the following estimate for a.e. $t \in (0,T)$.*

$$|u_1(t) - u_2(t)|_{L^1(\Omega)} \le |u_{0,1} - u_{0,2}|_{L^1(\Omega)} + \int_0^t |f_1(\tau) - f_2(\tau)|_{L^1(\Omega)}d\tau. \tag{3.30}$$

Once these theorems are proved, Theorem 3.3 follows easily.

*Proof of Theorem 3.3.* Let $u_i(i=1,2)$ be solutions of (E) satisfying $u_i \in L^\infty(0,T_0; L^1(\Omega) \cap L^\infty(\Omega)) \cap W^{1,2}([0,T_0]; H^{-1}(\Omega))$ and $|u_i|^{(\ell-2)/2}|\nabla u_i| \in L^2(Q_T)$. Put $f_i(x,t) = |u_i|^{q-2}u_i(x,t)$, then $f_i \in L^\infty(0,T_0; L^1(\Omega) \cap L^\infty(\Omega))$ and so Theorem 3.5 assures that $u_i$ give the unique solutions for $(E)_{f_i}$. Hence, by Theorem 3.6, we find that $w = u_1 - u_2$ satisfies

$$|w(t)|_{L^1(\Omega)} \le |w(0)|_{L^1(\Omega)} + \int_0^t \left||u_1|^{q-2}u_1(\tau) - |u_2|^{q-2}u_2(\tau)\right|_{L^1(\Omega)}d\tau$$

$$\le |w(0)|_{L^1(\Omega)} + \int_0^t (q-1)\left||u_1 + \theta(u_2 - u_1)|^{q-2}(\tau)\right|_{L^\infty(\Omega)}|w(\tau)|_{L^1(\Omega)}d\tau$$

$$\le |w(0)|_{L^1(\Omega)} + C\int_0^t |w(\tau)|_{L^1(\Omega)}d\tau \text{ a.e. } t \in (0,T),$$

which together with Gronwall's inequality concludes the uniqueness. $\qquad\square$

Thus it suffices to prove Theorems 3.5 and 3.6.

*Proof of Theorem* 3.5. In order to derive the uniqueness of the solution, we are going to work in the space $H^{-1}(\Omega)$ as in [3], where the case where $\Omega = \mathbb{R}^N$ is considered. Here we are concerned with general domains with Dirichlet boundary condition, so we need some modifications. It should be noted that the method to be introduced here works also for the other type of boundary conditions such as Neumann boundary condition.

For each $\varepsilon \in (0,1)$, consider the following problem.

$$(\mathrm{L})_\varepsilon \begin{cases} \varepsilon w_\varepsilon - \Delta w_\varepsilon = w & x \in \Omega, \\ w_\varepsilon = 0 & x \in \partial\Omega. \end{cases}$$

It is easy to see that for any $w \in H^{-1}(\Omega)$, $(\mathrm{L})_\varepsilon$ has a unique solution $w_\varepsilon \in H_0^1(\Omega)$.

Put $B_\varepsilon : w \mapsto w_\varepsilon = B_\varepsilon w$, then $B_\varepsilon$ gives a homeomorphism from $H^{-1}(\Omega)$ onto $H_0^1(\Omega)$. Hence ${}_{H_0^1}\!< B_\varepsilon u, v >_{H^{-1}}$ can give the inner product of $H^{-1}(\Omega)$ for any $\varepsilon > 0$.

Let $u_i (i = 1, 2)$ be two solutions of $(\mathrm{E})_f$, then $w = u_1 - u_2$ satisfies

$$w_t = \Delta h, \quad h = \left(\varphi(u_1) - \varphi(u_2)\right), \quad \varphi(u) = \frac{1}{\ell - 1}|u|^{\ell - 2}u.$$

Here, since $w \in W^{1,2}([0,T]; H^{-1}(\Omega))$, we note that $g_\varepsilon(t) = < B_\varepsilon w(t), w(t) >$ is well defined and that $< w_t, B_\varepsilon w(t) >= \frac{1}{2}\frac{dg_\varepsilon(t)}{dt}$. Then, multiplying the equation above by $B_\varepsilon w(t)$, with the aid of integration by parts, we obtain

$$\frac{d}{dt}g_\varepsilon(t) = 2\langle \Delta h(t), B_\varepsilon w(t)\rangle = 2\langle h(t), \Delta B_\varepsilon w(t)\rangle$$

$$= 2\langle h(t), \varepsilon B_\varepsilon w(t) - w(t)\rangle \le 2\langle h(t), \varepsilon B_\varepsilon h(t)\rangle \quad \text{a.e. } t \in [0,T],$$

where we used the fact that $h(t) \in H_0^1(\Omega)$. Hence, we get

$$g_\varepsilon(t) \le 2\int_0^t \left(h(s), \varepsilon B_\varepsilon w(s)\right)ds = I_\varepsilon(t). \tag{3.31}$$

Then in order to prove the uniqueness, it suffices to show that $\lim_{\varepsilon \to 0} I_\varepsilon(t) = 0$. Indeed, (3.31) implies that $g_\varepsilon(t) \to 0$ as $\varepsilon \to 0$, so plugging the relation $\varepsilon B_\varepsilon w - \Delta B_\varepsilon w = w$ into $g_\varepsilon$, we have

$$g_\varepsilon(t) = \left(B_\varepsilon w(t), \varepsilon B_\varepsilon w(t) - \Delta B_\varepsilon w(t)\right) = \varepsilon |B_\varepsilon w(t)|_{L^2}^2 + |\nabla(B_\varepsilon w)|_{L^2}^2,$$

whence follows $\varepsilon B_\varepsilon w(t) \to 0$ and $\nabla(B_\varepsilon w(t)) \to 0$ strongly in $L^2(\Omega)$ as $\varepsilon \to 0$. Therefore $\varepsilon B_\varepsilon w - \Delta B_\varepsilon w = w \to 0$ in $\mathcal{D}'(\Omega)$, which yields $w(t) = 0$ for a.e. $t \in [0,T]$.

In order to prove $\lim_{\varepsilon \to 0} I_\varepsilon(t) = 0$, we need the following lemma.

**Lemma 3.7.** *Assume that* $w \in L^2(Q_T) \cap L^s(Q_T)$ $(1 < s < \infty)$, *then*

$$\varepsilon B_\varepsilon w = \varepsilon w_\varepsilon \to 0 \qquad \text{weakly in } L^2(Q_T) \text{ and } L^s(Q_T). \tag{3.32}$$

Indeed, the fact $\lim_{\varepsilon \to 0} I_\varepsilon(t) = 0$ follows directly from Lemma 3.7 with $s = \ell$ and the fact that $h \in L^{\frac{\ell}{\ell-1}}(Q_T)$. $\qquad\square$

Thus we can complete the proof of Theorem 3.3 by verifying Lemma 3.7.

**Proof of Lemma** 3.7

We first prepare the following estimates.

**Lemma 3.8.** *Assume that $w \in L^2(\Omega)$, then $w_\varepsilon = B_\varepsilon w$ satisfies*

$$|\varepsilon w_\varepsilon|_{L^r(\Omega)} \le |w|_{L^r(\Omega)} \quad \forall w \in L^r(\Omega) \cap L^2(\Omega) \quad \forall r \in [1, \infty]. \qquad (3.33)$$

*Proof.* **Case 1 :** $r \ge 2$. Since $[w_\varepsilon]_n = g_n(w_\varepsilon) \in L^\infty(\Omega) \cap L^2(\Omega)$ (see (2.6)), we can multiply $(L)_\varepsilon$ by $|[w_\varepsilon]_n|^{r-2}[w_\varepsilon]_n$ to get

$$\varepsilon|[w_\varepsilon]_n|_{L^r}^r + (r-1)\int |[w_\varepsilon]_n|^{r-2}|\nabla w_\varepsilon|^2 g_n'(w_\varepsilon)dx \le |w|_{L^r}|[w_\varepsilon]_n|_{L^r}^{r-1},$$

whence follows $\varepsilon|[w_\varepsilon]_n|_{L^r} \le |w|_{L^r}$. Since $|[w_\varepsilon]_n(x)| \uparrow |w(x)|$ for a.e. $x \in \Omega$, letting $n \to \infty$, we obtain (3.33) for $2 \le r < \infty$. As for the case where $r = \infty$, noting $|w|_{L^r} \le |w|_{L^\infty}^{\frac{r-2}{r}}|w|_{L^2}^{\frac{2}{r}}$ and letting $r \to \infty$ in (3.33), by Lemma 2.1, we get (3.33) with $r = \infty$.

**Case 2 :** $1 \le r < 2$. Multiplying $(L)_\varepsilon$ by $(w_\varepsilon^2 + \frac{1}{n})^{\frac{r-2}{2}}w_\varepsilon \in H_0^1(\Omega)$, we get

$$\varepsilon\int\left(w_\varepsilon^2 + \frac{1}{n}\right)^{\frac{r-2}{2}}w_\varepsilon^2 dx + \int\left(w_\varepsilon^2 + \frac{1}{n}\right)^{\frac{r-2}{2}}|\nabla w_\varepsilon|^2 dx$$

$$+ (r-2)\int\left(w_\varepsilon^2 + \frac{1}{n}\right)^{\frac{r-4}{2}}w_\varepsilon^2|\nabla w_\varepsilon|^2 dx$$

$$\le \int\left(w_\varepsilon^2 + \frac{1}{n}\right)^{\frac{r-2}{2}}w_\varepsilon w\, dx.$$

Hence

$$\varepsilon\int\left(w_\varepsilon^2 + \frac{1}{n}\right)^{\frac{r-2}{2}}w_\varepsilon^2 dx \le \int\left(w_\varepsilon^2 + \frac{1}{n}\right)^{\frac{r-2}{2}}w_\varepsilon w\, dx$$

$$\le |w|_{L^r}\left(\int\left(|w_\varepsilon|\left(w_\varepsilon^2 + \frac{1}{n}\right)^{\frac{r-2}{2}}\right)^{\frac{r}{r-1}}dx\right)^{\frac{r-1}{r}}$$

$$\le |w|_{L^r}\left(\int\left(|w_\varepsilon|^{\frac{2(r-1)}{r}}\left(w_\varepsilon^2 + \frac{1}{n}\right)^{\frac{(r-1)(r-2)}{2r}}\right)^{\frac{r}{r-1}}dx\right)^{\frac{r-1}{r}}$$

$$\le |w|_{L^r}\left(\int\left(w_\varepsilon^2 + \frac{1}{n}\right)^{\frac{r-2}{2}}w_\varepsilon^2\, dx\right)^{\frac{r-1}{r}}.$$

Then

$$\varepsilon\left(\int\left(w_\varepsilon^2 + \frac{1}{n}\right)^{\frac{r-2}{2}}w_\varepsilon^2\, dx\right)^{\frac{1}{r}} \le |w|_{L^r}.$$

Thus, since $(w_\varepsilon^2 + \frac{1}{n})^{\frac{r-2}{2}}w_\varepsilon^2 \uparrow |w_\varepsilon|^r$ as $n \uparrow \infty$, we get (3.33) for $r \in [1, 2)$. $\qquad\square$

**Proof of Lemma** 3.7**(continued):**

Since $w \in L^2(Q_T)$, we have

$$
\begin{aligned}
|w|^2_{L^2(Q_T)} &= |\varepsilon w_\varepsilon - \Delta w_\varepsilon|^2_{L^2(Q_T)} \\
&= |\varepsilon w_\varepsilon|^2_{L^2(Q_T)} + |\Delta w_\varepsilon|^2_{L^2(Q_T)} + 2\varepsilon|\nabla w_\varepsilon|^2_{L^2(Q_T)},
\end{aligned}
\tag{3.34}
$$

whence follows

$$
|\varepsilon \nabla w_\varepsilon|_{L^2(Q_T)} \le \sqrt{\frac{\varepsilon}{2}}|w|_{L^2(Q_T)}.
\tag{3.35}
$$

Here we recall the following embedding (see [1] or [5]):

$$
|u|^2_{L^\infty(\Omega)} \le 2|u|_{L^2(\Omega)}|\nabla u|_{L^2(\Omega)} \quad \text{for } N = 1,
\tag{3.36}
$$

$$
|u|^4_{L^4(\Omega)} \le 2|u|^2_{L^2(\Omega)}|\nabla u|^2_{L^2(\Omega)} \quad \text{for } N = 2,
\tag{3.37}
$$

$$
|u|_{L^{2^*}(\Omega)} \le C \, |\nabla u|_{L^2(\Omega)} \quad 2^* = \frac{2N}{N-2} \quad \text{for } N \ge 3.
\tag{3.38}
$$

Hence, by (3.33) and (3.35), there exists some $r \in (2, \infty]$ such that $\varepsilon w_\varepsilon \to 0$ strongly in $L^2(0, T; L^r(\Omega))$. On the other hand, the fact that $w \in L^2(Q_T) \cap L^s(Q_T)$ together with Lemma 3.8 assures that $\varepsilon w_\varepsilon$ is bounded in $L^2(Q_T) \cap L^s(Q_T)$. Therefore there exists a subsequence $\{w_{\varepsilon_k}\}$ of $\{w_\varepsilon\}$ such that $\varepsilon_k w_{\varepsilon_k} \to \chi$ weakly in $L^2(Q_T) \cap L^s(Q_T)$ as $k \to \infty$, and hence $\chi = 0$. Since this argument does not depend on the choice of subsequences, (3.32) is verified. $\qquad\square$

**Proof of Theorem** 3.6

We first introduce the following approximate equations for $(E)_{f_i}$ $(i = 1, 2)$.

$$
(E)_{i,n} \quad
\begin{cases}
(u_i)_t = \Delta\left(\left(\frac{1}{n} + \varphi_n(u_i)\right) u_i\right) + f_{i,n}(x, t) \quad (x, t) \in Q_{n,T} = \Omega_n \times (0, \infty), \\
u_i|_{\partial\Omega_n} = 0 \ \ t \in (0, \infty), \qquad u_i(x, 0) = u^n_{i,0}(x) \quad x \in \Omega_n,
\end{cases}
$$

where $\varphi_n(u) = \frac{1}{\ell-1}\left(u^2 + \frac{1}{n}\right)^{\frac{\ell-2}{2}}$ and $u^n_{i,0}$, $f_{i,n}$ are sufficiently smooth such that $\hat{u}^n_{i,0} \to u_{i,0}$ strongly in $L^1(\Omega) \cap L^{\ell^*}(\Omega)$ and $\hat{f}_{i,n} \to f$ strongly in $L^1(0, T; L^1(\Omega) \cap L^{\ell^*}(\Omega))$ as $n \to \infty$. (Here $\hat{u}^n_{i,0}$ and $\hat{f}_{i,n}$ denote the zero extensions of $u^n_{i,0}$ and $f_{i,n}$ into $\Omega$ and $Q_T$.)

   Then the standard theory for quasilinear parabolic equations assures the existence of unique classical solutions $(u_i)_n$ $(i = 1, 2)$ of $(E)_{i,n}$ dented by $u_i$ (see, e.g., [6]). Now repeating the same arguments as in the proof of Theorem 3.2, we easily find that there exists a constant $C_0$ depending on $|u_{i,0}|_{L^1(\Omega) \cap L^{\ell^*}(\Omega)}$ and

$|f_i|_{L^1(0,T;L^1(\Omega)\cap L^{\ell^*}(\Omega))}$ but not on $n$ such that

$$|u_i|_{L^\infty(0,T;L^r(\Omega_n))} \leq C_0 \qquad\qquad \forall r \in [1, \ell^*]\,, \qquad (3.39)$$

$$|\nabla(\varphi_n(u_i)u_i)|_{L^2(0,T;L^2(\Omega_n))} \leq C_0\,, \qquad (3.40)$$

$$|\Delta(\varphi_n(u_i)u_i)|_{L^2(0,T;H^{-1}(\Omega_n))} \leq C_0\,, \qquad (3.41)$$

$$|\tfrac{1}{n}\Delta u_i|_{L^2(0,T;H^{-1}(\Omega_n))} \leq C_0 \frac{1}{\sqrt{n}}\,, \qquad (3.42)$$

$$|(u_i)_t|_{L^2(0,T;H^{-1}(\Omega_n))} \leq C_0\,. \qquad (3.43)$$

Then again by Lemma 3.4, we can extract a subsequence $\{(u_i)_{n_k}\}$ denoted by $\{u_i^k\}$ such that

$$\forall n,\ \exists k_0(n) \quad \text{s.t.} \quad \hat{u}_i^k = u_i^k \quad \text{in} \quad Q_{n,T} \qquad \forall k \geq k_0(n)\,, \qquad (3.44)$$

$$\hat{u}_i^k \to u_i \quad \text{weakly star in} \quad L^\infty(0,T;L^r(\Omega)) \qquad \forall r \in (1, \ell^*]\,, \qquad (3.45)$$

$$\hat{u}_i^k \to u_i \text{ strongly in} \quad L^p(0,T;L^p(\Omega_n)) \quad \forall p \in [1, \ell^*] \quad \forall n \in \mathbb{N}\,, \qquad (3.46)$$

$$\hat{u}_i^k \to u_i \quad \text{a.e. in} \quad Q_{n,T}\,, \qquad (3.47)$$

$$\varphi_{n_k}(u_i^k)u_i^k = \left((\hat{u}_i^k)^2 + \frac{1}{n_k}\right)^{\frac{\ell-2}{2}} \hat{u}_i^k \to |u_i|^{\ell-2}u_i \quad \text{weakly in } L^{\frac{\ell}{\ell-1}}(Q_T)\,, \quad (3.48)$$

$$\nabla\left(\left((\hat{u}_i^k)^2 + \frac{1}{n_k}\right)^{\frac{\ell-2}{2}} \hat{u}_i^k\right) \to \nabla(|u_i|^{\ell-2}u_i) \quad \text{weakly in} \quad L^2(0,T;L^2(\Omega))\,, \qquad (3.49)$$

where $\hat{u}_i^k$ denotes the zero extension of $u_i^k$ into $Q_T$. Then as in the proof of Theorem 3.2, we find that $u_i$ gives a solution of $(E)_{f_i}$ with $u_i(0) = u_{i,0}$ satisfying $u_i \in L^1(Q_T) \cap L^{\ell^*}(Q_T) \cap W^{1,2}(0,T;H^{-1}(\Omega))$ and $|u_i|^{(\ell-2)/2}|\nabla u_i| \in L^2(Q_T)$. Hence, by Theorem 3.5, $u_i$ constructed above turn out to be the solutions given in Theorem 3.6.

Let $u_i = (u_i)_n$ be solutions of $(E)_{i,n}$ with $(u_i)_n(0) = u_{i,0}^n$, then $w = (u_1)_n - (u_2)_n$ satisfies

$$w_t = \Delta(h_n(u_1) - h_n(u_2)) + f_{1,n} - f_{2,n}\,, \qquad h_n(u) = \left(\frac{1}{n} + \varphi_n(u)\right)u\,. \qquad (3.50)$$

Put $\gamma_\varepsilon(u) = g_\varepsilon(u)/\varepsilon$ (see (2.6)), i.e. , $\gamma_\varepsilon(u) = u/\varepsilon$, for $|u| \le \varepsilon$ and $\gamma_\varepsilon(u) = 1(-1)$, for $u > \varepsilon(u < -\varepsilon)$. Multiplying (3.50) by $\gamma_\varepsilon(w(t))$, we get

$$\frac{d}{dt}\int_{\Omega_n} G_\varepsilon(w)dx + I_1(t) + I_2(t) \le \int_{\Omega_n} |f_{1,n}(t) - f_{2,n}(t)|dx\,,$$

$$I_1(t) = \int_{\Omega_n} \left(\left(\frac{1}{n} + \varphi_n(u_1)\right)\nabla u_1 - \left(\frac{1}{n} + \varphi_n(u_2)\right)\nabla u_2\right)\nabla w\ \gamma_\varepsilon'(w)dx\,, \quad (3.51)$$

$$I_2(t) = \int_{\Omega_n} (\varphi_n'(u_1)u_1\nabla u_1 - \varphi_n'(u_2)u_2\nabla u_2)\nabla w\ \gamma_\varepsilon'(w)dx\,,$$

where $G_\varepsilon(u) = \int_0^u \gamma_\varepsilon(v)dv$. Here

$$I_1(t) = \frac{1}{n}\int_{\Omega_n} |\nabla w|^2\gamma_\varepsilon'(w)dx + J_1(t) + R_1(t)\,,$$

$$J_1(t) = \int_{\Omega_n} \varphi_n(u_1)|\nabla w|^2\gamma_\varepsilon'(w)dx\,, \quad (3.52)$$

$$R_1(t) = \int_{\Omega_n} (\varphi_n(u_1) - \varphi_n(u_2))\nabla u_2\nabla w\ \gamma_\varepsilon'(w)dx\,,$$

$$I_2(t) = J_2(t) + R_2(t)\,,$$

$$J_2(t) = \int_{\Omega_n} \varphi_n'(u_1)u_1|\nabla w|^2\gamma_\varepsilon'(w)dx\,, \quad (3.53)$$

$$R_2(t) = \int_{\Omega_n} (\varphi_n'(u_1)u_1 - \varphi_n'(u_2)u_2)\nabla u_2\nabla w\ \gamma_\varepsilon'(w)dx\,.$$

Let $Z_t^n(\varepsilon) = \{x \in \Omega_n \times [0,t]; 0 < |w(x)| < \varepsilon\}$, then $|Z_t^n(\varepsilon)| \to 0$ as $\varepsilon \to 0$, since $w \in L^2(\Omega_n)$. Hence, by letting $\varepsilon \to 0$, we find

$$\int_0^t |R_1(\tau)|d\tau \le \int_{Z_\tau^n(\varepsilon)} |\varphi_n'(u_1 + \theta(u_2 - u_1))||\nabla u_2||\nabla w|dxd\tau \to 0\,,$$
$$\int_0^t |R_2(\tau)|d\tau \le \int_{Z_\tau^n(\varepsilon)} |(\varphi_n'(u)u)'(u_1 + \theta(u_2 - u_1))||\nabla u_2||\nabla w|dxd\tau \to 0\,. \quad (3.54)$$

Since $\varphi_n'(u)u \le \varphi_n(u)$ and $G_\varepsilon(u) \to |u|$ as $\varepsilon \to 0$, by integrating (3.51) over [0,t] and letting $\varepsilon \to 0$, we obtain

$$|\hat u_1(t) - \hat u_2(t)|_{L^1(\Omega)} \le |\hat u_{0,1}^n - \hat u_{0,2}^n|_{L^1(\Omega)} + \int_0^t |\hat f_{1,n}(\tau) - \hat f_{2,n}(\tau)|_{L^1(\Omega)}d\tau\,. \quad (3.55)$$

By virtue of the fact (3.47) and Fatou's Lemma, letting $n \to \infty$ in (3.55), we deduce (3.30).                                                                                                        □

**Remark 3.9.**

(1) Theorem 3.5 and the proof of Theorem 3.6 also assuere that for any $u_0 \in L^1(\Omega) \cap L^{\ell^*}(\Omega)$ and $f \in L^1(0,T; L^1(\Omega) \cap L^{\ell^*}(\Omega))$, $(E)_f$ admits a unique

solution $u$ satisfying $u \in L^\infty(0, T; L^1(\Omega) \cap L^{\ell^*}(\Omega)) \cap W^{1,2}(0, T; H^{-1}(\Omega))$ and $|u|^{(\ell-2)/2}|\nabla u| \in L^2(Q_T)$.

(2) With more carefull applications of L$^\infty$-Energy Method, we can extend Theorems 3.1, 3.2 and 3.3 to more general equations.

(3) In Theorems 3.2 and 3.3, as for the regularity of $\Omega$, we only assume that there exists a sequence of monotone increasing bounded domains $\Omega_n$ with smooth boundaries $\partial\Omega_n$ such that $\Omega_n \to \Omega$ as $n \to \infty$. This approximation procedure is needed not only for unbounded domains but also for bounded domains with non-smooth boundaries.

# References

[1] A. R. Adams, *Sobolev Spaces*, Academic Press, 1978.

[2] H. Brézis, *Opérateurs maximaux monotone et semi-groupes de contractions dans les espaces Hilbert*, North-Holland Math. Studies **5**, 1973.

[3] H. Brézis and M. G. Crandall, *Uniqueness of solutions of the intial value problem for $u_t - \Delta\varphi(u) = 0$*, J. Math. pures et appl., **58** (1979), 153–163.

[4] J. A. Dubinskii, *Convergence faible dans les équations elliptiques paraboliques non linéaires*, Mat. Sbornik, **67** (109), (1965), 609–642.

[5] O. A. Ladyzhenskaya, *The Mathematical Theory of Viscous Imcompressible Flow*, Gordon & Breach, New York, 1969.

[6] O. A. Ladyzhenskaya, V. A. Solonnikov and N. N. Ural'ceva, *Linear and Quasi-linear Equations of Parabolic Type*, Transl. Math. AMS. Providence, R.I., 1968.

[7] G. M. Lieberman, *Second Order Parabolic Differential Equations*, World Scientific, 1996.

[8] J. L. Lions, *Quelques Méthodes de Résolution des Problèmes aux Limites Non Linéaires*, Duno Gauthier-Villars, Paris, 1969.

[9] E. Minchev and M. Ôtani, *$L^\infty$-Energy Method for a parabolic system with convection and hysteresis effect*, preprint.

[10] O. A. Oleinik, A. S. Kalashinikov and C. Yui-Lin', *The Cauchy problemand boundary-value problems for equations of unsteady filtration type*, Izv. Acad. Nauk S.S.S.R., Ser. Mat., **22** (1958), 667–704.

[11] M. Ôtani, *Non-monotone perturbations for nonlinear parabolic equations associated with subdifferential operators, Cauchy problems*, J. Differential Equations, **46**, No.12 (1982), 268–299.

[12] M. Ôtani, *$L^\infty$-energy method and its applications*, GAKUTO International Series, Mathematical Sciences and Applications, Gakkotosho, Tokyo, **20** (2004), 505–516.

[13] M. Ôtani, *$L^\infty$-energy method and its applications to some nonlinear parabolic systems*, GAKUTO International Series, Mathematical Sciences and Applications, Gakkotosho, Tokyo, **22** (2005), 233–244.

[14] M. Ôtani, *$L^\infty$-energy method, Applications*, preprint.

[15] M. Ôtani and Y. Sugiyama, *A method of energy estimates in $L^\infty$ and its application to porous medium equations*, J. Math. Soc. Japan, **53**, No.4 (2001), 745–789.

[16] M. Ôtani and Y. Sugiyama, *Lipschitz continuous soultions of some doubly nonlinear parabolic equations.* Discrete and Continuous Dynamical Stystems, **8**, No.3, (2002), 647–670.

[17] K. Yosida, *Functional Analysis*, 2nd Edition, Springer, 1968.

Mitsuharu Ôtani
3-4-1 Okubo
Shijuku-ku
J-169-8555, Tokyo
Japan
e-mail: otani@waseda.jp

Progress in Nonlinear Differential Equations
and Their Applications, Vol. 75, 377–391

# On the Singular Set of Certain Potential Operators in Hilbert Spaces

Biagio Ricceri

*To Professors Arrigo Cellina and James A. Yorke*

## 1. Introduction and statement of the main results

Here and in the sequel, $(X, \langle \cdot, \cdot \rangle)$ is a real Hilbert space and $J : X \to \mathbf{R}$ is a $C^1$ functional. For each $\lambda > 0$, set

$$\Phi_\lambda(x) = x + \lambda J'(x)$$

for all $x \in X$. For brevity, we will also write $\Phi$ instead of $\Phi_1$ when $\lambda = 1$.

As usual, for a generic operator $T : X \to X$, we say that $T$ is a local homeomorphism at a point $x_0 \in X$ if there are a neighbourhood $U$ of $x_0$ and a neighbourhood $V$ of $T(x_0)$ such that the restriction of $T$ to $U$ is a homeomorphism between $U$ and $V$. If $T$ is not a local homeomorphism at $x_0$, we say that $x_0$ is a singular point of $T$.

The set of all singular points of $T$ is called the singular set of $T$ and we denote it by $S_T$. Clearly, the set $S_T$ is closed.

When the restriction of $T$ to some open set $A \subseteq X$ is of class $C^1$, we also denote by $\hat{S}_{T|A}$ the set of all $x_0 \in A$ such that the operator $T'(x_0)$ is not surjective. Since the set of all surjective operators is open in $\mathcal{L}(X, X)$, by the continuity of $T'$, the set $\hat{S}_{T|A}$ is closed too.

When $T$ is $C^1$ in $X$, $T$ is said to be a Fredholm operator if $T'(x)$ is a Fredholm linear operator for each $x \in X$. The function $x \to \text{index}(T'(x))$ is then constant and its value is called the index of $T$.

In this paper, we are interested in the size of $S_\Phi$, of $\hat{S}_\Phi$ and of $\hat{S}_{\Phi_\lambda}$ for suitable $\lambda$.

To introduce our results, we first recall two canonical situations where $S_\Phi = \emptyset$. They are when $J'$ is a contraction and when $J$ is convex. Actually, in both cases, due to classical results, the operator $\Phi$ turns out to be a global homeomorphism between $X$ and itself. In particular, note that if $J'$ is Lipschitzian, then it cannot be positively homogeneous of degree different from 1.

Our first result reads as follows:

**Theorem 1.1.** *Let $X$ be infinite-dimensional. Assume that $J$ is sequentially weakly lower semicontinuous, not quasi-convex, and positively homogeneous of degree $\alpha \neq 2$. If $\alpha > 2$ assume also that $J$ is non-negative. Finally, suppose that $\Phi$ is closed. Then, both the sets $S_\Phi$ and $\Phi(S_\Phi)$ are not $\sigma$-compact.*

As usual, a set in $X$ is said to be $\sigma$-compact if it is the union of an at most countable family of compact sets, while a functional on $X$ is said to be quasi-convex if its sub-level sets are convex.

As far as we know, Theorem 1.1 is the first result providing a *general* class of not (necessarily) differentiable potential operators in Hilbert spaces whose singular set is not $\sigma$-compact.

Here is a remarkable consequence of Theorem 1.1.

**Theorem 1.2.** *Let $X$ be infinite-dimensional. Assume that $J'$ is compact and that*

$$\lim_{\|x\| \to \infty} \|\Phi(x)\| = +\infty. \tag{1}$$

*Assume that $J$ is positively homogeneous of degree $\alpha \neq 2$ and that it is not quasi-convex. If $\alpha > 2$ suppose, in addition, that $J$ is non-negative. Finally, assume that there exists a closed, $\sigma$-compact set $B \subset X$ such that the restriction of $J$ to $X \setminus B$ is of class $C^2$. Then, both the sets $\hat{S}_{\Phi_{|(X \setminus B)}}$ and $\Phi(\hat{S}_{\Phi_{|(X \setminus B)}})$ are not $\sigma$-compact.*

Note that there is already a known result with the same conclusion as that of Theorem 1.2, when $B = \emptyset$. We allude to Theorem 4 of [1]. But the profound difference between the two results is that this latter deals with $C^1$ Fredholm operators of *positive* index, while, in Theorem 1.2, $\Phi$ is a Fredholm operator of index 0. Just because $\Phi$ is so, again if $B = \emptyset$, the set $\Phi(\hat{S}_\Phi)$ has also an empty interior, by the classical Sard-Smale theorem ([7]). To get this information for a Fredholm operator of positive index $p$, we should assume in addition that the operator is of class $C^{p+1}$.

Both infinite dimensionality of $X$ and positive homogeneity of $J$ are essential assumptions in Theorems 1.1 and 1.2. Out of those assumptions, we have the following result:

**Theorem 1.3.** *Let $\dim(X) \geq 3$. Assume that $J$ is of class $C^2$ and not quasi-convex, that $J'$ is compact and that*

$$\liminf_{\|x\| \to \infty} \frac{J(x)}{\|x\|^2} \geq 0.$$

*Finally, suppose that*

$$\lim_{\|x\| \to +\infty} \|\Phi_\lambda(x)\| = +\infty$$

*for all $\lambda > 0$. Then, there exists $\lambda^* > 0$ such that the set $\hat{S}_{\Phi_{\lambda^*}}$ contains at least one accumulation point.*

## 2. Proofs

The proof of Theorem 1.1 is based on combining some ideas from [5] with the following general result by R. S. Sadyrkhanov ([6]) which extends to non-differentiable operators a previous one by R. A. Plastock ([3]):

**Theorem 2.1 ([6], Theorem 2.1).** *If $X$ is infinite-dimensional, if $T : X \to X$ is a closed continuous operator and if $S_T$ is $\sigma$-compact, then the restriction of $T$ to $X \setminus S_T$ is a homeomorphism between $X \setminus S_T$ and $X \setminus T(S_T)$.*

As in [5], we will also use two other major tools: a recent, very precise best approximation result by I. G. Tsar'kov [8] and a mini-max theorem that we have established in [4].

**Theorem 2.2 ([8], Corollary 2).** *Let $A \subset X$ be a sequentially weakly closed and non-convex set. Then, for each convex set $V \subseteq X$ dense in $X$, there exists $x_0 \in V \setminus A$ such that the set*

$$\big\{ x \in A : \|x_0 - x\| = \operatorname{dist}(x_0, A) \big\}$$

*has at least two points.*

**Theorem 2.3 ([4], Theorem 1).** *Let $I$ be a real interval, and $f : X \times I \to \mathbf{R}$ a function satisfying the following conditions:*

$(a_1)$ *for every $x \in X$, the function $f(x, \cdot)$ is quasi-concave and continuous;*

$(a_2)$ *for every $\lambda \in I$, the function $f(\cdot, \lambda)$ is sequentially weakly lower semicontinuous and each of its local minima is a global minimum;*

$(a_3)$ *there exist $\rho > \sup_{\lambda \in I} \inf_{x \in X} f(x, \lambda)$ and $\lambda_0 \in I$ such that the set*

$$\big\{ x \in X : f(x, \lambda_0) \le \rho \big\}$$

*is bounded. Then*

$$\sup_{\lambda \in I} \inf_{x \in X} f(x, \lambda) = \inf_{x \in X} \sup_{\lambda \in I} f(x, \lambda).$$

The successful link between [5] and [6] is essentially provided by the following proposition:

**Proposition 2.4.** *If $X$ is infinite-dimensional and if $U \subset X$ is a $\sigma$-compact set, then there exists a convex cone $C \subset X$, dense in $X$, such $U \cap C = \emptyset$.*

*Proof.* We distinguish two cases. First, assume that $X$ is separable. Fix a countable base $\{A_n\}$ of open sets. We claim that there exists a sequence $\{x_n\}$ in $X$ such that, for each $n \in \mathbf{N}$,

$$x_n \in A_n$$

and

$$U \cap C_{(x_1,\dots,x_n)} = \emptyset$$

where

$$C_{(x_1,\dots,x_n)} = \left\{ \sum_{i=1}^{n} \lambda_i x_i : \lambda_i \ge 0, \sum_{i=1}^{n} \lambda_i > 0 \right\}.$$

We proceed by induction on $n$. Clearly, the set $\cup_{\lambda>0}\lambda U$ is $\sigma$-compact and so, since $X$ is infinite-dimensional, it does not contain $A_1$. Thus, if we take $x_1 \in A_1 \setminus \cup_{\lambda>0}\lambda U$, we have $U \cap C_{(x_1)} = \emptyset$. Now, assume that $x_1, ..., x_n$, with the desired properties, have been constructed. Consider the set $\cup_{\mu>0}\mu(U - \overline{C_{(x_1,...,x_n)}})$. One readily sees that it is $\sigma$-compact, and so it does not contain $A_{n+1}$. Choose $x_{n+1} \in A_{n+1} \setminus \cup_{\mu>0}\mu(U - \overline{C_{(x_1,...,x_n)}})$. Then, one has

$$U \cap C_{(x_1,...,x_{n+1})} = \emptyset \,.$$

Indeed, if there was $\hat{x} \in U \cap C_{(x_1,...,x_{n+1})}$, we would have $\hat{x} = \sum_{i=1}^{n+1} \lambda_i x_i$, with $\lambda_i \geq 0$ and $\sum_{i=1}^{n+1} \lambda_i > 0$. In particular, $\lambda_{n+1} > 0$, since $U \cap C_{(x_1,...,x_n)} = \emptyset$. Consequently, we would have

$$x_{n+1} = \frac{1}{\lambda_{n+1}}\left(\hat{x} - \sum_{i=1}^{n} \lambda_i x_i\right)$$

and so $x_{n+1} \in \cup_{\mu>0}\mu(U - \overline{C_{(x_1,...,x_n)}})$, against our choice. Thus, the claimed sequence $\{x_n\}$ does exist. Now, put

$$C = \bigcup_{n=1}^{\infty} C_{(x_1,...,x_n)} \,.$$

It is clear that $C$ is a convex cone which does not meet $U$. Moreover, $C$ is dense in $X$ since it meets each set $A_n$. Now, assume that $X$ is not separable. Denote by $V$ the orthogonal complement of $\overline{\text{span}(U)}$. Since this latter subspace is separable, $V$ is not separable. Let $\{e_\gamma\}_{\gamma \in \Gamma}$ be an orthonormal basis of $V$. Introduce in $\Gamma$ a total order $\leq$ with no greatest element and set

$$D = \left\{x \in V : \exists \beta \in \Gamma : \langle x, e_\beta \rangle > 0 \text{ and } \langle x, e_\gamma \rangle = 0 \; \forall \gamma > \beta\right\} \,.$$

Clearly, $D$ is a convex cone. Let $x \in \text{span}(\{e_\gamma : \gamma \in \Gamma\})$. So, $x = \sum_{\gamma \in I}\langle x, e_\gamma \rangle e_\gamma$ for some finite $I \subset \Gamma$. Let $\beta \in \Gamma$ be such that $\beta > \gamma$ for all $\gamma \in I$. Then, for each $n \in \mathbf{N}$, the point $y_n = x + \frac{1}{n}e_\beta$ belongs to $D$, and the sequence $\{y_n\}$ tends to $x$. This clearly implies that $D$ is dense in $V$. Finally, set

$$C = \overline{\text{span}(U)} + D \,.$$

So, $C$ is a convex cone, dense in $X$, which does not meet $U$.                $\square$

We now are in a position to prove Theorem 1.1.

*Proof of Theorem* 1.1. Arguing by contradiction, assume that $S_\Phi$ is $\sigma$-compact. Then, since $\Phi$ is continuous, the set $\Phi(S_\Phi)$ is $\sigma$-compact too, and, by Theorem 2.1, for each $y \in X \setminus \Phi(S_\Phi)$, the equation

$$\Phi(x) = y$$

has a unique solution. By Proposition 2.4, there is a convex cone $C \subset X$, dense in $X$, such that

$$C \cap \Phi(S_\Phi) = \emptyset \,. \tag{2}$$

By assumption, there is $r > \inf_X J$ such that $J^{-1}(]-\infty, r])$ is sequentially weakly closed and not convex. Consequently, by Theorem 2.2, there exist $y_0 \in C$ and two distinct points $y_1, y_2$ in $J^{-1}(]-\infty, r])$ in such a way that

$$\|y_0 - y_1\| = \|y_0 - y_2\| = \mathrm{dist}\left(y_0, J^{-1}(]-\infty, r])\right). \tag{3}$$

Now, define the function $f : X \times [0, +\infty[\to \mathbf{R}$ by setting

$$f(x, \lambda) = \frac{1}{2}\|x - y_0\|^2 + \lambda(J(x) - r)$$

for all $(x, \lambda) \in X \times [0, +\infty[$. Let us check that $f$ satisfies the hypotheses of Theorem 2.3. It is clear that $(a_1)$ and $(a_3)$ (with $\lambda_0 = 0$) are satisfied. So, fix $\lambda \in ]0, +\infty[$. Clearly, the functional $f(\cdot, \lambda)$ is sequentially weakly lower semicontinuous. Fix $\epsilon > 0$ so that $\frac{1}{2} - \epsilon\lambda > 0$. We claim that there is $\delta > 0$ such that

$$\inf_{\|x\| > \delta} \frac{J(x)}{\|x\|^2} \geq -\epsilon. \tag{4}$$

This is clear when $\alpha > 2$ since, by assumption, $J$ is non-negative. Suppose $\alpha < 2$. Note that $\alpha \geq 1$ since $J$ is $C^1$. If (4) was not true, we could find a sequence $\{x_n\}$ in $X$, with $\lim_{n\to\infty} \|x_n\| = +\infty$, such that

$$\frac{J(x_n)}{\|x_n\|^2} < -\epsilon$$

for all $n \in \mathbf{N}$. Choosing $\gamma \in ]0, 2 - \alpha[$ and multiplying by $\|x_n\|^\gamma$, we would have

$$J\left(\|x_n\|^{\frac{\gamma-2}{\alpha}} x_n\right) = \|x_n\|^{\gamma-2} J(x_n) < -\epsilon\|x_n\|^\gamma.$$

This would contradict the continuity of $J$, since the sequence $\{\|x_n\|^{\frac{\gamma-2}{\alpha}} x_n\}$ converges to 0. Hence, (4) holds and from it we get

$$f(x, \lambda) \geq \left(\frac{1}{2} - \epsilon\lambda\right)\|x\|^2 - \|y_0\|\|x\| + \frac{1}{2}\|y_0\|^2 - \lambda r$$

for all $x \in X$, with $\|x\| > \delta$. Consequently

$$\lim_{\|x\| \to +\infty} f(x, \lambda) = +\infty.$$

From this, we infer that $f(\cdot, \lambda)$ has has a global minimum. On the other hand, the critical points of $f(\cdot, \lambda)$ are exactly the solutions of the equation

$$x + \lambda J'(x) = y_0. \tag{5}$$

If $u$ is one of such solutions, since $J'$ is positively homogeneous of degre $\alpha - 1$ and $\alpha \neq 2$, the point $\lambda^{\frac{1}{\alpha-2}} u$ is a solution of the equation

$$\Phi(x) = \lambda^{\frac{1}{\alpha-2}} y_0. \tag{6}$$

But, since $C$ is a cone, we have $\lambda^{\frac{1}{\alpha-2}} y_0 \in C$ and so, by (2), $\lambda^{\frac{1}{\alpha-2}} y_0 \notin \Phi(S_\Phi)$. As we remarked at the beginning of the proof, this implies that the equation (6) has a unique solution, and hence also (5) has a unique solution. Therefore, this argument shows that $f(\cdot, \lambda)$ has a unique global minimum and no other local minimum. Hence, condition $(a_2)$ is satisfied. Therefore, Theorem 2.3 ensures that

$$\sup_{\lambda \geq 0} \inf_{x \in X} f(x, \lambda) = \inf_{x \in X} \sup_{\lambda \geq 0} f(x, \lambda). \tag{7}$$

Clearly, one has

$$\inf_{x \in X} \sup_{\lambda \geq 0} f(x, \lambda) = \frac{1}{2} \inf_{x \in J^{-1}(]-\infty, r])} \|x - y_0\|^2. \tag{8}$$

Now, observe that the function $\inf_{x \in X} f(x, \cdot)$ is upper semicontinuous in $[0, +\infty[$ and that $\lim_{\lambda \to +\infty} \inf_{x \in X} f(x, \lambda) = -\infty$, since $r > \inf_X J$. Hence, there is $\lambda^* \geq 0$ such that

$$\inf_{x \in X} f(x, \lambda^*) = \sup_{\lambda \geq 0} \inf_{x \in X} f(x, \lambda). \tag{9}$$

Furthermore, observe that if $y \in J^{-1}(]-\infty, r])$ is such that

$$\|y_0 - y\| = \mathrm{dist}\left(y_0, J^{-1}(]-\infty, r])\right)$$

then

$$J(y) = r.$$

Indeed, if $J(y) < r$, since $J$ is continuous and $J(y_0) > r$, there would exist a point $z$ in the line segment joining $y_0$ and $y$ such that $J(z) = r$. So, we would have

$$\|y_0 - z\| < \mathrm{dist}\left(y_0, J^{-1}(]-\infty, r])\right),$$

an absurd. Hence, recalling $(3), (7), (8)$ and $(9)$, we have

$$\inf_{x \in X} \left(\frac{1}{2}\|x - y_0\|^2 + \lambda^* J(x)\right) = \inf_{x \in J^{-1}(]-\infty, r])} \frac{1}{2}\|x - y_0\|^2 + \lambda^* r$$

$$= \frac{1}{2}\|y_1 - y_0\|^2 + \lambda^* J(y_1) = \frac{1}{2}\|y_2 - y_0\|^2 + \lambda^* J(y_2).$$

This contradicts the fact (seen above) that the functional $x \to \frac{1}{2}\|x - y_0\|^2 + \lambda^* J(x)$ has a unique global minimum. Therefore, the set $S_\Phi$ is not $\sigma$-compact. To complete the proof, observe that, by Theorem 1.1 of [6], the operator $\Phi$ is proper, that is $\Phi^{-1}(K)$ is compact for each compact set $K \subset X$. Consequently, if $\Phi(S_\Phi)$ was $\sigma$-compact, $\Phi^{-1}(\Phi(S_\Phi))$ would be so, and hence, since $S_\Phi$ is closed and $S_\Phi \subseteq \Phi^{-1}(\Phi(S_\Phi))$, $S_\Phi$ would be $\sigma$-compact too, which is impossible. The proof is complete. $\square$

*Proof of Theorem* 1.2. The fact that $J'$ is compact implies that $J$ is sequentially weakly continuous and, jointly with $(1)$, that $\Phi$ is closed. Therefore, $J$ satisfies the hypotheses of Theorem 1.1. Consequently, the set $S_\Phi$ is not $\sigma$-compact. The compactness of $J'$ again implies that, for each $x \in X \setminus B$, the linear operator $J''(x)$ is compact. Consequently, $y \to y + J''(x)(y)$ is a Fredholm operator of index 0, and so it is invertible if and only if it is surjective. Now, observe that if $x \in X \setminus (\hat{S}_{\Phi_{|(X \setminus B)}} \cup B)$, then, by the inverse function theorem, $\Phi$ is a local homeomorphism at $x$, and so $x \notin S_\Phi$. Hence, we have

$$S_\Phi \subseteq \hat{S}_{\Phi_{|(X \setminus B)}} \cup B.$$

We then infer that $\hat{S}_{\Phi_{|(X \setminus B)}}$ is not $\sigma$-compact since, otherwise, $\hat{S}_{\Phi_{|(X \setminus B)}} \cup B$ would be so, and hence also $S_\Phi$ would be $\sigma$-compact being closed. Finally, the fact that $\Phi(\hat{S}_{\Phi_{|(X \setminus B)}})$ is not $\sigma$-compact follows as in the final part of the proof of Theorem 1.1, taking into account that $\hat{S}_{\Phi_{|(X \setminus B)}}$ is closed. $\square$

The proof of Theorem 1.3 comes directly out from a joint application of the two following results:

**Theorem 2.5 ([5], Theorem 2).** *Let $J$ satisfy the assumptions of Theorem 1.3. Then, there exist $\lambda^* > 0$ and $y^* \in X$ such that the equation*

$$\Phi_{\lambda^*}(x) = y^*$$

*has at least three solutions.*

See also [2] for an extension of Theorem 2.5.

**Theorem 2.6 ([3], Theorem 5).** *If $\dim(X) \geq 3$, if $T : X \to X$ is a $C^1$ proper Fredholm operator of index 0, and if $\hat{S}_T$ is discrete, then $T$ is a homeomorphism.*

*Proof of Theorem 1.3.* By Theorem 2.5, there exists $\lambda^* > 0$ such that the operator $\Phi_{\lambda^*}$ is not a homeomorphism. Since $\Phi_{\lambda^*}$ is a compact perturbation of the identity and $\lim_{\|u\| \to \infty} \|\Phi_{\lambda^*}(u)\| = +\infty$, it is a proper Fredholm operator of index 0, and so, by Theorem 2.6, the set $\hat{S}_{\Phi_{\lambda^*}}$ contains an accumulation point, as claimed. $\square$

## 3. Remarks and applications

First, note the following consequence of Theorem 2.6:

**Theorem 3.1.** *Let $\dim(X) \geq 3$ and let $T : X \to X$ be a $C^1$ proper Fredholm operator of index 0 which is not a homeomorphism. Also, assume that the set $T(\hat{S}_T)$ is discrete. Then, there exists $y^* \in T(\hat{S}_T)$ such that the set $T^{-1}(y^*)$ contains an accumulation point.*

*Proof.* By Theorem 2.6, the set $\hat{S}_T$ contains an accumulation point, say $x^*$. Choose $y^* = T(x^*)$. Since $T(\hat{S}_T)$ is discrete, there is a neighbourhood $V$ of $y^*$ such that $V \cap T(\hat{S}_T) = \{y^*\}$. By continuity, $T^{-1}(V)$ is a neighbourhood of $x^*$ and so, for every neighbourhood $U$ of $x^*$, there exists $\hat{x} \in T^{-1}(V) \cap U \cap \hat{S}_T$ with $\hat{x} \neq x^*$. Thus, we necessarily have $T(\hat{x}) = T(x^*)$ and so $x^*$ is an accumulation point of $T^{-1}(y^*)$. $\square$

**Remark 3.2.** On the basis of Theorem 3.1, it would be interesting to know classes of functionals $J$ satisfying the assumptions of Theorem 1.3 and such that, for each $\lambda > 0$, the set $\Phi_\lambda(\hat{S}_{\Phi_\lambda})$ is discrete.

It is also worth noticing the following consequence of Theorem 1.3:

**Theorem 3.3.** *Let* $\dim(X) \geq 3$. *Assume that $J$ is of class $C^2$ and not quasi-convex, that $J'$ is compact and that*

$$\lim_{\|x\| \to \infty} \|\Phi(x)\| = +\infty.$$

*Assume also that $J$ is positively homogeneous of degree $\alpha \neq 2$. If $\alpha > 2$ suppose, in addition, that $J$ is non-negative. Then, the set $\hat{S}_\Phi$ contains at least one accumulation point.*

*Proof.* In the proof of Theorem 1.1, we have already observed that from the present assumptions we get

$$\liminf_{\|x\| \to +\infty} \frac{J(x)}{\|x\|^2} \geq 0.$$

Clearly, for each $\lambda > 0, x \in X$, we have

$$\lambda^{\frac{1}{\alpha-2}} \Phi_\lambda(x) = \Phi\left(\lambda^{\frac{1}{\alpha-2}} x\right), \tag{10}$$

and so

$$\Phi'_\lambda(x) = \Phi'\left(\lambda^{\frac{1}{\alpha-2}} x\right). \tag{11}$$

From (10), we infer

$$\lim_{\|x\| \to +\infty} \|\Phi_\lambda(x)\| = +\infty$$

for all $\lambda > 0$. Therefore, all the assumptions of Theorem 1.3 are satisfied. Consequently, for some $\lambda^* > 0$, the set $\hat{S}_{\Phi_{\lambda^*}}$ contains an accumulation point. But, due to (11), we have

$$\hat{S}_\Phi = \lambda^{*\frac{1}{\alpha-2}} \hat{S}_{\Phi_{\lambda^*}}$$

and hence the set $\hat{S}_\Phi$ contains an accumulation point, as claimed. $\square$

**Remark 3.4.** It is obvious, but meaningful, to note that, when $\Phi_\lambda$ is a Fredholm operator of index 0, we have $x \in \hat{S}_{\Phi_\lambda}$ if and only if $-\frac{1}{\lambda}$ is an eigenvalue of the linear operator $J''(x)$.

**Remark 3.5.** It is interesting to apply the previous remark to Theorem 1.3 jointly with a classical characterization of $C^2$ convex functions. Namely, assume that $J$ is $C^2$ and not convex. Then, by the above quoted result ([9], Theorem 2.1.11), there exist some $x, y \in X$ such that $\langle J''(x)(y), y \rangle < 0$. Assuming that $J'$ is compact, this, in turn, implies that the operator $J''(x)$ has a negative eigenvalue (in general, depending on $x$). For not quasi-convex functionals $J$, Theorem 1.3 is able to ensure a much stronger conclusion: that is, there exists some negative number $\mu$ such that the set of all $x \in X$ for which $\mu$ is an eigenvalue of $J''(x)$ contains an accumulation point.

We now present some applications of Theorems 1.1 and 1.3 to the Dirichlet problem

$$\begin{cases} -\Delta u = f(x, u) \text{ in } \Omega \\ \\ u = 0 \text{ on } \partial\Omega. \end{cases} \tag{$P_f$}$$

So, in the sequel $\Omega \subset \mathbf{R}^n$ is a bounded domain, with smooth boundary, and $H_0^1(\Omega)$ is the usual Sobolev space, with the scalar product

$$\langle u, v \rangle = \int_\Omega \nabla u(x) \nabla v(x) dx$$

and the norm

$$\|u\| = \left( \int_\Omega |\nabla u(x)|^2 dx \right)^{\frac{1}{2}}.$$

Recall that

$$\int_\Omega |u(x)|^2 dx \leq \frac{1}{\lambda_1} \int_\Omega |\nabla u(x)|^2 dx$$

for all $u \in H_0^1(\Omega)$, $\lambda_1$ being the first eigenvalue of the problem

$$\begin{cases} -\Delta u = \lambda u \text{ in } \Omega \\ \\ u = 0 \text{ on } \partial\Omega. \end{cases}$$

As usual, if $f : \Omega \times \mathbf{R} \to \mathbf{R}$ is a Carathéodory function, a weak solution of problem $(P_f)$ is any $u \in H_0^1(\Omega)$ such that

$$\int_\Omega \nabla u(x) \nabla v(x) dx = \int_\Omega f\big(x, u(x)\big) v(x) dx$$

for all $v \in H_0^1(\Omega)$.

We denote by $\mathcal{A}$ the class of all Carathéodory functions $f : \Omega \times \mathbf{R} \to \mathbf{R}$ such that

$$\sup_{(x,\xi) \in \Omega \times \mathbf{R}} \frac{|f(x, \xi)|}{1 + |\xi|} < +\infty.$$

For each $f \in \mathcal{A}$ and $u \in H_0^1(\Omega)$, we put

$$I_f(u) = \int_\Omega \left( \int_0^{u(x)} f(x, \xi) d\xi \right) dx.$$

So, by classical results, the functional $I_f$ is (well defined and) continuously Gâteaux differentiable on $H_0^1(\Omega)$, its derivative is compact, and one has

$$I_f'(u)(v) = \int_\Omega f\big(x, u(x)\big) v(x) dx$$

for all $u, v \in H_0^1(\Omega)$.

The following propositions will be useful in the sequel.

**Proposition 3.6.** *Let $\beta \in L^\infty(\Omega)$, with ess $\sup_\Omega \beta > 0$, and let $g : \mathbf{R} \to \mathbf{R}$ be a continuous not quasi-convex function, with $g(0) = 0$, such that $\beta(\cdot)g(u(\cdot)) \in L^1(\Omega)$ for all $u \in H_0^1(\Omega)$. Set*

$$I(u) = \int_\Omega \beta(x)g\big(u(x)\big)dx$$

*for all $u \in H_0^1(\Omega)$. Then, the functional $I$ is not quasi-convex.*

*Proof.* Since $g$ is not quasi-convex, there are $\xi_0, \xi_1, \xi^* \in \mathbf{R}$, with $\xi_0 < \xi^* < \xi_1$, such that

$$\max\big\{g(\xi_0), g(\xi_1)\big\} < g(\xi^*).$$

Since ess $\sup_\Omega \beta > 0$, there is a compact set $C \subset \Omega$ such that

$$r := \int_C \beta(x) > 0.$$

Now, fix $\gamma$ satisfying

$$r \max\big\{g(\xi_0), g(\xi_1)\big\} < \gamma < rg(\xi^*)$$

and then choose an open set $A \subset \Omega$, with $C \subset A$, in such a way that

$$\|\beta\|_{L^\infty(\Omega)} \max_{|\xi| \le |\xi_0| + |\xi_1|} |g(\xi)|\mathrm{meas}(A \backslash C) < \min\big\{\gamma - r \max\big\{g(\xi_0), g(\xi_1)\big\}, rg(\xi^*) - \gamma\big\}.$$
$$(12)$$

Also, for $i = 0, 1$, fix a function $u_i \in H_0^1(\Omega)$ such that

$$u_i(x) = \xi_i$$

for all $x \in C$,

$$u_i(x) = 0$$

for all $x \in \Omega \setminus A$ and

$$|u_i(x)| \le |\xi_i|$$

for all $x \in \Omega$. Since $g(0) = 0$, in view of (12), we clearly have

$$I(u_i) = \int_{A \backslash C} \beta(x)g\big(u_i(x)\big)dx + \int_C \beta(x)g\big(u_i(x)\big)dx$$
$$< rg(\xi_i) + \|\beta\|_{L^\infty(\Omega)} \max_{|\xi| \le |\xi_0| + |\xi_1|} |g(\xi)|\mathrm{meas}(A \setminus C) < \gamma.$$

On the other hand, if $\lambda \in ]0, 1[$ is such that $\lambda\xi_0 + (1 - \lambda)\xi_1 = \xi^*$, by (12) again, we have

$$I\big(\lambda u_0 + (1 - \lambda)u_1\big) = \int_{A \backslash C} \beta(x)g\big(\lambda u_0(x) + (1 - \lambda)u_1(x)\big)dx + rg(\xi^*)$$
$$> -\|\beta\|_{L^\infty(\Omega)} \max_{|\xi| \le |\xi_0| + |\xi_1|} |g(\xi)|\mathrm{meas}(A \setminus C) + rg(\xi^*) > \gamma.$$

This shows that the sublevel set $I^{-1}(] - \infty, \gamma])$ is not convex, and the proof is complete. $\qquad\square$

**Proposition 3.7.** *Let $f \in \mathcal{A}$ satisfy*

$$\lim_{|\xi| \to +\infty} \sup_{x \in \Omega} \left| \frac{f(x, \xi)}{\xi} \right| = 0.$$

*Then, for each $\lambda \in \mathbf{R}$, one has*

$$\lim_{\|u\| \to +\infty} \|u + \lambda I'_f(u)\| = +\infty.$$

*Proof.* First, note that

$$\lim_{r \to +\infty} \frac{\sup_{(x,\xi) \in \Omega \times [-r,r]} |f(x,\xi)|}{r} = 0. \tag{13}$$

Indeed, arguing by contradiction, assume that

$$\limsup_{r \to +\infty} \frac{\sup_{(x,\xi) \in \Omega \times [-r,r]} |f(x,\xi)|}{r} > 0.$$

Then, there exist $\gamma > 0$, a sequence $\{r_k\}$, with $\lim_{k \to \infty} r_k = +\infty$, and a sequence $\{(x_k, \xi_k)\}$, such that

$$(x_k, \xi_k) \in \Omega \times [-r_k, r_k]$$

and

$$|f(x_k, \xi_k)| > \gamma r_k \tag{14}$$

for all $k \in \mathbf{N}$. But, by assumption, there is $\delta > 0$ such that

$$\sup_{x \in \Omega} \left| \frac{f(x, \xi)}{\xi} \right| < \gamma$$

provided $|\xi| > \delta$. Consequently, we have $|\xi_k| \leq \delta$ for all $k \in \mathbf{N}$. Then, since $f \in \mathcal{A}$, we have

$$\sup_{k \in \mathbf{N}} |f(x_k, \xi_k)| < +\infty$$

which contradicts (14). Since

$$\|u + \lambda I'_f(u)\| \geq \|u\| \left( 1 - |\lambda| \frac{\|I'_f(u)\|}{\|u\|} \right),$$

to prove the thesis it is enough to show that

$$\lim_{\|u\| \to +\infty} \frac{\|I'_f(u)\|}{\|u\|} = 0. \tag{15}$$

Since

$$\|I'_f(u)\| = \sup_{\|v\|=1} \int_{\Omega} |f(x, u(x))| |v(x)| dx,$$

we clearly have

$$\|I'_f(u)\| \leq \lambda_1^{-\frac{1}{2}} \left( \int_{\Omega} |f(x, u(x))|^2 dx \right)^{\frac{1}{2}} \tag{16}$$

for all $u \in H_0^1(\Omega)$. Now, fix $\epsilon > 0$ and, thanks to (13), choose $\eta > 0$ so that

$$\sup_{(x,\xi) \in \Omega \times [-r,r]} |f(x,\xi)| < \left( \frac{\epsilon}{2(\text{meas}(\Omega) + 1)} \right)^{\frac{1}{2}} r$$

for all $r > \eta$. Set

$$M_\eta = \sup_{\int_\Omega |u(x)|^2 dx \leq \eta^2} \int_\Omega |f(x, u(x))|^2 dx .$$

Clearly, $M_\eta < +\infty$. Fix $u \in H_0^1(\Omega)$, with $\|u\|^2 > \frac{M_\eta}{\epsilon}$. We distinguish two cases. If $\int_\Omega |u(x)|^2 dx \leq \eta^2$, we clearly have

$$\int_\Omega |f(x, u(x))|^2 dx \leq \frac{M_\eta}{\epsilon} \epsilon < \epsilon \|u\|^2 . \tag{17}$$

Now, suppose that $\int_\Omega |u(x)|^2 dx > \eta^2$. Set

$$D = \left\{ x \in \Omega : |u(x)|^2 \leq \int_\Omega |u(y)|^2 dy \right\} .$$

So, we have

$$|f(x, u(x))|^2 < \frac{\epsilon}{2\text{meas}(\Omega)} \int_\Omega |u(y)|^2 dy$$

for all $x \in D$, as well as

$$|f(x, u(x))|^2 < \frac{\epsilon}{2} |u(x)|^2$$

for all $x \in \Omega \setminus D$. Consequently

$$\int_\Omega |f(x, u(x))|^2 dx = \int_D |f(x, u(x))|^2 dx + \int_{\Omega \setminus D} |f(x, u(x))|^2 dx$$

$$< \frac{\epsilon \int_\Omega |u(x)|^2 dx}{2\text{meas}(\Omega)} \text{meas}(D) + \frac{\epsilon}{2} \int_{\Omega \setminus D} |u(x)|^2 dx \leq \epsilon \int_\Omega |u(x)|^2 dx . \tag{18}$$

Putting (16), (17) and (18) together, we then have

$$\|I_f'(u)\|^2 \leq \max \{ \lambda_1^{-1}, \lambda_1^{-2} \} \epsilon \|u\|^2 .$$

So, (15) holds, and the proof is complete.                                   $\square$

As an application of Theorem 1.1, we now prove the following

**Theorem 3.8.** *Let $\beta \in L^\infty(\Omega)$, with ess $\sup_\Omega \beta > 0$, and let $q \in ]0,1[$. For each $\varphi \in H_0^1(\Omega)$, denote by $\Lambda_\varphi$ the set of all weak solutions of the problem*

$$\begin{cases} -\Delta u = \beta(x) |u + \varphi(x)|^{q-1} (u + \varphi(x)) & \text{in } \Omega \\ u = 0 & \text{on } \partial\Omega . \end{cases}$$

*Then, there exist two closed, not $\sigma$-compact sets $A, B \subset H_0^1(\Omega)$ with the following properties:*

(i) *for each $\varphi \in B$ there exist $w \in A$ and three sequences $\{u_k\}$, $\{v_k\}$ and $\{\varphi_k\}$ in $H_0^1(\Omega)$ such that*

$$\lim_{k \to \infty} u_k = \lim_{k \to \infty} v_k = w - \varphi, \qquad \lim_{k \to \infty} \varphi_k = \varphi$$

*and, for each $k \in \mathbf{N}$,*

$$u_k \neq v_k \quad and \quad u_k, v_k \in \Lambda_{\varphi_k} \, ;$$

(ii) *for each $\varphi \in H_0^1(\Omega) \setminus B$, the set $\Lambda_\varphi$ is non-empty, finite and disjoint from $A - \varphi$.*

*Proof.* Set

$$f(x, \xi) = \beta(x)|\xi|^{q-1}\xi$$

for all $(x, \xi) \in \Omega \times \mathbf{R}$. Clearly, $f \in \mathcal{A}$. Let us apply Theorem 1.1, taking $X = H_0^1(\Omega)$ and $J = -I_f$. So, the functional $J$ is positively homogeneous of degree $q + 1 < 2$, is not quasi-convex (by Proposition 3.6), has compact derivative, and satisfies (1) (by Proposition 3.7). Thus, Theorem 1.1 applies. Choose

$$A = S_\Phi$$

and

$$B = \Phi(S_\Phi) \, .$$

By Theorem 1.1, the closed sets $A, B$ are not $\sigma$-compact. Observe that if $u, \varphi \in X$, then $\Phi(u) = \varphi$ if and only if

$$\int_\Omega \nabla\big(u(x) - \varphi(x)\big)\nabla v(x)dx - \int_\Omega \beta(x)|u(x)|^{q-1}u(x)v(x)dx = 0$$

for all $v \in X$. This is equivalent to the fact that $u - \varphi \in \Lambda_\varphi$. In other words, we have

$$\Phi^{-1}(\varphi) = \varphi + \Lambda_\varphi \, . \tag{19}$$

Fix $\varphi \in B$ and let $w \in S_\Phi$ be such that $\Phi(w) = \varphi$. Let $U$ be any open neighbourhood of $w$. Since $\Phi$ is not a local homeomorphism at $w$, it follows that $\Phi_{|U}$ is not injective. Indeed, assume that $\Phi_{|U}$ is injective. Then, since $\Phi$ is a compact perturbation of the identity, in view of the invariance of domain theorem ([10], Theorem 16.C), it would follow that, for each open set $E \subset U$, the set $\Phi(E)$ would be open in $X$. This, in turn, would imply that $\Phi_{|U}$ is a homeomorphism between $U$ and the neighbourhood $\Phi(U)$ of $\varphi$, which is impossible. Consequently, we can clearly construct two sequences $\{\hat{u}_k\}, \{\hat{v}_k\}$ in $X$, both converging to $w$, such that $\hat{u}_k \neq \hat{v}_k$ and $\Phi(\hat{u}_k) = \Phi(\hat{v}_k)$ for all $k \in \mathbf{N}$. Hence, if we choose

$$\varphi_k = \Phi(\hat{u}_k), \quad u_k = \hat{u}_k - \varphi_k, \quad v_k = \hat{v}_k - \varphi_k,$$

the sequences $\{\varphi_k\}$, $\{u_k\}$ and $\{v_k\}$ satisfy (*i*), by (19). Now, let $\varphi \in X \setminus B$. Clearly, the functional $u \to \frac{1}{2}\|u - w\|^2 + J(u)$ is coercive (since $q < 1$) besides being sequentially weakly lower semicontinuous. So, it has a global minimum $\hat{u}$ in $X$. Therefore, $\Phi(\hat{u}) = \varphi$. By the already invoked Theorem 1.1 of [6], the operator

$\Phi$ is proper, and so $\Phi^{-1}(\varphi)$ is compact. But, since $\varphi \notin \Phi(S_\Phi)$, $\Phi^{-1}(\varphi)$ is also discrete, and so it is finite. Now, (ii) follows directly from (19).                    □

Here is the final result.

**Theorem 3.9.** *Let $\beta \in L^\infty(\Omega)$, with ess $\sup_\Omega \beta > 0$, and let $\psi : \mathbf{R} \to \mathbf{R}$ be a $C^1$ function such that $\xi \to \int_0^\xi \psi(t)dt$ is not quasi-concave and*

$$\lim_{|\xi| \to +\infty} \frac{\psi(\xi)}{\xi} = 0. \tag{20}$$

*Moreover, if $n \geq 2$, assume that*

$$\sup_{\xi \in \mathbf{R}} \frac{|\psi'(\xi)|}{1 + |\xi|^p} < +\infty$$

*where $p > 0$ and $p \leq \frac{4}{n-2}$ if $n \geq 3$. Then, there exists $\lambda^* > 0$ such that the set of all $u \in H_0^1(\Omega)$ for which the problem*

$$\begin{cases} -\Delta v = \lambda^* \beta(x)\psi'\big(u(x)\big)v \ \text{ in } \Omega \\ \\ v = 0 \ \text{ on } \partial\Omega . \end{cases}$$

*has a non-zero weak solution contains an accumulation point.*

*Proof.* Set

$$f(x, \xi) = \beta(x)\psi(\xi)$$

for all $(x, \xi) \in \Omega \times \mathbf{R}$. Clearly, $f \in \mathcal{A}$. Let us apply Theorem 1.3, taking $X = H_0^1(\Omega)$ and $J = -I_f$. Our assumptions imply that $J$ is of class $C^2$, with

$$\big\langle J''(u)(v), w \big\rangle = - \int_\Omega \beta(x)\psi'\big(u(x)\big)v(x)w(x)dx$$

for all $u, v, w \in X$. Thanks to (20), from the proof of Proposition 3.7, we know that

$$\lim_{\|u\| \to +\infty} \frac{I_f'(u)}{\|u\|} = 0. \tag{21}$$

We claim that

$$\lim_{\|u\| \to +\infty} \frac{I_f(u)}{\|u\|^2} = 0.$$

Arguing by contradiction, assume that there exist $\gamma > 0$ and a sequence $\{u_k\}$, with $\lim_{k \to \infty} \|u_k\| = +\infty$, such that

$$I_f(u_k) > \gamma\|u_k\|^2$$

for all $k \in \mathbf{N}$. By the mean value theorem, for each $k \in \mathbf{N}$, there is $t_k \in ]0, 1[$ so that $I_f(u_k) = \langle I_f'(t_k u_k), u_k \rangle$. So, we clearly have

$$\big\|I_f'(t_k u_k)\big\| > \gamma\|u_k\| . \tag{22}$$

Since $I'_f$ is bounded on each bounded subset of $X$, we then infer that, up to a subsequence, $\lim_{k\to\infty} \|t_k u_k\| = +\infty$. Consequently, if $k$ is large enough, in view of (21), we would have

$$\|I'_f(t_k u_u)\| < \gamma \|t_k u_k\|$$

which contradicts (22). So, in view of Propositions 3.6 and 3.7, all the assumptions of Theorem 1.3 are satisfied. The conclusion then follows directly from Theorem 1.3, taking Remark 3.4 into account. □

**Acknowledgment**

I would like to thank J. Saint Raymond for a useful correspondence.

# References

[1] M. S. Berger and R. A. Plastock, *On the singularities of nonlinear Fredholm operators of positive index*, Proc. Amer. Math. Soc. **79** (1980), 217–221.

[2] F. Faraci and A. Iannizzotto, *An extension of a multiplicity theorem by Ricceri with an application to a class of quasilinear equations*, Studia Math. **172** (2006), 275–287.

[3] R. A. Plastock, *Nonlinear Fredholm maps of index zero and their singularities*, Proc. Amer. Math. Soc. **68** (1978), 317–322.

[4] B. Ricceri, *A further improvement of a minimax theorem of Borenshtein and Shul'man*, J. Nonlinear Convex Anal. **2** (2001), 279–283.

[5] B. Ricceri, *A general multiplicity theorem for certain nonlinear equations in Hilbert spaces*, Proc. Amer. Math. Soc. **133** (2005), 3255–3261.

[6] R. S. Sadyrkhanov, *On infinite dimensional features of proper and closed mappings*, Proc. Amer. Math. Soc. **98** (1986), 643–648.

[7] S. Smale, *An infinite-dimensional version of Sard's theorem*, Amer. Math. J. **87** (1965), 861–866.

[8] I. G. Tsar'kov, *Nonunique solvability of certain differential equations and their connection with geometric approximation theory*, Math. Notes **75** (2004), 259–271.

[9] C. Zălinescu, *Convex analysis in general vector spaces*, World Scientific, 2002.

[10] E. Zeidler, *Nonlinear Functional Analysis and its Applications*, vol. I, Springer, 1986.

Biagio Ricceri
Department of Mathematics
University of Catania
Viale A. Doria 6
I-95125 Catania
Italy
e-mail: ricceri@dmi.unict.it

Progress in Nonlinear Differential Equations
and Their Applications, Vol. 75, 393–406
© 2007 Birkhäuser Verlag Basel/Switzerland

# Shape and Conley Index of Attractors and Isolated Invariant Sets

José M. R. Sanjurjo

*Dedicated to A. Cellina and J. Yorke*

**Abstract.** This article is an exposition of several results concerning the theory of continuous dynamical systems, in which Topology plays a key role. We study homological and homotopical properties of attractors and isolated invariant compacta as well as properties of their unstable manifolds endowed with the intrinsic topology. We also provide a dynamical framework to express properties which are studied in Topology under the name of *Hopf duality*. Finally we see how the use of the intrinsic topology makes it possible to calculate the Conley–Zehnder equations of a Morse decomposition of an isolated invariant compactum, provided we have enough information about its unstable manifold.

## 1. Introduction

The aim of this paper is to survey several results in which Topology plays an important role in the study of the properties of flows, attractors and isolated invariant sets. The properties studied are mainly of a homological and homotopical nature. Some of the results are stated in terms of Čech cohomology and also we make use of the notion of shape, which was introduced by K. Borsuk in 1968 ([8]) as a generalized homotopy type, which agrees with the usual one when applied to spaces with good topological properties, but which gives deeper geometric insight in the case of spaces with more complicated topological structure, like many attractors. We give here an example of a situation in which shape theory appears in a natural way: Suppose $K$ is a compactum contained in the interior of a manifold $M$ such that there exists a neighborhood basis $\{U_m \mid m = 0, 1, 2, \dots\}$ of $K$ in $M$ with $U_0 = M$ and each inclusion $U_{m+1} \to U_m$ a homotopy equivalence, that is, $U_m$ can be deformed into $U_{m+1}$. This situation appears very often in dynamical systems and exemplifies a particular instance in which the inclusion $i : K \to M$ is a *shape equivalence* (and, hence $K$ and $M$ have the same shape). There are more general situations and there is even a general notion of *shape of a space* (see Kapitanski and Rodnianski [21] or Robbin and Salamon [26] for an exposition suitable for nonspecialists with application to dynamical systems). The reader of this article should think of shape as a notion similar to homotopy type, but a little more general. In many cases the two notions agree: for instance two manifolds have the same shape if and only if they have the same homotopy type. In general, two spaces which have the same homotopy type have also the same shape but the converse statement is not always true. Another topological notion we shall use is the one of *Absolute Neighborhood Retract (ANR)*, also introduced by Borsuk. Manifolds are the most important examples of this notion but Absolute Neighborhood Retracts are spaces more general than manifolds (they can be infinite dimensional) although they share many of their good properties. Some good references are Borsuk [7] and Hu [20].

Another aim of this paper is to provide a dynamical framework to express duality properties of flows that correspond to those that are studied in Topology under the name of *Hopf duality*. This kind of duality refers to $(n-1)$-manifolds, $W$, embedded in the n-sphere $S^n$ and establishes homological relations between the two $n$-manifolds with boundary in which $S^n$ is decomposed by $W$ (see Steenrod and Epstein [35]). We consider here the situation of flows defined in a locally compact metric space $X$ possessing an attractor, $M$, which is an $n$-manifold satisfying some specific conditions. The attractor is endowed with a Morse decomposition $\{M_0, M_1, M_2\}$ where $M_0$ is an $(n-1)$-submanifold of $M$ decomposing $M$ into two manifolds with common boundary $M_0$. We discuss in the paper some Hopf duality properties of the homology and cohomology Conley indices of the Morse sets. We also study the more general situation in which $M$ is required only to be an isolated invariant set (not necessarily an attractor) and we get some homological properties of the Morse sets. The most general result in this direction, stated in Corollary 3.8, is presented in terms of the unstable manifold with its intrinsic topology, as defined by Robbin and Salamon, and formulated in the language of Čech homology.

We also survey in this paper some results on the Morse theory of flows. Kapitanski and Rodnianski developed in [21] an approach to the Morse theory of attractors based on shape theory and we studied in [32] the more general case of isolated invariant compacta. This approach presents several advantages over the classical one. First of all, it shows how to calculate, in many cases, the Conley–Zehnder equations for a Morse decomposition of an isolated invariant compactum without making use of index pairs. The evaluation of the coefficients is made in terms of the unstable manifolds of the Morse sets. To be more precise, we use a truncation of the unstable manifolds, i.e., the part of the flow wich evolves before reaching a section of the unstable manifold. This greatly simplifies the whole process of calculation. In addition to the simplicity, this approach allows to handle more general metric spaces and evolutions. Finally, the use of shape theory makes unnecessary, in the most important cases, the usual hypotheses of finiteness of the Betti numbers of the cohomological Conley index, since in this situation finiteness is an automatic consequence of shape theory for many of the spaces that we consider here. We see that there is a filtration of truncated unstable manifolds associated to the Morse decomposition of an isolated invariant set from which it is possible to obtain the Conley–Zehnder equations. The unstable manifolds are endowed here with their intrinsic topology. We study in detail the intrinsic topology of the unstable manifolds and find necessary and sufficient conditions for the intrinsic and the extrinsic topologies to agree with each other. These conditions are expressed in terms of internal properties of the unstable manifolds.

The reader is supposed to be familiar with the most elementary notions of Dynamical Systems and Topology. Good references are the books [4] and [34].

## 2. The shape of isolated invariant compacta

In 1979 H. M. Hastings developed in [18] and [19] an analogue of the Poincaré–Bendixson theorem in Euclidean $n$-space, and gave several examples. According to him, the usual proof breaks down in higher dimensions because the main tool, the Jordan curve theorem, cannot be extended. He circumvented this problem by studying the motion of the whole manifold $M$ through time (and not just the orbits of separate points), and replacing a geometric description of the invariant set $K$ by a description of its shape (given by means of a neighborhood base in $\mathbb{R}^n$)

**Theorem 2.1.** *Let $M$ be a compact $n$-dimensional submanifold of $\mathbb{R}^n$ with boundary. Let $\varphi : \mathbb{R}^n \times [0, \infty) \to \mathbb{R}^n$ be a semi-dynamical system such that the orbits through the boundary of $M$ enter $M$ for increasing $t$. Then there is an asymptotically stable attractor $K$ contained in the interior of $M$ such that the inclusion $i : K \to M$ is a shape equivalence.*

As a consequence of this result the attractor $K$ and the manifold $M$ have the same shape, which implies that they share many global topological properties, in particular they have the same Čech homology and cohomology. The attractor might be topologically complicated even in the 2-dimensional case, for instance Hastings gave an example where $M$ is a planar ring and $K$ is the Warsaw circle. The later is, however, shape equivalent to the standard circle (although not homotopically equivalent).

Motivated by Hasting's result several authors used shape theory to study the properties of attractors. The following result has been established by various authors at different levels of generality.

**Theorem 2.2.** *Let $\varphi : M \times \mathbb{R} \to M$ be a flow defined on a manifold, $M$, or, more generally, on a locally compact $ANR$. Suppose $K$ is an asymptotically stable attractor of $M$, then $K$ has polyhedral shape (i.e., $K$ has the shape of a compact polyhedron).*

As a consequence of Theorem 2.2 and Borsuk's theorem on the shape classification of plane continua ([9], see also [12] and [22]) according which two of them have the same shape if and only if they decompose the plane in the same number of connected components, we have the following result.

**Corollary 2.3.** *If a compact connected subset $K$ of $\mathbb{R}^2$ is an attractor of a flow $\varphi : \mathbb{R}^2 \times \mathbb{R} \to \mathbb{R}^2$ then $K$ has the shape of a point or of a finite bouquet of circles.*

A consequence of Corollary 2.3 is that 1-dimensional Čech homology and cohomology groups of planar connected attractors are free and finitely generated.

Theorem 2.2 was first proved by Bogatyi and Gutsu [6] for differentiable flows and later by Günther and Segal [17] for continuous flows in manifolds and by J. M. R. Sanjurjo ([30] and [31]) for (nonnecessarily finite-dimensional) $ANR's$. As a consequence of this result attractors of flows in manifolds have finitely generated Čech homology and cohomology which vanishes in higher dimensions. This means that Algebraic Topology can be used as an efficient tool for the study of attractors.

Several forms of this result were proved later in a more general context.

**Theorem 2.4.** *Assume that a continuous semi-dynamical system $\varphi : M \times \mathbb{R} \to M$, where $M$ is a complete metric space, possesses a compact global attractor $K$. Assume that the system has an equilibrium $z \in K$. Then the inclusion $i : (K, z) \to (M, z)$ induces a shape equivalence of pointed spaces.*

Theorem 2.4 was proved by Kapitanski and Rodnianski in [21]. A more general result was given by Giraldo, Morón, Ruiz del Portal and Sanjurjo in [15] when $M$ is a topological Hausdorff space. Another result, eliminating the requirement of existence of an equilibrium of the system in Theorem 2.4, is presented in [16]. The result is the following.

**Theorem 2.5.** *Let $M$ be a metric ANR and assume that $\varphi : M \times \mathbb{R} \to M$ is a continuous semi-dynamical system which possesses a compact global attractor $K$. Then for every $z \in K$ the inclusion $i : (K, z) \to (M, z)$ induces a strong shape equivalence of pointed spaces.*

Two corollaries can be deduced from Theorem 2.5:

**Corollary 2.6.** *Let $M$ be a metric ANR and assume that $\varphi : M \times \mathbb{R} \to M$ is a continuous semi-dynamical system which possesses a compact global attractor $K$. Then the shape groups and the homotopy pro-groups of $M$ and $K$ are isomorphic.*

and

**Corollary 2.7.** *Let $M$ be a metric ANR and assume that $\varphi : M \times \mathbb{R} \to M$ is a continuous semi-dynamical system which possesses a local attractor $K$ which is also an ANR. Then $K$ and its basin of attraction $\mathcal{A}(K)$ have the same homotopy type and, as a consequence, their Euler characteristics $\chi(K)$ and $\chi(\mathcal{A}(K))$ agree.*

The existence of global attractors has some implications on the topological properties of the positively invariant regions of the phase space. In the following result [16] we present an example.

**Theorem 2.8.** *Let $M$ be a metric ANR and assume that $\varphi : M \times \mathbb{R} \to M$ is a continuous semi-dynamical system which possesses a compact global attractor $K$. If $\dim(K) = n$ then every positively invariant closed set has the shape of a $m -$ dimensional compactum ($m \leq n$). As a consequence its Čech homology and cohomology groups vanish for dimensions higher than $n$.*

In the same paper [16] some results are presented which study the properties of the connected components of attractors and their relations with the components of the phase space. We have in particular

**Theorem 2.9.** *Let $M$ be a metric space and let $\varphi : M \times \mathbb{R} \to M$ be a continuous semi-dynamical system with a compact global attractor $K$. Then for every connected component $M_\alpha$ of $M$ there is exactly one component $K_\alpha$ of $K$ contained in $M_\alpha$ such that $K_\alpha$ is a global attractor of the semiflow restricted to $M_\alpha$. In particular if $M$ is connected then $K$ is connected.*

Another result studying the same kind of properties when local connectedness is assumed is the following.

**Theorem 2.10.** *Let $M$ be a metric space and let $\varphi : M \times \mathbb{R} \to M$ be a continuous semi-dynamical system with a compact global attractor $K$. Suppose that either $M$ or $K$ is locally connected. Then $M$ and $K$ have a finite number of components*

Attractors and repellers are the main examples of *isolated invariant compacta,* which are the class of sets studied in the Conley index theory [10]. A natural question, in view of the previous results, is whether all isolated invariant compacta have polyhedral shape. This is not always true as the following result shows (see [14]):

**Theorem 2.11.** *Any finite-dimensional compactum $K$ can be embedded in $\mathbb{R}^n$, for suitable $n$, in such a way that there is a flow in $\mathbb{R}^n$ having $K$ as an isolated invariant set.*

This result is proved by embedding $K$ as a subset of the diagonal of some $\mathbb{R}^{2n}$ and defining a translation flow $\varphi$ without fixed points in this space. We then use a theorem of Beck [3] to modify $\varphi$ to a new flow $\varphi'$ in such a way that all the orbits of $\varphi$ not containing a point of $K$ are preserved in $\varphi'$ while the orbits containing a point of $K$ are decomposed into two orbits together with that point of $K$. Then $K$ is an isolated invariant set for the flow $\varphi'$. As a consequence of this result we see that the shape of isolated invariant sets might be quite general. In order to get some control over their shapes we need to impose an additional condition on the invariant set, namely that of being non-saddle.

Let $\varphi : M \times \mathbb{R} \to M$ be a flow. A compact set $K \subset M$ is said to be a *saddle set* if there is a neighborhood $U$ of $K$ in $M$ such that every neighborhood $V \subset U$ of $K$ contains a point $x \in V$ with $\gamma^+(x) \not\subset U$ and $\gamma^-(x) \not\subset U$ (where $\gamma^+(x)$ and $\gamma^-(x)$ are the positive and negative semitrajectories respectively). We say that $K$ is *non-saddle* if it is not a saddle set, i.e., if for every neighborhood $U$ of $K$ there exists a neighborhood $V \subset U$ such that for every $x \in V$, $\gamma^+(x) \subset U$ or $\gamma^-(x) \subset U$. The following result generalizes the previous theorems to a wider context.

**Theorem 2.12.** *Let $K$ be an isolated non-saddle set of the flow $\varphi : M \times \mathbb{R} \to M$, where $M$ is a locally compact $ANR$. Then $K$ has the shape of a polyhedron and, hence, it has finitely generated Čech homology and cohomology.*

The following result shows that if we limit ourselves to flows in manifolds then the topological condition of shape triviality has strong dynamical consequences when imposed onto non-saddle sets.

**Theorem 2.13.** *Let $K$ be an isolated non-saddle set of the flow $\varphi : M \times \mathbb{R} \to M$, where $M$ is an $n$-manifold with $n > 1$. If $K$ has trivial shape then $K$ is an attractor or a repeller.*

The property of robustness of attractors for discrete dynamical systems was studied by J. Milnor in [25]. The idea is to study the behaviour of the attractor when the map $f$ which originates the system is perturbed. In [31] we study several properties of attractors of flows related to robustness. We show that, in certain circumstances, small perturbations of the flow $\varphi$ produce attractors whose global topological properties are comparable to those of the attractors of $\varphi$. This means that attractors are not only dynamically robust but also topologically so. We have in particular:

**Theorem 2.14.** *Let $\varphi_\lambda : M \times \mathbb{R} \to M$ be a parametrized family of flows where $\lambda \in [0, 1]$ and $M$ is a manifold. Let $K$ be a stable attractor of the flow $\varphi_0 : M \times \mathbb{R} \to M$. Then for every neighborhood $V$ of $K$ in $M$ there exists an $\varepsilon > 0$ such that for every $\lambda \in [0, \epsilon]$ there exists an attractor $K_\lambda \subset V$ of $\varphi_\lambda$ with $Sh(K_\lambda) = Sh(K_0)$.*

When $M$ is a 2-manifold we obtain the following consequence of Theorem 2.14:

**Corollary 2.15.** *Suppose that in Theorem 2.14 $M$ is a 2-manifold and $K \subset int\ D$, where $D$ is a topological disk in $M$. Then we can conclude the existence of attractors $K_\lambda$ such that every $K_\lambda$ has a finite number of components and every component has the shape of a finite bouquet of circles.*

In our following result we show that the global topological properties of minimal sets are basically determined by the closure of positive semiorbits in their region of attraction.

**Theorem 2.16.** *Let $K$ be a compact minimal set of the flow $\varphi : M \times \mathbb{R} \to M$, where $M$ is a manifold. Then for every $x$ belonging to the region of attraction of $K$ we have that $Sh(\overline{\gamma^+(x)}) = Sh(K)$.*

In [13] some results are presented about the global structure of (positively) invariant regions of flows with asymptotically stable attractors. We are interested in understanding to what extent the topological structure of the invariant regions is conditioned by the existence of the attractors. This structure is rather simple in several important cases. In particular, when the invariant regions contain an asymptotically stable global attractor, they have trivial shape provided that the ambient space is a Banach space or a contractible manifold. More generally, we have the following result valid for Absolute Retracts ($AR$-spaces).

**Theorem 2.17.** *Consider a dynamical system defined on $M \in AR$ (in particular on a Banach space or a contractible manifold). Suppose that the system has an asymptotically stable global attractor, $K$. Then*

1) *Every positively invariant compactum containing the attractor has trivial shape.*
2) *If $K$ is unidimensional, then every positively invariant continuum has trivial shape.*
3) *If $K$ is $n$-dimensional (with $n \geq 2$) then the cohomotopy set $\pi^m(L)$ (or the cohomotopy group when it is defined) is trivial for every positively invariant continuum $L$ and every $m \geq n$.*

## 3. Morse decompositions, Conley–Zehnder equations and Hopf duality

We consider now another situation in which there is an interplay between Shape Theory and other aspects of the theory of flows, mainly those related to the Morse decompositions of isolated invariant compacta. We assume as known the elementary notions of the Conley index theory of isolated invariant sets, in particular the notions of isolating neighborhood and index pair. Good references are Conley's monograph [10], the book [27] by Rybakowski and Salamon's article [28]. Let $\varphi : X \times \mathbb{R} \to X$ be a flow defined on a locally compact metric space $X$. If $K$ is an isolated invariant set of $X$, we shall say that an index pair $(N, L)$ for $K$ is proper if every point $x$ in the exit set $L$ immediately leaves the neighborhood $N$ (i.e., if $x[0, t] \not\subset N$ for every $t > 0$). By a result of McCord [23] proper index pairs always exist and they can be chosen with the additional property that there is a compact set $L' \subset N$ such that $(N, L')$ is a proper index set for $K$ in the reverse flow. In the sequel we shall limit ourselves to considering almost exclusively proper index pairs. If we take the quotient $N/L$ then the point corresponding to the equivalence class of $L$ will be denoted by $*$ (note that $N/\emptyset$ is obtained from $N$ by adjoining the isolated point $*$, i.e., $N/\emptyset = N \cup \{*\}$). We denote by $h(K)$ the pointed homotopy class of $(N/L, *)$ (i.e., the Conley index of $K$) and by $s(K)$ the (pointed) shape of $(N/L, *)$, which is known under the name of *shape index* of $K$. The shape index has been defined by Robbin and Salamon in [26]. The cohomology Conley index is the (Čech) cohomology group $H^*(N/L, *) = H^*(N, L)$ and similarly for the homology Conley index. Both indexes can be determined from the homotopical index as well as from the shape index.

The notation $N^-$ stands for the negative asymptotic set i.e., the set of all $x \in N$ such that $x[0, -\infty) \subset N$ and similarly for the positive asymptotic set $N^+$. On the other hand, we also use the following notation: $n^+ = N^+ \cap L'$ and $n^- = N^- \cap L$.

In the following result we provide a characterization of attractors and some duality properties relating the cohomology indexes of the forward flow and the reverse flow (see also [24]) for other duality properties.

**Theorem 3.1.**   a) $s(K) = Sh(N^-/n^-, \{*\})$,
  b) $K$ *is an attractor if and only if the inclusion* $i : K \cup \{*\} \to N/L$ *is a shape equivalence,*
  c) *If* $N^-$ *and* $N^+$ *are (topological) orientable manifolds of dimension* $d_1$ *and* $d_2$ *and with boundaries* $n^-$ *and* $n^+$ *respectively, then the* $k$*-dimensional cohomology Conley index of* $K$ *agrees with the* $(d_2 - d_1 + k)$*-dimensional cohomology index for the reverse flow (i.e., there is a duality between cohomology indices for the forward flow and the reverse flow).*

Robbin and Salamon defined in [26] a topology for the unstable manifold of the isolated invariant set $W^u(K)$ which they called intrinsic to distinguish it from the extrinsic topology that $W^u(K)$ inherits from the state space. The intrinsic topology is difficult to grasp intuitively because it is defined in terms of the limit of an inverse system whose bonding maps are the elements of a certain semigroup induced by the flow. More concretely, if $(N, L)$ is a proper index pair for $K$, they considered the inverse system $((N/L)_s, p_{st})$, where $(N/L)_s = N/L$ for every $s \in \mathbb{R}_+$ and if $s \leq t$ then $p_{st} : (N/L)_t \to (N/L)_s$ is defined by

$$p_{st}(x) = \begin{cases} x(t-s) & \text{if } x[0, t-s] \subset N - L \\ * & \text{otherwise}. \end{cases}$$

If we take the inverse limit $Z = \lim((N/L)_s, p_{st})$ and denote by $\star$ the point in $Z$ all of whose coordinates are the base point $* \in N/L$ then there is a natural map $h : Z - \{\star\} \to W^u(K)$ defined in the following way:

If $\mathbf{x} = (x_s) \in Z - \{\star\}$ take $t \in \mathbb{R}_+$ such that $x_t \in N - L$. Then $h(\mathbf{x}) = x_t t$. Robbin and Salamon proved that $h$ does not depend on the choice of $t$ and that $h$ is a continuous bijection. They considered the topology on $W^u(K)$ which makes $h$ a homeomorphism and they called it the intrinsic topology. They finally proved that this topology does not depend on the particular index pair $(N, L)$. As Robbin and Salamon pointed out, a nice feature of the intrinsic topology is that it provides an alternative procedure to determine the shape index, namely the shape index of an isolated invariant compactum agrees with the shape of the Alexandroff compactification of its unstable manifold endowed with the intrinsic topology. The results that we present below (see [32] for complete proofs) provide additional justification for the introduction of the intrinsic topology.

In the sequel we denote by $W^i(K)$ the unstable manifold of $K$ endowed with its intrinsic topology. If we restrict the flow to $W^i(K)$ we obtain a flow again. The following theorem presents some useful properties that are formulated in terms of the unstable manifold with its intrinsic topology. The first of them expresses the fact that every isolated invariant set $K$ is a global repeller of $W^i(K)$. Some characterizations of the shape index in terms of the unstable manifold are also presented here.

**Theorem 3.2.**    1) $K$ *is a global repeller in* $W^i(K)$.
2) *Let* $S$ *be a compact section of* $W^i(K) - K$ *(for instance* $S = n^-$*) and denote by* $W^\#$ *the truncated unstable manifold (with boundary* $S$*) consisting of all points* $x \in W^i(K)$ *such that* $x \in K$ *or there is* $t \geq 0$ *with* $xt \in S$. *Then* $s(K) = Sh(W^\#/\partial W^\#, *)$, *where we use the notation* $\partial W^\#$ *to denote the section* $S$ *and* $*$ *is the base point* $[\partial W^\#]$.
3) *If the shape of* $K$ *is trivial then* $s(K) = Sh(\sum(\partial W^\#), *)$, *where* $\sum(\partial W^\#)$ *denotes the suspension of* $\partial W^\#$ *and* $*$ *one of its vertices.*

**Remark 3.3.** The use of the intrinsic topology is essential in Theorem 3.2; an isolated invariant set is not, in general, a global repeller for the flow in its unstable manifold endowed with the extrinsic topology.

A nice characterization of the cohomology Conley index in terms of the unstable manifold $W^i(K)$ and a section of the flow restricted to $W^i(K)$ is obtained as a consequence of Theorem 3.2.

**Corollary 3.4.** *The cohomology Conley index of* $K$ *is* $\check{H}^q(W^\#, \partial W^\#)$.

It is important to identify situations in which the intrinsic and the extrinsic topology of isolated invariant sets agree. The property of $K$ being a repeller turns out to be characteristic for this to happen. The following result provides an interesting correspondence between dynamical and topological properties.

**Theorem 3.5.** *A necessary and sufficient condition for the extrinsic and the intrinsic topologies to agree is that* $W^u(K)$ *is locally compact and* $K$ *is a global repeller in* $W^u(K)$.

Now we present a general discussion of the Conley–Zehnder theory of Morse decompositions. If we have a filtration $N_0 \subset N_1 \subset \cdots \subset N_n$ of compact topological spaces then there is a standard method of obtaining an associated Morse equation. This method, which uses the axioms of elementary cohomology theory, is explained, for instance, in [11]. We summarize here how the Morse equation is established. First we define the formal power series

$$p(t, N_j, N_{j-1}) = \sum_{k \geq 0} r^k(N_j, N_{j-1}) t^k$$

and

$$q(t, N_j, N_{j-1}, N_0) = \sum_{k \geq 0} d^k(N_j, N_{j-1}, N_0) t^k,$$

where

$$r^k(N_j, N_{j-1}) = \text{rank of } \check{H}^k(N_j, N_{j-1}) (\check{\text{C}}\text{ech cohomology}) \text{ and}$$

$$d^k(N_j, N_{j-1}, N_0) = \text{rank of the image of the coboundary operator}$$

$$\delta^k : \check{H}^k(N_{j-1}, N_0) \to \check{H}^{k+1}(N_j, N_{j-1})$$

in the long cohomology sequence of the triple $(N_j, N_{j-1}, N_0)$. All the cohomology groups are assumed to be of finite rank in the former expressions.

Under these conditions, Conley and Zehnder [11] proved that the following equation holds

$$\sum_{j=1}^{n} p(t, N_j, N_{j-1}) = p(t, N_n, N_0) + (1+t)Q(t),$$

where $Q(t) = \sum_{j=2}^{n} q(t, N_j, N_{j-1}, N_0)$. This is the Morse equation associated to the filtration $N_0 \subset N_1 \subset \cdots \subset N_n$.

If $(M_1, M_2, \ldots, M_n)$ is a Morse decomposition of $K$ and $M_1 = A_1 \subset A_2 \subset \cdots \subset A_n = K$ is the corresponding sequence of attractors (i.e. $A_j = \{x \in K \mid \exists i \leq j \text{ with } \omega^*(x) \subset M_i\}$, where $\omega^*(x)$ is the alpha-limit set of $x$), Conley and Zehnder proved that there is a filtration $N_0 \subset N_1 \subset \cdots \subset N_n$ of compact spaces in $X$ such that $(N_j, N_{j-1})$ is an index pair for $M_j$ and $(N_j, N_0)$ is an index pair for $A_j$. Hence, in this particular case, the coefficients of the Morse equation are

$$r^k(N_j, N_{j-1}) = \text{rank } \check{H}^k(N_j, N_{j-1})$$
$$= \text{rank } \check{H}^k(h(M_j)) \text{ (cohomology Conley index of } M_j)$$

and

$$d^k(N_j, N_{j-1}, N_0) = \text{rank of the image of } \delta^k : \check{H}^k(h(A_{j-1})) \to \check{H}^{k+1}(h(M_j))$$

and it makes sense to adopt the notation $p(t, h(M_j))$ instead of $p(t, N_j, N_{j-1})$ (which reflects the fact that $p(t, N_j, N_{j-1})$ depends only on the homotopical Conley index $h(M_j)$ and not on the particular index pair that we have used).

With this notation the Morse equation of the decomposition $(M_1, M_2, \ldots, M_n)$ takes the form

$$\sum_{j=1}^{n} p(t, h(M_j)) = p(t, h(K)) + (1+t)Q(t),$$

where $Q(t)$ has been defined before (observe that the coefficients of $Q$ are non-negative integers). We shall refer to this equation as the *Conley–Zehnder–Morse equation* in the statement of Theorem 3.6. This equation relates the cohomology Conley index of $K$ to the cohomology Conley indices of a Morse decomposition of $K$. It can be viewed as a generalization of Morse theory for flows other than gradient flows on spaces other than manifolds.

We shall now summarize our results from [32] where we develop an alternative method of calculation of the Morse equations. Let $S$ be a compact section of $W^i(K)$ and let us consider the corresponding truncated manifold $W^\#$ with $\partial W^\# = S$. We can assume, without loss of generality, that $(W^\#, \partial W^\#) = (N^- n^-)$. We mean by $W_j^\#$ the subspace of $W^\#$ consisting of all points $x \in W^\#$ with $\omega^*(x) \in A_j$. $W_j^\#$ is a compact set and it should be noted that it can be interpreted as the truncated unstable manifold of $A_j$ with $\partial W_j^\# = \partial W^\# \cap W^i(A_j)$. With this notation the following result is proved in [32].

**Theorem 3.6.** *Let $K$ be an isolated invariant set and let $M_1, \ldots, M_n$ be a Morse decomposition of $K$ with associated sequence of attractors $A_1 \subset \cdots \subset A_n = K$.*

*Consider the filtration of truncated unstable manifolds*

$$\partial W^{\#} \subset W_1^{\#} \cup \partial W^{\#} \subset \cdots \subset W_{n-1}^{\#} \cup \partial W^{\#} \subset W_n^{\#} = W^{\#}$$

*and assume the groups $\check{H}^q(W_i^{\#} \cup \partial W^{\#}, \partial W^{\#})$ and $\check{H}^q(W_j^{\#} \cup \partial W^{\#}, W_{j-1}^{\#} \cup \partial W^{\#})$ to be of finite rank for $q \geq 0$, $1 \leq i \leq n$ and $2 \leq j \leq n$. Then the Morse equation associated to this filtration agrees with the Conley–Zehnder–Morse equation of the decomposition. The condition of rank finiteness is automatically fulfilled in the following two cases: 1) $\varphi$ is a $C^1$-flow on a manifold and 2) $\varphi$ is a continuous flow on a locally compact ANR and $K$ is a global attractor.*

J. J. Sánchez-Gabites has shown in [29] how to determine the intrinsic topology of the unstable manifold of an isolated invariant set without using index pairs. As a consequence Theorem 3.6 provides a way of obtaining the Conley–Zehnder–Morse equations when we have sufficient knowledge about the unstable manifold of $K$. In [32] it is also proved that a large part of the theory previously exposed is also valid for semi-dynamical systems. This aspect of the theory can be used for the determination of the critical groups of isolated critical points, which are important in connection with the Morse inequalities under the Palais–Smale condition.

Some applications of the ideas presented before have been given by K. Athanassopoulos in [2], where he studied the complexity of the flow in the region of attraction of isolated invariant sets using the intrinsic topology of the stable manifold. All results concerning the unstable manifold can be dualized for the stable manifold. In particular, Theorem 3.5 has been used by Athanasssopoulos to define the *instability depth,* which is an ordinal and measures how far an isolated invariant set is from being asymptotically stable within its region of attraction. He has provided lower and upper bounds of the instability depth in certain cases.

A consequence of Theorem 3.5 and a result of Athanassopoulos in [1] is that if $K$ is an isolated 1-dimensional compact minimal set and its region of attraction is an $ANR$ with respect to the intrinsic topology then $K$ must be a periodic orbit. Partially motivated by this he remarks that an interesting problem is to find conditions under which the region of attraction of a general isolated set is an $ANR$ with the intrinsic topology, since this could lead to Poincaré–Bendixon type theorems for flows on higher-dimensional phase spaces.

We study now some properties of flows that are related to the topological situation of Hopf duality. This situation arises when $(n-1)$-manifolds, $W$, embedded in the n-sphere $S^n$, induce homological relations between the two $n$-manifolds with boundary in which $S^n$ is decomposed by $W$ (see Steenrod and Epstein [35]). We present here a much more general situation, applicable to a Morse decomposition $\{M_0, M_1, M_2\}$ of a manifold $M$ which is an isolated invariant set of a flow $\phi : X \times \mathbb{R} \to X$, where $M_0$ is an $(n-1)$-submanifold of $M$ decomposing $M$ into two manifolds with common boundary $M_0$, and $M_1$ and $M_2$ are general Morse sets (not necessarily manifolds). We recall that connected Morse decomposition means a decomposition where all the Morse sets are connected.

**Theorem 3.7.** *Let $\phi : X \times \mathbb{R} \to X$ be a flow defined on a locally compact metric space $X$. Let $M \subset X$ be an orientable, compact, connected $n$-dimensional manifold which is an attractor of $\phi$. Suppose that $H^k(M) = H^{k+1}(M) = \{0\}$ for a given index $k$. Let $\{M_0, M_1, M_2\}$ be a connected Morse decomposition of $M$, where $M_0$ is an $(n-1)$-submanifold of $M$, decomposing $M$ into two manifolds with common boundary $M_0$. We then have the following relations involving the homological and cohomological Conley indices: 1) $CH^{k+1}(M_1) = CH_{n-k}(M_2)$ and 2) $CH_{n-k-1}(M_0) = CH^{k+1}(M_1) \oplus CH^{k+1}(M_2)$.*

*If we only assume that $M$ is an isolated invariant set of $\phi$ (not necessarily an attractor) then, with the same hypotheses as above, we have the following relations involving Čech homology and cohomology of the Morse sets: 1) $\check{H}^k(M_1) = \check{H}_{n-k-1}(M_2)$ and 2) $\check{H}^k(M_0) = \check{H}^k(M_1) \oplus \check{H}^k(M_2)$.*

Theorem 3.7 has been proved in [33]. The second part of this theorem admits a different, more general, version, in which the homological hypothesis is placed on the unstable manifold of $M$ with its intrinsic topology. This can be obtained as the following consequence of the theorem:

**Corollary 3.8.** *Let $\phi : X \times \mathbb{R} \to X$ be a flow defined on a locally compact metric space $X$. Let $M \subset X$ be an orientable, compact, connected $n$-dimensional manifold which is an isolated invariant set of $\phi$. Suppose that $\check{H}^k(W^i(M)) = \check{H}^{k+1}(W^i(M)) = \{0\}$ for a given index $k$. Let $\{M_0, M_1, M_2\}$ be a connected Morse decomposition of $M$, where $M_0$ is an $(n-1)$-submanifold of $M$, decomposing $M$ into two manifolds with common boundary $M_0$. Then, we have the following relations involving Čech homology and cohomology of the Morse sets: 1) $\check{H}^k(M_1) = \check{H}_{n-k-1}(M_2)$ and 2) $\check{H}^k(M_0) = \check{H}^k(M_1) \oplus \check{H}^k(M_2)$.*

The isolated invariant set $M$ can be represented as the intersection of a nested sequence of isolating neighborhoods $N_i$ such that every $N_i$ has an exit set $L_i$ with $(N_i, L_i)$ a proper index pair. It can be proved that the unstable manifold $W^i$ with its intrinsic topology is homotopically equivalent to $N_i^-$ for every $i$. Since $M$ is also the intersection of the nested sequence of the $N_i^-$ we have, by the continuity property of Čech cohomology, that $\check{H}^k(M) = \check{H}^k(W^i(M)) = \{0\}$ and $\check{H}^{k+1}(M) = \check{H}^{k+1}(W^i(M)) = \{0\}$. Hence the corollary is a consequence of Theorem 3.7.

# References

[1] K. Athanassopoulos, *Cohomology and asymptotic stability of 1-dimensional continua.* Manuscripta Math. **72** (1991), no. 4, 415–423.

[2] K. Athanassopoulos, *Remarks on the region of attraction of an isolated invariant set.* Colloq. Math. **104** (2006), no. 2, 157–167.

[3] A. Beck, *On invariant sets.* Ann. of Math. (2) **67** (1958) 99–103.

[4] N. P. Bhatia and G. P. Szegő, *Stability theory of dynamical systems.* Grundlehren der Mat. Wiss. **161**, Springer-Verlag, Berlin-Heidelberg-New York, 1970.

[5] N. P. Bhatia, A. Lazer and G. P. Szegö, *On global weak attractors in dynamical systems.* J. Math. Anal. Appl. **16** (1966) 544–552.

[6] S. A. Bogatyi and V. I. Gutsu, *On the structure of attracting compacta.* (Russian) Differentsialnye Uravneniya **25** (1989), no. 5, 907–909

[7] K. Borsuk, *Theory of Retracts.* Monografie Matematyczne, Warsaw, 1967

[8] K. Borsuk, *Concerning homotopy properties of compacta.* Fund. Math. **62** (1968), 223–254.

[9] K. Borsuk, *Theory of shape.* Monografie Matematyczne, Tom **59**, [Mathematical Monographs, Vol. **59**], PWN-Polish Scientific Publishers, Warsaw, 1975.

[10] C. Conley, *Isolated invariant sets and the Morse index.* CBMS Regional Conference Series in Mathematics, **38**, American Mathematical Society, Providence, R. I., 1978.

[11] C. Conley and E. Zehnder, *Morse-type index theory for flows and periodic solutions for Hamiltonian equations.* Comm. Pure Appl. Math. **37** (1984), 207–253.

[12] J. Dydak and J. Segal, *Shape theory. An introduction.* Lecture Notes in Mathematics **688**, Springer, Berlin. 1978.

[13] A. Giraldo and J. M. R. Sanjurjo, *On the global structure of invariant regions of flows with asymptotically stable attractors.* Mathematische Zeitschrift **232** (1999), 739–746.

[14] A. Giraldo, M. A. Morón, F. R. Ruiz Del Portal, J. M. R. Sanjurjo, *Some duality properties of non-saddle sets.* Topology Appl. **113** (2001), 51–59.

[15] A. Giraldo, M. A. Morón, F. R. Ruiz Del Portal, J. M. R. Sanjurjo, *Shape of global attractors in topological spaces.* Nonlinear Anal. **60** (2005), no. 5, 837–847.

[16] A. Giraldo, R. Jiménez, M. A. Morón, F. R. Ruiz Del Portal, J. M. R. Sanjurjo, *Pointed shape and global attractors for metrizable spaces.* Preprint.

[17] B. Günther and J. Segal, *Every attractor of a flow on a manifold has the shape of a finite polyhedron.* Proc. AMS **119** (1993), 321–329.

[18] H. M. Hastings, *Shape theory and dynamical systems.* in: N. G. Markley and W. Perizzo.: The structure of attractors in dynamical systems, Lecture Notes in Math. **668**, Springer-Verlag, Berlin 1978, pp. 150–160.

[19] H. M. Hastings, *A higher-dimensional Poincaré-Bendixson theorem.* Glas. Mat. Ser. III **14** (34) (1979), no. 2, 263–268.

[20] S. T. Hu, *Theory of Retracts.* Detroit, 1965.

[21] L. Kapitanski and I. Rodnianski, *Shape and Morse theory of attractors.* Comm. Pure Appl. Math. **53** (2000), 218–242.

[22] S. Mardešić and J. Segal, *Shape theory. The inverse system approach.* North-Holland Mathematical Library **26**, North Holland, Amsterdam-NewYork, 1982.

[23] C. McCord, *Mappings and homological properties in the Conley index theory.* Ergodic Theory and Dynamical Systems **8** (1988), Charles Conley Memorial Volume, 175–198.

[24] C. McCord, *Poincaré-Lefschetz duality for the homology Conley index.* Transactions AMS **329** (1992), 233–252.

[25] J. Milnor, *On the concept of Attractor.* Commun. Math. Phys. **99** (1985), 177–195.

[26] J. W. Robbin and D. Salamon, *Dynamical systems, shape theory and the Conley index.* Ergodic Theory and Dynamical Systems **8** (1988), Charles Conley Memorial Volume, 375–393.

406 J. M. R. Sanjurjo

[27] K. P. Rybakowski, *The homotopy index and partial differential equations.* Universitext, Springer-Berlin-New York, 1987.

[28] D. Salamon, *Connected simple systems and the Conley index of isolated invariant sets.* Transactions AMS **291** (1985), 1–41.

[29] J. J. Sánchez-Gabites, *A description without index pairs of the intrinsic topology of the ustable manifold of a compact invariant set.* Preprint.

[30] J. M. R. Sanjurjo, *Multihomotopy, Čech spaces of loops and shape groups.* Proc. London Math. Soc. **69** (1994), 330–344.

[31] J. M. R. Sanjurjo, *On the structure of uniform attractors.* J. Math. Anal. Appl. **192** (1995), no. 2, 519–528.

[32] J. M. R. Sanjurjo, *Morse equations and unstable manifolds of isolated invariant sets.* Nonlinearity **16** (2003), 1435–1448.

[33] J. M. R. Sanjurjo, *Lusternik-Schnirelmann category, Hopf duality, and isolated invariant sets.* Bol. Soc. Mat. Mexicana (3) **10** (2004), 487–494.

[34] E. H. Spanier, *Algebraic Topology.* McGraw-Hill, New York-Toronto-London, 1966.

[35] N. E. Stenrod and D. B. A. Epstein, *Cohomology operations.* Princeton University Press, Princeton 1962.

José M. R. Sanjurjo
Facultad de Matemáticas
Universidad Complutense
E-28040 Madrid
Spain
e-mail: jose_sanjurjo@mat.ucm.es

Progress in Nonlinear Differential Equations
and Their Applications, Vol. 75, 407–414
© 2007 Birkhäuser Verlag Basel/Switzerland

# Regularity of Solutions for the Autonomous Integrals of the Calculus of Variations

Moulay Rchid Sidi Ammi and Delfim F. M. Torres

*Dedicated to Arrigo Cellina and James Yorke*

**Abstract.** The search for appropriate conditions under which we have regularity of solutions is an important area of study in the calculus of variations. In this note we describe some recent regularity conditions for the autonomous problems of the calculus of variations with second-order derivatives. We prove that autonomous integral functionals of the calculus of variations with a Lagrangian having superlinearity partial derivatives with respect to the higher-order derivatives admit only minimizers with essentially bounded derivatives. This imply non-occurrence of the Lavrentiev phenomenon and validity of the classical necessary optimality conditions.

**Mathematics Subject Classification (2000).** Primary 49N60; Secondary 49K05.

**Keywords.** Optimal control, calculus of variations, higher order derivatives, regularity of solutions, non-occurrence of the Lavrentiev phenomenon.

## 1. Introduction and motivation

Let $\mathcal{L}(x^0, \ldots, x^m)$ be a given $C^1(\mathbb{R}^{(m+1)\times n})$ real valued function. The autonomous problem of the calculus of variations with high-order derivatives consists in minimizing an integral functional

$$J^m[x(\cdot)] = \int_a^b \mathcal{L}\left(x(t), \dot{x}(t), \ldots, x^{(m)}(t)\right) dt \qquad (P_m)$$

over a certain class $\mathcal{X}$ of functions $x : [a, b] \to \mathbb{R}^n$ satisfying the boundary conditions

$$x(a) = x_a^0, x(b) = x_b^0, \ldots, x^{(m-1)}(a) = x_a^{m-1}, x^{(m-1)}(b) = x_b^{m-1}. \qquad (1.1)$$

The authors were supported by FCT (the Portuguese Foundation for Science and Technology): Sidi Ammi through the postdoc project *Thermistor Problems and Optimal Control*, reference SFRH/BPD/20934/2004.

Often it is convenient to write $x^{(1)} = x'$, $x^{(2)} = x''$, and sometimes we revert to the standard notation used in mechanics: $x' = \dot{x}$, $x'' = \ddot{x}$. Such problems arise, for instance, in connection with the theory of beams and rods [20]. Further, many problems in the calculus of variations with higher-order derivatives describe important optimal control problems with linear dynamics [18].

Regularity theory for optimal control problems is a fertile field of research and a source of many challenging mathematical issues and interesting applications [4, 23, 24]. The essential points in the theory are: (i) existence of minimizers and (ii) necessary optimality conditions to identify those minimizers.

The first systematic approach to existence theory was introduced by Tonelli in 1915 [21], who showed that existence of minimizers is guaranteed in the Sobolev space $W_m^m$ of the class of functions which are absolutely continuous with their derivatives up to order $m - 1$ and have $m$th derivative belonging to $L^m$. The direct method of Tonelli proceeds in three steps: (i) smoothness and convexity with respect to the highest-derivative of the Lagrangian $\mathcal{L}$ guarantees lower semi-continuity, (ii) the coercivity condition (the Lagrangian $\mathcal{L}$ must grow faster than a linear function) insure compactness, (iii) by the compactness principle, one gets the existence of minimizers for the problem $(P_m)$. Typically, Tonelli's existence theorem for $(P_m)$ is formulated as follows [4, 8]: under hypotheses (H1)–(H3) on the Lagrangian $\mathcal{L}$,

(H1) $(x^0, \ldots, x^m) \to \mathcal{L}(x^0, \ldots, x^m)$ is a $C^1$ function;
(H2) $\mathcal{L}(x^0, \ldots, x^m)$ is convex as a function of the last argument $x^m$;
(H3) $\mathcal{L}(x^0, \ldots, x^m)$ is coercive in $x^m$, i.e., $\exists\, \Theta : [0, \infty) \to \mathbb{R}$ such that

$$\lim_{r \to \infty} \frac{\Theta(r)}{r} = +\infty\,,$$
$$\mathcal{L}(x^0, \ldots, x^m) \geq \Theta(|x^m|) \quad \text{for all } (x^0, \ldots, x^m)\,,$$

there exists a minimizer to problem $(P_m)$ in the class $W_m^m$.

The main necessary condition in optimal control is the famous Pontryagin maximum principle, which includes all the classical necessary optimality conditions of the calculus of variations [14]. For $(P_m)$, the Pontryagin maximum principle [14] is established assuming $x \in W_m^\infty \subset W_m^m$ so, a priori, the hypotheses (H1)–(H3) do not assure the applicability of the necessary optimality conditions, being required more regularity on the class of admissible functions.

In the case $m = 1$, extra information about the minimizers was proved, for the first time, by Tonelli himself [21].

**Theorem 1.1 (Tonelli–Morrey).** *Under the hypotheses* (H1)–(H3) *of smoothness, convexity and coercivity, if $x$ is a minimizer of $(P_1)$ then $\dot{x}$ is locally essentially bounded on an open subset $\Omega \subset [a, b]$ of full measure. Moreover, if*

$$\left| \frac{\partial \mathcal{L}}{\partial x} \right| + \left| \frac{\partial \mathcal{L}}{\partial \dot{x}} \right| \leq c|\mathcal{L}| + r\,, \tag{1.2}$$

*for some constants c and r, c > 0, then $\Omega = [a, b]$ ($\dot{x}(t)$ is essentially bounded in all points t of $[a, b]$, i.e., $x \in W_1^\infty$), and the Pontryagin maximum principle, or the necessary condition of Euler–Lagrange, hold.*

Condition (1.2) is now known in the literature as the Tonelli–Morrey regularity condition [5, 7, 18]. Since Tonelli and Morrey, several Lipschitzian regularity conditions were obtained for the problem $(P_1)$ – see [1,2,6,12,13,15,17,24] and references therein. It turns out that condition (1.2) is not necessary in Theorem 1.1:

**Theorem 1.2 ( [6]).** *Under the hypotheses* (H1)–(H3) *of Tonelli's existence theorem, all the minimizing trajectories x of the autonomous fundamental problem of the calculus of variations* $(P_1)$ *are Lipschitzian:* $x \in W_1^\infty \subset W_1^1$.

We refer the reader to the works [1–3] of Arrigo Cellina for several extensions of Theorem 1.2.

Results of Lipschitzian regularity for $m > 1$ are scarcer: we are aware of the results in [8, 10, 17, 22]. In 1997 A.V. Sarychev [16] proved that the second-order problems of the calculus of variations may show new phenomena non-present in the first-order case: under the hypotheses (H1)–(H3) of Tonelli's existence theory, problems $(P_2)$ may exhibit the Lavrentiev phenomenon [11]. This is not a possibility for $(P_1)$, as follows immediately from Theorem 1.2. Sarychev's result was recently extended by A. Ferriero [9] for the case $m > 2$ (different extensions of Sarychev's result are found in [25]). It is also shown in [9] that, under some standard hypotheses, the problems of the calculus of variations $(P_m)$ with Lagrangians only depending on two consecutive derivatives $x^{(\gamma)}$ and $x^{(\gamma+1)}$, $\gamma \geq 0$, do not exhibit the Lavrentiev phenomenon for any boundary conditions (1.1) (for $m = 1$ this is trivially true from Theorem 1.2). In the case in which the Lagrangian only depends on the higher-order derivative $x^{(m)}$, it is possible to prove more [17, Corollary 2]: when $\mathcal{L} = \mathcal{L}\left(x^{(m)}\right)$, all the minimizers predicted by the existence theory belong to the space $W_m^\infty \subset W_m^m$ and satisfy the Pontryagin maximum principle (regularity). As to whether this is the case or not for Ferriero's problem with Lagrangians only depending on consecutive derivatives $x^{(\gamma)}$ and $x^{(\gamma+1)}$, seems to be an open question.

The results of Sarychev [16] and Ferriero [9] on the Lavrentiev phenomenon show that the problems of the calculus of variations with higher-order derivatives are richer than the problems with $m = 1$, but also show, in our opinion, that the regularity theory for higher-order problems is underdeveloped. Here we present a new regularity condition for $(P_2)$. We show that for the autonomous second-order problems of the calculus of variations regularity follows by imposing a superlinearity condition to the partial derivatives $\frac{\partial \mathcal{L}}{\partial \ddot{x}_i}$ of the Lagrangian.

## 2. Generalized integral form of duBois–Reymond condition

We shall limit ourselves here to $(P_2)$, i.e., to the problem of minimizing

$$\int_a^b \mathcal{L}\left(x(t), \dot{x}(t), \ddot{x}(t)\right) dt \qquad (P_2)$$

for some given Lagrangian $\mathcal{L}(\cdot, \cdot, \cdot)$, assumed to be a $C^1$ function with respect to all arguments. In this case it is appropriate to choose the admissible functions $x$ to be twice continuously differentiable with derivatives $\dot{x}$ and $\ddot{x}$ in $L^2$, i.e. $\mathcal{X} = W_2^2$. In this section we formulate a generalized duBois–Reymond necessary condition in integral form, valid for $\mathcal{X} = W_2^2$ (the optimal solutions $x$ may have unbounded derivatives $\dot{x}$, $\ddot{x}$). Then, in Section 3, we obtain a regularity condition under which all the minimizers of $(P_2)$ are in $W_2^\infty \subset W_2^2$ and thus satisfy the classical necessary conditions. We assume the following hypotheses $(S_i)$, $i = 1, \ldots, n$:

$(S_i)$ There exists a nonnegative continuous function $G(\cdot, \cdot, \cdot)$, and some $\delta > 0$, such that the function $t \to G(x(t), x'(t), x''(t))$ is $L^2$-integrable on $[a, b]$, and

$$\left| \frac{\partial \mathcal{L}}{\partial x_i}(y, x', x'') \right| \le G(x, x', x''),$$

$$\left| \frac{\partial \mathcal{L}}{\partial \dot{x}_i}(x, y, x'') \right| \le G(x, x', x''),$$

$$\left| \frac{\partial \mathcal{L}}{\partial \ddot{x}_i}(x, x', y) \right| \le G(x, x', x''),$$

for all $x$, $x'$, $x'' \in \mathbb{R}^n$, $x = (x_1, \ldots, x_n) \in \mathbb{R}^n$, $y = (y_1, \ldots, y_n) \in \mathbb{R}^n$, $y_j = x_j^{(k)}(t)$ for $j \ne i$, $\left| y_i - x_i^{(k)}(t) \right| \le \delta$, $i = 1, \ldots, n$ and $k = 0, 1, 2$, where $x_i^{(k)}(t)$ is the $i^{th}$ component of the $k^{th}$ derivative with the convention $x_i^{(0)}(t) = x_i(t)$.

Let $s$ be the arc length parameter on the curve $C_0 : x = x(t)$, $a \le t \le b$, so that the Jordan length of $C_0$ is $s(t) = \int_a^t \sqrt{1 + |x'(\tau)|^2} d\tau$ with $s(a) = 0$, $s(b) = l$ and $s(t)$ is absolutely continuous with $s'(t) \ge 1$ a.e. Thus $s(t)$ and its inverse $t(s)$, $0 \le s \le l$, are absolutely continuous with $t'(s) > 0$ a.e. in $[0, l]$. If $X(s) = x(t(s))$, $0 \le s \le l$, then $t(s)$ and $X(s)$ are Lipschitzian of constant one in $[0, l]$. By change of variable,

$$I[x] = \int_a^b \mathcal{L}\left(x(t), \dot{x}(t), \ddot{x}(t)\right) dt$$

$$= \int_0^l \mathcal{L}\left(X(s), \frac{X'(s)}{t'(s)}, \frac{1}{t'^2(s)}\left(X''(s) - \frac{X'(s)}{t'(s)} t''(s)\right)\right) t'(s) ds.$$

Setting $F(x, t', x', t'', x'') = \mathcal{L}\left(x, \frac{x'}{t'}, \frac{1}{t'^2}(x'' - \frac{x'}{t'} t'')\right) t'$, we have:

$$I[x] = J[C] = J[X] = \int_0^l F\left(X(s), t'(s), X'(s), t''(s), X''(s)\right) ds.$$

The following necessary condition is useful to prove our regularity theorem.

**Theorem 2.1.** *Under hypotheses* $(S_i)_{1 \leq i \leq n}$, *if* $x(\cdot) \in W_2^2$ *is a minimizer of problem* $(P_2)$, *then the following integral form of duBois–Reymond necessary condition holds:*

$$\phi_0(s) = \frac{\partial F}{\partial t''}(\theta(s)) - \int_0^s \frac{\partial F}{\partial t'}(\theta(\sigma))d\sigma = c_0, \quad 0 \leq \tau \leq s \leq l, \quad (2.1)$$

*where* $\partial F/\partial t''$ *and* $\partial F/\partial t'$ *are evaluated at* $\theta(s) = (X(s), t'(s), X'(s), t''(s), X''(s))$ *and* $c_0$ *is a constant.*

Theorem 2.1 is a direct corollary of a more general result proved in [19] for non-autonomous problems. The proof is done by contradiction, using conditions $(S_i)$ to justify the usual rule of differentiation under the sign of the integral.

## 3. Main result

We give now a regularity result for $(P_2)$ under an additional requirement on the Lagrangian $\mathcal{L}$.

**Theorem 3.1.** *In addition to the hypotheses* $(S_i)_{1 \leq i \leq n}$, *if* $\partial\mathcal{L}/\partial\ddot{x}$ *is superlinear, i.e. there exist constants* $a > 0$ *and* $b > 0$ *such that*

$$a|w| + b \leq \left|\frac{\partial\mathcal{L}}{\partial\ddot{x}}(s, v, w)\right| \quad \text{for all } (s, v, w) \in \mathbb{R}^n \times \mathbb{R}^n \times \mathbb{R}^n, \quad (3.1)$$

*then every minimizer* $x \in W_2^2$ *of the problem is on* $W_2^\infty$.

**Example 3.2.** A trivial example of a Lagrangian satisfying all the conditions $(S_i)_{1 \leq i \leq n}$ and (3.1) is $\mathcal{L}(x, \dot{x}, \ddot{x}) = \mathcal{L}(\ddot{x}) = a\ddot{x}^2 + b\ddot{x}$ with $a$ and $b$ strictly positive constants (one can choose $G(x, \dot{x}, \ddot{x}) = 2a|\ddot{x}| + b \in L^2$ in $(S_i)$). It follows from Theorem 3.1 that all minimizers of the problem

$$I[x(\cdot)] = \int_{t_0}^{t_1} \left[a\ddot{x}(t)^2 + b\ddot{x}(t)\right] dt \longrightarrow \min$$

$$x(\cdot) \in W_2^2, \quad a, b > 0$$

$$x(t_0) = \alpha, \quad x(t_1) = \beta$$

are $W_2^\infty$ functions.

*Proof.* Since $F(x, t', x', t'', x'') = \mathcal{L}\left(x, \frac{x'}{t'}, \frac{1}{t'^2}(x'' - \frac{x'}{t'}t'')\right)t'$, we have

$$\frac{\partial F}{\partial t'} = \mathcal{L} - \frac{\partial\mathcal{L}}{\partial\dot{x}}\frac{\dot{x}}{t'} + \frac{1}{t'^2}\frac{\partial\mathcal{L}}{\partial\ddot{x}}\left(\frac{-2\ddot{x}}{t'} + \frac{3t''}{t'}\dot{x}\right),$$

$$\frac{\partial F}{\partial t''} = -\frac{1}{t'^2}\frac{\partial\mathcal{L}}{\partial\ddot{x}}\dot{x},$$

and using (2.1) we get

$$\frac{1}{t'^2}\frac{\partial\mathcal{L}}{\partial\ddot{x}}\dot{x} + \int_0^s \left\{\mathcal{L} - \frac{1}{t'}\frac{\partial\mathcal{L}}{\partial\dot{x}}\dot{x} + \frac{1}{t'^2}\frac{\partial\mathcal{L}}{\partial\ddot{x}}\left(\frac{-2\ddot{x}}{t'} + \frac{3t''}{t'}\dot{x}\right)\right\} = c_0.$$

Therefore,

$$\frac{1}{t'^2}\frac{\partial\mathcal{L}}{\partial\ddot{x}}\dot{x} = c_0 - \int_0^s \left\{ \mathcal{L} - \frac{1}{t'}\frac{\partial\mathcal{L}}{\partial\dot{x}}\dot{x} + \frac{1}{t'^2}\frac{\partial\mathcal{L}}{\partial\ddot{x}}\left(\frac{-2\ddot{x}}{t'} + \frac{3t''}{t'}\dot{x}\right) \right\}$$

$$= c_0 - \int_0^s \mathcal{L} + \int_0^s \frac{1}{t'}\frac{\partial\mathcal{L}}{\partial\dot{x}}\dot{x} + 2\int_0^s \frac{1}{t'^3}\frac{\partial\mathcal{L}}{\partial\ddot{x}}\ddot{x} - \int_0^s \frac{3t''}{t'^3}\frac{\partial\mathcal{L}}{\partial\ddot{x}}\dot{x}.$$

Applying the Holder's inequality, we obtain

$$\left|\frac{1}{t'^2}\frac{\partial\mathcal{L}}{\partial\ddot{x}}\dot{x}\right| \leq |c_0| + \|\mathcal{L}\|_1 + k_1\left\|\frac{\partial\mathcal{L}}{\partial\dot{x}}\right\|_2 \|\dot{x}\|_2 + k_2\left\|\frac{\partial\mathcal{L}}{\partial\ddot{x}}\right\|_2 \|\ddot{x}\|_2 + \int_0^s \left|\frac{3t''}{t'}\right|\left|\frac{1}{t'^2}\frac{\partial\mathcal{L}}{\partial\ddot{x}}\dot{x}\right|,$$

where $k_1, k_2$ are positive constants. Then, using the fact that $\mathcal{L} \in C^1, \mathcal{L}, \frac{\partial\mathcal{L}}{\partial\dot{x}}, \frac{\partial\mathcal{L}}{\partial\ddot{x}} \in L^2$ and $x \in W_2^2$ (in other terms, $x, \dot{x}, \ddot{x} \in L^2$), it follows that $\frac{1}{t'^2}\frac{\partial\mathcal{L}}{\partial\ddot{x}}\dot{x}$ satisfies a condition of the form

$$\left|\frac{1}{t'^2}\frac{\partial\mathcal{L}}{\partial\ddot{x}}\dot{x}\right| \leq k_3 + \int_0^s \left|\frac{3t''}{t'}\right|\left|\frac{1}{t'^2}\frac{\partial\mathcal{L}}{\partial\ddot{x}}\dot{x}\right|,$$

for a certain positive constant $k_3$. Now, Gronwall's Lemma leads to the following uniform bound:

$$\left|\frac{1}{t'^2}\frac{\partial\mathcal{L}}{\partial\ddot{x}}\dot{x}\right| \leq k_4$$

with a positive constant $k_4$. Since $t' \leq 1$, we deduce that $\frac{\partial\mathcal{L}}{\partial\ddot{x}}\dot{x}$ is uniformly bounded. Besides, since $\left|\frac{\partial\mathcal{L}}{\partial\ddot{x}}\right|$ verifies (3.1), we have

$$|\dot{x}|\,(a|\ddot{x}| + b) \leq \left|\frac{\partial\mathcal{L}}{\partial\ddot{x}}\dot{x}\right| \leq k_4 \quad (b > 0).$$

Therefore we get for a positive constant $k_5$

$$|\dot{x}| \leq \frac{k_4}{a|\ddot{x}| + b} \leq k_5.$$

Then $\frac{\partial\mathcal{L}}{\partial\ddot{x}}$ is uniformly bounded. Since $\frac{\partial\mathcal{L}}{\partial\ddot{x}}(s, v, w)$ goes to $+\infty$ with $|w|$ (by superlinearity), this implies a uniform bound on $|\ddot{x}|$ which leads to the intended conclusion that $\ddot{x}$ is essentially bounded. $\square$

As an immediate corollary to our Theorem 3.1, we obtain conditions of non-occurrence of the Lavrentiev phenomenon for the autonomous second-order variational problems.

**Corollary 3.3.** *Under the hypotheses of Theorem 3.1, problems* $(P_2)$ *do not admit the Lavrentiev gap* $W_2^2 - W_2^\infty$:

$$\inf_{x(\cdot)\in W_2^2}\int_a^b \mathcal{L}\left(x(t), \dot{x}(t), \ddot{x}(t)\right) dt = \inf_{x(\cdot)\in W_2^\infty}\int_a^b \mathcal{L}\left(x(t), \dot{x}(t), \ddot{x}(t)\right) dt.$$

Theorem 3.1 admit a generalization for problems of an order higher than two. This is under study and will be addressed in a forthcoming paper.

# References

[1] A. Cellina, The classical problem of the calculus of variations in the autonomous case: relaxation and Lipschitzianity of solutions, Trans. Amer. Math. Soc. **356** (2004), no. 1, 415–426. MR2020039 (2004k:49087)

[2] A. Cellina and A. Ferriero, Existence of Lipschitzian solutions to the classical problem of the calculus of variations in the autonomous case, Ann. Inst. H. Poincaré Anal. Non Linéaire **20** (2003), no. 6, 911–919. MR2008683 (2004f:49069)

[3] A. Cellina, A. Ferriero and E. M. Marchini, Reparametrizations and approximate values of integrals of the calculus of variations, J. Differential Equations **193** (2003), no. 2, 374–384. MR1998639 (2004e:49013)

[4] F. H. Clarke, *Methods of dynamic and nonsmooth optimization*, SIAM, Philadelphia, PA, 1989. MR1085948 (91j:49001)

[5] F. H. Clarke, Necessary conditions in dynamic optimization, Mem. Amer. Math. Soc. **173** (2005), no. 816, x+113 pp. MR2117692 (2006i:49027)

[6] F. H. Clarke and R. B. Vinter, Regularity properties of solutions to the basic problem in the calculus of variations, Trans. Amer. Math. Soc. **289** (1985), no. 1, 73–98. MR0779053 (86h:49020)

[7] F. H. Clarke and R. B. Vinter, Regularity of solutions to variational problems with polynomial Lagrangians, Bull. Polish Acad. Sci. Math. **34** (1986), no. 1–2, 73–81. MR0850317 (87j:49042)

[8] F. H. Clarke and R. B. Vinter, A regularity theory for variational problems with higher order derivatives, Trans. Amer. Math. Soc. **320** (1990), no. 1, 227–251. MR0970266 (90k:49006)

[9] A. Ferriero, The approximation of higher-order integrals of the calculus of variations and the Lavrentiev phenomenon, SIAM J. Control Optim. **44** (2005), no. 1, 99–110. MR2176668 (2006e:49007)

[10] H. Frankowska and E. M. Marchini, Lipschitzianity of optimal trajectories for the Bolza optimal control problem, Calc. Var. Partial Differential Equations **27** (2006), no. 4, 467–492. MR2263674 (2007f:49021)

[11] M. Lavrentiev, Sur quelques problèmes du calcul des variations, Ann. Mat. Pura Appl. **4** (1927), 7–28.

[12] G. Dal Maso and H. Frankowska, Autonomous integral functionals with discontinuous nonconvex integrands: Lipschitz regularity of minimizers, DuBois–Reymond necessary conditions, and Hamilton–Jacobi equations, Appl. Math. Optim. **48** (2003), no. 1, 39–66. MR1977878 (2004c:49079)

[13] A. Ornelas, Lipschitz regularity for scalar minimizers of autonomous simple integrals, J. Math. Anal. Appl. **300** (2004), no. 2, 285–296. MR2098209 (2005f:49089)

[14] L. S. Pontryagin, V. G. Boltyanskiĭ, R. V. Gamkrelidze, E. F. Mishchenko, *Selected works. Vol. 4*, The mathematical theory of optimal processes. Translated from the Russian by K. N. Trirogoff, Translation edited by L. W. Neustadt, Reprint of the 1962 English translation, Gordon & Breach, New York, 1986. MR0898009 (90a:01108)

[15] M. Quincampoix and N. Zlateva, On Lipschitz regularity of minimizers of a calculus of variations problem with non locally bounded Lagrangians, C. R. Math. Acad. Sci. Paris **343** (2006), no. 1, 69–74. MR2241962 (2007b:49074)

[16] A. V. Sarychev, First- and second-order integral functionals of the calculus of variations which exhibit the Lavrentiev phenomenon, J. Dynam. Control Systems **3** (1997), no. 4, 565–588. MR1481627 (98m:49011)

[17] A. V. Sarychev and D. F. M. Torres, Lipschitzian regularity of minimizers for optimal control problems with control-affine dynamics, Appl. Math. Optim. **41** (2000), no. 2, 237–254. MR1731420 (2000m:49048)

[18] A. V. Sarychev and D. F. M. Torres, Lipschitzian regularity conditions for the minimizing trajectories of optimal control problems, in *Nonlinear analysis and its applications to differential equations (Lisbon, 1998)*, 357–368, Birkhäuser, Boston, Boston, MA, 2001. MR1800636 (2001j:49062)

[19] M. R. Sidi Ammi and D. F. M. Torres, Regularity of solutions to second-order integral functionals in variational calculus, Int. J. Appl. Math. Stat. **13**, no. J08, June 2008 (in press)

[20] D. R. Smith, *Variational methods in optimization*, Prentice Hall, Englewood Cliffs, N.J., 1974. MR0346616 (49:11341)

[21] L. Tonelli, *Opere scelte. Vol II: Calcolo delle variazioni* (Italian), Ed. Cremonese, Rome, 1961, 289–333. MR0125743 (23:A3041)

[22] D. F. M. Torres, Lipschitzian regularity of the minimizing trajectories for nonlinear optimal control problems, Math. Control Signals Systems **16** (2003), no. 2–3, 158–174. MR2006825 (2004i:49047)

[23] D. F. M. Torres, The role of symmetry in the regularity properties of optimal controls, in *Symmetry in nonlinear mathematical physics. Part 1, 2, 3*, 1488–1495, Natsīonal. Akad. Nauk Ukraïni, Īnst. Mat., Kiev, 2004. MR2077966 (2005b:49051)

[24] D. F. M. Torres, Carathéodory equivalence Noether theorems, and Tonelli full-regularity in the calculus of variations and optimal control, J. Math. Sci. (N. Y.) **120** (2004), no. 1, 1032–1050. MR2099056 (2006c:49034)

[25] A. J. Zaslavski, Nonoccurrence of the Lavrentiev phenomenon for many optimal control problems, SIAM J. Control Optim. **45** (2006), no. 3, 1116–1146. MR2247728 (2007f:49011)

Moulay Rchid Sidi Ammi and Delfim F. M. Torres
Control Theory Group (cotg)
Centre for Research on Optimization and Control
Department of Mathematics, University of Aveiro
P-3810-193 Aveiro
Portugal
e-mail: sidiammi@mat.ua.pt
        delfim@ua.pt

Progress in Nonlinear Differential Equations
and Their Applications, Vol. 75, 415–424
© 2007 Birkhäuser Verlag Basel/Switzerland

# Multi-modal Periodic Trajectories in Fermi–Pasta–Ulam Chains

Susanna Terracini

*Dedicated to Arrigo Cellina and James Yorke*

**Abstract.** This paper deals with the problem of bifurcation of periodic trajectories in the Fermi–Pasta–Ulam chains of nonlinear oscillator.

## 1. The Fermi-Pasta-Ulam model

The aim of this paper is to summarize some recent results concerning the problem of bifurcation of periodic trajectories in the Fermi–Pasta–Ulam chains of nonlinear oscillator, obtained in the joint papers [2] and [13]. The Fermi–Pasta-Ulam model consists in $N$ coupled nonlinear oscillators, interacting only with their nearest neighbours:

$$m_i \ddot{q}_i = \Phi'(q_{i+1} - q_i) - \Phi'(q_i - q_{i-1}), \qquad i = 1, \ldots, N,$$

where $\Phi(x) = \frac{x^2}{2} + \alpha \frac{x^3}{3} + \beta \frac{x^4}{4}$. This will be called the $\alpha$–$\beta$ model.

The linearization at $q_i \equiv 0$, $i = 1, \ldots, N$, of $\Phi$ gives the system

$$m_i \ddot{q}_i = q_{i+1} - 2q_i + q_{i-1}. \qquad i = 1, \ldots, N.$$

of second order finite differences. Now we add spatial boundary conditiond (periodic, zero, symmetric) and we define $A$ the matrix associated with the second order finite differences operator and the boundary conditions, $\mu_j^2$ its eigenvalues:

$$\mu_j^2 = 4\sin^2\left(\frac{j\pi}{N}\right)$$

This research was supported by MIUR project "Variational Methods and Nonlinear Differential Equations".

and $e_j$ the corresponding eigenvectors; it is convenient to change coordinate system (sometimes referred in the literature as phononic coordinates)

$$x_j = \sum_{i=1}^{N} e^{\frac{2ij\pi J}{N}} q_i, \qquad j = 1, \ldots, N.$$

We assume, for simplicity, that

$$m_i = 1, \qquad \forall i = 1, \ldots, N.$$

Then the system writes

$$-\ddot{x}_j = \mu_j x_j + \frac{\partial W}{\partial x_j}(x_1, \ldots, x_N), \qquad j = 1, \ldots, N.$$

## 2. Energy partition

Fermi, Pasta and Ulam studied numerically the model in [7] with the aim of computing the time needed for the relaxation to equipartition of the distribution of energy among modes. The *harmonic energy of the kth mode* is

$$E_k = \frac{1}{2}|\dot{x}_k|^2 + \frac{\mu_k}{2}|x_k|^2.$$

Let us start with an initial datum concentrating all the energy only on one mode. For small values of the total energy, the contribution of the potential $W$ can be neglected and the sum of the $E_k$-s can be thought to be constant. If the system is ergodic then the energy should be diffused (at least in the average, in time) over all the phase space and therefore *should eventually spread among all modes.*

Surprisingly, their numerical experiment yielded the opposite result. Fermi, Pasta and Ulam observed on the contrary that the energy of the system remained confined among the first modes, instead of spreading towards all modes. In addition, it seemed to show a regular recurrence in time. This result started a large number of additional numerical and analytical investigations.

## 3. KAM theory and Toda lattices

When $\Phi(s) = e^s - 1$ then the system belongs to the class of *Toda lattices* and it is integrable ([17]). Hence all the motions take place on invariant tori and they are regular (quasi–periodic) in time. This fact can explain the peculiar behaviour of the solutions examined by Fermi Pasta ad Ulam. As the $\alpha = \beta = 1$ model is the Taylor expansion of the exponential, it belongs to the class of systems wich are perturbations of an integrable system, for small values of the energy. A natural question is whether KAM theory is applicable. Unfortunately, it does not apply directly, because of the many symmetries and resonances. On the other hand, the very same presence of symmetries allows suitable reductions, and after all the application of KAM theory has been succesfully done by Rink [15, 16]. Hence for a close–to–one measure set of small values of the total energy, which can be

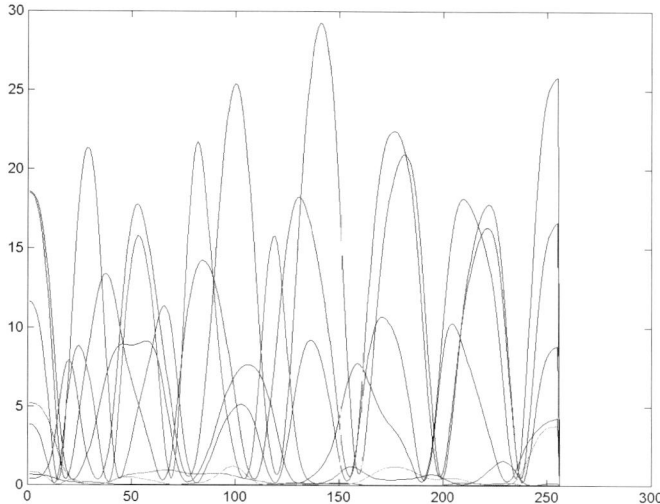

FIGURE 1. Energy partition for a genuinely multi–modal periodic trajectory (with large energy)

chosen as parameter, the motions are regular. Of course the energy threshold $E(N)$ depends on the number of degrees of freedom. A natural question concerns the the asymptotics of $E(N)$, when $N \to +\infty$? It appears that $E(N)$ vanishes as $N \to +\infty$ and its order of vanishing has been estimated (see [4] and references therein).

## 4. Lyapunov solutions

After the time scaling

$$x(t) = u(\omega t), \qquad \text{with} \qquad \omega = \frac{2\pi}{T},$$

the search for $T$–periodic solutions to the problem is equivalent to the boundary value problem

$$-\omega^2 \ddot{u} = Au + W'(u) \qquad u \quad 2\pi - periodic$$

the frequency $\omega$ being now a free (bifurcation) parameter. We denote by

$$\mu_1 < \cdots < \mu_N \qquad \text{and} \qquad e_1, \ldots, e_N$$

respectively the square roots of the eigenvalues of $A$ and their corresponding normalized eigenvectors. Consider the *characteristic frequencies*

$$\omega_{jk} = \frac{\mu_j}{k}, \qquad j = 1, \ldots, N, \quad k \in \mathbb{N}^+ .$$

Corresponding to $2\pi$–periodic solutions of the linearized system of the form

$$u_{jk}(t) = e_j \sin(kt) .$$

These solutions are called the *normal modes* of oscillation of the linearized FPU model and they correspond to periodic solutions of period $2\pi$ of the equation

$$-\omega^2 \ddot{u} = Ax \,.$$

By Lyapunov Center Theorem these solutions can be continued into the nonlinear problem, as an analytic branch, provided no resonances occur

$$\frac{\mu_j}{\mu_h} \notin \mathbb{Z}, \qquad \forall h \neq j \,.$$

We shall refer to such solutions as trivial solutions, for they do not mix the modes. Our problems are the following:

- Are there nontrivial periodic solutions?
- Are there secondary bifurcations from the Lyapunov branches?
- What is their variational characterization?

## 5. The $\beta$–model: A dual action principle

By homogeneity, if $\alpha = 0$, then the value of the parameter $\beta$ is irrelevant, since it can always be rescaled to 1. As $\omega$ decreases, the lagrangian action functional becomes negative definite on subspaces having a large dimension.

$$\mathcal{A}_\omega(u) = \int_0^{2\pi} \sum_{j=1}^{N} \left( \frac{\omega^2}{2} |\dot{u}_j|^2 - \frac{\mu_j}{2} |u_j|^2 \right) - W(u) \,.$$

We proceed as follows:

- introduce a dual functional which is suitable both for the minimization and for the numerical study the minimizers of the systems
- study the minimum of the dual functional
- numerically investigate some branches of Lyapunov solutions
- numerically investigate other branches of solution, bifurcating from the Lyapunov branches, and provide a bifurcation graph
- state a criterion based on the dual functional concerning the distribution of energy among modes and to provide a rigorous proof of the existence of many periodic solutions at some given frequency satisfying such criterion.

## 6. Comments

Since the $N$ positive eigenvalues $\mu_j^2$ of $A$ are explicitly known and satisfy the fundamental property $(Q)$

$$\frac{\mu_i}{\mu_j} \notin \mathbb{Q} \qquad \text{for all } i \neq j \,,$$

it is a standard fact from bifurcation theory that every frequency $\omega_{jk} = \mu_j/k$, $j = 1, \ldots, N$, $k \in \mathbb{N}^+$, is a bifurcation point for the system from the trivial line $(\omega, 0)$. (Crandall–Rabinowitz) The local primary branches that depart from these

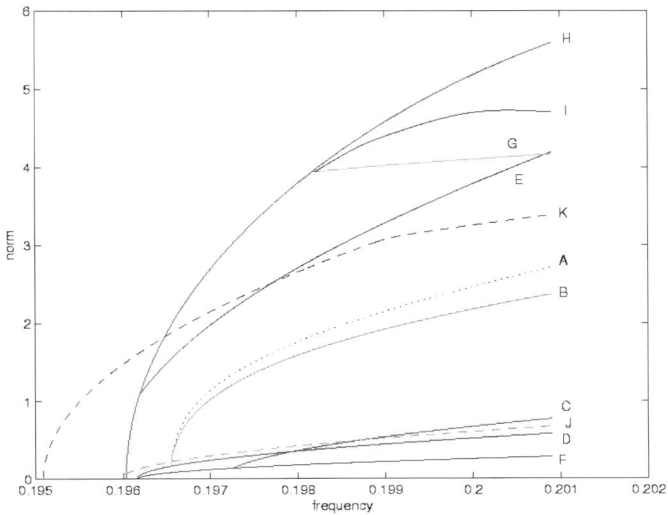

FIGURE 2. The bifurcation diagram in Arioli, Koch and Terracini, 2005

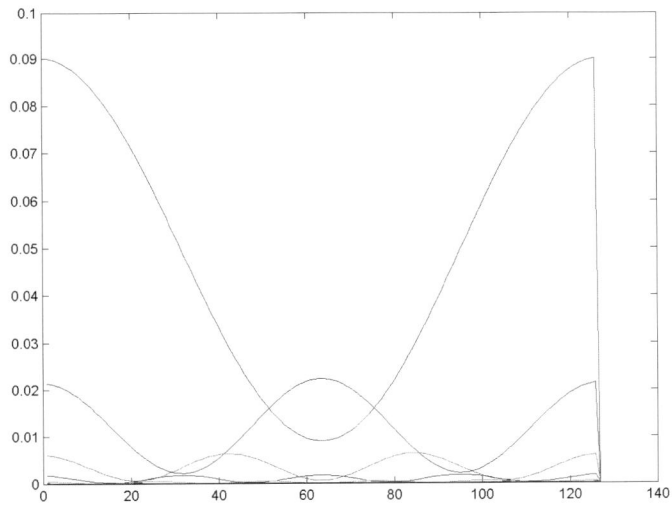

FIGURE 3. Energy partition for a multi–modal trajectory

points are small deformations of the normal modes of oscillation of the linearization at zero, and carry solutions with bounded minimal periods. Such is the case, in Figure 1, for the branches labelled by $F$, $H$, $J$ and $K$.

- The appearance of secondary bifurcation branches is related to anomalies in the distribution of the frequencies $\omega_{jk}$ on the real line. In the picture, for example, the bifurcation frequencies of the primary branches $F$, $H$, $J$ and $K$ appear as a tight cluster among all other frequencies, and those of $H$ and $J$ as an even tighter subcluster.
- A further analysis of the harmonic energy along the secondary branches shows that these, in contrast to the primary ones, carry solutions that "mix modes"; in this context, another role of the frequencies involved in the cluster is that they seem to decide which modes are mixed along a given secondary branch. The mixing however turns out to be quite asymmetric and prevalently a primary branch in a cluster interacts and mix modes with other branches located close to it and at its left.
- These considerations are the result of mainly numerical computations and call for a theoretical understanding. The main questions we address are the mechanism of formation of secondary bifurcations, the reasons for the asymmetry in the interactions, and the discussion of the notions we will introduce to prove the main results.

## 7. Assumptions

The $N \times N$ matrix $A$ and the potential $W : \mathbb{R}^N \to \mathbb{R}$ satisfy the assumptions

- *A is symmetric, positive definite and has only simple eigenvalues whose square roots are pairwise independent over $\mathbb{Q}$;*
- *W is an homogeneous polynomial of degree four and $W > 0$ except at zero.*

We denote by

$$\mu_1 < \cdots < \mu_N \qquad \text{and} \qquad e_1, \ldots, e_N$$

respectively the square roots of the eigenvalues of $A$ and their corresponding normalized eigenvectors. Moreover, we call *characteristic values* the numbers (frequencies)

$$\omega_{jk} = \frac{\mu_j}{k}, \qquad j = 1, \ldots, N, \quad k \in \mathbb{N}^+.$$

The same arguments as for the FPU problem show that from every $(\omega_{jk}, 0)$ there departs a *primary* branch $\Gamma_{jk}$ of *even* solutions to the problem.

## 8. Secondary bifurcations

The next theorems are our main results. We denote by $\omega^R$ and $\omega^L$ the nearest characteristic values respectively to the right and to the left of $\omega$.

**Bifurcation from the right branch.** *Under the assumptions $(A)$ and $(W)$, suppose moreover that*

$$6\mu_i^2 W(e_j) - \mu_j^2 W''(e_j) \cdot e_i^2 < 0$$

*for some for some $i \neq j$. Then there exists $\sigma > 0$ such that, if for some $h, k \in \mathbb{N}^+$ there results*

$$\omega_{ih} < \omega_{jk} \qquad \text{and} \qquad \frac{\omega_{jk} - \omega_{ih}}{\omega_{jk}^R - \omega_{jk}} < \sigma,$$

*then the primary branch $\Gamma_{jk}$ supports a secondary bifurcation in the interval $(\omega_{jk}, \omega_{jk}^R)$.*

**Bifurcation from the left branch.** *Under the assumptions $(A)$ and $(W)$, suppose moreover that*

$$6\mu_j^2 W(e_i) - \mu_i^2 W''(e_i) \cdot e_j^2 > 0$$

*for some for some $i \neq j$. Then there exists $\sigma > 0$ such that, if for some $h, k \in \mathbb{N}^+$ there results*

$$\omega_{ih} < \omega_{jk} \qquad \text{and} \qquad \frac{\omega_{jk} - \omega_{ih}}{\omega_{ih} - \omega_{ih}^L} < \sigma,$$

*then the primary branch $\Gamma_{ih}$ supports a secondary bifurcation in the interval $(\omega_{jk}, 2\,\omega_{ih} - \omega_{ih}^L)$.*


## 9. Asymptotic resonances and Morse indices

The proofs will make clear that secondary bifurcations appear on $\Gamma_{jk}$ and $\Gamma_{ih}$ at a distance from the respective feet $\omega_{jk}$ and $\omega_{ih}$ which is: much bigger than their mutual distance $\omega_{jk} - \omega_{ih}$, but much smaller than their "directional" distance to the remaining part of $\Omega$, namely

$$\omega_{jk}^R - \omega_{jk} \qquad \text{and} \qquad \omega_{ih} - \omega_{ih}^L$$

respectively. We say that $\mu_i$ is *right asymptotically resonant* with $\mu_j$, if there exist diverging sequences $h_n, k_n \in \mathbb{N}$ such that

$$\omega_{ih_n} < \omega_{jk_n} \quad \forall n \qquad \text{and} \qquad \lim_{n \to \infty} \frac{\omega_{jk_n} - \omega_{ih_n}}{\omega_{jk_n}^R - \omega_{jk_n}} = 0\,.$$

The presence of secondary bifurcations will be detected via a result by Kielhöfer, implying that along an analytic branch of solutions, every point of discontinuity for the Morse index of the solutions (seen as critical points of the usual action functional) is either a bifurcation or a turning point; the last possibility will be ruled out by trivial local arguments.

## 10.  A Birkoff–Lewis theorem

**Corollary 10.1.** *Under the assumptions* $(A)$ *and* $(W)$, *suppose moreover that*

$$W_{ij} = 6\mu_j^2 W(e_i) - \mu_i^2 W''(e_i) \cdot e_j^2 \neq 0$$

*for some for some* $i \neq j$. *If, according to* $W_{ij} < 0$ *or* $W_{ij} > 0$, *either* $\mu_i$ *is right asymptotically resonant with* $\mu_j$, *or* $\mu_j$ *is left asymptotically resonant with* $\mu_i$, *then there exists a sequence of periodic solutions, whose* $\mathcal{C}^2$ *norms tend to zero and whose minimal periods tend to infinity.*

The same conclusion could be obtained as a byproduct of the result by Rink, by showing that the fourth order Birkhoff normal form is nonresonant and nondegenerate (in the sense of Kolmogorov). We point out that this approach may fail in the more general framework identified by assumptions $(A)$ and $(W)$ because of two reasons:

– there might be  fourth order resonances among the $\mu_j$'s, preventing the normal form to be nonresonant; this fact seems to be unrelated to the notion of asymptotic resonance;

– even in the absence of resonances, the  Kolmogorov nondegeneracy condition needs not be fulfilled; although this condition holds generically, it is very hard to check, for one should know the explicit expression of the normal form, which rarely happens. This has to be compared with the simple and computable form of our nonlinear coupling conditions.

## 11.  Asymptotic resonance and nonlinear coupling in the FPU model

Now we turn to the analysis of conditions used in the main theorems and to the application of this analysis to the FPU problem. Here we use some  number theoretical arguments which may by unfamiliar to the analyst. This study involves:

– a complete study of the notion of asymptotic resonance shows that it is equivalent to a set of  Diophantine equations depending on the coefficients of the  ternary relations among the characteristic frequencies (null linear combinations with integer coefficients involving exactly three $\mu_j$'s) that may be present.

– as a conclusion, $\mu_i$ is proved to be asymptotically resonant with any other $\mu_j$, whenever the ternary relations involving them, if any exist, are of a special type.

– the complete solution of the abstract Diophantine equations, and the complete list of asymptotic resonances for every value of $N$ is obtained. As a particular case we obtain that $\mu_i$ is asymptotically resonant with any other $\mu_j$ as soon as $N + 1$ is not a multiple of 3.

– the analysis of the validity of the nonlinear coupling conditions for the FPU model. We obtain, for every $N$, that

$$W_{ij} < 0 \quad \text{if} \quad i+j \neq N+1 \quad \text{and} \quad W_{ij} > 0 \quad \text{if} \quad i+j = N+1 \,.$$

The relative preponderance of negative terms explains, we believe, the numerical results in Arioli, Koch and T.

# References

[1] G. Arioli, F. Gazzola and S. Terracini, *Periodic motions of an infinte lattice of particles with nearest neighbour interaction*, Nonlin. Anal. TMA (1996)

[2] G. Arioli , H. Koch and S. Terracini, *Two novel methods and Multi mode periodic solutions for the fermi–Pasta–Ulam model*, Comm. Math. Phys. (2005)

[3] A. Ambrosetti, G. Prodi, A Primer of Nonlinear Analysis. Cambridge studies in Adv. Math. 34. *Cambrige Univ. Press*, Cambridge, 1993.

[4] L. Berchialla, L. Galgani and A. Giorgilli, *Localization of energy in FPU chains*, Discrete Cont. Dynam. Systems A, (2004)

[5] B. Buffoni, J. Toland, Analytic theory of global bifurcation. An introduction. Princeton Series in Appl. Math. *Princeton Univ. Press*, Princeton, 2003.

[6] J. H. Conway, A. J. Jones, *Trigonometric Diophantine equations (on the vanishing sums of roots of unity)*, Acta Arith. **30** (1976), 3, 229–240.

[7] E. Fermi, J. Pasta and S. Ulam, *Studies of Nonlinear Problems*, Los Alamos Rpt. LA - 1940 (1955); also in Collected Works of E. Fermi University of Chicago Press, 1965, Vol II, p. 978.

[8] G. H. Hardy and E. M. Wright, An introduction to the theory of numbers, fifth ed., *The Clarendon Press Oxford University Press*, New York, 1979.

[9] P. C. Hemmer, *Dynamic and stochastic types of motion in the linear chain*, Ph.D. thesis, Trondheim, 1959.

[10] H. Kielhöfer, *A bifurcation theorem for potential operators*, J. Funct. Anal. **77** (1988), no. 1, 1–8.

[11] T. Y. Lam and K. H. Leung, *On vanishing sums of roots of unity*, J. Algebra **224** (2000), no. 1, 91–109.

[12] Henry B. Mann, *On linear relations between roots of unity*, Mathematika **12** (1965), 107–117.

[13] G. Molteni, E. Serra, M. Tarallo and S. Terracini, *Asymptotic resonance, interaction of modes and subharmonic bifurcation*, Arch. Rat. Mech. Anal (2006)

[14] W. Narkiewicz, Elementary and analytic theory of algebraic numbers, second ed., *Springer-Verlag*, Berlin, 1990.

[15] B. Rink, *Symmetri and resonance in periodic FPU chains*, Comm. Math. Phys. (2001)

[16] B. Rink, *Symmetric invariant manifolds in the Fermi-Pasta-Ulam lattice*, Physica D 175 (2003) 31–42.

[17] M. Toda, Theory of nonlinear lattices. Second edition. Springer Series in Solid-State Sciences, 20. *Springer-Verlag*, Berlin, 1989.

[18] L. C. Washington, *Introduction to cyclotomic fields*, second ed., Springer-Verlag, New York, 1997.

Susanna Terracini
Dipartimento di Matematica e Applicazioni
Università di Milano "Bicocca"
Via Cozzi 53,
I-20125 Milano,
Italy
e-mail: `susanna.terracini@gmail.com`

Progress in Nonlinear Differential Equations
and Their Applications, Vol. 75, 425–435
© 2007 Birkhäuser Verlag Basel/Switzerland

# Control of Transient Chaos Using Safe Sets in Simple Dynamical Systems

Samuel Zambrano and Miguel A. F. Sanjuán

*Dedicated to Arrigo Cellina and James Yorke*

**Abstract.** Transient chaos is nearly as ubiquitous as chaos itself, and it is a manifestation of the existence of a nonattractive chaotic set: a chaotic saddle. In some situations it might be desirable to keep the trajectories of a dynamical system with transient chaos far from the attractor and close to this set but its nonattractive nature, the complex dynamics associated with it and eventually the presence of noise may difficult this task. Assume, as an extra difficulty, that our action on the system is bounded and smaller than the action of noise. In such a situation this might seem impossible. However, we will show that in a variety of one dimensional maps this is possible indeed. The control strategy is based on the existence of a set, the safe set, with interesting properties that are due to the same conditions that imply the existence of a chaotic saddle in the system. An example of application of our control technique with the logistic map and some numerical simulations confirming our results are also presented in this work.

**Mathematics Subject Classification (2000).** Primary 37E05; Secondary 34C28.

**Keywords.** Control, transient chaos, maps.

## 1. Introduction

Some dynamical systems are not chaotic but they present a nonattractive invariant set where the dynamics is chaotic. A manifestation of the existence of that set, usually referred to as chaotic saddle, is the observation of chaotic transients: short periods of time in which the dynamics of a trajectory is chaotic, before it settles to an attractor [2]. Transient chaos is nearly as ubiquitous as chaos itself, and in different contexts [3] it might be desirable to keep the system close to the chaotic saddle in order to avoid the attractor.

Different techniques have been proposed in recent years to achieve this goal. A method inspired in the OGY chaos control scheme [5], based on stabilization of the system around one of the unstable periodic orbits that lie in the chaotic saddle, has shown its effectiveness [8]. Other authors have proposed a method based on applying small perturbations to the return map [3] of continuous-time dynamical systems.

The nonattracting nature of the chaotic saddle, and the erratic behavior of the trajectories that pass nearby, is the main difficulty for the control task. If the system is also affected by noise, staying close to the chaotic saddle might be even more difficult. Imagine, as an extra difficulty, that our action on the system is limited to be smaller than the action of noise. Then, it would seem that it is impossible to remain close to the nonattracting chaotic set. However, in a recent paper Aguirre *et al.* [1] showed that this is indeed possible for the simpler dynamical system with a chaotic saddle and escapes to infinity: the slope three tent map.

The aim of this work is to generalize the results obtained in [1] to a more general class of one dimensional maps presenting a chaotic saddle. We are going to show that, as in [1], paradoxically the same geometry giving rise to the existence of a chaotic saddle will help us to design a strategy to keep the trajectory close to the nonattracting chaotic set by using a control smaller than noise.

The structure of the paper is the following. In Section 2 we state the problem in a precise way and we enounce the main result of this work as a theorem. In Section 3 we present this particular set of points that will help us to design our control strategy, the safe set, and we give as a proposition its main properties. Once we have defined this set and its properties, in Section 4 we prove our main result. Finally, in Section 5 we show an example of application of our technique with the well known logistic map and in Section 6 we draw the main conclusions of our work.

## 2. Problem statement and main result

First we will define in a precise way the class of dynamical systems that we deal with. We consider one dimensional maps $x_{n+1} = f(x_n)$ where $f : \mathbb{R} \to \mathbb{R}$ is a map that satisfies the following conditions.

(i) There is an interval $I = [a, b]$ such that $I \subset f(I)$. The interval $I$ can be divided in three subintervals $A_1 = [a, x_-]$, $A_0 = (x_-, x_+)$, and $A_2 = [x_+, b]$ such that $f(A_1) = f(A_2) = I$ and $f(A_0) \notin I$.
(ii) The map $f$ is continuous and differentiable in $A_1 \cup A_2$ and for all $x_0 \in A_1 \cup A_2$, $|f'(x_0)| > 1$.
(iii) For all $x_0 \notin I$, $|f^n(x_0)| \to \infty$ as $n \to \infty$.

From conditions (i)–(iii) it can be proved (see [6]) that there is a nonattractive Cantor-like set $\Lambda \subset A_1 \cup A_2$ where the dynamics is topologically equivalent to a shift on two symbols, that is, there is a chaotic saddle. We must point out that

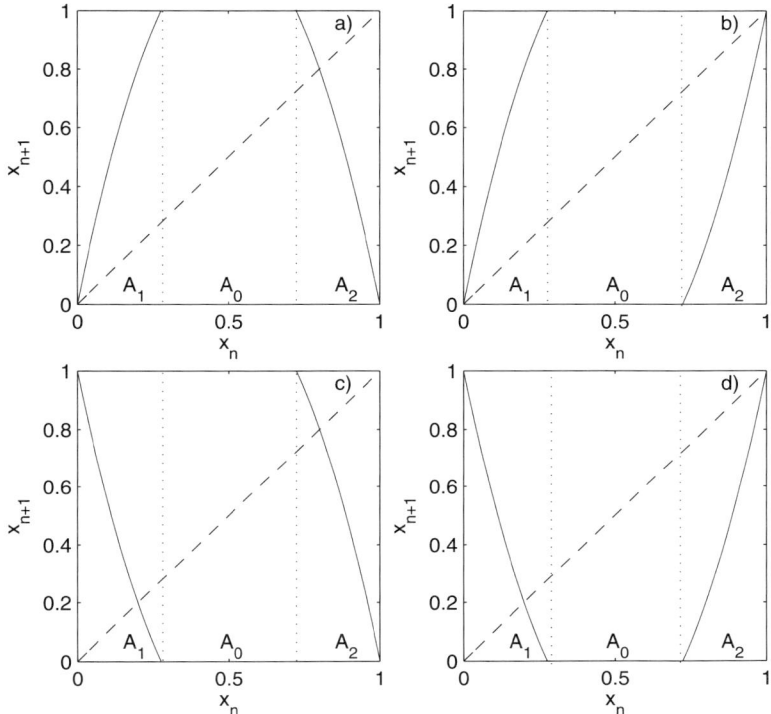

FIGURE 1. Four possible configurations of a map $x_{n+1} = f(x_n)$ satisfiying conditions (i)–(iii). Note that each point in $I$ has just one preimage in $A_1$ and other in $A_2$

condition (ii) is not a necessary condition for the existence of a chaotic saddle. However it makes both the proof of the existence of $\Lambda$ (see [6]) and the calculations needed in this paper much easier. Note that the slope three tent map $x_{n+1} = 3(1 - |x_n|) - 1$ studied in [1] satisfies these conditions. For this particular case, $\Lambda$ is like the classical middle-third Cantor set constructed using as the starting segment the $[-1, 1]$ interval.

On the other hand, condition (iii) does also imply that for all $x_0 \notin \Lambda$ $|f^k(x_0)| \to \infty$ for $k \to \infty$. If we would have established that all trajectories starting out of $I$ settle to any other type of attractor out of $I$ the existence of a chaotic saddle $\Lambda$ in $A_1 \cup A_2$ could be established in the same manner. However, we have opted to fix condition (iii) both for simplicity and to make a certain analogy with some chaotic scattering problems, a context in which control of transient chaos is important. For this kind of problems it is well-known [9] that all trajectories except those starting either on a zero-measure invariant set or in its stable manifold diverge from the scattering region to infinity. The same thing applies to the system that we are dealing with: in absence of control only the trajectories starting in $\Lambda$ will not diverge to infinity under iterations of $f$.

Once we have defined the type of dynamical system that we will deal with in this letter, we can define in a more precise way the type of situation that we want to control. When controlling a certain dynamical system, specially a physical system, there are two main ingredients that must be considered: first, the deterministic component of its dynamics, and second the eventual presence of a random deviation from the expected deterministic dynamics (the noise). We consider here systems where the deterministic part of the dynamics is modeled by a map $f : \mathbb{R} \to \mathbb{R}$ that satisfies conditions (i)–(iii), so we are considering a system that presents transient chaos. Thus, starting from a point $x_n$ the dynamics of the system takes it to

$$x' = f(x_n).$$

Now we introduce in our model an additive perturbation playing the role of noise, $u_n$, that deviates the trajectory from its deterministic path, taking it to

$$x'' = x' + u_n = f(x_n) + u_n.$$

We assume that $u_n$ is a random number such that $|u_n| \leq u_0$, where $u_0 > 0$.

As we said before, in the system considered nearly all the trajectories (except those lying in $\Lambda$) will diverge to infinity in absence of noise. With noise, it is clear that all the trajectories will diverge to infinity. Our objective here is to avoid such divergence to infinity. To do this, we can apply a small perturbation $r_n$ each iteration to control the system's dynamics. Thus, the final state of the system is the result of the action of the deterministic dynamics, modeled by $f$, of noise, modeled by $u_n$, and of the small control applied to the system, modeled by $r_n$. The control $r_n$ is also bounded by a positive constant $r_0$, so $|r_n| < r_0$ for all $n$. Thus, each time step, the evolution of the system is given by

$$x_{n+1} = x'' + r_n = f(x_n) + u_n + r_n. \tag{2.1}$$

Our aim here is to show that, contrary to what intuition may say, there is a way to keep the trajectories in $A_1 \cup A_2$ (or "close" to the chaotic saddle $\Lambda$) even if $r_0 < u_0$. Or, speaking in physical terms, if the control is smaller than noise. To do this, the only thing that we need is to fix the initial condition $x_0$ accurately. After this, by applying a wisely chosen perturbation $r_n$ each time step, with $|r_n| < r_0 < u_0$ trajectories can be kept bounded *ad infinitum*. This is the main result of this work and it can be stated as follows:

**Theorem 2.1 (Main result).** *For a dynamical system like the one given by eq. (2.1), where $f$ satisfies conditions* (i)–(iii), *for all $u_0 > 0$ there is a $0 < r_0 < u_0$ such that $x_n \in A_1 \cup A_2$ for all $n$.*

This theorem was proved in the particular case of $f(x)$ being the slope three tent map in [1]. Here we prove this theorem for a wider class of one-dimensional maps. As in [1], the key element for this control strategy is the existence of a set with very interesting properties: the safe set. In the next section we will define this set and we will show that its properties can be derived from the same conditions (i)–(iii) that implied the existence of a chaotic saddle. Thus, the main idea of this work is that, contrarily to what intuition might say, the existence of a chaotic saddle does not add an extra degree of difficulty but, instead, it can be of great help for the control task.

## 3. The safe set and its structure

In this section we define the safe set and explain and prove its main properties. We first define the following maps:

**Definition 3.1.** *Let $f$ be a map satisfying conditions* (i)–(iii). *Then $F_i \equiv f^{-1}(x) \cap A_i$ for $i = 1, 2$.*

These two maps have a first essential property that will allow us to define the safe sets:

**Proposition 3.2.** *The map $F_i : I \to A_i$ is a one-to-one map for $i = 1, 2$.*

*Proof.* From conditions (i) and (ii) it is clear that $f$ is invertible both in $A_1$ and in $A_2$, so $F_i : I \to A_i$ is one to one. In other words, any point in $I$ has just one preimage in $A_1$ and just one preimage in $A_2$. This can be clearly observed in the examples shown in Fig. 1. □

Thus, given a point $z \in I$, its preimage in $A_1$ is $F_1(z)$, and its preimage in $A_2$ is $F_2(z)$. We can now define the safe sets of order $k$.

**Definition 3.3.** *Let $x_1^0$ be the middle point of $A_0$, that will be called the safe point of order 0. Thus, we can define inductively the safe points of order $k$, $\{x_i^k\}_{i=1}^{2^k}$ as:*

$$\{x_i^k\}_{i=1}^{2^k} \equiv f^{-k}\left(x_1^0\right) \cap I = \cup_{i=1}^{2^{k-1}} \cup_{j=1}^{2} F_j\left(x_i^{k-1}\right) \tag{3.1}$$

*The set of safe points of order $k$ is called the safe set of order $k$. The sub index $i \in \{1...2^k\}$ of $x_i^k$ is assigned in such a way that $i < j \leftrightarrow x_i^k < x_j^k$ .*

From the definition given above, it might seem paradoxical to call these sets the "safe sets" of order $k$, as long as it is clear from eq. (3.1) that all the elements of this set fall out of $I$ after $k+1$ iterations, after which they will diverge to infinity. However, we will show now that the properties of this set justify this denomination.

A first main property of this set, that can be easily deduced from this definition, is the following: given a point $z$ that belongs to the safe set of order $k$, then $f(z)$ will belong to a safe set of order $k - 1$. This simple property, together with the two following ones, that will be presented as a proposition, are the properties that make these sets of points play a key role for our control strategy.

**Proposition 3.4.** *Consider the safe sets of order $k$ of a map $f$ satisfying conditions* (i)–(iii). *Then, for all $k \geq 0$ and for all $i$, $1 \leq i \leq 2^k$:*

– *The safe points of order $k$ and the safe points of order $k+1$ satisfy:*

$$x_{2i-1}^{k+1} < x_i^k < x_{2i}^{k+1}. \tag{3.2}$$

– *Consider the maximum and minimum distance between a safe point of order $k$ and the two adjacent safe points of order $k+1$:*

$$\delta_{max}^k = \max_i \left\{ |x_i^k - x_{2i-1}^{k+1}|, |x_i^k - x_{2i}^{k+1}| \right\} \tag{3.3}$$

$$\delta_{min}^k = \min_i \left\{ |x_i^k - x_{2i-1}^{k+1}|, |x_i^k - x_{2i}^{k+1}| \right\}, \tag{3.4}$$

*then*

$$\lim_{k\to\infty} \delta_{max}^k = \lim_{k\to\infty} \delta_{min}^k = 0. \tag{3.5}$$

*Proof.* First we must remember that in the definition of $f$ we assumed that the interval $A_1$ is to the left of the interval $A_2$. Thus, for those $x_i^{k+1}$ with $i = 1, ..., 2^k$ and certain $j$ that will depend on $i$, $x_i^{k+1} = F_i(x_j^k)$. Analogously, for those $x_i^{k+1}$ with $i = 2^k + 1, ..., 2^{k+1}$ and certain $j'$ that will depend on $i$, $x_i^{k+1} = F_i(x_{j'}^k)$. On the other hand we proved that both $F_1$ and $F_2$ are monotonous in $I$. The type of monotonicity will depend on whether $f$ is an increasing or a decreasing function in $A_1$ and in $A_2$. In this proof we will assume that given two points $z_1, z_2 \in I$ such that $z_1 < z_2$, then $F_1(z_1) < F_1(z_2)$ and $F_2(z_1) > F_2(z_2)$. This is the case of a map as the one shown in Fig. 1 (a). For the remaining configurations of the map shown in Fig. 1, the proof of this proposition is analogous.

The key observation now is that the only relation between the safe points of order $k+1$ and those of order $k$ that holds with our assumptions is $x_i^{k+1} = F_1(x_i^k)$ for $i = 1, ..., 2^k$ and $x_i^{k+1} = F_2(x_{2^{k+1}+1-i}^k)$ for $i = 2^k + 1, ..., 2^{k+1}$. Considering these relations, the proof of this proposition is easy.

We first prove eq. (3.2) inductively. The $k = 0$ case is simple as long as $A_1$ is to the left of $A_2$ and $A_0$ is between these intervals. Thus, $x_1^1 = F_1(x_1^0) < x_1^0 < F_2(x_1^0) = x_2^1$. Assuming that the eq. (3.2) is true for $k$, we will show that it is true for $k+1$. All we need is to apply $F_1$ and $F_2$ to this equation. Equation 3.2 and our assumption on $F_1$ implies that $F_1(x_{2i-1}^{k+1}) < F_1(x_i^k) < F_1(x_{2i}^{k+1})$ for $i = 1, ..., 2^k$ so, considering the relation given between the safe points of order $k$ and those of order $k+1$, this means that $x_{2i-1}^{k+2} < x_i^{k+1} < x_{2i}^{k+2}$ for $i = 1, ..., 2^k$.

Analogously, to complete the proof of eq. (3.2) we apply $F_2$ to eq. (3.2) and we have that $F_2(x_{2i-1}^{k+1}) > F_2(x_i^k) > F_2(x_{2i}^{k+1})$ for $i = 1, ..., 2^k$. Considering our observation, this is equivalent to $x_{2^{k+2}-2i+2}^{k+2} > x_{2^{k+1}+1-i}^{k+1} > x_{2^{k+2}+1-2i}^{k+2}$ and, by making the change of index $j = 2^{k+1}+1-i$, it is equivalent to $x_{2j-1}^{k+2} < x_j^{k+1} < x_{2j}^{k+2}$ with $j = 2^k + 1, ...2^{k+1}$. This completes the proof of eq. (3.2).

The proof of eq. (3.5) is also quite simple. We assumed that for all $x_0 \in A_1 \cup A_2$, $|f'(x_0)| > 1$. Thus there are two positive constants $L_{max} > 1, L_{min} > 1$

such that $L_{min} \leq |f'(x_0)| \leq L_{max}$. Then:

$$\delta^k_{max} = \max_i \left\{ |x^k_i - x^{k+1}_{2i-1}|, |x^k_i - x^{k+1}_{2i}| \right\}$$

$$= \max \left\{ |F_n(x^{k-1}_j) - F_n(x^k_{2j-1})|, |F_n(x^{k-1}_j) - F_n(x^k_{2j})| \right\}$$

for certain $j$ and for certain $n = 1, 2$. Thus, using the mean value theorem and the bound of the derivative given above:

$$\delta^k_{max} = \max \left\{ |F_n(x^{k-1}_j) - F_n(x^k_{2j-1})|, |F_n(x^{k-1}_j) - F_n(x^k_{2j})| \right\}$$

$$\leq \frac{1}{L_{min}} \max \left\{ |x^{k-1}_j - x^k_{2j-1}|, |x^{k-1}_j - x^k_{2j}| \right\} \leq \frac{\delta^{k-1}_{max}}{L_{min}},$$

so $\delta^k_{max} \leq \dfrac{\delta^0_{max}}{(L_{min})^k}$ and eq. (3.5) follows. $\qquad\square$

**Remark 3.5.** According to equation 3.2, a safe point of order $k$ has two adjacent safe points of order $k+1$ that are closer to it than any other safe point of order $k$. Thus, a trajectory lying in a safe point of order $k+1$ is mapped to a point that has a safe point of order $k+1$ to its left and another one to its right. This property is probably the most important one of the safe sets, and it will play a key role in our control strategy.

Once that we have given the key properties of the safe sets, we can now explain our control strategy, which completes the proof of Theorem 2.1.

## 4. Proof of the main result

Considering the properties given above, we can now give a demonstration of our main result.

*Proof of the main result.* The only thing that we have to do to control the system with $r_0 < u_0$ is to put the initial condition on a safe point of an accurately chosen order. To find it, we first have to chose $k$ in such a way that $u_0 > \delta^k_{max}$ which, by eq. (3.5) is always possible if $k$ is sufficiently big.

Considering this, we just have to put the initial condition on a safe point of order $k+1$. After this, $f$ maps this point to a safe point of order $k$, say $x^k_i$. Then noise acts, and there are two possibilities, according to eq. (3.2):

- That $x^k_i + u_n$ is to the left of $x^{k+1}_{2i-1}$ or to the right of $x^{k+1}_{2i}$. In this case, considering that the minimum distance between a safe point of order $k$ and the two adjacent safe points of order $k+1$ is $\delta^k_{min}$, a correction $r_n$ such that $|r_n| \leq u_0 - \delta^k_{min}$ will make $x^k_i + u_n + r_n$ lie on a safe point of order $k+1$.
- That $x^k_i + u_n$ is between $x^k_i$ and $x^{k+1}_{2i-1}$ or between $x^k_i$ and $x^{k+1}_{2i}$. In this case, considering that the maximum distance between a safe point of order $k$ and the two adjacent safe points of order $k+1$ is $\delta^k_{max}$, a correction $r_n$ such that $|r_n| \leq \delta^k_{max}$ will make $x^k_i + u_n + r_n$ lie on a safe point of order $k+1$.

Thus, even if the perturbations $r_n$ are bounded by $r_0 = \max\{u_0 - \delta_{min}^k, \delta_{max}^k\} < u_0$, trajectories starting on a safe point of order $k+1$ can always be placed on a safe point of order $k+1$. This procedure can be repeated forever, which completes our proof. $\qquad\square$

**Remark 4.1.** This theorem does not say a word about which is the optimal $k$ that allows to minimize the ratio $r_0/u_0$. It is a simple exercise to show that the optimal ratios are bounded by the following quantities:

- The ratio $\dfrac{r_0}{u_0} \leq \dfrac{u_0 - \delta_{min}^k}{u_0}$  if $u_0 \in (\delta_{max}^k + \delta_{min}^k, \delta_{max}^{k-1} + \delta_{min}^k]$.

- The ratio $\dfrac{r_0}{u_0} \leq \dfrac{\delta_{max}^k}{u_0}$  if $u_0 \in (\delta_{max}^k + \delta_{min}^{k+1}, \delta_{max}^k + \delta_{min}^k]$.

Thus, we have shown that in a variety of dynamical systems the same geometrical conditions giving rise to transient chaos have allowed us to define a set, the safe set, with some very interesting properties which, on the other hand, allow to keep the trajectories in the vicinity of the chaotic saddle even if control is smaller than noise. In next section we are going to give an example of application of our control technique using the well-known logistic map.

## 5. An example of application: Control of transient chaos for the logistic map

In this section we are going to explore our technique in a simple situation, using the well known logistic map $x_{n+1} = \mu x_n(1 - x_n)$. Although it is well known that for $\mu > 4$ this map presents a chaotic saddle [4], which is formed after a boundary crisis, in [6] it is proved that this map satisfies conditions (i)–(iii) just for $\mu \geq 2 + \sqrt{5}$. In the numerical simulations carried out here we will focus on the $\mu = 5$ case.

For this map, $x_- = \dfrac{1}{2} - \dfrac{\sqrt{\mu^2 - 4\mu}}{2\mu}$ and $x_+ = \dfrac{1}{2} + \dfrac{\sqrt{\mu^2 - 4\mu}}{2\mu}$ and thus $x_0^1 = \dfrac{1}{2}$.

As an example, assume first that we perturb the system with a random perturbation that is bounded by $u_0 = 0.25$. We must first find a $k$ such that $u_0 \geq \delta_{max}^k$. We observe numerically that with $k = 1$, this condition is fulfilled. The safe points of order 2 and those of order 1 are shown in Fig. 2 (a), and we can appreciate how they present the expected structure: each safe point of order 1 has two adjacent safe points of order 2.

In Fig. 2 (b) we can observe a controlled trajectory. As we said, the idea is to adjust $r_n$ in such a way that the resulting $x_{n+1} = f(x_n) + u_n + r_n$ lies always on a safe point of order two. The trajectory is kept bounded in 75 iterations and it could be bounded forever. Note that, in absence of perturbations (even of noise), considering that the initial condition lies on a safe point of order two, after three iterations the trajectory would lie out of $[0, 1]$, and then go to infinity. In Fig. 2 (c) we also show the value of the correction applied each iteration, showing that

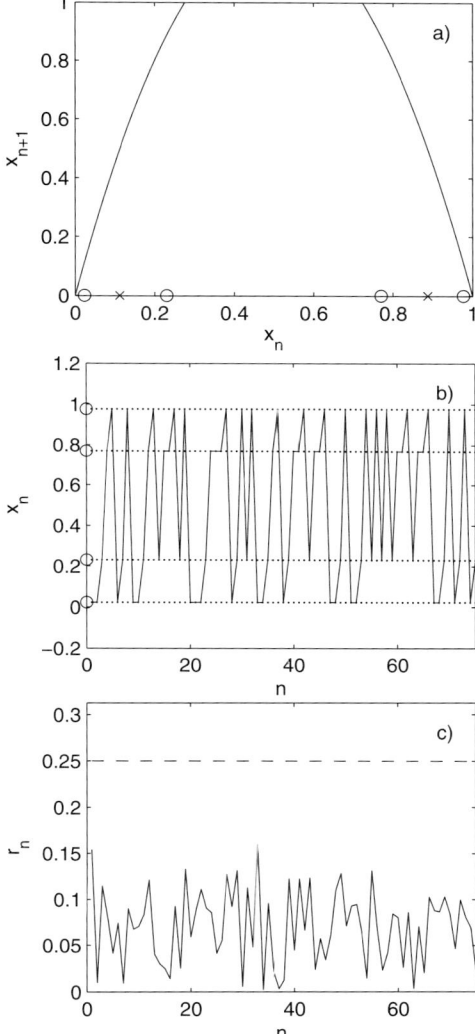

FIGURE 2. The safe points of order 2 ('o') and of order 1 ('×') plotted in the $x_n$ axis together with the curve of the logistic map $x_{n+1} = 5x_n(1 - x_n)$. A trajectory controlled with $u_0 = 0.25$, which is always kept in the safe points of order 2 (marked by 'o' in the $x_n$ axis) (b). The correction applied each iteration, which is always smaller than the maximum perturbation applied $u_0 = 0.25$, marked with a dashed line (c)

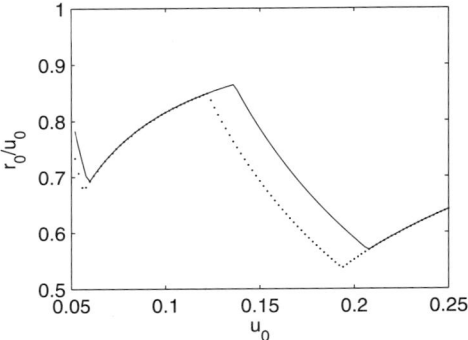

FIGURE 3. The ratio $r_0/u_0$ obtained for different values of $u_0$
numerically ('$\cdots$') and from the analytical expressions ('——'). Note
that this ratio is always smaller than 1

the main result obtained in this paper is observed in this example, as expected.
The correction applied in these iterations is always smaller than $u_0 = 0.25$. In
fact, we observe that $\max_{n}(|r_n|)/u_0 \approx 0.15/0.25 = 0.6$, so with a control that is
approximately 60 % of the noise the trajectories are kept bounded.

Finally, in Fig. 3 we have shown the bounds of the optimal ratios $r_0/u_0$ that
allow to keep the trajectories bounded ad infinitum, obtained analytically from
the expressions given in Remark 4.1, and their numerical estimations, which were
obtained by computing the maximum $|r_n|$ necessary to control a trajectory of
10000 time steps. Note that these ratios are always smaller than one, but their
value depend on the value of $u_0$.

## 6. Conclusions

In this paper we have shown a way to control transient chaos in one dynamical
systems using a very particular set of points: the safe sets. A main advantage of
this type of control is that, by accurately choosing the initial condition, we can
stabilize the system applying perturbations even smaller than the perturbation on
the dynamics induced by the presence of noise.

This is due to the very interesting properties of these sets, which themselves
can be derived from the same mathematical conditions from which the existence
of the chaotic saddle in the dynamical system can be inferred. These conditions
are intimately related with the typical "stretching and folding" processes associated with chaotic dynamics and transient chaos. It is well known that this type
of process is also present in higher dimensional dynamical systems, like in the
paradigmatic Smale horseshoe map [6] and we have recently proved [7] that safe

sets also arise in this kind of structures, which are themselves present in a variety of situations. All this makes us think that considering the global geometrical properties of a dynamical system can be useful from a control point of view, not only to control transient chaos but also to control other dynamical situations that involve this type of "stretching and folding" of the phase space.

# References

[1] J. Aguirre, F. d'Ovidio and M. A. F. Sanjuán, *Controlling chaotic transients: Yorke's game of survival*, Physical Review E **69** (2004), 016203.

[2] K. T. Alligood, T. D. Sauer and J. A. Yorke, *Chaos, An introduction to dynamical systems*. 1st Edition, Springer-Verlag, 1996.

[3] M. Dhamala and Y. C. Lai, *Controlling transient chaos in deterministic flows with applications to electrical power systems and ecology*, Physical Review E **59** (1999), 1646.

[4] C. Grebogi, E. Ott and J. A. Yorke, *Chaotic Attractors in Crisis*, Physical Review Letters (1982), 1507–1510.

[5] C. Grebogi, E. Ott and J. A. Yorke, *Controlling Chaos*, Physical Review Letters **64** (1990), 1196.

[6] C. Robinson, *Dynamical Systems. Stability, Symbolic Dynamics, and Chaos*. 2nd Edition, CRC Press, 1999.

[7] M. A. F. Sanjuán, J. A. Yorke, and Samuel Zambrano, In preparation (2006).

[8] T. Tél, *Controlling Transient Chaos*, Journal of Physics A: Mathematical and General **3** (1993), 417–425.

[9] T. Tél and E. Ott, *Chaotic Scattering: An introduction*, Chaos, (1993), 417–425.

## Acknowledgement

This research has been supported by the Spanish Ministry of Science and Technology under Project Number BFM2003-03081 and FIS2006-08525.

Samuel Zambrano and Miguel A. F. Sanjuán
Nonlinear Dynamics and Chaos Group,
Departamento de Física,
Universidad Rey Juan Carlos,
Tulipán s/n,
E-28933 Móstoles, Madrid,
Spain
e-mail: samuel.zambrano@urjc.es
       miguel.sanjuan@urjc.es

# Progress in Nonlinear Differential Equations and Their Applications (PNLDE)

Edited by

**Haim Brezis**, Université Pierre et Marie Curie, Paris, France and Rutgers University, New Brunswick, N.J., USA

*Progress in Nonlinear Differential Equations and Their Applications* is a book series that lies at the interface of pure and applied mathematics. Many differential equations are motivated by problems arising in diversified fields such as mechanics, physics, differential geometry, engineering, control theory, biology and economics. This series is open to both the theoretical and applied aspects, hopefully stimulating a fruitful interaction between the two sides. It will publish monographs, polished notes arising from lectures and seminars, graduate level texts, and proceedings of focused and refereed conferences.

**BIRKHÄUSER**

**PNLDE 75: Staicu, V.** (Ed.)
Differential Equations, Chaos and Variational Problems (2007).
ISBN 978-3-7643-8581-4

**PNLDE 74: Berti, M.**
Nonlinear Oscillations of Hamiltonian PDEs (2007).
ISBN 978-0-8176-4680-6

**PNLDE 73: Pucci, P. / Serrin, J.**
The Maximum Principle (2007).
ISBN 978-3-7643-8144-8

**PNLDE 72: Tarantello, G.**
Self-Dual Gauge Field Vortices. An Analytical Approach (due 2007).
ISBN 978-0-8176-4310-2

**PNLDE 71: Kichenassamy, S.**
Fuchsian Reduction. Applications to Geometry, Cosmology and Mathematical Physics (2007). ISBN 978-0-8176-4352-2

**PNLDE 70: Sandier, E. / Serfaty, S.**
Vortices in the Magnetic Ginzburg-Landau Model (2006). ISBN 978-0-8176-4316-4

**PNLDE 69: Bove, A. / Colombini, F. / Del Santo, D.** (Eds.) Phase Space Analysis of Partial Differential Equations (2006).
ISBN 978-0-8176-4511-3

**PNLDE 68: Dal Maso, G. / DeSimone, A. / Tomarelli, F.** (Eds.)
Variational Problems in Materials Science (2006). ISBN 978-3-7643-7564-5

**PNLDE 67: Aftalion, A.**
Vortices in Bose–Einstein Condensates (2006). ISBN 978-0-8176-4392-8

**PNLDE 66: Cazenave, T. / Costa, D. / Lopes, O. / Manásevich, R. / Rabinowitz, P. / Ruf, B. / Tomei, C.** (Eds.) Contributions to Nonlinear Analysis. A Tribute to D.G. de Figueiredo (2005).
ISBN 978-3-7643-7149-4

**PNLDE 65: Bucur, D. / Buttazzo, G.**
Variational Methods in Shape Optimization Problems (2005). ISBN 978-0-8176-4359-1

**PNLDE 64: Chipot, M. / Escher, J.** (Eds.)
Nonlinear Elliptic and Parabolic Problems. A Special Tribute to the Work of Herbert Amann (2005). ISBN 978-3-7643-7266-8

**PNLDE 63: Bandle, C. / Berestycki, H. / Brighi, B. / Brillard, A. / Chipot, M. / Coron, J.-M. / Sbordone, C. / Shafrir, I. / Valente, V. / Vergara Caffarelli, G.** (Eds.)
Elliptic and Parabolic Problems. A Special Tribute to the Work of Haim Brezis (2005).
ISBN 978-3-7643-7249-1

**PNLDE 62: Suzuki, T.**
Free Energy and Self-Interacting Particles (2005). ISBN 978-0-8176-4302-7

**PNLDE 61: Rodrigues, J.F. / Seregin, G. / Urbano, J.M.** (Eds.)
Trends in Partial Differential Equations of Mathematical Physics (2005). ISBN 978-3-7643-7165-4